HANDBOOK OF CHEMICAL EQUILIBRIA IN ANALYTICAL CHEMISTRY

ELLIS HORWOOD SERIES IN ANALYTICAL CHEMISTRY
Series Editors: Dr. R. A. CHALMERS and Dr. MARY MASSON
University of Aberdeen

APPLICATION OF ION SELECTIVE MEMBRANE ELECTRODES IN ORGANIC ANALYSIS
G. E. BAIULESCU and V. V. COSOFRET
EDUCATION AND TEACHING IN ANALYTICAL CHEMISTRY
G. E. BAIULESCU, C. PATROESCU and R. A. CHALMERS
HANDBOOK OF PRACTICAL ORGANIC MICROANALYSIS
S. BANCE
ORGANIC TRACE ANALYSIS
K. BEYERMANN
FOUNDATIONS OF CHEMICAL ANALYSIS
O. BUDEVSKY
INORGANIC REACTION CHEMISTRY Volume 1: Systematic Chemical Separation
D. T. BURNS, A. TOWNSEND and A. G. CATCHPOLE
INORGANIC REACTION CHEMISTRY Reactions of the Elements and their Compounds Volume 2, Part A: Alkali Metals to Nitrogen and Volume 2, Part B: Osmium to Zirconium
D. T. BURNS, A. TOWNSHEND and A. H. CARTER
AUTOMATIC CHEMICAL ANALYSIS
J. K. FOREMAN and P. B. STOCKWELL
FUNDAMENTALS OF ELECTROCHEMICAL ANALYSIS
Z. GALUS
LABORATORY HANDBOOK OF PAPER AND THIN LAYER CHROMATOGRAPHY
J. GASPARIC and J. CHURACEK
HANDBOOK OF ANALYTICAL CONTROL OF IRON AND STEEL PRODUCTION
T. S. HARRISON
ORGANIC ANALYSIS USING ATOMIC ABSORPTION SPECTROMETRY
SAAD S. M. HASSAN
HANDBOOK OF ORGANIC REAGENTS IN INORGANIC CHEMISTRY
Z. HOLZBECHER, L. DIVIS, M. KRAL, L. SUCHA and F. VLACIL
GENERAL HANDBOOK OF ON-LINE PROCESS ANALYSERS
DAVID HUSKINS
QUALITY MEASURING INSTRUMENTS IN ON-LINE PROCESS ANALYSIS
DAVID HUSKINS
ANALYTICAL APPLICATIONS OF COMPLEX EQUILIBRIA
J. INCZEDY
PARTICLE SIZE ANALYSIS
Z. K. JELINEK
OPERATIONAL AMPLIFIERS IN CHEMICAL INSTRUMENTATION
R. KALVODA
METHODS OF PROTEIN ANALYSIS
Edited by I. KERESE
ATLAS OF METAL-LIGAND EQUILIBRIA IN AQUEOUS SOLUTION
J. KRAGTEN
GRADIENT LIQUID CHROMATOGRAPHY
C. LITEANU and S. GOCAN
STATISTICAL THEORY AND METHODOLOGY OF TRACE ANALYSIS
C. LITEANU and I. RICA
SPECTROPHOTOMETRIC DETERMINATION OF ELEMENTS
Z. MARCZENKO
LABORATORY HANDBOOK OF CHROMATOGRAPHIC AND ALLIED METHODS
O. MIKES
STATISTICS FOR ANALYTICAL CHEMISTRY
MILLER and MILLER
SEPARATION AND PRECONCENTRATION METHODS IN INORGANIC TRACE ANALYSIS
J. MINCZEWSKI, J. CHWASTOWSKA and R. DYBCZYNSKI
HANDBOOK OF ANALYSIS OF ORGANIC SOLVENTS
V. SEDIVEC and J. FLEK
ELECTROCHEMICAL STRIPPING ANALYSIS
F. VYDRA, K. STULEK and B. JULAKOVA
ANALYSIS WITH ION-SELECTIVE ELECTRODES
J. VESELY, D. WEISS and K. STULIK

HANDBOOK OF CHEMICAL EQUILIBRIA IN ANALYTICAL CHEMISTRY

Stanislav Kotrlý
and
Ladislav Šůcha

ELLIS HORWOOD LIMITED
Publishers • Chichester

Halsted Press: a division of
JOHN WILEY & SONS
New York • Brisbane • Chichester • Toronto

First published in 1985 by
ELLIS HORWOOD LIMITED
Market Cross House, Cooper Street, Chichester, West Sussex, P019 1EB, England

The publisher's colophon is reproduced from James Gillison's drawing of the ancient Market Cross, Chichester.

Distributors:
Australia, New Zealand, South-east Asia:
Jacaranda-Wiley Ltd., Jacaranda Press,
JOHN WILEY & SONS INC.,
G.P.O. Box 859, Brisbane, Queensland 40001, Australia

Canada:
JOHN WILEY & SONS CANADA LIMITED
22 Worcester Road, Rexdale, Ontario, Canada

Europe, Africa:
JOHN WILEY & SONS LIMITED
Baffins Lane, Chichester, West Sussex, England

North and South America and the rest of the world:
Halsted Press: a division of
JOHN WILEY & SONS
605 Third Avanue, New York, N.Y. 10016, U.S.A.
© 1985 Kotrlý, Šůcha/Ellis Horwood Ltd.

 British Library Cataloguing in Publication Data

Kotrlý, S.
Handbook of chemical equilibria in analytical chemistry.—(Ellis Horwood series in analytical chemistry)
1. Chemistry, Analytic—Tables
2. Chemical equilibriums—Tables
I. Title   II. Šůcha, L.
543′.00212   QD78

ISBN 0-85312-107-9 (Ellis Horwood Limited)
ISBN 0-470-27479-4 (Halsted Press)

Published in co-edition with
SNTL — Publishers of Technical Literature, Prague, Czechoslovakia

All Rights reserved. No part of this publication may be reproduced, stored in a retrieval system, or transmitted, in any form or by any means, electronic, mechanical, photocopying, recording or otherwise, without the permission of Ellis Horwood Limited, Market Cross House, Cooper Street, Chichester, West Sussex, England
Printed in Czechoslovakia

Contents

Foreword 9

Chapter 1 — Chemical Reactions and Equilibria in Analytical Chemistry 11

Chapter 2 — Chemical Equilibria and Equilibrium Constants 16
  2.1 Definitions of equilibrium constants 16
  2.2 The effect of temperature on the equilibrium constant 18
  2.3 The effect of ionic strength on the value of the equilibrium constant 20
    2.3.1 Activity coefficients of non-electrolytes 20
    2.3.2 Activity coefficients of electrolytes 22
    2.3.3 Interconversion of thermodynamic and concentration constants 29
  2.4 Types of chemical equilibria in solution and expressions for equilibrium constants 30
    2.4.1 Protonation (acid–base) equilibria 31
    2.4.2 Complexation equilibria 34
    2.4.3 Oxidation–reduction equilibria 35
    2.4.4 Liquid–liquid partition equilibria 38
    2.4.5 Partition equilibria on ion-exchangers 39
    2.4.6 Precipitation equilibria 40

Chapter 3 — Chemical Reactions Affected by Side-Equilibria 43
  3.1 Formulation of conditional equilibrium constants and side-reaction coefficients 44
    3.1.1 Complex-formation equilibria influenced by side-reactions 44
    3.1.2 Conditional precipitation equilibria 51
    3.1.3 Conditional oxidation–reduction potentials 53
    3.1.4 Conditional equilibria in liquid–liquid partition systems . 57

3.1.5 Conditional equilibria on ion-exchangers 59
3.2 Calculation of complicated equilibria with the aid of a computer 59
    3.2.1 Program COMICS 61
    3.2.2 Program HALTAFALL 62

Chapter 4 – Diagrams for Consecutive Equilibrium Systems 67
4.1 Distribution diagrams 67
4.2 Logarithmic diagrams 69
    4.2.1 Logarithmic concentration diagrams for protonation equilibria 70
    4.2.2 Logarithmic solubility diagram for a complexation reaction with a common ion 74
    4.2.3 Logarithmic activity-ratio diagrams 77
4.3 Predominance-area diagrams 79
    4.3.1 Construction of a predominance-area diagram in the absence of solid phase 81
    4.3.2 Construction of a predominance-area diagram in the presence of a solid compound 84
    4.3.3 Significance of predominance-area diagrams 86

Chapter 5 – Constants of Protonation Equilibria 87

Chapter 6 – Stability Constants of Complexes with Inorganic Ligands 109

Chapter 7 – Stability Constants of Complexes with Organic Ligands 142

Chapter 8 – Solubility Products of Sparingly Soluble Salts 209

Chapter 9 – Oxidation-Reduction Potentials 221

Chapter 10 – Side-Reaction Coefficients for Protonation Equilibria . . 251

Chapter 11 – Side-Reaction Coefficients for Complexation Equilibria of Metal Ions 278

Chapter 12 – Diagrams of Precipitation Equilibria 339
12.1 Graphs of conditional solubility products 339
12.2 Logarithmic concentration diagrams for precipitation equilibria 343

    12.2.1 pH-Dependence solubility diagram for boric acid . . . 343
    12.2.2 Logarithmic solubility diagram for iron(III) hydroxide,
         showing pH-dependence 346
    12.2.3 Logarithmic solubility diagram of indium(III) hydroxide, showing dependence on pH and concentration of chloride 348
    12.2.4 Predominance-area diagram for the equilibrium system of indium(III) hydroxo- and chloro-complexes 351

Chapter 13 — Diagrams for Oxidation-Reduction Equilibria 356
13.1 Logarithmic activity-ratio diagrams of some redox systems 356
    13.1.1 Logarithmic activity-ratio diagram of redox forms of iron in aqueous solution 357
    13.1.2 Simplified logarithmic activity-ratio diagram for redox forms of iron 357
    13.1.3 Logarithmic activity-ratio diagram for the redox system of iron in presence of hydrolysis products 359
13.2 Predominance-area diagrams 370
    13.2.1 Diagram of redox, complexation and precipitation equilibria of indium in aqueous solution, showing the dependence on pH 370
    13.2.2 Diagram for redox equilibria in a system of oxygen, hydrogen peroxide and water, showing pH-dependence 376
    13.2.3 Dependence of redox equilibria of cadmium in aqueous solution on the concentration of chloride ions 378

Chapter 14 — Examples of Applications of Tables and Graphs 382
14.1 Calculation of the pH in an isopolyacid equilibrium system in aqueous solution 382
14.2 Treatment of side-reactions in the calculation of masking efficiency 384
14.3 Consideration of the course of a redox reaction by use of a predominance-area diagram 386
14.4 Hydrolysis of metal ions in aqueous solution 390
    14.4.1 Use of the $\log \alpha_{M(OH)}$–pH relationship for the determination of the extent of hydrolysis of a metal ion in aqueous solution 390
    14.4.2 Precipitation of sparingly soluble hydroxides 398
    14.4.3 Use of plots of $pM' = f(pH)$ 399

Index 401

Foreword

It is a pleasure and a privilege to write an introduction to this book. The authors are both well known for their care and thoroughness in teaching the theory of analytical chemistry to students and applying it in research. They are also noteworthy for their ability to write a clear and accurate account of the ways in which stability constants and other physicochemical data can be used to elucidate the nature of the equilibria existing in solutions that are of interest to analysts and to research workers in inorganic and co-ordination chemistry. In this book they show the reader how to steer a safe course between the Scylla of inadequate chemical insight and the Charybdis of mathematical complexity. The various tables given cover most of the systems likely to be found in ordinary chemical practice, and will save the reader many excursions to the library and the reference books. The data given have been extracted from many sources and conscientiously checked against the original literature, and some of them have been recalculated *ab initio*; in this search for the 'best' values, Martell and Smith's invaluable compilation *Critical Stability Constants* (Plenum Press) was, of course, the yardstick used, and it was found that their selection of values could not be faulted. Professors Martell and Smith (and Plenum Press) have very generously given permission to use the values given in that collection.

The compilation of the tables was practically all done by Ladislav Šůcha, and Stanislav Kotrlý felt that Šůcha's name should be the first on the title page, but Šůcha would not agree to this. Šůcha's sad death during the final stages of preparation of the book precluded further argument, and his wishes have been respected, but Stanislav Kotrlý will join with me in saying that this book is a fitting and lasting monument to a very good friend and a fine and talented researcher and teacher.

R. A. CHALMERS
Aberdeen

CHAPTER 1

Chemical reactions and equilibria in analytical chemistry

Development of instrumentation, which has so profoundly changed science and technology over the past 30 years, has resulted in the introduction of numerous new instrumental methods and techniques for chemical analysis. Some of these methods of analytical determination are based on interaction between energy and the matter of the component to be determined in the sample; the need for a chemical reaction with an analytical reagent is thus eliminated from the finish of the analysis. Automated spectrometry (quantometry), which is widely used in metallurgy, is a typical example of such an instrumental method. However, most analytical methods and procedures currently used are still based on chemical reactions.

Reaction chemistry is important, especially for the following stages of chemical analysis. (1) Treatment of the sample in order to obtain a form suitable for further analysis; for many sample materials, dissolution in acids or fusion is necessary in order to convert the components into compounds soluble in water or another solvent. (2) Separation of individual components of the sample material. (3) The determination itself.

As can be seen, chemical reactions will remain indispensable for chemical analysis for the foreseeable future, and new inspiring applications that help to achieve higher selectivity or lower the limit of determination, often in combination with instrumental techniques, will be found.

Analytical chemistry is thus concerned with many aspects of the kinetics and thermodynamics of chemical reactions. From the view-point of chemical kinetics all chemical reactions applicable in analytical chemistry should proceed rapidly (best within the mixing time of reactants) and without formation of side-products, which implies a one-way reaction mechanism. Kinetic methods of chemical analysis represent an exception in this respect: determination of the concentration of a component is in this case based on a measurement influenced by the rate of the reaction.

Decrease in concentration of one of the reactants or increase in concentration of the reaction product may be monitored.

The thermodynamics of a chemical reaction is concerned with the difference in energy between the initial state (the reactants) and the final state (the reaction products). With reactions proceeding at constant temperature and pressure (which is typical for reactions in solution) the energy difference is given by the change in Gibbs free energy ΔG, in J/mole, which is the "driving force" of chemical reaction. The value of ΔG is given by the algebraic sum of the contributions of the reacting species. For example, for a reaction of the type

$$a\text{A} + b\text{B} \rightleftharpoons m\text{M} + n\text{N} \tag{1.1}$$

the change in Gibbs free energy is

$$\Delta G = m\bar{G}_\text{M} + n\bar{G}_\text{N} - a\bar{G}_\text{A} - b\bar{G}_\text{B} \tag{1.2}$$

where $\bar{G}_\text{A} = \mu_\text{A}$, $\bar{G}_\text{B} = \mu_\text{B}$, etc., represent partial molal (Gibbs) free energies of the reacting species, which are equal by definition to chemical potentials, and a, b, etc. are amounts of substance of individual reacting species. The value of μ_i for the ith component is related to its activity a_i by

$$\mu_i = \mu_i^\circ + RT \ln a_i \tag{1.3}$$

where the quantity μ_i° represents the standard chemical potential. It refers to a unit activity of the component i ($a_i = 1$ at constant temperature), which is attained in the standard state of the substance (see further Chapter 2, p. 16).

The expressions in Eqs. (1.2) and (1.3) indicate that the change in Gibbs free energy, characterizing the reaction system, depends both on the kind of reacting compounds and on their concentrations (more exactly on the activities). If the value of ΔG, calculated for a certain composition of the system, has a negative sign, the reaction proceeds in the direction of formation of reaction products; conversely, with a positive value of ΔG the reverse reaction takes place. A zero value of ΔG corresponds to an equilibrium state when the two reactions proceed simultaneously in opposing directions to such an extent that the composition of the system is not changed. An equilibrium state is attained by all reaction systems after a certain time, provided no further changes are introduced from outside, for example, by addition or removal of a reaction species, by changing the temperature of the system, etc.

For analytical applications the equilibrium state should be shifted as much towards the formation of reaction products as possible. The reaction should proceed quantitatively; this means that the unreacted concentration of the component to be determined should not be detectable by the method or device (sensor) used for the determination.

When the expression for the chemical potential in Eq. (1.3) is substituted for the partial molal Gibbs free energies in Eq. (1.2) and the condition for equilibrium, $\Delta G = 0$, is applied, the following equation is obtained

$$-\Delta G^\circ = RT \ln \left[\frac{a_M^m \cdot a_N^n}{a_A^a \cdot a_B^b} \right]_{equil} = RT \ln K \qquad (1.4)$$

which defines the equilibrium constant K for the reaction in Eq. (1.1) in terms of the standard Gibbs free energy ΔG° for the reaction. As can be seen, the value of ΔG° is equal to the sum of the standard chemical potentials of all reacting species. The relation of ΔG° to other thermodynamic quantities is defined by

$$\Delta G^\circ = \Delta H^\circ - T\Delta S^\circ. \qquad (1.5)$$

The values for the standard enthalpy change ΔH° and the standard entropy change ΔS° can be found from tables and used to calculate ΔG° and hence the equilibrium constant for the reaction. More often the equilibrium concentrations (or activities in some cases) are determined experimentally and used for the determination of the equilibrium constant. The value of the equilibrium constant from a tabulation can be used for the calculation of the composition of a system when the equilibrium is reached; thus completeness of the chemical reaction can be estimated. If further information is available, it is also possible to evaluate the salt effect, the change of equilibrium position with temperature and, if necessary, to estimate the influence of various side-reactions on the equilibrium during certain stages of the analytical procedure.

In Chapters 5–9 of this book, a series of tables of equilibrium constants gives data for protonation, complexation, precipitation, and oxidation–reduction equilibria in aqueous solutions; for protonation equilibria some data are given for mixed or non-aqueous solvents. For more detailed or further information the reader should consult the comprehensive tabulations, e.g. [1–6].

In the second part of this book, the selected values of constants are used for evaluation of side-reaction coefficients, and construction of various diagrams and graphs which elucidate the composition of equilibrium systems of some interest or importance.

The data presented in various forms in this book were selected with reference to applicability, and to the importance of particular chemical equilibria in analytical chemistry, electrochemistry, biochemistry, etc. and, of course, in various fields of technology such as corrosion of metals, etc. The final chapter gives several examples that illustrate various possible applications of the data and supplement the graphical material which has been compiled in this monograph.

If some other problems concerning chemical equilibria in solution not covered by this book have to be solved, the necessary information may be found in a selection of monographs [7–18] which can also be recommended for further detailed study in this field.

REFERENCES

[1] L. G. Sillén and A. E. Martell, *Stability Constants of Metal-Ion Complexes*, 2nd Ed., The Chemical Society, London, 1964.
[2] L. G. Sillén and A. E. Martell, *Stability Constants of Metal-Ion Complexes, Supplement No. 1*, The Chemical Society, London, 1971.
[3] J. J. Christensen, D. J. Eatough and R. M. Izatt, *Handbook of Metal Ligand Heats*, 2nd Ed., Dekker, New York, 1975.
[4] G. Kortüm, W. Vogel and K. Andrussow, *Dissociation Constants of Organic Acids in Aqueous Solution*, Butterworths, London, 1961.
[5] D. D. Perrin, *Dissociation Constants of Organic Bases in Aqueous Solution*, Butterworths, London, 1965.
[6] A. E. Martell and R. M. Smith, *Critical Stability Constants*, Vols. 1–5, Plenum Press, New York, 1974–1982.
[7] G. Hägg, *Die theoretischen Grundlagen der analytischen Chemie*, Birkhäuser, Basel, 1950.
[8] I. M. Kolthoff and P. J. Elving (eds.), *Treatise on Analytical Chemistry*, Vol. I, Part 1, Interscience, New York, 1959.
[9] A. Ringbom, *Complexation in Analytical Chemistry*, Interscience, New York, 1963.
[10] H. Freiser and Q. Fernando, *Ionic Equilibria in Analytical Chemistry*, Wiley, New York, 1963.
[11] J. N. Butler, *Ionic Equilibrium*, Addison-Wesley, Reading, Massachusetts, 1964.
[12] G. M. Fleck, *Equilibria in Solution*, Holt, Rinehart and Winston, New York, 1966.
[13] L. Šůcha and S. Kotrlý, *Solution Equilibria in Analytical Chemistry*, Van Nostrand Reinhold, London, 1972.

[14] C. F. Baes, Jr. and R. E. Mesmer, *The Hydrolysis of Cations*, Wiley, New York, 1976.
[15] J. Inczédy, *Analytical Applications of Complex Equilibria*, Horwood, Chichester, 1976.
[16] J. Kragten, *Atlas of Metal-Ligand Equilibria in Aqueous Solution*, Horwood, Chichester, 1978.
[17] Z. Holzbecher, L. Diviš, M. Král, L. Šůcha and F. Vláčil, *Handbook of Organic Reagents in Inorganic Analysis*, Horwood, Chichester, 1976.
[18] O. Budevsky, *Foundations of Chemical Analysis*, Horwood, Chichester, 1979.

CHAPTER 2

Chemical equilibria and equilibrium constants

2.1 DEFINITIONS OF EQUILIBRIUM CONSTANTS

The thermodynamic equilibrium constant of a chemical reaction [cf. Eq. (1.1), p. 12] is defined in terms of the activities of all the reacting species. In the overall product, each particular activity is raised to the power of the relevant stoichiometric coefficient, with a positive sign for a reaction product and a negative sign for a reactant. The expression in Eq. (1.4) (p. 13) relates the thermodynamic equilibrium constant to the standard Gibbs free energy of the reaction (for constant temperature and pressure). After insertion of the values for the universal gas constant and the thermodynamic temperature (here 25° C) the following expression is obtained

$$-\Delta G° = 5.71 \times 10^3 \log K \quad [\text{J/mole}] \tag{2.1}$$

which allows interconversion of the two quantities.

The activity of an equilibrium species in solution is related to the concentration scale used. For chemical equilibria in solution, especially in applications typical for analytical chemistry, concentrations are expressed on the molar scale.
Then

$$a_i = c_i y_i \tag{2.2}$$

where both the activity of the species i and the concentration c_i are in units of mole/litre (symbol M). The activity coefficient y_i can be taken as a correction factor representing the deviation of the substance from ideal behaviour.

The standard state, to which the activities of solutes are referred for reactions in solutions, is represented by a hypothetical $1M$ solution where $a_i = c_i = 1$; this means that $y_i = 1$. In such an ideal solution the activity coefficient of a solute is unity for all concentrations. A solvent of defined composition and salt content is chosen for achieving the standard state, and the activity coefficient y_i approaches unity for very low concentrations of the species i, i.e. a real solution then behaves as an ideal one. Thus the standard state can be realized arbitrarily. For example, a pure solvent, a solvent mixture, or a solution of an inert background salt (e.g. $1M\ NaClO_4$) in a given solvent may be taken as the medium for realization of the standard state. It is evident that this selected medium (called the reference state to differentiate it from the standard state) should be explicitly specified for all data which refer to the particular standard state; unfortunately, this is not always done with published constants. Thus the standard state is a useful abstraction: an ideal solution (of the same nature as an ideal gas, etc.) to which all real solutions can be referred under given conditions to define the deviations from ideality. For all solutes involved in the reaction equilibrium, the standard state should be realized in the same way; it then corresponds to unit concentration of all solutes in this ideal solution, which is the advantage of this concept.

If the activity of a solute i refers to the molality m_i (in units of mole/kg), as is common in physical chemistry, an expression analogous to Eq. (2.2) defines the molal activity coefficient:

$$a_i = m_i \gamma_i. \tag{2.3}$$

It is useful to know the relationship between the two concentration scales:

$$c_i = \frac{m_i \varrho}{1 + 10^{-3} m_i M_i}; \quad \varrho\ [\text{g/cm}^3] \tag{2.4a}$$

or

$$m_i = \frac{c_i}{\varrho - 10^{-3} M_i c_i} \tag{2.4b}$$

where M_i is the molecular weight of species i, and ϱ is the density of the solution. For dilute solutions (concentration less than $10^{-2}M$ or 10^{-2} mole/kg) the terms $10^{-3} c_i M_i$ or $10^{-3} m_i M_i$ become negligible with respect to unity, and the following simplified expression may be used:

$$c_i = m_i \varrho. \tag{2.5}$$

The density ϱ_S of the solvent may also be used instead of that of a dilute solution; thus for very dilute aqueous solutions at room temperature there is no substantial difference between the molal and molar concentration scales.

If molar concentrations are substituted for activities, the thermodynamic equilibrium constant for the reaction in Eq. (1.1) can be written as

$$K = \frac{[M]^m [N]^n}{[A]^a [B]^b} \frac{y_M^m y_N^n}{y_A^a y_B^b} = K_c \frac{y_M^m y_N^n}{y_A^a y_B^b} \qquad (2.6)$$

where square brackets are used to denote equilibrium (free) molar concentrations of the species A, B, etc., to differentiate them from the total (initial) concentrations of the reactants denoted by the symbols c_A, c_B. To differentiate a concentration (stoichiometric) equilibrium constant, expressed in terms of equilibrium concentrations, from the corresponding thermodynamic constant defined by activities, the subscript c will be used.

2.2 THE EFFECT OF TEMPERATURE ON THE EQUILIBRIUM CONSTANT

The change in activity with temperature is small for a solute if the activity is expressed on the molal concentration scale, which is essentially independent of temperature. The activity of a substance is always related to the standard state valid for the given temperature. Thus both the chemical potential and the standard chemical potential change with temperature, but the change in activity is small and can be expressed, for the two concentration scales, as the dependence of a particular activity coefficient on temperature (cf. Section 2.3.2, p. 22). However, the molar concentration of a solute is affected by temperature and so the activity expressed on this concentration scale is also influenced by temperature.

The temperature dependence of the equilibrium constant of a chemical reaction results from the fact that a change in temperature brings about a shift in the equilibrium state, according to the sign and value of the enthalpy change ΔH of the reaction. This is expressed by the van't Hoff equation

$$\frac{d \ln K}{dT} = \frac{\Delta H^\circ}{RT^2} . \qquad (2.7)$$

If ΔH° is practically constant within the temperature range T_1–T_2, integration yields the expression

$$\log K_{T_2} = \log K_{T_1} + \frac{\Delta H^\circ}{19.145} \left(\frac{1}{T_1} - \frac{1}{T_2} \right) \qquad (2.8)$$

The Effect of Temperature on the Equilibrium Constant

Table 2.1

Values of the term $\dfrac{1}{19.14}\left(\dfrac{1}{T_1} - \dfrac{1}{T_2}\right) \times 10^5$ in Eq. (2.8) at various temperatures in the range 0–90°C

Temperature T_1, °C	Temperature T_2, °C								
	10	20	30	40	50	60	70	80	90	100
0	0.6754	1.305	1.892	2.443	2.959	3.444	3.901	4.332	4.739	5.125
10		0.6293	1.217	1.767	2.283	2.769	3.226	3.657	4.064	4.449
20			0.5878	1.138	1.654	2.139	2.596	3.027	3.435	3.820
30				0.5502	1.066	1.552	2.009	2.440	2.847	3.232
40					0.5162	1.001	1.458	1.889	2.297	2.682
50						0.4852	0.9421	1.373	1.780	2.166
60							0.4569	0.8880	1.295	1.681
70								0.4310	0.8383	1.224
80									0.4073	0.7928
90										0.3855

where T is thermodynamic temperature in K and ΔH° should be in units of J/mole. Table 2.1 lists the values of the term $(1/19.14)(1/T_1 - 1/T_2)$ in Eq. (2.8) for the temperature range 0–90°C. For temperature intervals typical for aqueous solutions, only ΔH° values greater than about 40 kJ/mole (i.e. about 10 kcal/mole) can be considered to be constant. For reactions with a smaller enthalpy change, ΔH° depends on temperature:

$$\Delta H^\circ_{T_2} = \Delta H^\circ_{T_1} + \Delta C^\circ_P(T_2 - T_1) \tag{2.9}$$

where the change in heat capacity (at constant pressure) ΔC°_P, can usually be considered to be constant, even though the values of the molar heat capacities C°_P of the reacting substances vary with temperature, since the changes in C°_P for the reaction products and the reactants cancel out in most cases. To calculate the equilibrium constant for a reaction which has an enthalpy change ΔH° smaller than 40 kJ/mole, the following more complicated expression should be used:

$$\log K_{T_2} = \log K_{T_1} + \frac{\Delta H^\circ_{T_1}}{19.14}\left(\frac{1}{T_1} - \frac{1}{T_2}\right)$$
$$+ \Delta C^\circ_P\left[\frac{1}{19.14}\left(\frac{T_1}{T_2} - 1\right) - \frac{1}{8.314}\log\frac{T_1}{T_2}\right]. \tag{2.10}$$

The values of the term in square brackets, calculated for temperatures in the range 0–90°C, are listed in Table 2.2.

Table 2.2

Values of the term $\left[\dfrac{1}{19.44}\left(\dfrac{T_1}{T_2}-1\right)-\dfrac{1}{8.314}\log\dfrac{T_1}{T_2}\right]\times 10^4$ in Eq. (2.10) at various temperatures in the range 0–100°C

Temperature T_1, °C	Temperature T_2, °C									
	10	20	30	40	50	60	70	80	90	100
0	0.3336	1.274	2.740	4.663	6.983	9.649	12.616	15.847	19.307	22.968
10		0.3110	1.189	2.562	4.366	6.547	9.057	11.857	14.910	18.184
20			0.2906	1.113	2.401	4.096	6.150	8.519	11.164	14.053
30				0.2721	1.044	2.254	3.851	5.789	8.027	10.530
40					0.2553	0.9806	2.121	3.627	5.458	7.576
50						0.2401	0.9232	1.999	3.422	5.155
60							0.2262	0.8706	1.887	3.234
70								0.2134	0.8224	1.784
80									0.2017	0.7781
90										0.1910

2.3 THE EFFECT OF IONIC STRENGTH ON THE VALUE OF THE EQUILIBRIUM CONSTANT

The thermodynamic equilibrium constant for a reaction proceeding in solution does not depend on the concentrations of the reacting substances. In contrast, the value of the concentration constant varies to an extent that is controlled by the deviations of the reacting substances from ideality. This is shown by Eq. (2.6). The activity coefficient of an ion in solution depends considerably on the salt content, but for an uncharged species the salt effect is relatively small for the usual range of possible salt concentrations. For both cases, the effect depends on the concentrations of all the ions and on their charges (z), as expressed by the ionic strength I of the solution, defined by

$$I_\mathrm{c} = 0.5 \sum_i z_i^2 c_i \quad \text{or} \quad I_\mathrm{m} = 0.5 \sum_i z_i^2 m_i \qquad (2.11\mathrm{a, b})$$

in accordance with the concentration scale used.

2.3.1 Activity coefficients of non-electrolytes

In the absence of salts the behaviour of non-electrolytes is almost ideal in aqueous solutions, at least up to concentrations of about $1M$ or

Sec. 2.3] The Effect of Ionic Strength on the Value of the Equilibrium Constant

1 mole/kg; the values of the activity coefficients are within the range 1 ± 0.05. For higher concentrations, deviations from ideality are greater.

In the presence of a salt, the activity coefficient γ_N of a non-electrolyte N can be calculated by means of the equation

$$\log \gamma_N = kI \qquad (2.12)$$

where the constant k (the so-called salting-out coefficient) depends on the properties of the non-electrolyte, the solvent, and on the kinds of ions present. Its value is determined by measuring the solubility of a given non-electrolyte for various concentrations of a salt; it is assumed that any decrease in solubility can be interpreted as resulting from an increase in activity of the non-electrolyte, shown as a higher value of γ_N, and vice versa.

Table 2.3

Values of salting-out coefficients k for the calculation of activity coefficients of non-electrolytes in aqueous solutions of some electrolytes [cf. Eq. (2.12)]

Electrolyte	Gases			Non-electrolytes					
	H_2	O_2	CO_2	Ethyl acetate	Acetic acid	Chloro-acetic acid	Benzoic acid	o-Nitro-benzoic acid	Salicylic acid
LiCl	0.066			0.088	0.075				
NaCl	0.094	0.132		0.166	0.066	0.088	0.191	0.180	0.196
KCl	0.078			0.143	0.033	0.026	0.152		0.140
KBr			0.059	0.105	0.014	0.008			
KI			0.045	0.032	0.037				
NaNO$_3$	0.080		0.053	0.074			0.064		−0.078
KNO$_3$	0.061		0.025	0.060	−0.020	−0.043	0.025		−0.006
BaCl$_2$			0.060	0.080		0.056	0.100	0.105	
CH$_3$COONa					−0.014				

In most cases salting-out coefficients have positive signs; thus the activity coefficient of a non-electrolyte in solution increases with increasing ionic strength — the non-electrolyte is "salted-out". A negative sign of the coefficient means that in the presence of the salt, the non-electrolyte is "salted-in", i.e. its solubility is increased because of a decrease in the activity coefficient caused by the higher salt content.

Some values of salting-out coefficients are listed in Table 2.3.

2.3.2 Activity coefficients of electrolytes

Activity coefficients of electrolytes in solutions are determined experimentally, for example by potentiometric or osmometric methods; mean activity coefficients are obtained. For an electrolyte of general formula $M_m N_n$ the mean activity coefficient is defined (for the molar concentration scale) by

$$y_\pm = (y_+^m y_-^n)^{1/(m+n)}. \tag{2.13}$$

An analogous expression is written for the coefficient $\gamma_\pm$ for the molal scale. Mean activity coefficients for some common electrolytes are available in various tabulations (for example see refs. [1–7]), usually for limited ranges of concentration and temperature.

If a value is not found in a table for given conditions, the activity coefficient can be estimated from an expression based on the Debye–Hückel theory. Thus, the mean activity coefficient $y_\pm$ of an electrolyte which yields M^{z+} and N^{z-} ions can be calculated from the expression

$$-\log y_\pm = \frac{A|z_+ \cdot z_-|I_c^{1/2}}{1 + BaI_c^{1/2}} \tag{2.14}$$

which holds satisfactorily for low values of ionic strength (on the molar concentration scale), $I_c \leqq 0.1$ (in mole/l), evaluated by means of Eq. (2.11a). The numerical values of the parameters A and B depend on the thermodynamic temperature and on the relative permittivity of the solvent:

$$\begin{aligned} A &= 1.8246 \times 10^{-6}(\varepsilon T)^{-3/2} \quad [\text{mole}^{-1/2} \cdot l^{1/2} \cdot K^{3/2}] \\ B &= 502.9(\varepsilon T)^{-1/2} \quad\quad\quad\quad [\text{nm}^{-1} \cdot \text{mole}^{-1/2} \cdot l^{1/2} \cdot K^{1/2}]. \end{aligned} \tag{2.15}$$

The values of these parameters are much influenced by the accuracy with which ε is available for the given temperature. If the ionic strength is expressed as a molality, the values of the two parameters also involve the square root of the density of the solvent [cf. Eqs. (2.5), (2.11) and (2.15)]: $A_m = A_c \varrho_s^{1/2}$; $B_m = B_c \varrho_s^{1/2}$. Since the density of water approaches 1 at room temperature (0.99707 g/ml for 25°C), the numerical values of these parameters on the two concentration scales differ negligibly. For such solutions $c \sim m$ (cf. p. 17) and it is possible to approximate $\gamma_\pm \sim y_\pm$ and so Eq. (2.14) is also valid for activity coefficients on the molal scale. Calculated values of the two parameters for the temperature range 0–100° C are summarized in Table 2.4.

Sec. 2.3] The Effect of Ionic Strength on the Value of the Equilibrium Constant

Table 2.4
Constants of the extended Debye–Hückel expression for 0–100°C (for ion-size parameters a in nm)

Temperature, °C	Ionic strength of aqueous solution expressed as			
	I_c, mole/l		I_m, mole/kg	
	A_c	B_c	A_m	B_m
0	0.4918	3.248	0.4918	3.248
5	0.4952	3.256	0.4952	3.256
10	0.4989	3.264	0.4988	3.264
15	0.5028	3.273	0.5026	3.272
20	0.5070	3.282	0.5066	3.279
25	0.5115	3.291	0.5108	3.286
30	0.5161	3.301	0.5150	3.294
35	0.5211	3.312	0.5196	3.302
40	0.5262	3.323	0.5242	3.310
45	0.5317	3.334	0.5291	3.318
50	0.5373	3.346	0.5341	3.326
55	0.5432	3.358	0.5393	3.334
60	0.5494	3.371	0.5448	3.343
65	0.5558	3.384	0.5504	3.351
70	0.5625	3.397	0.5562	3.359
75	0.5695	3.411	0.5623	3.368
80	0.5767	3.426	0.5685	3.377
85	0.5842	3.440	0.5750	3.386
90	0.5920	3.456	0.5817	3.396
95	0.6001	3.471	0.5886	3.404
100	0.6086	3.488	0.5958	3.415

According to Eq. (2.14), the mean activity coefficient is a function of the mean ionic diameter a (in nm), which is regarded as the minimum distance of approach of the two solvated ions in solution. In reality, a is an experimental parameter which is obtained from measurements of activity coefficients. In order to estimate the activity coefficient for a solvated ion i carrying a charge z_i, Eq. (2.14) is written in the form

$$-\log y_i = \frac{Az_i^2 I_c^{1/2}}{1 + Ba_i I_c^{1/2}} \tag{2.16}$$

where a_i is an estimate of the effective diameter of the particular ion. In order to facilitate activity-coefficient calculations, Kielland [8] assigned empirical values of the parameter a_i to common individual ions (see Table 2.5). Kielland's values may be used for approximate evaluation

Table 2.5

Values of the ion-size parameter a_i for calculation of individual activity coefficients by Kielland [cf. Eq. (2.16)]

a, nm	Ions
1.1	$Ce^{4+}, Sn^{4+}, Th^{4+}, Zr^{4+}$
0.9	H^+
	$Al^{3+}, Ce^{3+}, Cr^{3+}, Fe^{3+}, In^{3+}, La^{3+}, Sc^{3+}$
0.8	Be^{2+}, Mg^{2+}
0.7	$(C_3H_7)_3NH^+$
	$(CH_2)_n(COO)_2^{2-}$ $(n = 5, 6)$
0.6	$Li^+, (C_2H_5)_4N^+$
	$C_6H_5COO^-, C_6H_4(OH)COO^-, C_6H_5CH_2COO^-$
	$Ca^{2+}, Co^{2+}, Cu^{2+}, Fe^{2+}, Mn^{2+}, Ni^{2+}, Sn^{2+}, Zn^{2+}$
	$C_6H_4(COO)_2^{2-}$
0.5	$Ba^{2+}, Cd^{2+}, Hg^{2+}, Ra^{2+}, Sr^{2+}$
	$CHCl_2COO^-, CCl_3COO^-$
	$(C_2H_5)_3NH^+$
	$CH_2(COO)_2^{2-}, (CH_2COO)_2^{2-}, (CHOHCOO)_2^{2-}, S^{2-}, S_2O_4^{2-}, WO_4^{2-}$
	$Fe(CN)_6^{4-}, C_6H_5O_7^{3-}$ (citrate)
0.45	$CH_3COO^-, CH_2ClCOO^-, (CH_3)_4N^+, NH_2CH_2COO^-$
	$Pb^{2+}, CO_3^{2-}, MoO_4^{2-}, SO_3^{2-}$
	$C_2O_4^{2-}, HC_6H_5O_7^{2-}$ (citrate)
0.4—0.45	$Na^+, CdCl^+$
	$H_2AsO_4^-, HCO_3^-, ClO_2^-, IO_3^-, H_2PO_4^-, HSO_3^-$
0.4	$Hg_2^{2+}, {}^+NH_3CH_2COOH, (CH_3)_3NH^+, C_2H_5NH_3^+$
	$CrO_4^{2-}, HPO_4^{2-}, SO_4^{2-}, S_2O_3^{2-}, S_2O_6^{2-}, S_2O_8^{2-}, SeO_4^{2-}$
	$Fe(CN)_6^{3-}, PO_4^{3-}$
0.35	$ClO_3^-, ClO_4^-, BrO_3^-, IO_4^-, F^-, OH^-, MnO_4^-$
	$HCOO^-, H_2C_6H_5O_7^-$ (citrate), $CH_3NH_3^+, (CH_3)_2NH_2^+$
0.3	K^+
	$Cl^-, Br^-, I^-, CN^-, NO_2^-, NO_3^-$
2.5	$Ag^+, Cs^+, NH_4^+, Rb^+, Tl^+$

of the activity coefficients of various ions, but if a certain ion is not found among those listed in Table 2.5, a value for a_i is selected according to its ionic charge and any chemical similarity to other ions. As can be seen from the values in Table 2.5, the size of an aquated ion is between 0.3 and 1.1 nm. The lower value 0.3 was proposed by Güntelberg [9] to approximate the value of an unknown mean activity coefficient; hence

$$-\log \gamma_\pm \sim -\log y_\pm = A|z_+ z_-| \frac{I^{1/2}}{1 + I^{1/2}} \qquad (2.17)$$

Sec. 2.3] The Effect of Ionic Strength on the Value of the Equilibrium Constant

where z_i^2 should be inserted instead of the product of the ionic charges to calculate the activity coefficient of an individual ion. The values calculated by this equation tend to be low in comparison with experimental data, even at low ionic strength. Guggenheim therefore added a linear term to account for the contribution from all specific effects referring to individual properties of the electrolyte [10]:

$$-\log y_\pm = A |z_+ z_-| \frac{I^{1/2}}{1 + I^{1/2}} - CI. \qquad (2.18)$$

Davies proposed a modified version of this expression [4]:

$$-\log y_\pm = 0.5 |z_+ z_-| \left(\frac{I^{1/2}}{1 + I^{1/2}} - 0.3I \right) \qquad (2.19)$$

which can conveniently be used for calculation up to ionic strength 0.5 if the values of the parameters a and C are not known for the electrolyte. This expression yields values of activity coefficients for electrolytes of 1:1 stoichiometry (ionic charges from 1 to 5), which are in agreement with experimental results within the limits $\pm 1\%$ for $I = 0.05$, $\pm 3\%$ for $I = 0.1$, $\pm 10\%$ for $I = 0.5$. The calculation of activity coefficients by Eq. (2.17) or (2.19) can easily be done with the aid of the values of the expressions $I^{1/2}/(1 + I^{1/2})$ or $I^{1/2}/(1 + I^{1/2}) - 0.3I$, given in Table 2.6.

Table 2.6
Values of functions for calculation of activity coefficients by the Debye–Hückel, Davies and Pitzer equations, respectively, for various ionic strengths

I	$I^{0.5}/(1 + I^{0.5})$	$[I^{0.5}/(1 + I^{0.5})] - 0.3I$	$F(I)$†
0	0.0000	0.0000	—
0.001	0.0306	0.0303	0.9488
0.002	0.0428	0.0422	0.9284
0.003	0.0519	0.0510	0.9131
0.004	0.0595	0.0583	0.9004
0.005	0.0660	0.0645	0.8893
0.006	0.0719	0.0701	0.8795
0.007	0.0772	0.0751	0.8705
0.008	0.0821	0.0797	0.8623
0.009	0.0866	0.0839	0.8546
0.010	0.0909	0.0879	0.8474
0.020	0.1239	0.1179	0.7918
0.030	0.1476	0.1386	0.7518

Table 2.6 (Continued)

I	$I^{0.5}/(1 + I^{0.5})$	$[I^{0.5}/(1 + I^{0.5})] - 0.3I$	$F(I)$†
0.040	0.1667	0.1547	0.7199
0.050	0.1827	0.1677	0.6929
0.060	0.1968	0.1788	0.6695
0.070	0.2092	0.1882	0.6487
0.080	0.2205	0.1965	0.6300
0.090	0.2308	0.2038	0.6130
0.100	0.2402	0.2102	0.5974
0.150	0.2792	0.2342	0.5340
0.200	0.3090	0.2490	0.4863
0.250	0.3333	0.2583	0.4482
0.300	0.3539	0.2639	0.4166
0.350	0.3717	0.2667	0.3898
0.400	0.3874	0.2674	0.3666
0.450	0.4015	0.2665	0.3462
0.500	0.4142	0.2642	0.3281
0.550	0.4258		0.3119
0.600	0.4365		0.2972
0.650	0.4464		0.2840
0.700	0.4555		0.2718
0.750	0.4641		0.2607
0.800	0.4721		0.2504
0.850	0.4797		0.2409
0.900	0.4868		0.2321
0.950	0.4936		0.2238
1.000	0.5000		0.2162
1.100	0.5119		0.2022
1.200	0.5228		0.1899
1.300	0.5328		0.1789
1.400	0.5420		0.1691
1.500	0.5505		0.1602
1.600	0.5585		0.1521
1.700	0.5659		0.1448
1.800	0.5730		0.1381
1.900	0.5796		0.1319
2.000	0.5858		0.1263
2.500	0.6126		0.1036
3.000	0.6340		0.0873
3.500	0.6517		0.0752
4.000	0.6667		0.0659
4.500	0.6796		0.0586
5.000	0.6910		0.0526

† $F(I) = [1 - (1 + 2I^{0.5} - 2I)\exp(-2I^{0.5})]/4I$

Sec. 2.3] The Effect of Ionic Strength on the Value of the Equilibrium Constant

For Eqs. (2.17–2.19) it is assumed that the ionic strength gives an adequate expression of the concentration effects of all ions present in solution, irrespective of their nature. However, changes in concentration or in the type of ions can bring about rearrangement of the ionic cloud and thus affect the value of the parameter a. With high concentrations even the relative permittivity of the medium undergoes a change; it is also necessary to consider that a certain amount of the solvent is bound in the solvation spheres of the ions. These factors have to be taken into account in the calculation of activity coefficients from the Debye–Hückel expression for $I > 0.1$. This is done by means of additional terms which can be evaluated empirically [cf. Eqs. (2.18) and (2.19)]. Robinson and Stokes [2] assumed that the additional term represents a correction for the change in real concentration of solvated ions, arising from the considerable amount of solvent bound in the solvation sheaths. They derived values for the parameter a for electrolytes of the type M^+N^- and $M^{2+}(N^-)_2$, within the range 0.35–0.62 nm. The calculated values of $\gamma_\pm$ agree well with experimental data up to quite high molalities (cf. also ref. [11]); however, the Robinson and Stokes equation is not generally applicable.

In a mixture of electrolytes, the parameter a, expressing the effective size of the ion for which the activity coefficient is to be calculated, is changed according to the kind and concentration of all electrolytes present. According to Harned's rule, the Debye–Hückel expression is applicable in the form shown in Eq. (2.17) if an additional term involving the interaction coefficients is added. The interaction coefficient expresses the effect of all the ions present, on the value of the activity coefficient. The effect of each ion is proportional to the contribution of the particular electrolyte to the total concentration. An analogous approach has been recommended by Baes and Mesmer [1], who used, on the basis of work published by Pitzer et al. [13, 14], the following expressions to calculate activity coefficients for a mixture of electrolytes. If a mixture of electrolytes of the type M^+N^- has a common ion M^+, the molal activity coefficient is calculated as

$$-\log \gamma_M = \frac{A z_M^2 I^{1/2}}{1 + I^{1/2}} - \sum_N B_{MN} m_N. \qquad (2.20)$$

Analogously the activity coefficient for a common anion in such a mixture of electrolytes is given by

$$-\log \gamma_N = \frac{A z_N^2 I^{1/2}}{1 + I^{1/2}} - \sum_M B_{MN} m_M. \qquad (2.21)$$

Table 2.7

Interaction coefficients for some electrolytes, allowing calculation of activity coefficients according to Pitzer;† ionic strength should be expressed as molality units (mole/kg)

Cation M	Anion N	Electrolyte	$B_{MN}(0)$	$B_{MN}(\infty)$
H^+	Cl^-	HCl	0.274	0.1351
H^+	ClO_4^-	$HClO_4$	0.228	0.1515
H^+	NO_3^-	HNO_3	−0.304	−0.0217
Li^+	Cl^-	$LiCl$	0.245	0.1178
Li^+	ClO_4^-	$LiClO_4$	0.382	0.1523
Li^+	OH^-	$LiCl$	0.424	0.334
Na^+	Cl^-	$NaCl$	0.159	0.0486
Na^+	ClO_4^-	$NaClO_4$	0.163	0.0240
Na^+	NO_3^-	$NaNO_3$	−0.463	−0.0913
Na^+	OH^-	$NaCl$	0.339	0.093
K^+	Cl^-	KCl	0.099	0.0187
K^+	OH^-	KCl	0.279	0.105
Mg^{2+}	Cl^-	$MgCl_2$	1.16	0.217
Mg^{2+}	ClO_4^-	$Mg(ClO_4)_2$	1.33	0.451
Ca^{2+}	Cl^-	$CaCl_2$	1.09	0.179
Ca^{2+}	ClO_4^-	$Ca(ClO_4)_2$	1.34	0.291
Sr^{2+}	Cl^-	$SrCl_2$	1.11	0.152
Ba^{2+}	Cl^-	$BaCl_2$	0.99	0.110
Ba^{2+}	OH^-	$BaCl_2$	0.79	−0.140
$SrOH^+$	Cl^-	$SrCl_2$	0.08	0.08
$BaOH^+$	Cl^-	$BaCl_2$	0.11	0.11
Mn^{2+}	Cl^-	$MnCl_2$	1.09	0.165
Fe^2	Cl^-	$FeCl_2$	1.05	0.187
Co^{2+}	Cl^-	$CoCl_2$	1.04	0.203
Ni^{2+}	Cl^-	$NiCl_2$	1.09	0.203
Cu^{2+}	Cl^-	$CuCl_2$	0.98	0.124
Zn^{2+}	ClO_4^-	$Zn(ClO_4)_2$	1.38	0.358
Pb^{2+}	ClO_4^-	$Pb(ClO_4)_2$	1.22	0.184
Al^{3+}	Cl^-	$AlCl_3$	4.10	0.356
Sc^{3+}	Cl^-	$ScCl_3$	3.70	0.331
La^{3+}	Cl^-	$LaCl_3$	3.77	0.254
La^{3+}	ClO_4^-	$La(ClO_4)_3$	4.75	0.489
Th^{4+}	Cl^-	$ThCl_4$	9.28	0.300

† See ref. [12], p. 440.

The interaction coefficient B_{MN} has a different value for every electrolyte and is a function of ionic strength. It can be calculated from the limiting values of $B_{MN}(0)$ and $B_{MN}(\infty)$ which correspond to the two extreme values of ionic strength ($I \to 0$ and $I \to \infty$), by the equation

$$B_{MN} = B_{MN}(\infty) + [B_{MN}(0) - B_{MN}(\infty)] F(I) \tag{2.22}$$

where

$$F(I) = [1 - (1 + 2I^{1/2} - 2I)\exp(-2I^{1/2})]/4I. \tag{2.23}$$

A list of interaction coefficients $B_{MN}(0)$ and $B_{MN}(\infty)$ is given in Table 2.7.

2.3.3 Interconversion of thermodynamic and concentration constants

Concentration equilibrium constants, which are the type of constant most often obtained experimentally, can be converted into thermodynamic constants by means of Eq. (2.6) rearranged according to the type of reaction. Tabulated or calculated values of activity coefficients are inserted in such an equation. However, it is more rigorous to determine a set of concentration equilibrium constants for the widest possible range of ionic strength and to extrapolate the relationship $\log K_c = f(I)$ to $I \to 0$. Conversely, the thermodynamic equilibrium constants found in tabulations (sometimes also as the values of $\Delta G°$) can be used to calculate concentration constants with the aid of calculated or experimentally determined activity coefficients.

If the evaluation of activity coefficients can be based on the expression in Eq. (2.17), the concentration equilibrium constant for a reaction in which ions participate can be expressed as

$$\log K_c = \log K + \frac{0.51 \Delta z^2 I^{1/2}}{1 + I^{1/2}} \quad \text{(for an aqueous solution and 25°C)} \tag{2.24}$$

where Δz^2 is the algebraic sum of the squares of the charges of the ions taking part in the reaction. The squares are assigned a positive sign for the products, and a negative sign for the reactants (see examples in Section 2.4).

For a mixture of electrolytes it is more correct to convert the constants according to Eqs. (2.20) and (2.21); hence

$$\log K_c = \log K + \frac{A \Delta z^2 I^{1/2}}{1 + I^{1/2}} + \Delta BI \quad \text{(for an aqueous solution and 25°C).} \tag{2.25}$$

The parameter ΔB is calculated from the values of the interaction coefficients of the reacting substances for the given ionic strength.

The calculation of the composition of an equilibrium system, the assessment of the completeness of a reaction when equilibrium is reached, etc., should best be based on concentration constants determined under conditions similar to those chosen for the reaction. The dependence of $-\log y_i$ on ionic strength [cf. Eq. (2.16)] indicates that the values of concentration equilibrium constants change most when $I < 0.1$. As shown in Fig. 2.1 by the family of curves for the dependence of $\Delta \log K = \log K_c - \log K = f(I)$, calculated with the aid of the Davies equation (2.19) for various types of reactions, the concentration constants are not much affected by ionic strength at higher salt concentrations ($0.2 < I < 0.5$). However, the higher the charges of the ions participating in the reaction, the stronger is the dependence of the concentration equilibrium constant on ionic strength.

2.4 TYPES OF CHEMICAL EQUILIBRIA IN SOLUTION AND EXPRESSIONS FOR EQUILIBRIUM CONSTANTS

2.4.1 Protonation (acid–base) equilibria

Acid–base equilibria in solutions are commonly described in terms of dissociation processes with the reaction scheme

$$HB^z \rightleftharpoons H^+ + B^{z-1} \tag{2.26}$$

and defined by the dissociation constant

$$K_a = \frac{a_{H^+} a_{B^{z-1}}}{a_{HB^z}}. \tag{2.27}$$

Dissociation of an acid can proceed, however, only in the presence of a substance able to accept the released proton; therefore, the dissociation reaction should be written as

$$HB^z + S \rightleftharpoons SH^+ + B^{z-1}. \tag{2.28}$$

Hence the equilibrium constant should be considered as the protolysis constant of the acid HB^z in the solvent S:

$$K_{HB^z,S} = \frac{a_{SH^+} a_{B^{z-1}}}{a_{HB^z} a_S}. \tag{2.29}$$

In dilute solutions the activity of the solvent, a_S, remains constant. A common convention is to take pure solvent for the standard state; then $a_S = 1$, and if a simplified notation for the solvated proton is used, Eq. (2.29) becomes identical with the equation for the dissociation constant.

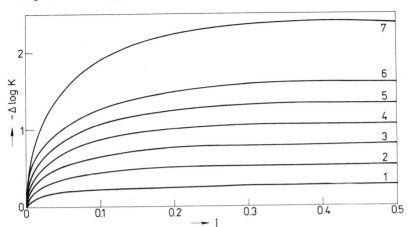

Fig. 2.1 — Conversion of thermodynamic equilibrium constant into the corresponding concentration constant, based on the Davies expression and with the assumption that the activity coefficient of an electrically uncharged species is equal to unity,

$$\Delta \log K = \log K_c - \log K = z^2\, 0.5 \left(\frac{I^{0.5}}{1 + I^{0.5}} - 0.3 I \right),$$

for the various types of ionic reactions. $1 - M^+ + N^- = MN$; $2 - M^{2+} + N^- = MN^+$, or $M^+ + N^{2-} = MN^-$; $3 - M^{2+} + 2 N^- = MN_2$, or $2 M^+ + N^{2-} = M_2N$; $4 - M^{2+} + N^{2-} = MN$; $5 - M^{3+} + 2 N^- = MN_2^+$; $6 - M^{3+} + N^{2-} = MN^+$; $7 - M^{3+} + N^{3-} = MN$.

The numerical value of $K_{HB^z,S}$ is also the same as for the constant K_a, which actually represents a conventional dissociation constant of the acid HB^z in the solvent S.

An acid–base equilibrium can also be considered in terms of protonation of the Brønsted base B^{z-1}:

$$B^{z-1} + H^+ \rightleftharpoons HB^z \qquad (2.30)$$

characterized by the protonation constant

$$K_H = \frac{a_{HB^z}}{a_{B^{z-1}} a_{H^+}}. \qquad (2.31)$$

The protonation reaction can again be simplified by neglecting the solvent molecule from which the proton is transferred to the base. The protonation constant K_H includes the activity of the solvent, which is conventionally unity for very dilute solution. The protonation and dissociation constants are related by

$$K_H = 1/K_a. \qquad (2.32)$$

The advantage of considering the acid–base equilibrium as a protonation process instead of as a dissociation process becomes evident in the case of a polybasic species which is protonated consecutively to yield a polyprotic acid H_nB:

$$B^{n-} + H^+ \rightleftharpoons HB^{1-n}; \quad K_{H1}$$
$$HB^{1-n} + H^+ \rightleftharpoons H_2B^{2-n}; \quad K_{H2}$$
$$\text{etc.} \tag{2.33}$$
$$H_{n-1}B^- + H^+ \rightleftharpoons H_nB; \quad K_{Hn}$$
$$H_nB + H^+ \rightleftharpoons H_{n+1}B^+; \quad K_{Hn+1}$$
$$\text{etc.}$$

The subscript to K_H indicates the ordinal number of the protonation step. Direct formation of each protonated species from the base B^{n-} can be expressed by the overall reaction

$$B^{n-} + kH^+ \rightleftharpoons H_kB^{k-n} \tag{2.34}$$

and by the overall constant

$$\beta_{Hk} = K_{H1}K_{H2} \ldots K_{Hk} \tag{2.35}$$

where k denotes the number of protons involved in the overall protonation.

If a polyprotic equilibrium of the acid H_nB is instead characterized by a series of stepwise dissociation constants, the subscript n for the dissociation constant K_{an} gives the number of protons released up to the particular dissociation step, but the symbol for the dissociation constant does not give direct information about the composition of the acid. For further protonation of the acid H_nB it is necessary to devise another notation. A further advantage of using protonation constants is the analogy between protonation and complex-formation reactions that then becomes apparent.

The calculation of a concentration protonation constant from the thermodynamic value can be based on Eq. (2.24) where

$$\Delta z^2 = z^2_{H_nB} - z^2_{H_{n-1}B} - 1 = 2z_{H_{n-1}B} \tag{2.36}$$

since $z_{H_nB} = z_{H_{n-1}B} + 1$. Thus for a concentration protonation constant a simplified expression is obtained (for an aqueous solution and 25°C):

$$\log(K_{Hn})_c = \log K_{Hn} + z_{H_{n-1}B} \frac{I^{1/2}}{1 + I^{1/2}}. \tag{2.37}$$

This equation indicates that the thermodynamic and concentration constants for protonation of an uncharged molecule should have the same value, and that the concentration constant should be independent of ionic strength. In reality such protonation constants do vary with ionic strength, but to a much smaller extent than the constants for protonation of charged bases.

For protonation reactions realized at constant ionic strength, so-called "mixed constants" are defined as

$$K^*_{Hn} = \frac{[H_n B]}{[H_{n-1} B] a_{H^+}}. \tag{2.38}$$

These constants are found in experiments where pH values are measured with glass and calomel electrodes, standardized with the buffers constituting the practical pH(S) scale recommended internationally.† In this case pH(S) = pa_H and so 10^{-pH} is nearly equal to the activity of the hydrogen ion in solution. The value of the ratio of the concentrations of the two equilibrium species of the acid, e.g. $[H_n B]/[H_{n-1} B^-]$, is determined by some other approach. The analysis of a set of log K^*_H values determined for a sufficiently wide range of ionic strength gives both the parameters of the Debye–Hückel expression and the thermodynamic value of the protonation constant (obtained by extrapolation to zero ionic strength if a pure solvent is chosen to realize the standard state). For example, if $n = 1$, $z_{H_{n-1}B} = -1$, and the extended Debye–Hückel equation [cf. Eq. (2.18)] is used, an expression similar to that in Eq. (2.37) is obtained:

$$\log K^*_H = \log K_H - \frac{AI^{1/2}}{1 + aBI^{1/2}} + CI \tag{2.39}$$

where $C = C_{HB} - C_B$. The ionic strength represents an independent variable which is far less subject to experimental error. When a sufficiently large experimental set of pairs of values of $\log(K^*_H)_i$, I_i is obtained, it is possible to use least-squares regression analysis of the non-linear function $\log(K^*_H)_i = f(I; \log K_H, a, C)$ to determine the three parameters $\log K_H$, a and C. Various computer programs have been written for such numerical analysis; for example, the program DH-LETAG [15] uses a minimizing subroutine not based on derivative-determination, which was adapted from the successful LETAGROP programs of Sillén et al. [16–18].

---

† The pH(S) scale was realized for molal concentrations; hence pH(S) = p$(a_H)_m$ = p$(a_H)_c$ + log ϱ_S, where ϱ_S is the density of the solvent. For aqueous solutions and temperatures up to 35°C this correction is less than 0.003 pH unit.

The computed thermodynamic constant and the values of the Debye–Hückel parameters are then used for calculation of the mixed constant for any given ionic strength, by Eq. (2.39).

A mixed constant can also be converted into the concentration constant by

$$(K_{Hn})_c = K^*_{Hn} y_{H^+}. \tag{2.40}$$

2.4.2 Complexation equilibria

The complex-formation equilibrium of a metal ion M (central ion) with the complex-forming species L (ligand) generally proceeds in successive steps

$$\begin{aligned} M + L &\rightleftharpoons ML; & K_1 \\ ML + L &\rightleftharpoons ML_2; & K_2 \\ &\vdots \\ ML_{i-1} + L &\rightleftharpoons ML_i; & K_i \\ &\vdots \\ ML_{n-1} + L &\rightleftharpoons ML_n; & K_n \end{aligned} \tag{2.41}$$

The equilibrium of the ith step is characterized by a stepwise (consecutive) stability constant, K_i.

The overall equilibrium, for example

$$M + kL \rightleftharpoons ML_k \tag{2.42}$$

is expressed by the overall stability constant

$$\beta_k = \frac{a_{ML_k}}{a_M a_L^k} = K_1 K_2 \ldots K_k. \tag{2.43}$$

For a polynuclear complex the formation equilibrium is

$$mM + nL \rightleftharpoons M_m L_n \tag{2.44}$$

and the overall stability constant is defined as

$$\beta_{nm} = \frac{a_{M_m L_n}}{a_M^m a_L^n}. \tag{2.45}$$

The calculation of stability constants for solutions differing in ionic strength is a difficult problem, because species involved in complex-formation equilibria are often complicated in nature. Thus it is preferred to determine the stability constants in solutions with fairly high concentration ($I > 0.5$) of an inert background electrolyte. The equilibrium concentration constants are then not much affected by change in ionic strength. Such constants should be used under experimental conditions close to those used for their determination, but they are preferred in general practice.

Experimental study of complex formation within a particular pH region may lead to a constant for an equilibrium in which hydrogen ions participate. For example, for stepwise formation

$$ML_{n-1} + HL \rightleftharpoons ML_n + H^+; \quad *K_n \tag{2.46}$$

or for the overall reaction

$$M + nHL \rightleftharpoons ML_n + nH^+; \quad *\beta_n \tag{2.47}$$

the equilibrium constants are written as

$$*K_n = \frac{a_{ML_n} a_{H^+}}{a_{ML_{n-1}} a_{HL}}; \quad *\beta_n = \frac{a_{ML_n} a_{H^+}^n}{a_M a_{HL}^n}. \tag{2.48a, b}$$

The actual stability constants are then calculated from

$$K_n = *K_n K_{H1}; \quad \beta_n = *\beta_n K_{H1}^n. \tag{2.49}$$

2.4.3 Oxidation–reduction equilibria

A general oxidation–reduction (redox) reaction

$$n_B A_{red} + n_A B_{ox} \rightleftharpoons n_B A_{ox} + n_A B_{red} \tag{2.50}$$

consists of two partial redox reactions (redox couples) both of which are conventionally written as reductions:

$$A_{ox} + n_A e^- \rightleftharpoons A_{red} \tag{2.51}$$

$$B_{ox} + n_B e^- \rightleftharpoons B_{red}. \tag{2.52}$$

For an overall chemical reaction, the redox couples are combined so that the electrons provided by the oxidation of one reactant, i.e. A_{red}, are consumed in the reduction of the other reacting species, B_{ox}. In a spontaneous process, the transfer of electrons from A_{red} to B_{ox} brings about release of energy which is the useful work that can be done by the reaction system. If the temperature and pressure are constant, this work is given by the Gibbs free energy of the reaction. Thus the following expression holds for equilibrium:

$$-\Delta G° = nFE° = RT \ln K \tag{2.53}$$

where K is the equilibrium constant of the redox reaction

$$K = \frac{a_{A_{ox}}^{n_B} a_{B_{red}}^{n_A}}{a_{A_{red}}^{n_B} a_{B_{ox}}^{n_A}} \tag{2.54}$$

$n = n_A n_B$ is the total number of electrons exchanged, F is the Faraday constant (96487 C/equivalent), and $E°$ the standard e.m.f. of a cell corresponding to the chemical reaction in Eq. (2.50).

For an electrochemical cell the partial redox reactions correspond to the half-reactions representing the electrode processes. The tendency of an oxidized form to accept electrons is represented by the oxidation–reduction (redox) potential E, which depends on the activities of the two redox forms, as expressed by the Nernst–Peters equation, e.g. for the redox pair A_{ox}, A_{red}:

$$E = E°_{A_{ox},A_{red}} + \frac{RT}{n_A F} \ln \frac{a_{A_{ox}}}{a_{A_{red}}} \tag{2.55}$$

The Nernst–Peters equation can be derived from the expression for the Gibbs free energy ΔG of the chemical reaction which takes place in a cell consisting of a standard hydrogen electrode and the redox electrode representing the redox half-reaction under consideration, e.g. $A_{ox} + n_A \rightleftharpoons A_{red}$. In the overall reaction of this cell, the oxidized form of the substance, A_{ox}, reacts with hydrogen to yield the reduced form A_{red} and hydrogen ions. For 1 atm pressure of hydrogen and $a_{H^+} = 1$, the e.m.f. of the cell represents the redox potential of the pair A_{ox}, A_{red}. As the standard potential of the standard hydrogen electrode is equal to zero by convention for all temperatures, the e.m.f. of this cell is equal to the standard redox potential of the redox couple, $E = E°_{A_{ox},A_{red}}$, if the activities of the two redox forms are equal. Thus the standard redox potential is a characteristic quantity for every redox couple.

On combination of Eq. (2.54) with the Nernst–Peters equations for the two redox couples, the expression for the equilibrium constant of the overall reaction in Eq. (2.50) is obtained:

$$\log K = \frac{n_A n_B F}{2.303 RT} (E°_{B_{ox},B_{red}} - E°_{A_{ox},A_{red}}). \tag{2.56}$$

On comparison with Eq. (2.53) it can be seen that the standard potential $E°$ is equal to the difference between the standard redox potentials of the two redox couples forming the overall redox reaction.

Although all other types of chemical equilibria (both for homogeneous and heterogeneous reaction systems) are characterized by equilibrium constants, for oxidation–reduction processes it is convenient to continue to consider separately the two redox couples involved in the overall redox reaction. To each redox pair it is also possible to assign a corresponding ΔG° value from

$$-\Delta G^\circ = 96487 n E^\circ \quad [\text{J/mole}] \tag{2.57}$$

where the standard redox potential is expressed in volts.

Any oxidation–reduction half-reaction (redox couple) can also be characterized, according to a proposal by Sillén [19], by a partial equilibrium constant K; for example, for the redox couple A_{ox}, A_{red} the constant K is written as

$$K = \frac{a_{A_{red}}}{a_{A_{ox}} a_e} \tag{2.58}$$

where a_e denotes the hypothetical activity of the electron. This equilibrium constant can be calculated from the standard redox potential as

$$\log K = \frac{n_A F}{2.303 RT} E^\circ_{A_{ox}, A_{red}} \tag{2.59}$$

Further, in analogy to the definition of $\text{pH} = -\log a_{H^+}$, the electrode potential can be expressed in terms of the quantity pe which is dimensionless and is defined† as

$$\text{p}e = -\log a_e = \frac{EF}{RT \ln 10}. \tag{2.60a}$$

For a system where all the chemical substances are at unit activity we can define a quantity

$$\text{p}e^\circ = \frac{E^\circ F}{RT \ln 10} = \log K. \tag{2.60b}$$

Hence at 25°C, $\log K = nE^\circ/0.059$; i.e. p$e^\circ = E^\circ/0.059$ and p$e = E/0.059$ [see also Eqs. (4.16) and (4.17)].

The use of such half-reaction equilibrium constants instead of the corresponding standard (or formal) redox potentials is of some advantage for calculating concentrations of reacting species and especially for formulating equilibrium constants when other types of reactions are also involved.

---

† Sillén [19] used pE, not pe, but we think this is confusing, since it uses the same symbol (E) for two different quantities.

2.4.4 Liquid–liquid partition equilibria

In the transfer of a substance A dissolved in one liquid (e.g. in water) into another liquid (immiscible with the first) a distribution (extraction) equilibrium is attained, and this is characterized by the partition constant

$$(A)_{aq} \rightleftharpoons (A)_{org}; \quad K_{D,A}^0 = \frac{(a_A)_{org}}{(a_A)_{aq}}. \tag{2.61}$$

The thermodynamic value $K_{D,A}^0$, defined by the activities of A in the two phases, can be recalculated to a concentration partition constant defined for a particular concentration (here molar) scale

$$K_{D,A} = \frac{[A]_{org}}{[A]_{aq}} = K_{D,A}^0 \frac{(y_A)_{aq}}{(y_A)_{org}}. \tag{2.62}$$

The partition constant can represent the distribution equilibrium correctly only if the substance extracted does not take part in any futher reactions in either of the two phases (such as protonation, formation of aggregates, complexes, etc.).

For analytical practice it is important to consider the overall distribution of a substance regardless of the actual forms in which it is present in the two phases. This is expressed by the extraction (or distribution) coefficient, defined either for total (analytical) molar concentrations in the two phases (D_c) or analogously for molalities (D_m):

$$D_c = c_{A,org}/c_{A,aq} \tag{2.63}$$

$$D_m = m_{A,org}/m_{A,aq}. \tag{2.64}$$

The dependence of the extraction coefficient on equilibria occurring in the aqueous phase will be dealt with in Chapter 3.

A further important characteristic quantity, especially for partition equilibria of extractable complexes, is the extraction constant, which is defined for each particular extraction reaction. For example, for the overall reaction combined with distribution between the two phases

$$M_{aq}^{n+} + nHL_{org} \rightleftharpoons ML_{n,org} + nH_{aq}^+ \tag{2.65}$$

the extraction constant is written as

$$K_{ex}^{HL} = \frac{[ML_n]_{org}[H^+]_{aq}^n}{[M]_{aq}[HL]_{org}^n}. \tag{2.66}$$

In practical applications it is usually convenient to express this equilibrium constant in terms of concentrations; its value then holds for constant ionic strength.

The extraction reaction in Eq. (2.65) includes distribution and dissociation of the reagent HL (protonation constant K_H) and formation of the complex ML_n (stability constant β_{ML_n}), which is extracted into the organic phase. Hence it can be written as

$$K_{ex}^{HL} = K_{D,M}\beta_n(K_{D,L}K_H)^{-n} \tag{2.67}$$

where $K_{D,M}$ and $K_{D,L}$ are the partition constants of the complex ML_n and of the extracting agent, respectively.

If both free metal M and reagent L are present in the aqueous phase, the extraction is represented by the following equilibrium

$$M_{aq}^{n+} + nL_{aq}^- \rightleftharpoons ML_{n,org} \tag{2.68}$$

and the extraction constant is expressed by

$$K_{ex} = \frac{[ML_n]_{org}}{[M]_{aq}[L]_{aq}^n} = K_{D,M}\beta_n = K_{ex}^{HL}(K_{D,L}\beta_{HL})^n. \tag{2.69}$$

2.4.5 Partition equilibria on ion-exchangers

An ion-exchange process between the aqueous and the ion-exchanger phase (index s) can be exemplified by the following reaction scheme for a cation-exchanger in the sodium form:

$$M_{aq}^{n+} + nNaR_{(s)} \rightleftharpoons MR_{n(s)} + nNa_{aq}^+. \tag{2.70}$$

The ion-exchange reaction for an anion-exchanger is analogous. The equilibrium constant for this reaction, the selectivity coefficient, can be defined in the following two ways

$$K_{nNa}^{M} = \frac{[M]_s[Na]_{aq}^n}{[M]_{aq}[Na]_s^n} \tag{2.71a}$$

or

$$K_{Na}^{M/n} = \frac{[M]_s^{1/n}[Na]_{aq}}{[M]_{aq}^{1/n}[Na]_s}. \tag{2.72b}$$

The values of selectivity coefficients hold for particular values of ionic strength. Recalculations to thermodynamic values are virtually impossible, because little is known about activity coefficients in the exchanger phase.

As with extraction equilibria, the distribution of an ion between the aqueous and ion-exchanger phases can be expressed by a distribution coefficient; for example, the partition of the metal ion M^{n+} between water and a cation-exchanger is characterized by

$$D_M = \frac{[M]_s}{[M]_{aq}}. \tag{2.73}$$

From consideration of the conditions, equilibrium concentrations may be replaced by total (analytical) concentrations (here c_M), and this is more suitable for the ion-exchanger phase. Effects that influence the concentrations of separated ions in the aqueous phase will be discussed in Chapter 3.

The distribution coefficient and the selectivity coefficient are related by the expression

$$D_M = K_{nNa}^M \frac{[Na]_s^n}{[Na]_{aq}^n} = \left(K_{Na}^{M/n} \frac{[Na]_s}{[Na]_{aq}}\right)^n \tag{2.74}$$

which is written for a cation-exchanger in Na^+-form [cf. Eq. (2.70)]. For low concentrations of the metal M this expression can be simplified to

$$D_M = K_{nNa}^M [Na]_{aq}^{-n} = (K_{Na}^{M/n}[Na]_{aq}^{-1})^n \tag{2.75}$$

since the concentration $[Na^+]_s$ may be considered to be constant until about 1% of the capacity of the exchanger is utilized.

2.4.6 Precipitation equilibria

The heterogeneous equilibrium which takes place between the solid phase of a substance and a liquid phase consisting of the saturated solution of that substance, is followed by dissociation of the solute to form ions:

$$M_m A_{n(s)} \rightleftharpoons M_m A_{n(soln)} \rightleftharpoons mM^{n+} + nA^{m-}. \tag{2.76}$$

The activity (and also the concentration) of the undissociated form of the solute is constant at constant temperature. In polar solvents (e.g. water) the concentration of an undissociated strong electrolyte is negligible in most cases, so the equilibrium can be represented by the direct transfer of the ions from the solid phase into the solution. The corresponding equilibrium constant, the solubility product, is defined by

$$K_{s(M_mA_n)} = a_M^m a_A^n. \tag{2.77}$$

In terms of equilibrium concentrations the solubility product is written as

$$(K_s)_c = [M]^m [A]^n = K_s y_M^{-m} y_A^{-n}. \tag{2.78}$$

Sparingly soluble substances often form soluble complexes with common ions, as illustrated by the following general reaction scheme

$$\frac{a}{m} M_m A_n \rightleftharpoons M_a A_{b(soln)} + \left(\frac{an}{m} - b\right) A; \qquad K_{sba}. \qquad (2.79)$$

The equilibrium constant for this heterogeneous equilibrium includes the stability constant of the complex $M_a A_b$ and the solubility product K_s for the substance $M_m A_n$:

$$K_{sba} = \beta_{ba} K_s^{a/m}. \qquad (2.80)$$

An example of such a precipitation equilibrium is solid HgI_2 in the presence of excess of iodide in solution:

$$\begin{aligned}
HgI_{2(s)} &\rightleftharpoons Hg^{2+} + 2I^-; & K_s &= [Hg^{2+}][I^-]^2 \\
HgI_{2(s)} &\rightleftharpoons HgI^+ + I^-; & K_{s11} &= \beta_1 K_s = [HgI^+][I^-] \\
HgI_{2(s)} &\rightleftharpoons HgI_{2(soln)}; & K_{s21} &= \beta_2 K_s = [HgI_2] \qquad (2.81)\\
HgI_{2(s)} + I^- &\rightleftharpoons HgI_3^-; & K_{s31} &= \beta_3 K_s = [HgI_3^-]/[I^-] \\
HgI_{2(s)} + 2I^- &\rightleftharpoons HgI_4^{2-}; & K_{s41} &= \beta_4 K_s = [HgI_4^{2-}]/[I^-]^2.
\end{aligned}$$

The effects of various side-reactions such as protonation or hydrolysis of common ions, complex formation with foreign ions, etc., will be discussed in Chapter 3.

REFERENCES

[1] H. S. Harned and B. B. Owen, *The Physical Chemistry of Electrolyte Solutions*, 3rd Ed., Reinhold, New York, 1958.
[2] R. A. Robinson and R. H. Stokes, *Electrolyte Solutions*, 2nd Ed., Butterworths, London, 1959.
[3] *Spravochnik khimika (Chemist's Handbook)*, Vol. 3, Goskhimizdat, Leningrad–Moscow, 1952.
[4] C. W. Davies, *Ion Association*, Butterworths, London, 1962; *J. Chem. Soc.*, **1938**, 2093.
[5] D'Ans-Lax, *Taschenbuch für Chemiker und Physiker*, Vol. 1, Springer, Berlin, 1967.
[6] *Solute–Solvent Interactions*, J. F. Coetzee and C. D. Ritchie, (eds.), Dekker, New York, 1969.
[7] V. Sýkora and V. Zátka, *Příruční tabulky pro chemiky (Concise Tables for Chemists)*, SNTL, Prague, 1967.
[8] J. Kielland, *J. Am. Chem. Soc.*, 1937, **59**, 1675.

[9] E. Güntelberg, *Z. Phys. Chem.*, (*Leipzig*), 1926, **123**, 199.
[10] E. A. Guggenheim, *Phil. Mag.*, 1935, **19**, 588.
[11] J. Dvořák, J. Koryta and V. Boháčková, *Elektrochemie*, 2nd Ed., p. 49, Academia, Prague, 1975.
[12] C. F. Baes, Jr. and R. E. Mesmer, *The Hydrolysis of Cations*, p. 19. Wiley, New York, 1976.
[13] K. S. Pitzer, *J. Phys. Chem.*, 1973, **77**, 268.
[14] K. S. Pitzer and G. Mayorga, *J. Phys. Chem.*, 1973, **77**, 2300.
[15] M. Meloun and S. Kotrlý, *Collection Czech. Chem. Commun.*, 1977, **42**, 2115.
[16] L. G. Sillén, *Acta Chem. Scand.*, 1962, **16**, 159.
[17] N. Ingri and L. G. Sillén, *Arkiv Kemi*, 1964, **23**, 97.
[18] L. G. Sillén and B. Warnquist, *Arkiv Kemi*, 1969, **31**, 377.
[19] L. G. Sillén and A. E. Martell, *Stability Constants of Metal-Ion Complexes*, p. xv, The Chemical Society, London, 1964.

CHAPTER 3

Chemical reactions affected by side-equilibria

Chemical reactions in solution are often influenced by side-reactions in which various components of the solution participate. Solvation of reactants can usually be included as part of the main reaction process; it is possible to assume that any equilibrium is always affected by solvation to the same extent irrespective of the reactant concentrations, because in a fairly dilute solution there is a great excess of solvent, and its concentration and influence on the reaction therefore remain constant.

Various side-reactions affect the main reaction more seriously. A great variety of species can take part in side-reactions, such as the ions of an amphiprotic solvent, the components of a buffer, masking agents etc. In these side-reactions, certain proportions of the reactants are bound and thus removed from the main reaction. The value of the equilibrium constant for the main reaction [expressed either in terms of activities (the thermodynamic constant) or in terms of equilibrium concentrations (the stoichiometric constant)] remains unchanged, but if total equilibrium concentrations are used in the equilibrium quotient expression, the value obtained is found to change with the reaction conditions. It is thus convenient to define an equilibrium constant which takes account of all the side-reactions in a defined solution. This constant is called the *conditional* (or effective) constant, and it is denoted by a symbol marked with a prime, i.e. K'. The extents of the side-reactions in which the species X and Y affect the main reactants M and N respectively, are characterized by the values of the side-reaction coefficients, $\alpha_{M(X)}$ and $\alpha_{N(Y)}$. Values for the side-reaction coefficients can be calculated if the equilibrium concentrations of the species X and Y and all the necessary equilibrium constants for the side-reactions with the components M and N are known.

A great advantage of the conditional constant concept is that the constant can be used to characterize the equilibrium exactly for given conditions. Values of conditional constants are obtained experimentally in studies of ionic equilibria in which side-reactions are involved. Once the value of a conditional constant has been determined, further calculation based on the main equilibrium is usually quite simple. This advantage was pointed out by Schwarzenbach, who introduced the concept of conditional constants in an early edition of his monograph on complexometric titrations [1]. Ringbom further developed the idea for more complicated equilibria of various types [2], and demonstrated convincingly by many examples the practical importance of these calculations for analytical chemistry [3].

In this chapter, various types of side-reactions will be discussed with reference to complexation, precipitation, oxidation–reduction and solvent–solvent partition equilibria, and for separations on ion-exchangers. The conversion of an equilibrium constant into a conditional constant will be demonstrated.

3.1 FORMULATION OF CONDITIONAL EQUILIBRIUM CONSTANTS AND SIDE-REACTION COEFFICIENTS

3.1.1 Complex-formation equilibria influenced by side-reactions

The main equilibrium in the formation of a complex

$$M + L \rightleftharpoons ML; \quad \beta_{ML} = \frac{[ML]}{[M][L]} \tag{3.1}$$

may be influenced by side-reactions of the metal M, the ligand L, and also of the complex ML with other species present in solution. If the concentrations of the two reactants, which have not reacted to yield the products of the main complexation reaction, are denoted by the symbols [M'] and [L'], and if the sum of the concentrations of all the forms of the complex is denoted by [ML'], the equilibrium can be characterized by the expression

$$\beta'_{ML} = \frac{[ML']}{[M'][L']} \tag{3.2}$$

which defines the conditional stability constant for the complex ML. The conditional concentrations of the three main-reaction components are

$$[M'] = c_M - [ML'] = [M] + [MX] + [MX_2] + \ldots + [MX_Q] \tag{3.3}$$

Sec. 3.1]   Formulation of Conditional Equilibrium Constants

$$[L'] = c_L - [ML'] = [L] + [LY] + [LY_2] + \ldots + [LY_P] \quad (3.4)$$
$$[ML'] = [ML] + [MXL] + [MX_2L] + \ldots$$
$$+ [MYL] + [MY_2L] + \ldots \quad (3.5)$$

For each component the ratio of the conditional to the equilibrium concentration defines the side-reaction coefficient:

$$\alpha_{M(X)} = [M']/[M] = 1 + \beta_{MX}[X] + \beta_{MX_2}[X]^2 + \ldots + \beta_{MX_Q}[X]^Q \quad (3.6)$$
$$\alpha_{L(Y)} = [L']/[L] = 1 + \beta_{LY}[Y] + \beta_{LY_2}[Y]^2 + \ldots + \beta_{LY_P}[Y]^P \quad (3.7)$$
$$\alpha_{ML(X,Y)} = [ML']/[ML] = 1 + \beta_{MXL}[X] + \beta_{MX_2L}[X]^2 + \ldots$$
$$+ \beta_{MYL}[Y] + \beta_{MY_2L}[Y]^2 + \ldots \quad (3.8)$$

The stability constants for the mixed complexes MX_rL (and similarly for MY_sL) in Eq. (3.8) are defined as

$$\beta_{MX_rL} = \frac{[MX_rL]}{[ML][X]^r}. \quad (3.9)$$

Calculation of side-reaction coefficients

The value for a side-reaction coefficient can be calculated if the stability constants for the side-reaction complexes are known and the equilibrium concentrations of the side-reaction ligands can be estimated. The approach to calculation is indicated in Eqs. (3.6–3.8). If it is required to evaluate the side-reaction coefficient over a wide range of concentrations of a ligand, it is useful to arrange the calculation in a tabular form, as exemplified in Table 3.1. The evaluation of side-reaction coefficients is usually the most time-consuming step in calculations with conditional constants, so it is convenient to use the tabulations or graphs of $\log \alpha = f(\log [X])$ available in the literature for the commonest ions and ligands. Ringbom presented some such graphs in his monograph [3], and others are given by Kragten [4].

The formation of hydroxo-complexes of the metal ion, protonation of the ligand and formation of mixed proton- or hydroxo-complexes of the type MH_zL† are the most frequent side-reactions in aqueous solution. In order to calculate the side-reaction coefficients $\alpha_{M(OH)}$ and $\alpha_{L(H)}$, it is necessary to know the stability constants of hydroxo-complexes, the protonation (or dissociation) constants of the ligand [cf. Eqs. (2.30–2.35), p. 31] and the pH of the solution, which is usually controlled

---

† For a proton-complex $z > 0$, whereas $z < 0$ represents a hydroxo-complex; it is assumed that all species are more or less hydrated in an aqueous solution.

Table 3.1
Arrangement of tabulation for functions and coefficients characteristic for zinc–ammine complexes

pNH$_3$	$\log \beta_1[NH_3]$	δ_1 (%)	$\log \beta_2[NH_3]^2$	δ_2 (%)	$\log \beta_3[NH_3]^3$	δ_3 (%)	$\log \beta_4[NH_3]^4$	δ_4 (%)	$\alpha_{Zn(NH_3)}$	$\log \alpha_{Zn(NH_3)}$	$\bar{n}$
4.000	−1.730	1.83	−3.390	0.04	−4.990	0.001	−6.940	—	1.019	0.008	0.02
3.398	−1.130	6.86	−2.190	0.60	−3.190	0.06	−4.540	0.003	1.081	0.034	0.08
3.000	−0.730	15.04	−1.390	3.29	−1.990	0.83	−2.940	0.09	1.238	0.093	0.24
2.699	−0.429	22.76	−0.788	9.96	−1.087	5.00	−1.736	1.12	1.636	0.214	0.62
2.550	−0.280	24.41	−0.490	15.05	−0.640	10.66	−1.140	3.37	2.150	0.332	1.00
2.398	−0.128	22.26	−0.186	19.48	−0.184	19.57	−0.531	8.78	3.346	0.525	1.55
2.300	−0.030	18.78	0.010	20.59	0.110	25.92	−0.140	14.58	4.970	0.696	1.96
2.222	0.048	15.35	0.166	20.15	0.344	30.34	0.172	20.42	7.277	0.862	2.28
2.097	0.173	9.91	0.416	17.34	0.719	34.83	0.672	31.26	1.503 × 10	1.176	2.74
2.000	0.270	6.50	0.610	14.22	1.010	35.71	1.060	40.07	2.865 × 10	1.457	3.02
1.850	0.420	3.05	0.910	9.42	1.460	33.41	1.660	52.96	8.632 × 10	1.936	3.34
1.699	0.571	1.30	1.212	5.68	1.913	28.56	2.264	64.10	2.866 × 10^2	2.457	3.55
1.398	0.872	0.20	1.814	1.78	2.816	17.85	3.468	80.12	3.668 × 10^3	3.565	3.78
1.222	1.048	0.06	2.166	0.85	3.344	12.81	4.172	86.24	1.723 × 10^4	4.237	3.85
1.097	1.173	0.03	2.416	0.50	3.719	9.96	4.672	89.51	5.254 × 10^4	4.720	3.89
1.000	1.270	0.015	2.610	0.32	4.010	8.15	5.060	91.47	1.255 × 10^5	5.099	3.91
0.000	2.270	—	4.610	0.004	7.010	0.88	9.060	99.14	1.158 × 10^9	9.064	3.99

The calculations of coefficients are indicated by the following expressions:

$$\delta_{Zn} = \frac{100}{\alpha_{Zn(NH_3)}} \%; \qquad \delta_k = \frac{\beta_k[NH_3]^k}{\alpha_{Zn(NH_3)}} 100\%$$

$$\alpha_{Zn(NH_3)} = 1 + 10^{\log \beta_1} \times 10^{\log[NH_3]} + 10^{\log \beta_2} \times 10^{2\log[NH_3]} + 10^{\log \beta_3} \times 10^{3\log[NH_3]} + 10^{\log \beta_4} \times 10^{4\log[NH_3]}$$

$$\bar{n} = \delta_1 + 2\delta_2 + 3\delta_3 + 4\delta_4.$$

$$pNH_3 = -\log[NH_3]$$

The following overall stability constants were used in the calculations (20°C, $I = 0.1$): $\log \beta_1 = 2.27$, $\log \beta_2 = 4.61$, $\log \beta_3 = 7.01$, $\log \beta_4 = 9.06$.

during the experiment. As equilibrium concentrations and stoichiometric (concentration) constants are frequently used in calculations, it is essential to pay attention to the actual meaning of the pH value (i.e. to consider the pH scale used in standardizing the pH-meter). If necessary, the pH value should be converted to the concentration scale (or vice versa) with the aid of an activity coefficient valid for the conditions.

If various other species present in solution (e.g. a masking agent or components of a buffer) take part in side-reactions, only the total concentrations are usually available and the equilibrium concentrations needed for calculation of the side-reaction coefficient are not known. Quite commonly such a complexing component, X, is present in excess, so it is possible to set the equilibrium concentration [X] as approximately equal to the total concentration c_X (if $c_X \gg c_M$), or, if MX_q is the prevailing complex in a side-reaction, [X] may be set equal to $c_X - qc_M$ (if $c_X > qc_M$). If there is only a small excess of the ligand X it is necessary to use a more rigorous approach, and a correction for the amount of the ligand consumed in the side-reaction is essential. This can be done by considering the conditional concentration of the metal ion [M'], since $[X] \sim c_X - q[M']$. Then the conditional concentration [M'] can be estimated by successive approximations. If more than one complex is present in significant amount, the approach by successive approximations becomes a necessity.

A further complication in formulation of the side-reactions results from protonation of the interfering ligand X, with the corresponding decrease in the equilibrium concentration [X]. The side-reaction coefficient $\alpha_{M(X/H)}$ is then written as

$$\alpha_{M(X/H)} = 1 + \beta'_{MX}[X'] + \beta'_{MX_2}[X']^2 + \ldots + \beta'_{MX_Q}[X']^Q \qquad (3.10)$$

where

$$\beta'_{MX_q} = \frac{[MX_q]}{[M][X']^q} \qquad (3.11)$$

and [X'] is the conditional, i.e. the total concentration of the remaining uncomplexed interfering ligand:

$$[X'] = [X] + [HX] + [H_2X] + \ldots = [X]\alpha_{X(H)} \qquad (3.12)$$

which implies that

$$[X'] = c_X - \sum_{q=0}^{Q} q[MX_q]. \qquad (3.13)$$

If $c_X \gg [M']$, then $[X'] \sim c_X$. The coefficient $\alpha_{X(H)}$ is calculated from the protonation constants for the interfering ligand X and for a known value of pH:

$$\alpha_{X(H)} = 1 + \beta_{HX}[H^+] + \beta_{H_2X}[H^+]^2 + \ldots \tag{3.14}$$

The dependence of the side-reaction coefficient $\alpha_{M(X/H)}$ on pH is calculated, therefore, for a given conditional concentration of free complexing agent. Ringbom [3] arranged the values of log $\alpha_{M(X/H)}$ for various important ligands in a table for integer values of pH and two total concentrations, 0.01 and 0.1M, of ligand. A large collection of graphs of log $\alpha_{M(X/H)}$ at various concentration levels of the ligand is available in the book by Kragten [4].

An analogous complicated side-reaction system is encountered when some interfering metal ion, which also forms a complex with the ligand L of the main reaction, hydrolyses to form soluble hydroxo-complexes.

When polynuclear complexes of general stoichiometry $M_m X_n$ are formed in the side-reaction, the value of the coefficient $\alpha_{M(X)}$ is a function of the two equilibrium concentrations [X] and [M]. Thus it depends on the total (conditional) concentration of the metal M. The calculation of the side-reaction coefficient is again based on successive approximations.

It is therefore useful to make a simple estimation of whether polynuclear complexes can occur to an appreciable extent in the system concerned. This can readily be done in the following way. Consider that a mononuclear complex ML_n prevails in the presence of lower concentrations of consecutive polynuclear complexes in the following equilibrium system: ML_n, M_2L_n, $M_3L_n \ldots M_mL_n$. The overall contribution of polynuclear complexes can be neglected if

$$\frac{[ML_n]}{\sum_{i=2}^{m} i[M_iL_n]} \geq 100. \tag{3.15}$$

As the formation of a polynuclear species M_iL_n can be characterized by a stability constant of the type

$$\beta(ML_n + (i-1)M = M_iL_n) = \frac{[M_iL_n]}{[ML_n][M]^{i-1}} \tag{3.16}$$

it is possible to write the condition in Eq. (3.15) as

$$(\sum_{i=2}^{m} i\beta[M]^{i-1})^{-1} \geq 100. \tag{3.17}$$

Usually only one polynuclear species prevails in such a system, so

$$\beta^{-1} \geqq 100i[M]^{i-1}. \tag{3.18}$$

If two or more complexing species take part in a side-reaction with one of the main components [e.g., see Eq. (3.5)], the overall side-reaction coefficient can be expressed as indicated in Eq. (3.8). Thus an overall side-reaction coefficient involving simultaneous side-reactions of the metal M with the ligands X, Y and Z is given by

$$\alpha_{M(X,Y,Z)} = \alpha_{M(X)} + \alpha_{M(Y)} + \alpha_{M(Z)} - 2 \tag{3.19}$$

or for the general case of p such ligands:

$$\alpha_{M(X,Y...A)} = \alpha_{M(X)} + \alpha_{M(Y)} + \ldots + \alpha_{M(A)} - p + 1 \tag{3.19a}$$

which indicates an easy evaluation of the overall side-reaction coefficient from the values of the individual side-reaction coefficients (cf. Chapters 10 and 11). Commonly, one of the contributing side-reaction coefficients predominates: that is, within a particular region of conditions, the main complexation reaction is influenced by only one of the side-reactions.

Formulation of a conditional stability constant

For solution of the systems of complexation equilibria that are important in analytical practice, it is essential first to define the main equilibrium and to characterize it by a stability constant value which is valid for the given solvent composition, ionic strength, and temperature. Next, all the important side-reactions arising from the presence of complexing species added as buffers or masking agents should be considered. Last, but not least, all possible protonation and/or hydrolytic equilibria should be taken into account.

It is evident that detailed information should be available on the complexation system, including the values of all the equilibrium constants needed for calculation of side-reaction coefficients. In many cases an approximate result is acceptable; it is then possible to read the values of the side-reaction coefficients from graphs without having to worry that the values of the constants used to construct the graphs are not strictly correct for the conditions chosen for the actual experiment. A compilation of side-reaction coefficients, such as that presented in the following chapters, can thus save much time and effort in analytical calculations.

Once values of the side-reaction coefficients are known, it is easy to calculate the conditional stability constant. For example, the following simple expression relates the conditional stability constant [cf. Eq. (3.2)] to the stoichiometric constant of the main complexation reaction [Eq. (3.1)]:

$$\beta'_{ML} = \beta_{ML} \frac{\alpha_{ML(X,Y)}}{\alpha_{M(X)}\alpha_{L(Y)}}. \qquad (3.20)$$

This equation is derived by combination of Eqs. (3.2) and (3.6–3.8). Because values for side-reaction coefficients are greater than unity, a conditional stability constant is smaller than the true equilibrium constant if the central ion M and/or the ligand L can participate in side-reactions. However, the formation of mixed complexes in addition to the main complex ML increases the value of the conditional stability constant.

For a conditional stability constant defined for the formation of a polynuclear complex of the type $M_m L_n$ [cf. Eq. (2.44), p. 34] an analogous expression may be written:

$$\beta'_{nm} = \beta_{nm} \frac{\alpha_{M_m L_n}}{\alpha^m_{M(X)}\alpha^n_{L(Y)}}. \qquad (3.21)$$

In this case the conditional equilibrium concentration $[M_m L'_n]$ may be defined to involve all consecutive complexes formed by the ligand L with the metal ion M; this is done if the overall concentration of the metal bound in various complex species is important; for example, when a masking procedure is investigated. If the formation of only a particular complex species such as $M_m L_n$ is of interest, all other consecutive complexes are considered as products of a side-reaction and should be included in the conditional concentration $[M']$ and considered in estimation of $[L']$.

Figure 3.1 illustrates the pH-dependence of the conditional stability constant of the EDTA complex of iron(III), which is influenced not only by protonation of the Y^{4-} ion and by hydrolysis of Fe^{3+} to hydroxo-complexes [and polynuclear hydroxo-complexes which may be formed at higher concentrations of iron(III)] but also by formation of a protonated chelate FeHY or of hydroxo-complexes $Fe(OH)Y^{2-}$ and $Fe(OH)_2 Y^{3-}$.

The formulation of a complicated reaction system, such as encountered quite commonly when an analytical procedure is modified or developed to satisfy the needs of practice, requires a certain amount of experience, and this can only be gained by solving problems. Some illustrative problems, including the use of various diagrams, are given in Chapter 14. The reader is further referred to Ringbom [3] and to other recent books where more detailed information on complexation and other conditional systems may be found [5–7].

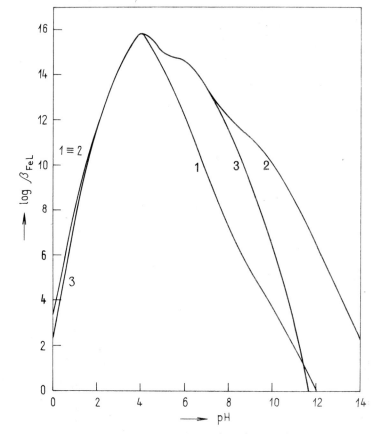

Fig. 3.1 — Dependence of the conditional stability constant of the iron(III)–EDTA complex on pH, showing the influence of the following side-reactions and the protonation of the ligand. 1 — Formation of the mixed proton-complex Fe(Hedta), the hydroxo-complexes Fe(OH)(edta)$^{2-}$ and Fe(OH)$_2$(edta)$^{3-}$, and the formation of iron(III) hydroxo-complexes, including the polynuclear species ([Fe^{3+}] = 10^{-5} M); 2 — formation of the mixed complexes Fe(Hedta), Fe(OH)(edta)$^{2-}$ and only mononuclear iron(III) hydroxo-complexes; 3 — effect of formation of only mononuclear hydroxo-complexes.

3.1.2 Conditional precipitation equilibria

In aqueous solution the solubility of a sparingly soluble electrolyte of the type M$_m$A$_n$ is often influenced by hydrogen or hydroxide ions. In acidic medium some of the anion A^{m-} is protonated. At a higher pH, hydroxo-complexes may be formed. Other complexing side-reactions of M^{n+} and A^{m-} will occur when other compounds are present in solution. Soluble

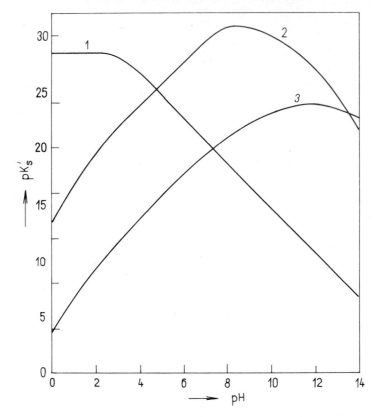

Fig. 3.2 — Dependence of the logarithmic values of the conditional solubility products on pH ($I \to 0$, $20°C$), $pK'_s = f(pH)$. 1 — HgI_2; 2 — $Zn_3(PO_4)_2$; 3 — $Mg_3(PO_4)_2$. (The solubility of these precipitates is very low and so the effect of polynuclear complexes is negligible.)

complexes of the metal ion M^{n+} with the anion A^{m-} may also be formed (cf. p. 41).

A conditional solubility product for use in calculation of the solubility of a given substance M_mA_n in water, is defined in terms of conditional concentrations as

$$K'_s = [M']^m [A']^n = (K_s)_c \alpha^m_{M(X)} \alpha^n_{A(Y)}. \tag{3.22}$$

The side-reaction coefficients are calculated by means of equations analogous to Eqs. (3.6), (3.7) or (3.19). The degree of protonation of the anions of weak acids which participate in side-reaction complexation with the metal ion M^{n+} depends on the pH, so Eq. (3.10) has to be used in calculations. The effect of complex formation with a common ion of the precipitate can also be taken into account by means of the appropriate total side-reaction coefficient α_M.

The equilibrium concentrations of particular common-ion complexes may be calculated with advantage from expressions for the constants K_{sba} [cf. Eqs. (2.79) and (2.80), p. 41]. For example, the following equations can be written for the problem given on p. 41.

$$K'_s = K_s \alpha_{Hg(I)} = [Hg'][I^-]^2.$$

Iodide is not involved in a further side-reaction and is not protonated, so $\alpha_I = 1$.

$$[Hg'] = [Hg^{2+}] + [HgI^+] + [HgI_2] + [HgI_3^-] + [HgI_4^{2-}].$$

Thus

$$\alpha_{Hg(I)} = 1 + \beta_1[I^-] + \beta_2[I^-]^2 + \beta_3[I^-]^3 + \beta_4[I^-]^4.$$

Hence the solubility of mercury(II) iodide can be calculated for a given concentration of free iodide ions as

$$[Hg'] = K'_s/[I^-]^2 = K_s([I^-]^{-2} + \beta_1[I^-]^{-1} + \beta_2 + \beta_3[I^-] + \beta_4[I^-]^2).$$

The same expression is obtained from the equation for $[Hg']$ when the solubility product and the expressions for the constants K_{s11}–K_{s41} (p. 41) are substituted for the concentrations of the complexes HgI_n^{2-n} ($n = 0$–4).

Figure 3.2 shows the pH-dependence of the conditional solubility products for three different sparingly soluble electrolytes.

3.1.3 Conditional oxidation–reduction potentials

The Nernst–Peters equation for the oxidation–reduction (redox) potential can be written to involve, in addition to the salt effect, the influence of pH or formation of a complex or a precipitate.

The influence of hydrogen ions

The pH of the solution affects the redox potential for processes in which hydrogen ions take part in the redox half-reaction; for example

$$A_{ox} + n_A e + mH^+ \rightleftharpoons A_{red} + \frac{m}{2} H_2O. \tag{3.23}$$

The corresponding Nernst–Peters equation is

$$E = E_A^\circ - \frac{RT}{n_A F} \ln \frac{a_{A_{red}}}{a_{A_{ox}} \cdot a_{H^+}^m} \tag{3.24}$$

and can be rewritten as

$$E = E^{o\prime}_{A(I,\text{pH})} - 2.303 \frac{RT}{n_A F} \log \frac{[A_{\text{red}}]}{[A_{\text{ox}}]}. \tag{3.25}$$

The symbol $E^{o\prime}_{A(I,\text{pH})}$ denotes the conditional redox potential of the couple A_{ox}, A_{red} for given values of ionic strength and pH:

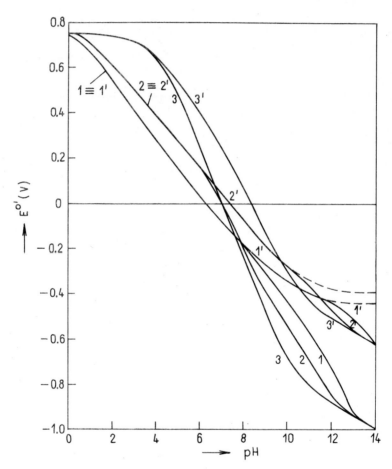

Fig. 3.3 — Dependence of conditional redox potential for the couple Fe^{3+}/Fe^{2+} on pH at a constant concentration of a complex-forming reagent (5-sulphosalicylic acid, H_2L). 1 — $[L'] = 0.1M$; 1' — $[L'] = 0.1M$, formation of polynuclear hydroxo-complexes is not considered; 2 — $[L'] = 0.01M$; 2' — absence of polynuclear hydroxo-complexes; 3 — $[L'] = 0$; 3' — absence of polynuclear species. Dashed parts of the curves show pH-dependence if formation of hydroxo-complexes is neglected.

$$E^{\circ\prime}_{A(I,\text{pH})} = E^{\circ}_A - 2.303 \frac{RT}{n_A F} \log \frac{y_{A_{\text{red}}}}{y_{A_{\text{ox}}}} - 2.303 \frac{mRT}{n_A F} \text{pH} \qquad (3.26)$$

where pH $= -\log a_{H^+}$ and y denotes the activity coefficients.

If either or both of the redox forms can be protonated (or hydrolysed to yield hydroxo-complexes), the value of the conditional potential is influenced by hydrogen ions in an additional way which can be expressed by including the side-reaction coefficients $\alpha_{A_{\text{ox}}(H)}$ and $\alpha_{A_{\text{red}}(H)}$ in the expression for the conditional redox potential, as indicated below.

The pH-dependence is illustrated in Fig. 3.3.

Effect of complexation

The value of a redox potential may also be considerably changed if one or both constituents of a redox pair can be complexed with a complexing agent present in the solution. Consider that the two redox forms take part in formation of consecutive complexes with a ligand X. Then the redox potential may be simply written as

$$E = E^{\circ\prime}_{A(I,X)} - \frac{RT}{n_A F} \ln \frac{[A'_{\text{red}}]}{[A'_{\text{ox}}]} \qquad (3.27)$$

where

$$E^{\circ\prime}_{A(I,X)} = E^{\circ}_A - 2.303 \frac{RT}{n_A F} \log \frac{y_{A_{\text{red}}}}{y_{A_{\text{ox}}}} - 2.303 \frac{RT}{n_A F} \log \frac{\alpha_{A_{\text{ox}}(X)}}{\alpha_{A_{\text{red}}(X)}}. \qquad (3.28)$$

A value for this conditional redox potential holds for a given ionic strength and given equilibrium concentration of the ligand X. This definition of conditional redox potential has the advantage that the conditional (i.e. the total) concentration of all the species in a particular oxidation state can be used for further calculation—there is no need to specify the equilibrium concentrations of all the individual species involved. The influence of the ligand concentration on the conditional redox potential is illustrated in Fig. 3.4 for two different pH values.

The effect of precipitation of a redox form

If a sparingly soluble compound formed by one constituent of a redox couple is present in contact with a solution, the activity (concentration) of this redox form is constant. For example, consider the precipitation side-reaction

$$A_{\text{red}} X_{m(s)} \rightleftharpoons A_{\text{red}} + mX; \qquad K_s = a_{A_{\text{red}}} a_X^m. \qquad (3.29)$$

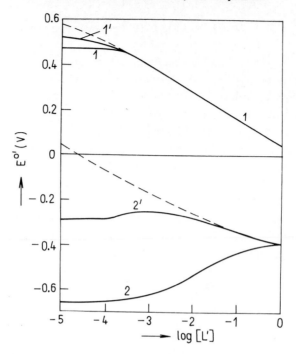

Fig. 3.4 — Dependence of conditional redox potential for the couple Fe^{3+}/Fe^{2+} on the concentration of a complexing agent (5-sulphosalicylic acid) at a given constant pH. 1 — pH = 5; 1' — pH = 5, formation of polynuclear hydroxo-complexes is not considered; 2 — pH = 10; 2' — pH = 10, effect of polynuclear hydroxo-complexes is not included; dashed parts of the curves show pH-dependence if formation of hydroxo-complexes is not considered.

The redox potential is then written as

$$E = E^{o'}_{A(I,K_s)} + \frac{RT}{n_A F} \ln [A_{ox}][X]^m \tag{3.30}$$

where

$$E^{o'}_{A(I,K_s)} = E^{o}_{A} - 2.303 \frac{RT}{n_A F} \log \frac{y_{A_{red}}}{y_{A_{ox}} y_X^m} - 2.303 \frac{RT}{n_A F} \log K_s. \tag{3.31}$$

For constant concentration of the precipitant X Eq. (3.30) may be simplified to

$$E = E^{o'}_{A(I,K_s,c_X)} + \frac{RT}{n_A F} \ln [A_{ox}] \tag{3.32}$$

and then

$$E^{\circ\prime}_{A(I,K_s,c_X)} = E^{\circ}_A - 2.303 \frac{RT}{n_A F} \log \frac{y_{A_{red}}}{y_{A_{ox}} y_X^m}$$
$$- 2.303 \frac{RT}{n_A F} \log K_s + 2.303 \frac{RT}{n_A F} \log c_X^m. \quad (3.33)$$

The conditional redox potential in Eq. (3.33) is written to include the concentration of the species X forming the precipitate. The expressions in Eqs. (3.30–3.33) hold for constant ionic strength.

3.1.4 Conditional equilibria in liquid–liquid partition systems

The side-reactions that affect extraction equilibria occur mainly in the aqueous phase (most often they are protonation and complexation equilibria), but side-reactions in the organic solvent are also encountered (formation of higher aggregates of the extracted complex or association-complexes with other compounds present in the organic phase). A conditional constant may again be formulated to take account of these and characterize the partition equilibrium for the given conditions.

In the extraction of the metal ion M^{n+} with the extraction agent HL according to Eq. (2.68) both the metal ion and the predominant reacting form of the extractant L^- are present in the aqueous phase. The conditional extraction constant is then written as

$$K'_{ex} = \frac{[ML'_n]_{org}}{[M']_{aq}[L']^n_{aq}} = K_{ex} \frac{\alpha_{M,org}}{\alpha_M \alpha_L^n} \quad (3.34)$$

where the side-reaction coefficients are calculated by means of expressions analogous to Eq. (3.6) or (3.7).

The coefficient α_M involves all side-reactions of the metal ion M^{n+} in the aqueous phase: the formation of hydroxo-complexes, complexes with components of buffers and/or complexes with masking agents.

The conditional concentration $[ML'_n]_{org}$ represents the sum of the concentrations of all forms of the metal present in the organic phase, i.e. the extracted complex ML_n, together with any higher aggregates, and aggregates with extractant or other compounds in the organic solvent. The calculation of this side-reaction coefficient requires knowledge of the particular association constants and concentrations of the reacting components in the organic phase, but this information is seldom available. If aggregates of the type $(ML_n)_p$ are formed, the value of $\alpha_{M,org}$ depends also on the concentration of the metal M.

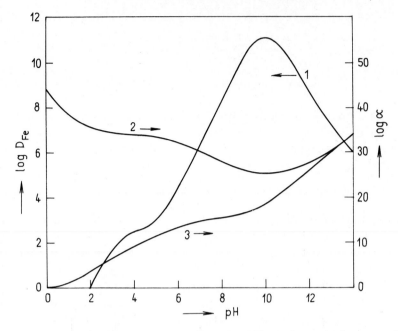

Fig. 3.5 — The influence of pH on the extraction of an iron(III) 8-hydroxyquinolinato-complex into chloroform from a solution containing citric acid ($I = 0.1$, 25°C, $V_{aq} = V_{org}$, [Hox'] = $10^{-2}M$, [H_3 cit'] = $0.1M$). 1 — $\log D_{Fe} = f(pH)$; 2 — $\log \alpha_{Hox} = f(pH)$; 3 — $\log \alpha_{Fe(cit,OH)} = f(pH)$. The concentration of free Fe^{3+} is too low for polynuclear iron(III) hydroxo-complexes to have any influence.

The conditional concentration [L'] represents the sum of all free forms of the extracting agent (i.e. not co-ordinated to the metal to be extracted) calculated as if they are all present in the aqueous phase, of the extraction system, and includes the complexing agent (e.g. HL) in the organic phase. The conditional concentration and the side-reaction coefficient are easily calculated [3, 5] from the volumes of the phases, the pH, and the partition constant of HL. For example, if HL can be protonated as well as deprotonated (e.g. 8-hydroxyquinoline):

$$[L'] = [H_2L^+]_{aq} + [HL]_{aq} + [L^-]_{aq} + [HL]_{org} \frac{V_{org}}{V_{aq}}. \quad (3.35)$$

The side-reaction coefficient α_L is calculated from

$$\alpha_L = 1 + \left(1 + \frac{V_{org}}{V_{aq}} K_{D,HL}\right) \beta_{H1}[H^+] + \beta_{H2}[H^+]^2 \quad (3.36)$$

where $K_{D,HL}$ is the partition constant for the form HL and β_{H1} and β_{H2} are the two protonation constants of the ligand L. Figure 3.5 shows the influence of pH on an extraction.

The extraction coefficient of the metal M can be calculated from the conditional extraction constant defined by means of Eq. (3.34), as indicated by the simple relationship

$$D_M = K'_{ex}[L']^n. \qquad (3.37)$$

In calculations, the conditional concentration of the extracting agent $[L']$ can usually be replaced by the original concentration in the organic phase recalculated for partition into the aqueous layer, as the extracting agent is commonly present in great excess with respect to the metal. If a solution of the extractant HL in an organic solvent is used in excess to extract the metal ion from the aqueous phase, it is more convenient to express the extraction coefficient by means of the conditional extraction constant $K_{ex}^{HL'}$. On rearrangement, Eq. (2.66), p. 38, yields the following relationship for the ratio of the metal concentrations in the two phases

$$D_M = K_{ex}^{HL'}([HL]_{org}/[H^+]_{aq})^n. \qquad (3.38)$$

3.1.5 Conditional equilibria on ion-exchangers

Sorption of metal ions on an ion-exchanger is influenced by equilibria in which the ion being separated can participate in aqueous solution (cations can form complexes with various ligands, anions can be protonated). Such processes decrease the completeness of sorption.

When a complexing agent is present in the exchange system, the conditional distribution coefficient for the metal M is written as

$$D'_M = \frac{[M]_s}{[M']_{aq}} = \frac{[M]_s}{[M]_{aq}\alpha_M} = \frac{D_M}{\alpha_M}. \qquad (3.39)$$

If the value of the side-reaction coefficient α_M is sufficiently high, practically all the metal ion remains in the aqueous phase. That is, the metal ion is masked and no sorption takes place.

3.2 CALCULATION OF COMPLICATED EQUILIBRIA WITH THE AID OF A COMPUTER

The computation of the concentrations of all the species participating in a complicated reaction, involving a number of simple reactions, has to be based on detailed information about the reaction system.

First, the composition of the solution of interest and, if necessary, of other phases present must be known; i.e. it is essential to find the exact nature of all the compounds and ions participating in the reaction, and to know the actual concentrations of the starting components. Secondly, the values of the constants for all the individual reactions involved should be known with sufficient accuracy. Commonly, the only equilibrium concentration of a species in solution that is actually known is the hydrogen-ion concentration, from the pH value. The calculation of the equilibrium concentrations of all the species is done by successive approximations, best with the aid of an electronic computer. The reliability of the results and the agreement with experiment depend mainly on the input data, especially the equilibrium constants. However, the speed of an electronic computer makes it possible to study the various sources of error which may affect calculated data; for example, it is possible to test the effect on the equilibrium of a change in ionic strength (which affects the values of the equilibrium constants).

Various programs have been written for the calculation of the composition of complicated reaction mixtures. The programs COMICS [8] and HALTAFALL [9, 10], which are particularly useful for various applications in practice, will be introduced here. The principles of these programs and the scope of their applications will be explained in detail, and illustrative examples for reaction systems involving complexation and protolytic equilibria will be given.

The compositions of complexes and species formed in related protolytic equilibria, by the ligand L with the metal ion M and hydrogen or hydroxide ion, can be represented by the general formula $M_m L_n H_z$. The stoichiometric coefficients m and n may have positive integral values or be equal to zero, and z may take positive or negative integral values, or be equal to zero. Thus the general formula can include free metal, free ligand and its protonated forms and, of course, various mononuclear and/or polynuclear complex species together with protonated or hydroxo forms. A negative value of the coefficient z denotes either a hydroxo-complex of the metal M (for $n = 0$) or a mixed hydroxo-complex with the ligand L. (Molecules of water are assumed to be present in the hydration sphere or in various aquo-complexes, but this is not indicated in the general formula). The equilibrium concentration of any species occurring in the system can thus be expressed by the equation:

$$c_j = [M_m L_n H_z] = \beta_j [M]^m [L]^n [H]^z \tag{3.40}$$

where β_j is the overall equilibrium constant for the formation of the species which is characterized by the given combination of stoichiometric coefficients:

$$\beta_j = \frac{[M_m L_n H_z]}{[M]^m [L]^n [H]^z}. \tag{3.41}$$

The values of the equilibrium constants must be known for the conditions of the experiment, in addition to the total concentrations of the reactants which were mixed to yield the reaction system. The total (analytical) concentrations are represented by the mass balances, which are made up according to the composition of the reaction system; for example,

$$c_M = [M] + \sum_{j=1}^{j=k} p_j c_j \tag{3.42}$$

where k is the number of all complex species of the metal M present in the system and p_j the number of M ions present in the jth species. Analogous expressions are written for c_L and c_H. The equilibrium concentrations of all the species in the equilibrium are then calculated by means of successive approximations.

3.2.1 Program COMICS

In the program COMICS, and in its major subroutine COGS (concentrations of generalized species) by Perrin [8] the calculation begins with an initial assumption that complex formation is negligible with respect to the total concentrations of reactants. Thus the preliminary estimates of the equilibrium concentrations of the free metal ion $[M]'$, free ligand $[L]'$, and hydrogen ion $[H^+]'$, (if a pH value on the concentration scale is not given in the input data) are approximated as

$$[M]' = c_M; \quad [L]' = c_L; \quad [H^+]' = c_{H^+}. \tag{3.43}$$

These preliminary estimates of equilibrium concentrations are substituted in the expressions for the calculation of equilibrium concentrations of individual species in the reaction system [equations of the type indicated in Eq. (3.40)]. In this way the first estimates of the concentrations c'_j for individual species are obtained; these are used to evaluate the total concentrations [cf. Eq. (3.42)]. The calculated values $(c_M)_{calc}$, $(c_L)_{calc}$, etc. differ considerably, of course, from the total concentrations c_M, c_L, etc. In the next step, corrected values for the equilibrium concentrations of the species M, L, etc. are calculated from

$$[M]'' = [M]' [c_M/(c_M)_{calc}]^{1/2} \tag{3.44}$$

$$[L]'' = [L]' [c_L/(c_L)_{calc}]^{1/2} \tag{3.45}$$

etc.

The values of the corrected estimates $[M]''$, $[L]''$, etc. are used to calculate corrected values of equilibrium concentrations c_j''. Subsequently new approximations to the total concentrations of the reaction components are calculated.

The iteration cycle is repeated until

$$c_M - (c_M)_{calc} \leqq \varepsilon c_M \tag{3.46}$$

$$c_L - (c_L)_{calc} \leqq \varepsilon c_L \tag{3.47}$$

where ε is a selected small number, usually between 0.001 and 0.0001.

The program COMICS is written in FORTRAN IV [8]. According to the authors it can be used for complexation equilibria in solutions, without any limitation on the number or type of components. The authors have successfully calculated the equilibrium composition of a system which consisted of 10 metals and 10 ligands with 195 individual equilibria. In another example, the composition of a solution in which 2 different metal ions reacted with 3 ligands to yield 21 different simple and mixed complexes was determined.

3.2.2 Program HALTAFALL

The program HALTAFALL from the Sillén school [9, 10] allows the calculation of equilibrium concentrations of species in many equilibrium systems of interest to the analytical chemist. The program can be used for complicated equilibria of various types, including normal single-phase solution equilibria, equilibria between solutions and solid phases, liquid–liquid (extraction) equilibria and liquid–gas equilibria.

The initial data for the calculation include the total concentrations of the reactants, the stoichiometry of the species formed in the equilibrium system, and the values of all the corresponding equilibrium constants.

HALTAFALL can be used to solve a wide range of problems. It is possible to calculate the equilibrium concentration of hydrogen ions for a mixture of acids and conjugate bases, the pH-dependence of the buffer capacity of a buffer solution of given composition, the full titration curve for titrations of various types, the equilibrium composition of a system of complexes of several different metals and ligands and the partition of an extractable metal complex between an aqueous and an organic phase. The array sizes may be adjusted so that the number of reaction components and complexes is not limited, but each additional component or equilibrium species increases the computation time. The composition of an equilibrium system can be calculated for various conditions (pH, concentration of the ligand, temperature). The output data can be used for construction of a distribution or logarithmic diagram, for plotting of an extraction curve, etc.

The algorithmic features of the HALTAFALL program will be outlined briefly here and illustrated by a simple example [10]. Consider a complexation equilibrium in which two metals M and N and two ligands L and R form the complexes MR, ML, NL, NL_2, N_2L. The experiment is as follows. A certain volume V_0 ml of solution of known molar concentrations of the reactants M, N, and R, denoted as $(c_M)_0$, etc., is titrated with a standard solution of the complexing agent L of concentration $(c_L)_{\text{titrant}}$ (the additions are v ml).

The equilibrium concentrations of the complex species can be evaluated from the overall stability constants [cf. Eq. (3.40)]:

$$[MR] = \beta_{MR}[M][R] \tag{3.48}$$

$$[ML] = \beta_{ML}[M][L] \tag{3.49}$$

$$[NL] = \beta_{NL}[N][L] \tag{3.50}$$

$$[NL_2] = \beta_{NL_2}[N][L]^2 \tag{3.51}$$

$$[N_2L] = \beta_{N_2L}[N]^2[L]. \tag{3.52}$$

The total concentrations of the four reactants, c_M, c_N, etc. in the solution at each titration point are calculated from the initial concentration and the volumes used to prepare the solution and are compared with the total concentrations calculated from the mass-balance equations for the equilibrium system.

$$(c_M)_{\text{calc}} = [M](1 + \beta_{MR}[R] + \beta_{ML}[L]) \tag{3.53}$$

$$(c_N)_{\text{calc}} = [N](1 + \beta_{NL}[L] + 2\beta_{N_2L}[N][L] + \beta_{NL_2}[L]^2) \tag{3.54}$$

$$(c_R)_{\text{calc}} = [R](1 + \beta_{MR}[M]) \tag{3.55}$$

$$(c_L)_{\text{calc}} = [L](1 + \beta_{ML}[M] + \beta_{NL}[N] + \beta_{N_2L}[N]^2 + 2\beta_{NL_2}[N][L]). \tag{3.56}$$

At the beginning of the calculation the reaction components are sorted into a sequence for computation according to the number of components with which they do not react. Thus the component M forms complexes with R and L, but it does not react with N; so its number is 1. The same holds for the component L. The components R and N react with only one of the three remaining components and so their number is 2. A second criterion for the computation sequence is the formation of mono- or polynuclear complexes. The calculations then proceed in decreasing order of the numbers of the components; those forming mononuclear species are treated first. Thus the component R is sorted out first in the sequence of calculation, then comes the component N which forms a polynuclear complex, etc.

The cycles of approximations start with the estimated values of equilibrium concentrations, which are given with the input data. First a better value is sought for the equilibrium concentration of R, the first component in the computation sequence, while other equilibrium concentrations are kept constant. The calculated value $(c_R)_{calc}$ is then compared with the actual total concentration. If a tolerance limit $(tol)_R$ given in the input data for c_R is exceeded, i.e. if

$$|c_R - (c_R)_{calc}| = w_R > (tol)_R \tag{3.57}$$

a new value of $[R]$ is calculated. As indicated by Eq. (3.55), the values of $[R]$ and c_R are directly proportional, so the calculated total concentration of the component $[R]$ can be used to evaluate a new approximation to the value of $[R]$:

$$[R]_{new} : c_R = [R]_{guess} : (c_R)_{calc}. \tag{3.58}$$

Component N comes next, and again the difference between a calculated and a known value of c_N is compared with a given tolerance $(tol)_N$. The calculation of a new value of $[N]$ is then based on Eq. (3.54). First, it is necessary to find a limited range within which a value $[N]_{new}$ can be found. This is done by varying $\ln [N]$ in steps:

$$\text{if } (c_N)_{calc} < c_N, \quad \text{then} \quad \ln [N]_{new} = \ln [N] + 2;$$

otherwise

$$\text{if } (c_N)_{calc} > c_N, \quad \text{then} \quad \ln [N]_{new} = \ln [N] - 2.$$

This stepwise variation of $[N]$ is repeated until the inequality is no longer fulfilled. Then $\ln [N]$ is varied again in the opposite direction, with half the value of the previous step, (the next step is 1, then 0.5 etc.) until the magnitude of the step is less than the value specified in the input data. Thus, two values $[N]_1$ and $[N]_2$ for a final range are obtained, between which an improved value of $[N]_3$ can be found.

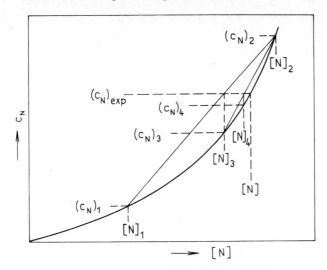

Fig. 3.6 — Evaluation of an improved value of the equilibrium concentration of the component N by successive approximations by the program HALTAFALL. The curve represents Eq. (3.54) for a constant value of [L].

The approach to subsequent numerical calculation is indicated in Fig. 3.6. A chord is drawn between the points $[N]_1, (c_N)_1$ and $[N]_2, (c_N)_2$ on the graph of Eq. (3.54) (the value of $[L]_{guess}$ is now taken as constant). The abscissa of the point of intersection of the chord with the line $c_N = (c_N)_{exp}$ yields the next approximation to the equilibrium concentration $[N]_3$ which is used to evaluate $(c_N)_3$. A new chord is then drawn between the points $[N]_3, (c_N)_3$ and $[N]_2, (c_N)_2$ to yield a yet better approximation $[N]_4$. The approximation procedure is continued until $w_N < (tol)_N$. Then an improved estimate of $[M]$ is calculated by Eq. (3.53) on inserting the value of $[R]_{new}$ and $[L]_{guess}$, and an improved value of $[L]$ is found by using the last estimates of $[M]$ and $[N]$ and the approximation procedure indicated above.

The computer then returns to the beginning of the cycle to calculate a better approximation for $[R]$ and then for the other variables. The cycles are repeated until the differences between the calculated and experimental values of the total concentrations of all components are smaller than the allowed tolerances.

It is worth adding a note on the reliability of calculation of an equilibrium system with the aid of a computer. The agreement between the output data and experiment is influenced not only by the reliability of the input data, especially by the values of the constants, but it depends crucially on the correct formulation of the equilibrium itself. It is essential to check whether all possible reaction species of the components in the mixture have been included. Also, since equilibrium constants are used in the calculations, it is essential that the actual reaction system should reach equilibrium. Large discrepancies between observed and calculated values may be due to equilibrium not having been reached.

REFERENCES

[1] G. Schwarzenbach and H. Flaschka, *Die komplexometrische Titration*, 5th Ed., Enke, Stuttgart, 1965; 2nd English Ed., Methuen, London, 1969.

[2] A. Ringbom, *Complexation Reactions*, Chap. 14, Part I, Vol. 1. in *Treatise on Analytical Chemistry*, I. M. Kolthoff and P. J. Elving (eds.), Interscience, New York, 1959.

[3] A. Ringbom, *Complexation in Analytical Chemistry*, Interscience, New York, 1963.

[4] J. Kragten, *Atlas of Metal–Ligand Equilibria in Aqueous Solution*, Horwood, Chichester, 1978.

[5] L. Šůcha and S. Kotrlý, *Solution Equilibria in Analytical Chemistry*, Van Nostrand Reinhold, London, 1972.

[6] J. Inczédy, *Analytical Applications of Complex Equilibria*, Horwood, Chichester, 1976.

[7] O. Budevsky, *Foundations of Chemical Analysis*, Horwood, Chichester, 1979.

[8] D. D. Perrin and I. G. Sayce, *Talanta*, 1967, **14**, 833; errata: *Talanta*, 1968, **15**, No. 3, p. xi.

[9] N. Ingri, W. Kakołowicz, L. G. Sillén and B. Warnqvist, *Talanta*, 1967, **14**, 1261.

[10] D. Dyrssen, D. Jagner and F. Wengelin, *Computer Calculation of Ionic Equilibria and Titration Procedures: With Specific Reference to Analytical Chemistry*, p. 66. Almqvist and Wiksell, Stockholm; Wiley, New York, 1968.

CHAPTER 4

Diagrams for consecutive equilibrium systems

It is usually not easy to calculate the composition of a solution in which the reaction proceeds in overlapping consecutive steps, or which is influenced by complicated side-reactions. In such situations it is useful to consider which species will predominate under given conditions in the solution and therefore virtually determine its properties. This is especially the case in analytical practice.

The graphical representation of a complicated equilibrium system gives objective information on the composition of the system for a large range of values of the controlled variable, usually the pH or pOH of the solution, the logarithm of the concentration of the ligand or precipitation reagent, or the redox potential. Distribution or logarithmic concentration diagrams, and in some cases predominance-area diagrams, can be constructed for such purposes, and may be used in a variety of situations. Various applications of these diagrams are discussed in many monographs and text-books; for further reading references [1–6] may be useful.

4.1 DISTRIBUTION DIAGRAMS

A distribution diagram presents a family of curves showing the relative proportions of the individual species taking part in an equilibrium system of consecutive complexes of the type ML_n or of the various protonated forms of a polyprotic acid H_nB. The distribution curves of the diagram are defined by distribution coefficient functions, $\delta_n = [ML_n]/c_M = f(\log[L])$ and $\delta_{H_nB} = [H_nB]/c_B = f(pH)$, respectively. A typical simple distribution diagram is shown in Fig. 4.1. For calculation of distribution coefficients, the expressions used are:

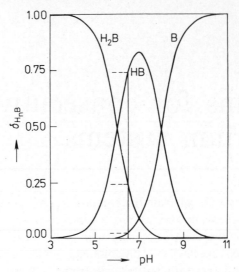

Fig. 4.1 — Distribution diagram of the protonated forms of a diacidic Brønsted base B (log $K_{H1} = 8.00$; log $K_{H2} = 6.00$). The diagram shows, for example, that the following values of particular distribution coefficients are found at pH 6.5: $\delta_B = 0.02$; $\delta_{HB} = 0.74$; $\delta_{H_2B} = 0.24$.

for stepwise complexation

$$\delta_n = \beta_n[L]^n/(1 + \sum_{n=1}^{N} \beta_n[L]^n) \qquad (4.1)$$

and for a similar protonation equilibrium

$$\delta_{H_nB} = \beta_{Hn}[H^+]^n/(1 + \sum_{n=1}^{N} \beta_{Hn}[H^+]^n). \qquad (4.2)$$

For graphical construction of a distribution diagram, it is convenient to calculate the values of distribution coefficients for integer logarithmic values of the controlled variable. It is also useful to determine δ_n for $\log[L] = \frac{1}{2}\log K_n K_{n+1}$, where the particular distribution curve reaches its maximum. If the ratios of successive stability constants K_n/K_{n+1} are at least 4×10^4, the distribution curves for $\delta_1 - \delta_{N-1}$ have identical bell-shapes: a particular distribution coefficient increases sigmoidally from zero to a maximum (of unity). If the ratio of the two consecutive constants is lower than the value given above, then $\delta_{max} < 1$. The curves for δ_0 and δ_N differ in shape from the rest: the curve δ_0 is a sigmoid decreasing from 1 to 0; on the other side of the diagram the curve δ_N is the mirror image of δ_0, increasing from 0 to 1. When the values of equilibrium constants for consecutive steps differ sufficiently, the distribution curves occur in the diagram with the intersection of two successive curves, i.e. $\delta_n = \delta_{n+1}$, located at $\log[L] = \log K_{n+1}$.

When polynuclear complexes are present, the shape of the distribution curves is influenced by the concentration of the metal ion. The distribution coefficient for a complex species $M_m L_n$ is defined by

$$\delta_{nm} = \frac{m[M_m L_n]}{c_M} = \frac{m\beta_{nm}[M]^{m-1}[L]^n}{1 + \sum_{m=1}^{M} m \sum_{n=1}^{N} \beta_{nm}[M]^{m-1}[L]^n} \qquad (4.3)$$

if the distribution of the metal ion in the system of complexes is to be demonstrated. Alternatively, the coefficient is defined as

$$\delta_{nm} = \frac{[M_m L_n]}{c_M} = \frac{\beta_{nm}[M]^{m-1}[L]^n}{1 + \sum_{m=1}^{M} m \sum_{n=1}^{N} \beta_{nm}[M]^{m-1}[L]^n} \qquad (4.4)$$

for representation of the individual complex species in the equilibrium.

4.2 LOGARITHMIC DIAGRAMS

A logarithmic concentration diagram is a plot of the logarithms of the concentrations of equilibrium species *vs.* the logarithm of the concentration (or activity) of the species representing the independent variable. For a polyprotic equilibrium the logarithms of concentrations of the various protonated forms are plotted against pH. For a complexation equilibrium the logarithmic diagram gives plots of the logarithms of the concentrations of the metal ion and the various complex species against pL or $\log[L]$. The presence of a solid phase formed by a precipitation reaction between the species M and L can also be included. Logarithmic concentration diagrams can also be used to illustrate the composition of an oxidation–reduction (redox) system, e.g. for a multivalent metal, as a function of the redox potential. The logarithm of the ratio of the activity (or concentration) of a particular redox form to that of a suitable reference can also be plotted.

The main importance of logarithmic diagrams lies in providing graphical information on acid–base systems and complexation equilibria in the presence of a precipitate with an ion involved in the complexation reaction. The construction of such diagrams is easy. Often no further calculations of the composition of the system are necessary. On the other hand, the construction of a diagram for an equilibrium of consecutive complexes cannot normally be done without the precise calculation of a series of distribution coefficients.

4.2.1 Logarithmic concentration diagrams for protonation equilibria

The equilibrium concentrations of the species participating in the protonation equilibrium of a polyprotic Brønsted base can be expressed as functions of pH by the equation

$$\log [H_nB] = \log c_B + \log \delta_{H_nB} = f(\text{pH}) \tag{4.5}$$

which results from the definition of the distribution coefficient [cf. Section 4.1 and Eq. (4.2)].

The non-linear function obtained can be represented by linear relationships for wide regions of pH, but not near to the pK_a values for the system, i.e. for $\text{pH} = \log K_{Hn} \pm 1.3$. The error in drawing the diagram is at most $\pm 0.05 c_B$ (i.e. about 5%) for the species predominating in the equilibrium within a certain pH region. In view of this limit and the simplifications involved, it may also be assumed that the values of equilibrium constants and hence the values of distribution coefficients will not be significantly influenced by changes of ionic strength over a wide range of pH, e.g. from 0 to 14.

The logarithmic diagram is thus made up from $N + 1$ plots of the functions $\log [H_nB] = f(\text{pH})$ (n varying from 0 to N). On each curve, linear sections are connected by short curves in regions where the pH values are in the intervals $\log K_{Hn} \pm 1.3$. The curves can readily be drawn, except for systems with overlapping protonation steps, i.e. those for which $\log K_{Hn} - \log K_{H(n+1)} < 3.2$.

Construction of a diagram will be illustrated by the following example of a diacidic Brønsted base. This protonation system will be considered for two different ratios of protonation constants.

4.2.1.1 Examples of construction of logarithmic concentration diagrams

Consider the equilibrium system of a diacidic Brønsted base ($\log K_{H1} = 6.0$, $\log K_{H2} = 2.0$, $c_B = 0.1M$) which is shown in Fig. 4.2.

First of all, the "system points", i.e. the points with co-ordinates $S_1(2.0, -1.0)$ and $S_2(6.0, -1.0)$ are plotted on the diagram together with

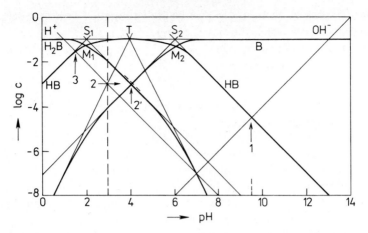

Fig. 4.2 — Logarithmic concentration diagram for a diacidic Brønsted base ($\log K_{H1} = 6.00$; $\log K_{H2} = 2.00$; $\log c_B = 0.1 M$).

the points $T(4.0, -1.0)$, $M_1(2.0, -1.3)$ and $M_2(6.0, -1.3)$. The points S_1, T, and S_2 lie on a line parallel to the x-axis and intersecting the y-axis at $\log c_B$, which represents the highest attainable concentration value for any species of the acid–base equilibrium. The heavy sections of this line denote the regions for $\text{pH} \leq (\log K_{H2} - 1.3)$, $(\log K_{H2} + 1.3) \leq \text{pH} \leq (\log K_{H1} - 1.3)$, and $\text{pH} \geq (\log K_{H1} + 1.3)$, which correspond to the concentrations of the particular species predominating in the equilibrium within the given pH ranges. In these regions, plots of the logarithms of concentration against pH for all other equilibrium species are also linear. The slopes of these linear sections always have the values ± 1 or ± 2, depending on the difference in the number of protons possessed by the predominant species and the species considered. The lines with slopes ± 1 can be extrapolated to intersect at points S_1 and S_2; the point T represents a point of intersection for extrapolations of the lines with slopes ± 2.

Curves within the pH region given by $\log K_{Hn} \pm 1.3$ connect the linear sections with slopes of 0 and -1 or $+1$ and 0, respectively, and intersect at the points M_1 and M_2, since $\text{pH} = \log K_{H2}$ when $[H_2B] = [HB]$ and $\text{pH} = \log K_{H1}$ when $[HB] = [B] = c_B/2$. The curves can be drawn easily with the aid of these auxiliary points. As there are no further auxiliary points, the location of the curves that join the lines of slopes $+2$ and $+1$ is not so easy, but these parts of the curves correspond to nearly negligible concentration levels of the equilibrium species. The curved sections can thus be drawn roughly, with consideration of the ranges over which the slopes change.

Finally, the auxiliary lines $\log[H^+] = -pH$ and $\log[OH^-] = pH - 14$ are drawn to complete the diagram.

4.2.1.2 Reading concentrations of equilibrium species for an acid–base system

Composition of the system at a given pH. Logarithms of concentrations for individual species can be read directly from a line perpendicular to the x-axis at the relevant pH. The concentration of a predominant species is obtained more precisely by subtraction of the concentrations of minor components from the total concentration c_B.

For example, the following values for pH 3 can be read from Fig. 4.2: $\log[B] = -4.05$ ($[B] = 9.1 \times 10^{-5} M$), $\log[H_2B] = -2.05$ ($[H_2B] = 9.1 \times 10^{-3} M$). The concentration of the species HB is obtained by difference: $[HB] = c_B - [H_2B] - [B]$, so $[HB] = 9.1 \times 10^{-2} M$.

Estimation of the pH of a solution containing Brønsted base B or its protonated forms. The proton-balance equation (PBE) defines the hydrogen-ion concentration for a solution of a defined protonated form of a Brønsted base. For an aqueous solution of the base B the equation is

$$[H^+] + 2[H_2B] + [HB] = [OH^-]. \tag{4.6}$$

This is satisfied at point 1 in Fig. 4.2, where $[HB] = [OH^-]$, because the concentrations of the other species in Eq. (4.6) can be neglected. The pH corresponding to point 1 is 9.50.

For a solution of pure acid, H_2B, the proton balance takes the form

$$[H^+] = [HB] + 2[B] + [OH^-] \tag{4.7}$$

which is valid for point 3 where $[H^+] = [HB]$ and the concentrations of the other species are again negligible: the pH read from the diagram is 1.55.

For a solution of a pure acid salt with predominant species HB, the proton balance is

$$[H^+] + [H_2B] = [B] + [OH^-]. \tag{4.8}$$

This equation is approximately satisfied at point 2′, where $[H_2B] = [B]$ and the first estimate of pH (pH′) is 4.0; thus $[OH^-]$ is negligible. However, in this case the concentration of hydrogen ion cannot be neglected with respect to H_2B. A better approximation to the pH must be obtained by the following procedure. The value $[H_2B] = 1.0 \times 10^{-3} M$ is read from the diagram at pH', and a value for the left-hand side of the proton balance $[H^+] + [H_2B] = 1.1 \times 10^{-3} M$ is calculated. A parallel to the line segment for H_2B is then drawn through the point {pH', $\log([H^+] + [H_2B])$} (dashed line in Fig. 4.2). The point of intersection with the curve for B (point 2) gives the corrected reading of pH 4.05.

As illustrated by these examples, a proton balance takes into consideration on the one hand the number of protons gained by the predominant Brønsted base to yield the minor species of Brønsted acids and on the other the number of protons lost by the predominant acid species to yield the minor species of Brønsted bases. The equilibrium concentration of each species is thus multiplied by the number of protons gained or lost relative to the original compound, which does not itself appear in the PBE.

If summation of concentrations is necessary in a simplified PBE in order to make a correction, extrapolation with a parallel is easy in regions where the curves are linear. If the curve for a more significant species is non-linear at the location of summation, it can be assumed that the curve necessary for the corrected reading is nearly parallel to that of the predominant component.

A logarithmic diagram constructed for a particular total concentration c_B of a given acid–base system can also be used for another value of the total concentration. The lines on the diagram are simply shifted along the y-axis to place the points S_1 and S_2 at the ordinate corresponding to the new value of $\log c_B$. The location of the lines for H^+ and OH^- remains unchanged, of course. The logarithmic diagram can be used for another protonation system only if the ratio of the protonation constants remains the same, so that the distance between the system points is not changed. Then the lines of the diagram are just shifted along the x-axis. Otherwise a new diagram has to be constructed, as illustrated by that in Fig. 4.3 where the values of the protonation constants are much closer together than for Fig. 4.2.

4.2.1.3 Construction of a logarithmic concentration diagram for a protonation system with close values of $\log K_{Hn}$

Consider a diacidic Brønsted base with $\log K_{H1} = 8.0$, $\log K_{H2} = 6.0$ and $c_B = 0.01 M$. The composition of the system has to be calculated for pH $= \frac{1}{2}(\log K_{H1} + \log K_{H2}) = 7.0$, since the concentrations of the species H_2B and B cannot be neglected here with respect to the predominant

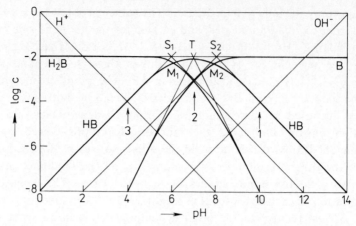

Fig. 4.3 — Logarithmic concentration diagram for a diacidic Brønsted base ($\log K_{H1} = 8.00$; $\log K_{H2} = 6.00$; $\log c_B = 0.01 M$).

species HB. From the distribution-coefficient expressions the following values are obtained: $\log[B] = \log[H_2B] = -3.08$. The concentration of HB is then found by subtraction from c_B: $\log[HB] = -2.08$. With the aid of these calculated values and the symmetry of the curves the diagram shown in Fig. 4.3 can readily be constructed in a way analogous to that used for Fig. 4.2.

4.2.2 Logarithmic solubility diagram for a complexation reaction with a common ion

This type of diagram shows the dependence of the logarithms of the concentrations of complex species (e.g. M_aB_b) on $\log[A]$ in the presence of a precipitate M_mA_n (cf. the system discussed in Section 2.4.6, p. 41). The construction of such diagrams is easy if constants of the type K_{sba} are used for the dissolution equilibria involved. Linear expressions of the form $\log[M_aB_b] = f(\log[A])$ are obtained. As an example, consider the equilibrium system of iodomercurates $HgI_b^{(2-b)+}$ ($b = 0$–4) in the presence of solid HgI_2. The equilibrium constants are as follows (for definitions see Section 2.4.6, p. 41): $\log K_s = -28.0$, $\log K_{s11} = -15.1$, $\log K_{s21} = -4.1$, $\log K_{s31} = -0.3$, $\log K_{s41} = 1.9$ (for $I = 0.5$ and $25°C$).

Construction of the diagram. The expressions for the equilibrium constants given on p. 41 for mercury(II) and its complex species present in a solution saturated with solid mercury(II) iodide yield the following linear relationships when expressed in logarithmic form (see Fig. 4.4):

$$\log [Hg^{2+}] = -2 \log [I^-] + \log K_s \qquad (4.9)$$
$$\log [HgI^+] = -\log [I^-] + \log K_{s11} \qquad (4.10)$$
$$\log [HgI_2] = \log K_{s21} \qquad (4.11)$$
$$\log [HgI_3^-] = \log [I^-] + \log K_{s31} \qquad (4.12)$$
$$\log [HgI_4^{2-}] = 2 \log [I^-] + \log K_{s41}. \qquad (4.13)$$

The slopes of the lines are given by the difference between the number of iodide ions in the particular complex species and in HgI_2 (the presence of which is not shown in the diagram, but is a necessary condition for validity of the diagram).

In addition to the lines corresponding to Eqs. (4.9)–(4.13) another two lines are drawn in the diagram: an auxiliary line for I^- (cf. the auxiliary lines for H^+ and OH^- in the acid–base plots) and the line of total solubility, which represents the solubility of HgI_2 under the given conditions, obtained by summation of the concentrations of all mercury species present in solution. In the region of predominant existence of particular complex species the solubility curve coincides with the species line. When two species participate in the equilibrium, the solubility curve has a bend, which can be drawn approximately by locating an auxiliary

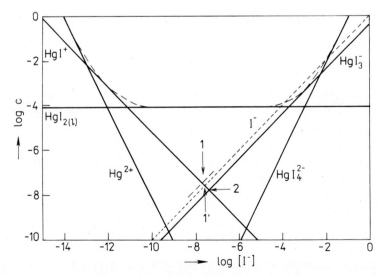

Fig. 4.4 — Logarithmic diagram for the dependence of the solubility of mercury(II) iodide on the concentration of iodide ions (see text for the equilibrium constants).

point above the intersection point of the two lines at a distance of 0.3 logarithmic unit. The bend is then drawn symmetrically to approach the line of a predominant species at a point where the concentration of the next most important species drops below 5% of the total solubility (i.e. where its line is located at about 1.3 logarithmic units below the line of the predominant species). It is seldom necessary to sum the concentrations of more than two significant equilibrium species.

4.2.2.1 Reading concentrations from a solubility concentration diagram

Composition of the system for a chosen value of $\log[I^-]$. As for acid–base systems, the logarithms of concentrations are read at intersection points on a perpendicular to the x-axis at the relevant value of $\log[I^-]$. The total concentration of mercury(II) in solution (i.e. the solubility of the precipitate of HgI_2) changes with iodide concentration, so the concentrations of all the soluble forms of mercury(II) have to be read from the diagram. For example, for $[I^-] = 10^{-5}M$ the following values are obtained (the concentration of HgI_2 remains unchanged at varied concentrations of free iodide): $\log[HgI_{2(soln)}] = -4.1$ ($8 \times 10^{-5}M$); $\log[HgI_3^-] = -5.4$ ($4 \times 10^{-6}M$); $\log[HgI_4^{2-}] = -8.2$ ($6 \times 10^{-9}M$); the concentrations of the other ionic species are negligibly small. Thus the total solubility of mercury(II) iodide is here $8.4 \times 10^{-5}M$.

Solubility of mercury(II) iodide in pure water. For a solution obtained by dissolving solid HgI_2 in pure water the following ligand balance holds:

$$2[Hg^{2+}] + [HgI^+] = [I^-] + [HgI_3^-] + 2[HgI_4^{2-}]. \tag{4.14}$$

This mass balance is based on the fact that the sum of the concentrations of free iodide ion and the iodide bound in complexes containing more iodide than the original solute, HgI_2, must be equal to the sum of the concentrations of species resulting from release of iodide. Two iodide ions are released to yield free Hg(II) ion and that is why the equilibrium concentration $[Hg^{2+}]$ has to be multiplied by 2 in Eq. (4.14).

It is evident from Fig. 4.4 that the ligand balance equation can be simplified to

$$[HgI^+] = [I^-] + [HgI_3^-] \tag{4.15}$$

since the concentrations of the other species are negligible. A summation line is then constructed (see Fig. 4.4) as for the acid–base system discussed above to give the intersection point 1 corresponding to $\log[I^-] = -7.65$.

In this region the solubility of mercury(II) iodide is determined only by the presence of undissociated HgI_2; the value found is $8 \times 10^{-5} M$.

Minimum solubility of HgI_2. The minimum of the solubility curve and the corresponding concentration of iodide can readily be determined by reading $\log[I^-]$ for the intersection point of the lines for HgI^+ and HgI_3^-. These complex species also contribute to the solubility of the precipitate in addition to the predominant complex HgI_2. Reading the diagram gives an iodide concentration of $4 \times 10^{-8} M$, which is only slightly higher than that corresponding to the solubility of HgI_2 in pure water, so in this case the minimum solubility and the solubility of mercury(II) iodide in pure water are practically equal. Both values are determined by the predominant complex HgI_2; however, this result does not represent the general situation. Quite often the minimum solubility may be significantly changed by adding a small excess of the precipitant.

4.2.3 Logarithmic activity-ratio diagrams

The relative amounts of species participating in a redox equilibrium can be shown graphically with the aid of a logarithmic activity-ratio diagram.

If the Nernst–Peters equation for the redox pair A_{ox}, A_{red} is re-arranged by division by the expression $2.303RT/F$ (approximately 0.06 at 25°C), and is written in terms of Sillén's notation $pe = E/0.06$, the following form results:

$$pe = pe^°_{A_{ox},A_{red}} + \frac{1}{n_A}\log\frac{a_{A_{ox}}}{a_{A_{red}}} = pe^°_A + \frac{1}{n_A}\log\frac{a_1}{a_2}. \qquad (4.16)$$

Hence the logarithm of the ratio of the activities of the two redox forms is expressed as

$$\log\frac{a_{A_{ox}}}{a_{A_{red}}} = n_A pe - n_A pe^°_A. \qquad (4.17)$$

In Cartesian co-ordinates, $\log(a_1/a_2)$, pe, the expression in Eq. (4.17) is represented by a straight line with slope n_A and intercept $-n_A pe^°_A$ on the ordinate. For construction of the diagram, the activity a_2 (i.e. $a_{A_{red}}$) is considered as unity; then the logarithmic ratio a_2/a_2 is represented by the x-axis.

When the diagram is constructed for the general case, for example, for the successive redox equilibria of a multivalent element, one of the intermediate redox terms is chosen as a reference form. The diagram then consists of straight lines with both positive and negative slopes, according to the number of electrons exchanged with respect to the reference redox form.

Instead of activities it is also possible to consider the concentrations of species involved. The values of the conditional redox potentials for the given conditions should be used for construction of the diagram.

The procedure for the construction of a logarithmic activity-ratio diagram will be illustrated with the redox systems Tl(III)/Tl(I) and Tl(I)/Tl. Further examples are given in Chapter 13.

4.2.3.1 Construction of the logarithmic activity-ratio diagram for the redox system Tl(III)–Tl(I)–Tl

The oxidation–reduction systems of thallium are characterized by the following standard redox potentials (for 25°C and ionic strength 0): $E°_{Tl(III),Tl(I)} = E°_{3,1} = 1.267$ and $E°_{Tl(I),Tl} = E°_{1,0} = -0.336$ V (cf. Table 9.1).

By rearrangement of the Nernst–Peters equation, as indicated above, the following expressions are obtained:

$$pe = pe°_{3,1} + \tfrac{1}{2}\log(a_{Tl(III)}/a_{Tl(I)})$$
$$= 21.30 + \tfrac{1}{2}\log(a_3/a_1) \tag{4.18}$$

$$pe = pe°_{1,0} + \log(a_{Tl(I)}/a_{Tl})$$
$$= -5.64 + \log(a_1/a_0). \tag{4.19}$$

If the activity of thallium(I) is chosen as reference for construction of the diagram, the logarithms of the activity ratios are expressed by the following linear equations with respect to pe (see Fig. 4.5):

Straight line expression	Notation for the straight line	
$\log a_0/a_1 = -pe - 5.64$	a_0	
$\log a_1/a_1 = 0$	a_1	(4.20)
$\log a_3/a_1 = 2pe - 42.60$	a_3	

It is evident that the point of intersection of lines a_0 and a_1 has the co-ordinates $pe°_{1,0}$, 0, and for the lines a_1 and a_3 it is $pe°_{3,1}$, 0. These points are needed for construction of the diagram.

The slope of a particular straight line always corresponds to the number of electrons released in the transition from the reference redox form to the one represented by the line. Thus, for oxidations the

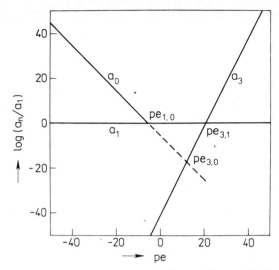

Fig. 4.5 — Logarithmic activity-ratio diagram for an oxidation–reduction system: thallium(III)–thallium(I)–metallic thallium (the values of the standard redox potentials are given in the text).

slopes are positive (e.g. line a_3) and a negative slope indicates reduction (e.g. line a_0).

Figure 4.5 illustrates the feasibility of successive oxidation of metallic thallium to Tl(I) and then to Tl(III). The ion Tl$^+$ is relatively stable in aqueous solution over a wide range of redox potentials. It is oxidized to Tl(III) only by strong oxidizing agents. The dotted parts of line a_0 may be taken to represent the activity of thallium in an alloy, e.g. HgTl, in equilibrium with Tl(I).

4.3 PREDOMINANCE-AREA DIAGRAMS

The composition of equilibrium systems is, in practice, often influenced by the concentrations of several reactants. Some typical examples are complex-formation and precipitation equilibria which are often strongly influenced by formation of mixed complexes, and various redox equilibria in which one or both redox forms are affected by acid–base, complexation, and precipitation reactions. If such an equilibrium is controlled by two independently variable concentrations of two reactants, the reaction system can be represented by a space distribution diagram. The relative predominance of equilibrium species is shown by distribution surfaces which are functions of the concentrations (or activities) of the two reactants, taken as controlled variables. Each point on a distribution surface corresponds to the concentration ratio of a given component of the equilibrium.

The space distribution diagram can be transformed into a planar diagram with co-ordinates carrying the scales of the two controlled variables. This planar diagram is essentially a projection of the distribution surfaces on to a plane base where the only parts visible are those which represent the regions of predominant existence of relevant species. The parts of the diagram are separated by straight lines or line segments which are projections of the intersection curves of neighbouring distribution surfaces. Thus a predominance-area diagram is actually the flat surface of a section through a solid model, perpendicular to the axis corresponding to the distribution coefficients. The intersection lines, which separate the regions of predominant existence of particular forms, correspond to equal ratios (i.e. to equal concentrations or activities) of neighbouring forms in the equilibrium. (These neighbouring forms can be interconverted by reaction with the species corresponding to the controlled variable). With acid–base (protonation) equilibria one of the controlled variables is pH; for redox reactions it is redox potential. If one of the compounds taking part in the equilibrium is present as a solid phase (i.e. in a pure form, conventionally considered to be at unit activity), its distribution surface is represented by a plane parallel to the plane of projection and at unit distance from it. The intersection lines of distribution surfaces of the species present in solution and in equilibrium with the solid component, also correspond to unit activities (concentrations).

Any species that has no region of predominant existence is not represented in the predominance-area diagram. If it is desired to indicate the existence of minor species in the diagram, the chosen value for the location of the intersection lines should be given; these lines are actually hidden below the distribution surfaces of successive components of the equilibrium system.

For equilibria involving a solid phase it is convenient to draw a boundary line between the solid compound and the solutes, corresponding to a lower value of activity (or concentration) of the solutes; for example, the concentration level $10^{-6}M$ is often chosen, because it may be considered as a practical limit for the beginning of dissolution or, on the other hand, for completeness of precipitation.

Construction of predominance-area diagrams will be illustrated by the complexation equilibrium of the ion M with two different ligands A and B.

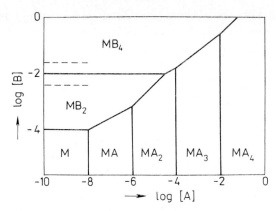

Fig. 4.6 — Predominance-area diagram for the system of complexes of metal M with ligands A and B (see text for the values of equilibrium constants).

No mixed complexes are formed in the system; the only species present are soluble complex species of the type MA_n and MB_n (n is 0–4) and a sparingly soluble compound MB in the solid phase. The following equilibrium constants will be used to characterize the system:

Species	Consecutive stability constants			
	$\log K_1$	$\log K_2$	$\log K_3$	$\log K_4$
Complexes MA_n	8	6	4	2
MB_n	4	4	1	2

Precipitate MB with solubility product $\log K_s = -10$.

Figure 4.6 is the diagram for a solution with such a small metal ion concentration that no precipitation of MB has occurred. The same equilibrium system in the presence of the solid phase MB is shown in Fig. 4.7.

4.3.1 Construction of a predominance-area diagram in the absence of solid phase

The diagram will be set up for axes graduated in units of $\log [B]$ and $\log [A]$. The boundary lines corresponding to the stepwise formation of complexes MA_n from metal M with ligand A are not influenced by changes in concentration of ligand B. Therefore they are represented by lines parallel to the $\log [B]$ axis. The equations for these lines can be derived by arrangement of the expressions for the consecutive stability constants of complexes MA_n: equilibrium concentrations of neighbouring species cancel since they are equal at the boundary lines.

For example, for the boundary line between the region of predominance of M and MA, $[M] = [MA]$ and the expression for the boundary line is obtained from $K_1 = [MA]/[M][A] = 10^8$, which gives $\log[A] = -8$. In a similar way other boundary lines are obtained from expressions for the constants K_2-K_4.

Boundary line equation	Species	
$\log[A] = -8$	M, MA	
$\log[A] = -6$	MA, MA_2	
$\log[A] = -4$	MA_2, MA_3	
$\log[A] = -2$	MA_3, MA_4	(4.21)

Analogously, the boundary lines separating the regions for the complexes MB_n do not depend on the concentration of the ligand A; thus they are formed by straight lines parallel to the $\log[A]$ axis. The following equations are obtained for this family of boundary lines:

Boundary line equation	Species	
$\log[B] = -4$	M, MB	
$\log[B] = -4$	MB, MB_2	
$\log[B] = -1$	MB_2, MB_3	
$\log[B] = -2$	MB_3, MB_4	(4.22)

In this case, the first two boundary lines coincide, and when the concentration of the ligand B is continuously increased, the fourth boundary line precedes the third one. It is thus evident that the complex species MB and MB_3 are never the predominating components in the equilibrium system: their boundaries are located under the distribution surfaces for M, MB_2 and MB_4. Therefore, only the boundary lines between the areas of predominance for the species M, MB_2, and MB_4 are drawn in the diagram. The equilibrium constants for the corresponding reactions are given by the product of consecutive constants for the neighbouring steps.

$$K(M + 2B \rightleftharpoons MB_2) = K_{MB}K_{MB_2} = 10^8 \qquad (4.23)$$

$$K(MB_2 + 2B \rightleftharpoons MB_4) = K_{MB_3}K_{MB_4} = 10^3. \qquad (4.24)$$

Thus the equations for the boundary lines are written as

$\log[B] = 4$	(boundary $M-MB_2$)	
$\log[B] = 1.5$	(boundary MB_2-MB_4).	(4.25)

In this system interconversion reactions between complexes, such as $MA_n \rightleftharpoons MB_n$, may also occur. These reactions are not influenced by the concentrations of the two ligands, so the corresponding boundaries are not parallel to either of the axes. In order to locate these boundaries it is useful to consider first the complexes of the simplest stoichiometry with adjoining areas of predominance. In this system, the complexes $[MA]$ and $[MB_2]$ come first. The equilibrium of the conversion is defined by the constant

$$K(MB_2 + A \rightleftharpoons MA + 2B) = K_{MA}/K_{MB}K_{MB_2} = 1. \tag{4.26}$$

Thus for $[MA] = [MB_2]$ the boundary line equation represents a straight line:

$$\log[B] = \tfrac{1}{2}\log[A] \tag{4.27}$$

which passes through the origin and has a slope of 0.5. The line starts at the point of intersection of the boundary lines M–MA and M–MB$_2$; thus only the slope need be known for its construction. The relevant slope is given by the ratio of the number of ligands accepted (A) to the number of ligands released (2 B) in the reaction, written in the sense of transformation of a species predominating close to the y-axis into one predominating in an area close to the x-axis.

The next boundary line in the diagram corresponds to conversion of MB_2 into MA_2; it is evident that the slope is 1. For a check, the corresponding equations can be derived:

$$K(MB_2 \rightleftharpoons MA_2) = K_{MA}K_{MA_2}/(K_{MB}K_{MB_2}) = 10^6 \tag{4.28}$$
$$\log[B] = \log[A] + 3.$$

The slope of the boundary line is indeed 1; extrapolation on the diagram confirms that the intercept on the y-axis is 3.

In a similar way the following equations for the remaining boundary lines in the diagram are obtained:

Boundary line equation	Species	
$\log[B] = \tfrac{1}{2}\log[A] + 0.75$	MB_4, MA_2	
$\log[B] = \tfrac{3}{4}\log[A] + 1.75$	MB_4, MA_3	
$\log[B] = \log[A] + 2.25$	MB_4, MA_4	(4.29)

In Fig. 4.6 the dashed boundary lines between the areas for the species MB_2–MB_3 and MB_3–MB_4 lie under the plane of projection.

4.3.2 Construction of a predominance-area diagram in the presence of a solid compound

If a precipitate of the compound MB is present in the solution, the diagram (see Fig. 4.7) differs considerably from the one just discussed, especially in the region of predominance of MB_n.

First, the boundary lines between the region for the solid phase and the predominance regions for the metal ion M and the complexes MB_n are drawn. Again, these boundary lines are not influenced by log [A] and thus are parallel to the log [A] axis. They correspond to unit concentration of M and the complexes MB_n, respectively, in a solution saturated with $MB_{(s)}$. Hence from the solubility product, for $[M] = 1M$ it follows that $\log [B] = \log K_s = -10$, which represents the boundary line M–$MB_{(s)}$.

In order to derive equations for the other boundary lines between the areas for the solid phase and the complexes MB_n, it is necessary to use expressions for the constants K_{sba} defined for equilibria involving

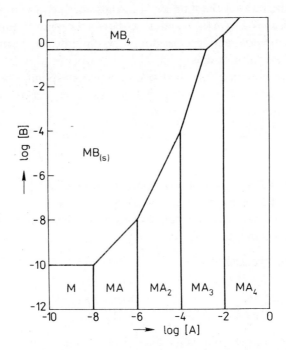

Fig. 4.7 — Predominance-area diagram for the system of complexes of metal ion M with ligands A and B in presence of a sparingly soluble compound $MB_{(s)}$.

the precipitate and the complexes with the common ion of the precipitant (see Section 2.4.6, p. 41). For the system under consideration these equilibrium constants are defined as

$$K_{s11} = [MB]_{(soln)} = K_{MB}K_s = 10^{-6} \qquad (4.30)$$

$$K_{s21} = [MB_2]/[B] = K_{MB}K_{MB_2}K_s = 10^{-2} \qquad (4.31)$$

$$K_{s31} = [MB_3]/[B]^2 = K_{MB}K_{MB_2}K_{MB_3}K_s = 10^{-1} \qquad (4.32)$$

$$K_{s41} = [MB_4]/[B]^3 = K_{MB}K_{MB_2}K_{MB_3}K_{MB_4}K_s = 10^{1}. \qquad (4.33)$$

The value of the constant K_{s11} indicates that the concentration of the species $MB_{(soln)}$ in a solution saturated with the precipitate is $10^{-6}M$; therefore, this species is not shown in the diagram, where the lowest concentration of the species plotted is $1M$. This concentration is attained for the other complex species at the following values of $\log[B]$: for MB_2, 2; MB_3, 0.5; MB_4, -0.33. Thus, only the last complex species is represented in the diagram. The boundary lines between predominance areas for the complexes MA_n are obtained similarly to those for Fig. 4.6.

The equations for the boundary lines between the region of existence of the solid $MB_{(s)}$ and the regions for the complexes MA_n are obtained in the same way as the lines for MB_n–MA_n. As they correspond to unit activities of the complex species MA_n, the following expressions and equations hold.

Boundary	Equilibrium constant	Species for which unit activity is assumed	Boundary line equation
$MB_{(s)}$–MA	$K(MB_{(s)} + A \rightleftharpoons MA + B)$ $= [MA][B]/[A] = K_{MA}K_s$ $= 10^{-2}$	MA	$\log[B]$ $= \log[A] - 2$
$MB_{(s)}$–MA_2	$K(MB_{(s)} + 2A \rightleftharpoons MA_2 + B)$ $= [MA_2][B]/[A]^2 = K_{MA}K_{MA_2}K_s$ $= 10^4$	MA_2	$\log[B]$ $= 2\log[A] + 4$
$MB_{(s)}$–MA_3	$K(MB_{(s)} + 3A \rightleftharpoons MA_3 + B)$ $= [MA_3][B]/[A]^3$ $= K_{MA}K_{MA_2}K_{MA_3}K_s = 10^8$	MA_3	$\log[B]$ $= 3\log[A] + 8$ (4.34)

At a higher concentration of B, the complex MA_3 is converted into the species MB_4; here the boundary line is the same as for the system in Fig. 4.6. The same holds for the equilibrium between the complexes MB_4 and MA_4.

4.3.3 Significance of predominance-Area diagrams

In such diagrams the predominance of certain species in a given equilibrium controlled by two variables can be seen at a glance. Thus they give a good idea of the composition of a solution at equilibrium for given concentrations of controlled variables and under given conditions. This type of diagram can also be used to estimate whether a precipitate may be formed in the presence of a complicated set of complexation equilibria. The information obtained can provide the first estimates of concentrations that are necessary for accurate calculations with the use of computer programs.

Further examples and applications of predominance-area diagrams will be given in the chapters on particular types of equilibria.

REFERENCES

[1] L. G. Sillén, in *Treatise on Analytical Chemistry*, I. M. Kolthoff and P. J. Elving (eds.), Part I, Vol. 1, Chap. 8, Interscience, New York, 1959.
[2] J. N. Butler, *Ionic Equilibrium*, Addison-Wesley, Reading, Massachusetts, 1964.
[3] Q. Fernando and H. Freiser, *Ionic Equilibria in Analytical Chemistry*, Wiley, New York, 1963.
[4] F. Seel, *Grundlagen der analytischen Chemie*, 4th Ed., Verlag Chemie, Weinheim, 1966.
[5] M. Pourbaix, *Atlas d'équilibres électrochimiques*, Gauthiers–Villars, Paris, 1963.
[6] L. Šůcha and S. Kotrlý, *Solution Equilibria in Analytical Chemistry*, Van Nostrand Reinhold, London, 1972.

CHAPTER 5

Constants of protonation equilibria

The literature on protonation (acid–base) equilibria is wide and extensive. Investigation of these equilibria and determination of protonation constants at present mainly depends on measurements by potentiometry and spectrophotometry (in the visible and ultraviolet region; Raman spectroscopy and NMR are also used). Other techniques are not so often used, but conductometric, kinetic, cryoscopic methods and various other approaches based on partition chromatography, extraction, electrophoresis, and solubility measurements should be mentioned. Principles and applicability, necessary instrumentation and apparatus, and critical evaluation of individual methods are discussed in many monographs (e.g. [1–10]) and in numerous original contributions.

The basic problem met in the determination of protonation constants is the definition of the pH-scale: practically all methods for protonation equilibria are directly or indirectly based on pH measurements. The definition adopted should be clearly stated.

In view of the international recommendation [11] and of the extent of applications of the conventional activity pH scale (often referred to as the practical pH scale) it is probably most convenient to realize the pH scale with the aid of standard buffers for the solvent in which the protonation equilibria are to be studied. If measurement of the other species in the protonation equilibrium (i.e. the acid form and the conjugated base) yields results in terms of concentration, the logarithmic value of the mixed protonation constant [as defined in Eq. (2.38), p. 33] can be converted into a concentration constant with the aid of the following values for the difference $\Delta \log K_H$ given by Smith and Martell [12]:

I	0.05	0.10	0.15	0.2	0.5	1.0	2.0	3.0
$-\Delta \log K_H$	0.09	0.11	0.12	0.13	0.15	0.14	0.11	0.07

The values of mixed or concentration protonation constants are valid for a given ionic strength and solvent.

When complexation equilibria are investigated at constant ionic strength, it is often preferable to define pH in terms of a concentration scale: $\mathrm{pH} \equiv pc_H = -\log[H^+]$ (mole/l) or $pm_H = -\log m_H$ (molality scale). Then standardization of the pH-cell can be based on the procedure recommended by Sillén [13] or by Irving et al. [14]. Alternatively, the standardization can be realized by emf measurements of both strong and weak acids in constant ionic medium [15]. The suggestion by Bates [16, 17] that standard reference buffer solutions of known hydrogen ion concentration be established should inspire further investigation leading to the recommendation of standard reference buffer solutions for the concentration pH scale.

Selected values for the protonation constants of inorganic and organic substances in aqueous solutions are listed in Tables 5.1 and 5.2, respectively. Some protonation constants for mixed and non-aqueous solvents are given in Tables 5.3 and 5.4. The data are logarithmic values of protonation constants determined at 25°C and ionic strength approaching zero [cf. the thermodynamic protonation constant defined by Eq. (2.31), p. 31]. For ionic strength $I = 0.1$ and 1.0, the concentration constants, defined as $K_H = [HB]/([H^+][B])$, are given. If a value refers to another temperature or ionic strength, a suitable note is added.

Inorganic compounds are arranged according to the position of the main element in the periodic table. For organic compounds the empirical formulae (Chemical Abstracts system) are taken as a basis for their arrangement.

The precision of a value of $\log K_{H_n}$ is indicated by the number of decimal places given. Data given as arithmetical means for various species or thought to be unreliable for various reasons are given in parentheses.

Equilibria in which a species of the system is present in a solid or gaseous phase are characterized by equilibrium constants defined for that particular equilibrium. Both the reaction products and the reactants are written in square brackets after the symbol for the constant; molecules of water or the solvent are not given. For example, the equilibrium constant $K[H_2CO_3/CO_{2(g)}]$ corresponds to the reaction

$$CO_{2(g)} + H_2O \rightleftharpoons H_2CO_3.$$

REFERENCES

[1] H. C. Brown, D. H. McDaniel and O. Häflinger, in *Determination of Organic Structures by Physical Methods*, E. A. Braude and F. C. Nachod (eds.), Chap. 14, Academic Press, New York, 1955.
[2] F. J. C. Rossotti and H. S. Rossotti, *The Determination of Stability Constants*, McGraw-Hill, New York, 1961.
[3] A. Albert and E. P. Serjeant, *Ionization Constants of Acids and Bases: A Laboratory Manual*, Methuen, London, 1962; 3rd Ed. Chapman and Hall, London, 1984.
[4] T. Sheldovsky, in *Physical Methods of Organic Chemistry*, A. Weissberger (ed.), Chap. 25, Interscience, New York, 1949.
[5] J. N. Butler, *Ionic Equilibrium*, Addison–Wesley, Reading, Massachusetts, 1964.
[6] D. Dyrssen, D. Jagner and F. Wengelin, *Computer Calculation of Ionic Equilibria and Titration Procedures*, Almqvist and Wiksell, Stockholm, 1968; Wiley, New York, 1968.
[7] H. F. Ebel, *Die Acidität der CH-Säuren*, Thieme, Stuttgart, 1969.
[8] J. Inczédy, *Analytical Applications of Complex Equilibria*, Horwood, Chichester, 1976.
[9] R. G. Bates, *Determination of pH*, 2nd Ed., Wiley, New York, 1973.
[10] C. F. Baes, Jr. and R. E. Mesmer, *The Hydrolysis of Cations*, Wiley, New York, 1976.
[11] *Manual of Symbols and Terminology for Physicochemical Quantities and Units*, IUPAC, Butterworths, London, 1970.
[12] R. M. Smith and A. E. Martell, *Critical Stability Constants*, Vol. 4, *Inorganic Complexes*, p. xii, Plenum Press, New York, 1976.
[13] L. G. Sillén and G. Biedermann, *Arkiv Kemi*, 1953, **5**, 425.
[14] H. M. Irving, M. G. Miles and L. D. Pettit, *Anal. Chim. Acta*, 1967, **38**, 475.
[15] W. A. E. McBryde, *Analyst*, 1969, **94**, 337; 1971, **96**, 739.
[16] R. G. Bates, Chap. 9. in ref. [9].
[17] R. G. Bates, in *Essays on Analytical Chemistry*, E. Wänninen (ed.), p. 23, Pergamon, Oxford, 1977.

Table 5.1
Protonation constants of inorganic compounds in aqueous medium

Brønsted base	Equilibrium constants Symbol for the constant, value of log K, $[I]$†	Ref. §
$B(OH)_4^-$	K_{H1} 9.236 [0]; 8.97 [0.1]; 8.88 [1.0]	[2]
$B_2O(OH)_5^-$	K_{H1} 9.36 [0]; 9.05 [1.0]	[2]
$B_3O_3(OH)_4^-$	K_{H1} 7.03 [0]; 6.75 [1.0]	[2]
$B_4O_5(OH)_4^{2-}$	$K_{H1}K_{H2}$ 16.3 [0]; 14.91 [1.0]	[2]
CN^-	K_{H1} 9.21 [0]; 9.01 [0.1]; 8.95 [1.0]	
CNS^-	K_{H1} 0.9 [0]	
CO_3^{2-}	K_{H1} 10.33 [0]; 10.00 [0.1]; 9.57 [1.0]	
	K_{H2} 6.35 [0]; 6.16 [0.1]; 6.02 [1.0]	
	$K[H_2CO_3/CO_{2(g)}]$ -1.46 [0]; -1.51 [1.0]	
CS_3^{2-}	K_{H1} 8.22, K_{H2} 2.68 [0]^a	[3]
$SiO_2(OH)_2^{2-}$	K_{H1} 13.1 [0]; 12.56 [0.5]; K_{H2} 9.86 [0]; 9.46 [0.5];	
	$K[H_4SiO_4/SiO_{2(s,am.)}]$ -2.74 [0]	
	$K[H_4SiO_4/SiO_{2(s,silica)}]$ -4.0 [0]	[1, 2]
$Si_4O_6(OH)_6^{2-}$	$K_{H1}K_{H2}$ 13.44 [0]	[2]
$Si_4O_8(OH)_4^{4-}$	$K_{H1}K_{H2}K_{H3}K_{H4}$ 35.80 [0]	[2]
$GeO_2(OH)_2^{2-}$	K_{H1} 12.6 [0]; 11.7 [1.0]; K_{H2} 9.3 [0]; 9.02 [1.0]	[1, 2]
$Ge_8O_{16}(H_2O)_5(OH)_3^{3-}$	$K_{H1}K_{H2}K_{H3}$ 14.24 [0]	[2]
NH_3	K_{H1} 9.244 [0]; 9.29 [0.1]; 9.40 [1.0]	
	$K[NH_4^+/H^+, NH_{3(g)}]$ 11.11 [1.0]	
NH_2NH_2	K_{H1} 7.98 [0]; 8.18 [1.0]; K_{H2} -0.9 [0]^a	
NH_2OH	K_{H1} 5.96 [0]; 6.06 [1.0]	
N_3^-	K_{H1} 4.65 [0]; 4.45 [0.1]; 4.44 [1.0]	
	$K[HN_3/HN_{3(g)}]$ 1.08 [0]	
NO_2^-	K_{H1} 3.15 [0]; 3.00 [1.0]	
NO_3^-	K_{H1} -1.34 [0]	[4]
$N_2O_2^{2-}$	K_{H1} 11.54 [0]; 10.85 [1.0]; K_{H2} 7.18 [0]; 6.75 [1.0]	
PO_4^{3-}	K_{H1} 12.35 [0]; 11.74 [0.1]; 10.79 [3.0]	
	K_{H2} 7.199 [0]; 6.72 [0.1]; 6.46 [1.0]	
	K_{H3} 2.148 [0]; 2.0 [0.1]; 1.70 [1.0]	
$P_2O_7^{4-}$	K_{H1} 9.40 [0]; 8.37 [0.1]; 9.00 [0.1]^b; 7.43 [1.0]	
	K_{H2} 6.70 [0]; 6.04 [0.1]; 6.19 [0.1]^b; 5.41 [1.0]	
	K_{H3} 2.2 [0]; 1.8 [0.1]; 2.0 [0.1]^b; 1.4 [1.0]	
	K_{H4} 0.8 [0]; 0.8 [0.1]^b; 0.8 [1.0]	
$P_3O_{10}^{5-}$	K_{H1} 9.25 [0]; 8.00 [0.1]; 8.70 [0.1]^b; 8.61 [1.0]^b	
	K_{H2} 6.54 [0]; 5.50 [0.1]; 5.90 [0.1]^b; 5.69 [1.0]^b	
	K_{H3} 2.5 [0]; 2.6 [0.1]; 2.2 [0.1]^b; 2.0 [1.0]^b	
	K_{H4} 1.0 [1.0]^b	
$P_4O_{13}^{6-}$	K_{H1} 8.34, K_{H2} (6.63), K_{H3} 2.23, K_{H4} 1.4 [1.0]^b	
$P_3O_9^{3-}$	K_{H1} 2.05 [0]; 1.35 [0.1]; 0.65 [1.0]^b	

† Temperature 25°C unless otherwise noted.
§ If no reference is given, the data were taken from [1].

Table 5.1 (Continued)

Brønsted base	Equilibrium constants Symbol for the constant, value of log K, $[I]$†	Ref. §
$P_4O_{12}^{4-}$	K_{H1} 2.76 [0]; 1.53 [1.0]^b	
PO_5^{3-}	K_{H1} 12.8; K_{H2} 5.5; K_{H3} 1.1 [0.1]	
PS_4^{3-}	K_{H1} 6.5; K_{H2} 3.4; K_{H3} 1.7 [0.5]^a	
HPO_3F^-	K_{H1} 5.12 [0]; 4.47 [1.0]	
AsO_4^{3-}	K_{H1} 11.50, K_{H2} 6.96, K_{H3} 2.24 [0]	[1, 2]
$As(OH)_4^-$	K_{H1} 9.29 [0]; 9.13 [0.1]; 9.11 [1.5] $K[As(OH)_3/0.25As_4O_{6(s)}]$ −0.68 [0]	[1, 2]
$Sb(OH)_6^-$	K_{H1} 2.72 [0]	[2]
OH^-	K_{H1} 13.997 [0]; 13.78 [0.1]; 13.79 [1.0]; 14.18 [3.0] K_{H1} 13.997 [0]; 13.78 [0.1]; 13.74 [0.5]; 13.80 [1.0]; 13.97 [2.0]; 14.20 [3.0]; 14.42 [4.0]	[2]
O_2H^-	K_{H1} 11.65 [0]; 12.13 [3.0]^c	
S^{2-}	K_{H1} 13.9 [0]; 13.8 [1.0]; K_{H2} 7.02 [0]; 6.83 [0.1]^a; 6.61 [1.0]; $K[H_2S/H_2S_{(g)}]$ −0.99 [0]	
S_4^{2-}	K_{H1} 6.3, K_{H2} 3.8 [0.1]^a	[5]
S_5^{2-}	K_{H1} 5.7, K_{H2} 3.5 [0.1]^a	[5]
SO_3^{2-}	K_{H1} 7.18 [0]; 6.34 [1.0]; K_{H2} 1.91 [0]; 1.37 [1.0] $K[H_2SO_3/SO_{2(g)}]$ 0.09 [0]; 0.03 [1.0]	
SO_4^{2-}	K_{H1} 1.99 [0]; 1.55 [0.1]; 1.10 [1.0]; K_{H2} −3.0 [0]	[1, 6]
HSO_5^-	K_{H1} 9.86 [0]^d	
$S_2O_3^{2-}$	K_{H1} 1.6 [0]; 1.3 [0.1]; 0.8 [1.0]; K_{H2} 0.6 [0]	
$NH_2SO_3^-$	K_{H1} 0.988, $K[H^+, H_2NSO_3^-/H_2NSO_3H_{(s)}]$ −0.92 [0]	
Se^{2-}	K_{H1} 15.0 [0]^e; (11.6) [1.0]; K_{H2} 3.89 [0]; 3.48 [1.0]	
SeO_3^{2-}	K_{H1} 8.5 [0]; 7.78 [1.0]; K_{H2} 2.75 [0]; 2.27 [1.0] $K[(H_2SeO_3)_2/2\,H_2SeO_3]$ 0.17 [1.0]	[1, 2, 7]
SeO_4^{2-}	K_{H1} 1.7 [0]	[1, 2]
Te^{2-}	K_{H1} 12.2, K_{H2} 2.6^g [dil.]	[2, 8, 9]
$TeO_2(OH)_2^{2-}$	K_{H1} 9.8 [0]; 9.36 [0.1]; K_{H2} 2.68 [0]; 2.46 [0.1] K_{H3} −3.13 [0]; −3.16 [0.1]; $K[Te(OH)_4/TeO_{2(s)}]$ (−4.7)	[1, 2, 10]
$TeO_3(OH)_3^{3-}$	K_{H1} (15.6), K_{H2} 11.00, K_{H3} 7.68 [0]; 7.32 [1.0]	[2]
CrO_4^{2-}	K_{H1} 6.51 [0]; 6.09 [0.1]; 5.74 [1.0]; 5.90 [3.0]; K_{H2} −0.20 [0]; −0.7 [1.0]; −1.09 [3.0]^a $K[Cr_2O_7^{2-}/2\,HCrO_4^-]$ 1.53 [0]; 1.97 [1.0]; 2.17 [3.0]	[1, 2, 11]
MoO_4^{2-}	K_{H1} 3.65; K_{H2} 3.55 [1.0]; $K_{H1}K_{H2}$ 7.75 [0.1]^a $K[Mo_7O_{24}^{6-}/8\,H^+, 7\,MoO_4^{2-}]$ 52.81 [1.0]; $K[Mo_7O_{23}(OH)^{5-}/9\,H^+, 7\,MoO_4^{2-}]$ 62.14, $K[Mo_7O_{22}(OH)_2^{4-}/10\,H^+, 7\,MoO_4^{2-}]$ 65.68, $K[Mo_7O_{21}(OH)_3^{3-}/11\,H^+, 7\,MoO_4^{2-}]$ 68.21 [3.0]	[2, 12]
WO_4^{2-}	K_{H1} (4.6); K_{H2} (3.5) [0.1]^a $K[W_{12}O_{39}^{6-}/18\,H^+, 12\,WO_4^{2-}]$ 132.51 [3.0]^f	[2] [15]

Table 5.1 (Continued)

Brønsted base	Equilibrium constants Symbol for the constant, value of log K, $[I]$†	Ref. §
F^-	K_{H1} 3.17 [0]; 2.92 [0.1]; 2.96 [1.0]; 3.30 [3.0] $K[HF_2^-/HF, F^-]$ 0.5 [0]; 0.59 [0.5, 1.0]; 0.86 [3.0]	
Cl^-	K_{H1} −7 [0]	[13]
ClO^-	K_{H1} 7.53 [0]	
ClO_2^-	K_{H1} 1.95 [0]	
ClO_3^-	K_{H1} −2.7 [0]	[13]
ClO_4^-	K_{H1} −7.3 [0]	[13]
Br^-	K_{H1} −9 [0]	[13]
BrO^-	K_{H1} 8.63 [0]	
I^-	K_{H1} −9.5 [0]	[13]
IO^-	K_{H1} 10.64 [0]	
IO_3^-	K_{H1} 0.77 [0]	
MnO_4^-	K_{H1} −0.3 [1.0]	[14]
$Fe(CN)_6^{4-}$	K_{H1} 4.30 [0]; 3.25 [0.2]; K_{H2} 2.6 [0]; 1.79 [0.2]	

[a] 20°C; [b] $(CH_3)_4NCl$ medium; [c] 5°C; [d] 19°C; [e] 22°C; [f] 50°C; [g] 18°C

REFERENCES

[1] R. M. Smith and A. E. Martell, *Critical Stability Constants*, Vol. 4, *Inorganic Complexes*, Plenum Press, New York, 1976.
[2] C. F. Baes, Jr. and R. E. Mesmer, *The Hydrolysis of Cations*, Wiley, New York, 1976.
[3] G. Gatow and B. Krebs, *Z. Anorg. Allgem. Chem.*, 1963, **323**, 13.
[4] O. Redlich, *Monatsh. Chem.*, 1955, **86**, 329.
[5] G. Schwarzenbach and A. Fischer, *Helv. Chim. Acta*, 1960, **43**, 1365.
[6] A. Kossiakoff and D. Harker, *J. Am. Chem. Soc.*, 1938, **60**, 2047.
[7] C. T. Kawassiades, G. E. Manoussakis and J. A. Tossidis, *J. Inorg. Nucl. Chem.*, 1967, **29**, 401.
[8] M. de Hlasko, *J. Chim. Phys.*, 1923, **20**, 167.
[9] A. J. Panson, *J. Phys. Chem.*, 1963, **67**, 2177.
[10] T. Sekine, H. Iwaki, M. Sakairi and F. Shimada, *Bull. Chem. Soc. Japan*, 1968, **41**, 1.
[11] O. Lukkari, *Suom. Kemistilehti*, 1970, **B43**, 347.
[12] Y. Sasaki and L. G. Sillén, *Acta Chem. Scand.*, 1964, **18**, 1014.
[13] J. C. McCoubrey, *Trans. Faraday Soc.*, 1955, **51**, 743.
[14] N. Bailey, A. Carrington, K. A. K. Lott and M. C. R. Symons, *J. Chem. Soc.*, **1960**, 290.
[15] J. Aveston and J. S. Johnson, *Inorg. Chem.*, 1964, **3**, 1051.

Table 5.2
Protonation constants for organic substances in aqueous solution†

Compound Emp. form. Name	Symbol for the acid	Protonation constants for Brønsted base L (Symbol for the constant, value of log K, [I§])	Ref.‡
CH_4N_2O urea	HL	K_{H1} 14.3 [0]; K_{H2} 0.1 [0]; 0.11 [0.1]	
CH_5N methylamine	L	K_{H1} 10.64 [0]; 10.67 [0.2]; 10.72 [0.5]	
CH_5N_3 guanidine	L	K_{H1} 13.54 [1.0][a]	
$C_2HCl_3O_2$ trichloroacetic acid	HL	K_{H1} 0.66 [0.1]	
$C_2H_2Cl_2O_2$ dichloroacetic acid	HL	K_{H1} 1.30 [0]; 0.87 [0.2]	
$C_2H_2O_3$ glyoxylic acid	HL	K_{H1} 3.46 [0]; 3.18 [0.1][b]; 2.91 [1.0]	
$C_2H_3BrO_2$ bromoacetic acid	HL	K_{H1} 2.902 [0]; 2.72 [0.1]; 2.69 [1.0]	
$C_2H_3FO_2$ fluoroacetic acid	HL	K_{H1} 2.586 [0]	
$C_2H_4N_2S_2$ dithio-oxamide	L	K_{H1} 10.89 [0]; 10.42 [1.0]	[2]
$C_2H_6N_2O$ glycine amide	L	K_{H1} 7.93 [0.1]; 8.19 [1.0]	
$C_2H_6S_2$ ethane-1,2-dithiol	H_2L	K_{H1} 10.43: K_{H2} 8.85 [0.1][c]	
C_2H_7N ethylamine	L	K_{H1} 10.636 [0]; 10.66 [0.5]	
C_2H_7N dimethylamine	L	K_{H1} 10.77 [0]; 10.80 [0.2]	
C_2H_7NS 2-aminoethanethiol	HL	K_{H1} 10.71 [0.1]; 10.69 [1.0][c]; K_{H2} 8.23 [0]; 8.21 [0.1]; 8.28 [1.0][c]	
$C_3H_3NOS_2$ rhodanine	L	K_{H1} 5.18 [0.1]	[3]
$C_3H_4N_2$ imidazole	HL	K_{H1} 14.44 [0]; 14.29 [0.5]; K_{H2} 6.993 [0]; 7.03 [0.18]; 7.31 [1.0]	
$C_3H_4O_3$ 2-oxopropionic acid	HL	K_{H1} 2.55 [0]; 2.26 [0.1]; 2.20 [0.5]	
$C_3H_6O_2$ propionic acid	HL	K_{H1} 4.874 [0]; 4.67 [0.1]; 4.67 [1.0]	

† Other organic substances and their protonation constants are listed in the tables of stability constants of complexes of organic ligands (Chapter 7)
§ Temperature is 25°C unless noted otherwise
‡ If no reference is given, the data were taken from [1]

Table 5.2 (Continued)

Compound Emp. form. Name	Symbol the acid	Protonation constants for Brønsted base L (Symbol for the constant, value of log K, [I§])	Ref.‡
$C_3H_6O_2S$ 3-mercaptopropionic acid	H_2L	K_{H1} 10.84 [0]; K_{H2} 4.34 [0]; 4.86 [0.1]^c	[4]
$C_3H_7NO_2$ sarcosine	HL	K_{H1} 10.200 [0]; 9.99 [0.1]; 10.08 [1.0]^b; K_{H2} 2.11 [0]; 2.20 [0.1]; 2.18 [1.0]^b	
$C_3H_7NO_2S$ L-cysteine	H_2L	K_{H1} 10.77 [0]; 10.29 [0.1]; 9.95 [1.0]; K_{H2} 8.36 [0]; 8.15 [0.1]; 8.07 [1.0]; K_{H3} 1.71 [0]; 1.88 [0.1]; 2.44 [3.0]	
$C_3H_7NO_3$ L-serine	HL	K_{H1} 9.209 [0]; 9.06 [0.1]; 9.12 [1.0]^b; K_{H2} 2.187 [0]; 2.13 [0.1]; 2.26 [1.0]^b	
$C_3H_7NO_3$ DL-isoserine	HL	K_{H1} 9.13 [0.18]	
$C_3H_8NO_6P$ L-phosphoserine	H_3L	K_{H1} 9.71 [0.18]; K_{H2} 6.19 [0]; 5.65 [0.18]; K_{H3} 2.07 [0.18]	
$C_4H_4O_4$ fumaric acid	H_2L	K_{H1} 4.494 [0]; 4.10 [0.1]; 3.91 [1.0]; K_{H2} 3.053 [0]; 2.85 [0.1]; 2.80 [1.0]	
$C_4H_7NO_3$ L-asparagine	HL	K_{H1} 8.72 [0.1]; 8.79 [1.0]^b; K_{H2} 2.14 [0.1]; 2.09 [1.0]^b	
C_4H_9N pyrrolidine	L	K_{H1} 11.305 [0]; 11.2 [0.2]	
$C_4H_9NO_3$ L-homoserine	HL	K_{H1} 9.28 [0.1]; K_{H2} 2.27 [0.1]	
$C_4H_9NO_3$ L-threonine	HL	K_{H1} 9.100 [0]; 8.97 [0.1]; 8.86 [1.0]^b; K_{H2} 2.088 [0]; 2.21 [0.1]; 2.34 [1.0]^b	
$C_4H_{10}N_2$ piperazine	L	K_{H1} 9.731 [0]; 9.71 [0.1]; K_{H2} 5.333 [0]; 5.59 [0.1]	
$C_4H_{11}N$ diethylamine	L	K_{H1} 10.933 [0]; 10.97 [0.5]	
$C_4H_{11}NO_2$ 2,2'-iminodiethanol	L	K_{H1} 8.883 [0]; 8.90 [0.1]; 9.00 [0.5]	
$C_4H_{12}N_2O$ 2-(2'-aminoethyl- amino)ethanol	L	K_{H1} 9.56 [0]; 9.59 [0.1]; 9.74 [0.5]; K_{H2} 6.34 [0]; 6.60 [0.1]; 6.85 [0.5]	
$C_4H_{12}N_2S$ thiobis(2-ethylamine)	L	K_{H1} 9.50, K_{H2} 8.70 [1.0]^c	
$C_4H_{13}N_3$ tris(aminomethyl)methane	L	K_{H1} 10.39 [0.1]^b; 10.51 [1.0]^d; K_{H2} 8.56 [0.1]^b; 8.86 [1.0]^d; K_{H3} 6.44 [0.1]^b; 6.90 [1.0]^d	
$C_4H_{13}NO_6P_2$ N-ethyliminodimethylene- diphosphonic acid	H_4L	K_{H1} 12.42; K_{H2} 5.92; K_{H3} 4.70 [1.0]	

Table 5.2 (Continued)

Compound Emp. form. Name	Symbol for the acid	Protonation constants for Brønsted base L (Symbol for the constant, value of $\log K$, $[I\S]$)	Ref.‡
$C_5H_4N_4$ purine	HL	K_{H1} 8.66; K_{H2} 2.41 [0.05]	[5]
$C_5H_5N_5$ 6-aminopurine	HL	K_{H1} 9.87 [0]; 9.67 [0.1]; 9.26 [0.15]^e; K_{H2} 4.20 [0]; 4.07 [0.1]; 3.84 [0.15]^e	
$C_5H_6N_2$ 2-aminopyridine	L	K_{H1} 6.70 [0.2]; 6.93 [1.0]	
$C_5H_6N_2$ 3-aminopyridine	L	K_{H1} 6.03 [0]; 6.06 [0.1]; 6.33 [1.0]	
$C_5H_6N_2$ 4-aminopyridine	L	K_{H1} 9.114 [0]; 9.14 [0.2]; 9.39 [1.0]	
$C_5H_9NO_2$ L-proline	HL	K_{H1} 10.640 [0]; 10.38 [0.1]; 10.52 [1.0]^b; K_{H2} 1.952 [0]; 1.90 [0.1]; 2.02 [1.0]^b	
$C_5H_9NO_3$ L-hydroxyproline	HL	K_{H1} 9.662 [0]; 9.46 [0.1]; 9.58 [1.0]^b; K_{H2} 1.818 [0]; 1.80 [0.1]; 1.93 [1.0]^b	
$C_5H_9NO_4$ L-glutamic acid	H_2L	K_{H1} 9.95 [0]; 9.59 [0.1]; 9.55 [1.0]^b; K_{H2} 4.42 [0]; 4.20 [0.1]; 4.21 [1.0]^b; K_{H3} 2.23 [0]; 2.18 [0.1]; 2.39 [1.0]^b	
$C_5H_9N_3$ histamine	L	K_{H1} 9.86 [0]; 9.83 [0.1]; 9.72 [1.0]^c; K_{H2} 5.96 [0]; 6.07 [0.1]; 6.24 [1.0]^c	
$C_5H_{10}N_2O_3$ L-glutamine	HL	K_{H1} 9.01 [0.1]; 9.64 [3.0]; K_{H2} 2.17 [0.1]; 2.72 [3.0]	
$C_5H_{11}N$ piperidine	L	K_{H1} 11.123 [0]; 11.01 [0.2]; 11.12 [0.5]	
$C_5H_{11}NO_2$ DL-norvaline	HL	K_{H1} 9.806 [0]; 9.65 [0.05]; K_{H2} 2.318 [0]; 2.31 [0.05]	
$C_5H_{11}NO_2$ L-valine	HL	K_{H1} 9.718 [0]; 9.49 [0.1]; 9.59 [1.0]^b; K_{H2} 2.286 [0]; 2.26 [0.1]; 2.38 [1.0]^b	
$C_5H_{11}NO_2S$ DL-methionine	HL	K_{H1} 9.05 [0.1]; 9.13 [1.0]; K_{H2} 2.20 [0.1]; 2.26 [1.0]	
$C_5H_{12}N_2O_2$ L-ornithine	HL	K_{H1} 10.755 [0]; 10.55 [0.1]; 10.46 [1.0]; K_{H2} 8.690 [0]; 8.74 [0.1]; 8.58 [1.0]; K_{H3} 1.705 [0]; 2.11 [1.0]^b	
$C_5H_{16}N_4$ tetra(aminomethyl)methane	L	K_{H1} 9.89, K_{H2} 8.17, K_{H3} 5.67, K_{H4} 3.03 [0.1]	
$C_6H_2O_6$ rhodizonic acid	H_2L	K_{H1} 3.58, K_{H2} 3.45 [3.0]	[6]
$C_6H_5NO_2$ nicotinic acid	HL	K_{H1} 4.81 [0]; 4.67 [0.1]; 4.70 [0.5]^b K_{H2} 2.05 [0]; 2.09 [0.5]^b	
C_6H_6O phenol	HL	K_{H1} 9.98 [0]; 9.82 [0.1]; 9.52 [3.0]	

Table 5.2 (Continued)

Compound Emp. form. Name	Symbol for the acid	Protonation constants for Brønsted base L (Symbol for the constant, value of log K, [I§])	Ref.‡
$C_6H_6N_2O_2$ cupferron	HL	K_{H1} 4.16 [0.1]	
C_6H_6S benzenethiol	HL	K_{H1} 6.615 [0]; 6.46 [0.1]; 6.43 [1.0]	
$C_6H_7AsO_3$ phenylarsonic acid	H_2L	K_{H1} 8.75 [0]; 8.25 [0.1]; K_{H2} 3.61 [0]; 3.39 [0.1]	
$C_6H_7AsO_4$ 2-hydroxyphenylarsonic acid	H_3L	K_{H1} 13.27 [0.1]; K_{H2} 7.92 [0]; 7.50 [0.1]; K_{H3} 4.02 [0]; 3.83 [0.1]	
$C_6H_7AsO_4$ 4-hydroxyphenylarsonic acid	H_3L	K_{H1} 10.24, K_{H2} 8.65, K_{H3} 3.85 [0]	
C_6H_7N aniline	L	K_{H1} 4.601 [0]; 4.65 [0.1]; 4.82 [1.0]	
C_6H_7N 2-picoline	L	K_{H1} 5.95 [0]; 5.95 [0.1]; 6.02 [0.5]	
C_6H_7N 3-picoline	L	K_{H1} 5.68 [0]; 5.76 [0.1]; 5.87 [0.5]	
C_6H_7N 4-picoline	L	K_{H1} 6.03 [0]; 6.04 [0.1]; 6.11 [0.5]	
$C_6H_7NO_3S$ 4-aminobenzenesulphonic acid	HL	K_{H1} 3.232 [0]; 3.01 [0.1]	
$C_6H_7NO_3S$ 3-aminobenzenesulphonic acid	HL	K_{H1} 3.738 [0]	
$C_6H_8O_6$ propane-1,2,3-tricarboxylic acid	H_3L	K_{H1} 6.38 [0]; 5.82 [0.1]; K_{H2} 4.87 [0]; 4.50 [0.1]; K_{H3} 3.67 [0]; 3.48 [0.1]	
$C_6H_9N_3O_2$ L-histidine	HL	K_{H1} 9.02 [0.1]; 9.63 [3.0]; K_{H2} 6.02 [0.1]; 6.97 [3.0]; K_{H3} 1.7 [0.1]; 2.28 [3.0]	
$C_6H_{10}N_2O_2$ cyclohexane-1,2-dione dioxime	H_2L	K_{H1} 12.5 [0]; 12.1 [0.1]; K_{H2} 10.7 [0]; 10.57 [0.1]	
$C_6H_{10}O_4$ adipic acid	H_2L	K_{H1} 5.42 [0]; 5.03 [0.1]; 4.92 [1.0]; K_{H2} 4.42 [0]; 4.26 [0.1]; 4.21 [1.0]	
$C_6H_{11}NO_4S$ N-(2-mercaptoethyl)imino-diacetic acid	H_2L	K_{H1} 10.79, K_{H2} 8.17, K_{H3} 2.14 [0.1][b]	
$C_6H_{11}NO_5$ N-(2-hydroxyethyl)imino-diacetic acid	H_2L	K_{H1} 8.66 [0.1]; 8.67 [1.0]; K_{H2} 2.20 [0.1]; 2.22 [1.0]	

Table 5.2 (Continued)

Compound Emp. form. Name	Symbol for the acid	Protonation constants for Brønsted base L (Symbol for the constant, value of log K, [I§])	Ref.‡
$C_6H_{11}N_3O_4$ glycylglycylglycine	HL	K_{H1} 8.09 [0]; 7.89 [0.1]; 8.01 [1.0]; K_{H2} 3.22 [0]; 3.20 [0.1]; 3.26 [1.0]	
$C_6H_{13}NO_2$ L-leucine	HL	K_{H1} 9.747 [0]; 9.57 [0.1]; 9.62 [1.0]; K_{H2} 2.329 [0]; 2.35 [0.1]; 2.37 [1.0]	
$C_6H_{13}NO_2$ L-isoleucine	HL	K_{H1} 9.754 [0]; 9.62 [0.1]; 9.55 [0.5]; K_{H2} 2.319 [0]; 2.25 [0.1]; 2.51 [0.5]	
$C_6H_{14}N_2O_2$ L-lysine	HL	K_{H1} 10.69 [0.1]; 10.56 [1.0]; K_{H2} 9.08 [0.1]; 9.05 [1.0]	
$C_6H_{14}N_4O_2$ L-arginine	HL	K_{H1} 8.991 [0]; 9.01 [0.1]; 9.21 [1.0]^b; K_{H2} 1.823 [0]; 2.05 [0.1]; 2.19 [1.0]^b	
$C_7H_5NO_5$ chelidamic acid	H_3L	K_{H1} 10.88, K_{H2} 3.11, K_{H3} 1.4 [0.1]^b	
$C_7H_7NO_2$ 3-aminobenzoic acid	HL	K_{H1} 4.76, K_{H2} 3.08 [0]	
$C_7H_7NO_2$ 4-aminobenzoic acid	HL	K_{H1} 4.87, K_{H2} 2.42 [0]	
C_7H_8O 3-methylphenol	HL	K_{H1} 10.09 [0]	
C_7H_8O 4-methylphenol	HL	K_{H1} 10.26 [0]	
C_7H_9N 2-methylaniline	L	K_{H1} 4.447 [0]; 4.57 [0.1]	
C_7H_9N 4-methylaniline	L	K_{H1} 5.084 [0]; 5.23 [0.1]	
C_7H_9N 2,3-dimethylpyridine	L	K_{H1} 6.57 [0]	
C_7H_9N 2,4-dimethylpyridine	L	K_{H1} 6.63 [0]	
C_7H_9N 2,5-dimethylpyridine	L	K_{H1} 6.40 [0]	
C_7H_9N 2,6-dimethylpyridine	L	K_{H1} 6.72 [0]	
C_7H_9N 3,4-dimethylpyridine	L	K_{H1} 6.46 [0]; 6.65 [0.5]	
C_7H_9N 3,5-dimethylpyridine	L	K_{H1} 6.17 [0]; 6.24 [0.5]	
C_7H_9N 4-ethylpyridine	L	K_{H1} 5.87 [0]; 6.33 [0.5]; 5.89 [1.0]	
$C_7H_{12}O_4$ pimelic acid	H_2L	K_{H1} 5.43 [0]; 5.08 [0.1]; 4.97 [1.0]; K_{H2} 4.49 [0]; 4.31 [0.1]; 4.28 [1.0]	

Table 5.2 (Continued)

Compound Emp. form. Name	Symbol for the acid	Protonation constants for Brønsted base L (Symbol for the constant, value of log K, [I§])	Ref.‡
$C_8H_6O_4$ phthalic acid	H_2L	K_{H1} 5.408 [0]; 4.93 [0.1]; 4.71 [1.0]; K_{H2} 2.950 [0]; 2.75 [0.1]; 2.66 [1.0]	
$C_8H_9NO_2$ N-phenylglycine	HL	K_{H1} 4.38, K_{H2} 2.0 [0.1]	
$C_8H_{14}N_4O_5$ glycylglycylglycylglycine	HL	K_{H1} 7.87 [0.1]; 7.88 [1.0]; K_{H2} 3.18 [0.1]; 3.21 [1.0]	
$C_8H_{16}N_2O_4$ ethyldi-iminodi-3-propionic acid	H_2L	K_{H1} 9.60, K_{H2} 6.87 [0.1]^c	
C_9H_7N quinoline	L	K_{H1} 4.81 [0]; 4.97 [0.1]	
$C_9H_8N_2$ 8-aminoquinoline	L	K_{H1} 4.04 [0.1]^b	
$C_9H_{11}NO_2$ DL-phenylalanine	HL	K_{H1} 9.31 [0]; 9.11 [0.05]; 9.18 [1.0]^b K_{H2} 2.20 [0]; 2.18 [0.05]; 2.21 [1.0]^b	
$C_9H_{11}NO_3$ DL-o-tyrosine	H_2L	K_{H1} 10.54, K_{H2} 8.48 [0.18]	
$C_9H_{11}NO_3$ L-tyrosine	H_2L	K_{H1} 10.47 [0]; 10.14 [0.1]; K_{H2} 9.19 [0]; 9.04 [0.1]; K_{H3} 1.48 [0]; 2.17 [0.1]	
$C_{10}H_8O$ 1-naphthol	HL	K_{H1} 9.34 [0]; 9.14 [0.1]; 9.05 [1.0]	[7]
$C_{10}H_8O$ 2-naphthol	HL	K_{H1} 9.51 [0]; 9.31 [0.1]; 9.25 [1.0]	[7]
$C_{10}H_{10}O_2$ benzoylacetone	HL	K_{H1} 8.89 [0.1]; 8.24 [1.0]	
$C_{10}H_{12}N_2O_4$ N-(2-pyridylmethyl)iminodiacetic acid	H_2L	K_{H1} 8.21, K_{H2} 2.8 [0.1]	
$C_{10}H_{13}N_5O_4$ adenosine	HL	K_{H1} 12.35, K_{H2} 3.50 [0]	
$C_{10}H_{14}N_5O_7$ adenosine-2'-monophosphoric acid	H_3L	K_{H2} 6.02, K_{H3} 3.72 [0.1]	
$C_{10}H_{14}N_5O_7P$ adenosine-3'-monophosphoric acid	H_3L	K_{H2} 5.83, K_{H3} 3.65 [0.1]	
$C_{10}H_{14}N_5O_7P$ adenosine-5'-mono- phosphoric acid	H_3L	K_{H1} 13.06 [0]; K_{H2} 6.67 [0]; 6.19 [0.1]; K_{H3} 3.80 [0.1]	
$C_{10}H_{15}N_5O_{10}P_2$ adenosine-5'-diphosphoric acid	H_4L	K_{H1} 7.20 [0]; 6.40 [0.1]; K_{H2} 3.96 [0.1]	
$C_{10}H_{16}N_5O_{13}P_3$ adenosine-5'-triphosphoric acid	H_5L	K_{H1} 7.68 [0]; 6.51 [0.1]; K_{H2} 4.06 [0.1]	

Constants of Protonation Equilibria

Table 5.2 (Continued)

Compound Emp. form. Name	Symbol for the acid	Protonation constants for Brønsted base L (Symbol for the constant, value of log K, [I§])	Ref. §
$C_{10}H_{17}NO_4$ N-cyclohexyliminodiacetic acid	H_2L	K_{H1} 10.81 [0.1]^b; 10.58 [0.5]; K_{H2} 2.15 [0.1]^b; 2.26 [0.5]; K_{H3} 1.49 [0.5]	
$C_{10}H_{20}N_6O_8$ ethylenedinitrilotetra-aceto-hydroxamic acid	H_4L	K_{H1} 11.1, K_{H2} 10.6, K_{H3} 7.23, K_{H4} 6.67, K_{H5} 6.05, K_{H6} 5.55 [0.1]^b	
$C_{11}H_{12}N_2O_2$ L-tryptophan	HL	K_{H1} 9.33 [0]; 9.43 [0.1]^b; K_{H2} 2.35 [0]; 2.39 [0.1]^b	

Measured at ^{a} 27°C; ^{b} 20°C; ^{c} 30°C; ^{d} 22°C; ^{e} 37°C.

REFERENCES

[1] A. E. Martell and R. M. Smith, *Critical Stability Constants*, Vols. 1–3, Plenum Press, New York, 1974–1976.
[2] R. P. Yaffee and A. F. Voigt, *J. Am. Chem. Soc.*, 1952, **74**, 2941.
[3] O. Navrátil and J. Kotas, *Collection Czech. Chem. Commun.*, 1965, **30**, 2736.
[4] M. Cefola, A. S. Fompa, A. V. Celiano and P. S. Gentile, *Inorg. Chem.*, 1962, **1**, 290.
[5] H. Reinert and R. Weiss, *Z. Physiol. Chem.*, 1969, **350**, 1310.
[6] D. Alexandersson and N. G. Vannerberg, *Acta Chem. Scand.*, 1972, **26**, 1909.
[7] O. Makitie, *Suom. Kemistilehti*, 1966, **B39**, 23.
[8] R. Karlíček and J. Majer, *Collection Czech. Chem. Commun.*, 1972, **37**, 151.

Table 5.3

Protonation constants of inorganic substances in mixed and non-aqueous media† (at 25°C)

Brønsted base	Medium	log K_{Hn}	Ref.
$B(OH)_4^-$	dioxane (45%), [$I \to 0$]	K_{H1} 11.406	[1]
	C_6H_6 (50%) + CH_3OH (50%)	K_{H1} 13.23	[2]
NH_3	CH_3OH (p%)	K_{H1} 9.044 (20%); 8.687 (50%); 8.571 (70%)	[3]
	C_2H_5OH	K_{H1} 10.5	[4]
	CH_3CN	K_{H1} 16.46	[5]
N_3^-	CH_3OH [$I \to 0$]	K_{H1} 9.20	[6]

† The solvent medium is named or given by formula; the composition of a mixed solvent (usually with water) is given in percentage by weight.

Table 5.3 (Continued)

Brønsted base	Medium	log K_{Hn}	Ref.
NO_3^-	C_2H_5OH [$I \to 0$]	K_{H1} 3.57	[7]
	CH_3COOH	K_{H1} 5.1	[8]
	CH_3CN	K_{H1} 8.9	[9]
	C_5H_5N	K_{H1} 4.06	[10]
PO_4^{3-}	CH_3OH (50%)	K_{H2} 8.443 (mole/kg)	[11]
	CH_3OH	K_{H3} 7.60	[12]
	C_2H_5OH (75%)	K_{H3} 4.17 (?)	[13]
	$CH_3CH_2CH_2OH$	K_{H3} 8.75	[12]
	CH_3COOH	K_{H3} 4.4	[8]
H_2O	CH_3OH	K_{H1} 0.64	[14–16]
	C_2H_5OH	K_{H1} 1.23	[15–18]
	CH_3COOH	K_{H1} 1.92	[19]
	CH_3CN	K_{H1} 2.15; β_2 3.43; $\beta_n[H(H_2O)_n^+/H^+, nH_2O]$ β_3 4.52; β_4 4.08	[20]
SO_3^{2-}	CH_3OH ($p\%$), [$I \to 0$]	K_{H2} 2.00 (20%); 2.30 (45%); 2.72 (70%)	[21]
SO_4^{2-}	C_2H_5OH (20%)	K_{H1} 2.62	[22]
	CH_3COOH	K_{H1} 7.24	[23]
	CH_3CN	K_{H1} 25.9; K_{H2} 7.8	[24]
	$(CH_3)_2SO$	K_{H1} 14.5	[24]
F^-	C_2H_5OH (20°)	K_{H1} 6.4–7.0	[25]
Cl^-	$R-OH$ ($R-$), [$I \to 0$]	K_{H1} 4.26 (CH_3-); 4.21 (C_2H_5-); 4.30 (C_3H_7-); 4.43 (($CH_3)_2CH-$); 4.39 (C_4H_9-)	[12]
	HCOOH, [$I \to 0$]	K_{H1} 1.96	[26]
	$R-COOH$ ($R-$)	K_{H1} 1.25 (H$-$); 5.40 (CH_3-)	[27, 28]
	CH_3CN	K_{H1} 8.9	[9]
	C_6H_5CN ($I \to 0$)	K_{H1} 3.6	[29]
	CH_3COCH_3 ($p\%$)	K_{H1} 1.3 (80%); 2.5 (90%); 8 (100%)	[30, 31]
	dioxane (70%)	K_{H1} 2.18	[32]
	$(ClC_2H_4)_2O$	$K_{H1} \sim 7$	[33–35]
	$C_6H_5NO_2$ (sat. with H_2O)	K_{H1} 7.53	[36]
ClO_4^-	CH_3COOH	K_{H1} 5.24	[37]
Br^-	CH_3OH	K_{H1} 0.42	[38]
	C_2H_5OH	K_{H1} 1.73	[39]
	CH_3COOH	K_{H1} 6.73; 4.10 [for $I \to 0$]	[40, 41]
	CH_3CN	K_{H1} 5.5	[9]
	$(ClC_2H_4)_2O$	K_{H1} 5.2	[34, 35]
	$C_6H_5NO_2$ (sat. with H_2O)	K_{H1} 7.53	[36]
I^-	C_5H_5N	K_{H1} 3.11	[37]

REFERENCES

[1] D. Feakins and D. J. Turner, *J. Chem. Soc.*, **1965**, 4986.
[2] V. V. Alexandrov, L. L. Spivak and L. K. Zakharchenko, *Zh. Fiz. Khim.*, 1965, **39**, 58.
[3] M. Paabo, R. G. Bates and R. A. Robinson, *J. Phys. Chem.*, 1966, **70**, 247.
[4] E. Larsson, *Thesis*, p. 77. Lund, 1924.
[5] J. F. Coetzee and G. R. Padmanabhan, *J. Am. Chem. Soc.*, 1965, **87**, 5005.
[6] R. G. Pearson, P. M. Henry and F. Basolo, *J. Am. Chem. Soc.*, 1957, **79**, 5379.
[7] A. J. Deyrup, *J. Am. Chem. Soc.*, 1934, **56**, 60.
[8] Yu. Ya. Fialkov and Yu. Ya. Borovikov, *Ukr. Khim. Zh.*, 1964, **30**, 119.
[9] I. M. Kolthoff, S. Bruckenstein and M. K. Chantooni, Jr., *J. Am. Chem. Soc.*, 1961, **83**, 3927.
[10] L. M. Mukherjee, J. J. Kelly, W. Baranetzky and J. Sica, *J. Phys. Chem.*, 1968, **72**, 3410.
[11] M. Paabo, R. A. Robinson and R. G. Bates, *J. Am. Chem. Soc.*, 1965, **87**, 415.
[12] A. P. Kreshkov, V. A. Drozdov and N. A. Kolchina, *Zh. Fiz. Khim.*, 1966, **40**, 2150.
[13] D. F. Peppard, G. W. Mason and C. M. Andrejasich, *J. Inorg. Nucl. Chem.*, 1965, **27**, 697.
[14] L. Thomas and E. Marum, *Z. Phys. Chem. (Leipzig)*, 1929, **143**, 191.
[15] P. K. Migal' and N. Kh. Grinberg, *Uch. Zap. Kishinev. Univ.*, 1960, **56**, 179.
[16] L. S. Guss and I. M. Kolthoff, *J. Am. Chem. Soc.*, 1940, **62**, 1494.
[17] H. Goldschmidt, *Z. Phys. Chem. (Leipzig)*, 1914, **89**, 129.
[18] V. V. Aleksandrov and N. A. Izmailov, *Zh. Fiz. Khim.*, 1958, **32**, 404.
[19] S. Bruckenstein and I. M. Kolthoff, *J. Am. Chem. Soc.*, 1957, **79**, 5915.
[20] I. M. Kolthoff and M. K. Chantooni, Jr., *J. Am. Chem. Soc.*, 1968, **90**, 3320.
[21] E. N. Deikhman, G. V. Rodicheva and L. S. Krysina, *Zh. Neorgan. Khim.*, 1966, **11**, 2237.
[22] W. D. Bale, E. W. Davies, D. B. Morgans and C. B. Monk, *Discuss. Faraday Soc.*, 1957, **24**, 94.
[23] S. Bruckenstein and I. M. Kolthoff, *J. Am. Chem. Soc.*, 1956, **78**, 2974.
[24] I. M. Kolthoff and M. K. Chantooni, Jr., *J. Am. Chem. Soc.*, 1968, **90**, 5961.
[25] E. Nachbaur, *Monatsh. Chem.*, 1960, **91**, 749.
[26] T. C. Wehman and A. I. Popov, *J. Phys. Chem.*, 1968, **72**, 4031.
[27] A. M. Shkodin, *Zh. Fiz. Khim.*, 1962, **36**, 595.
[28] H. Zannovich-Tesarin, *Z. Phys. Chem. (Leipzig)*, 1896, **19**, 251.
[29] G. J. Janz, I. Ahmad and H. V. Venkattasetty, *J. Phys. Chem.*, 1964, **68**, 889.
[30] V. M. Atkins and C. B. Monk, *J. Chem. Soc.*, **1961**, 1817.
[31] D. H. Everett and S. E. Rasmussen, *J. Chem. Soc.*, **1954**, 2812.
[32] E. L. Purlee and E. Grunwald, *J. Chem. Soc.*, 1957, **79**, 1366.
[33] A. M. Poskanzer, R. J. Dietz, Jr., E. Rudzitis, J. W. Irvine, Jr. and C. D. Coryell, *Proc. Intern. Conf. Radioisotopes, Sci. Res. (Paris)*, 1957, **2**, 518.
[34] R. J. Dietz, Jr., *Thesis*, Massachusetts Inst. Technol., 1958.
[35] J. S. Mendez-Salchi, *Thesis*, Massachusetts Inst. Technol., 1959.
[36] R. H. McCorkell, M. M. Sein and J. W. Irvine, Jr., *J. Inorg. Nucl. Chem.*, 1968, **30**, 1155.
[37] P. Alcais, F. Rothenberg and J.-E. Dubois, *J. Chim. Phys.*, 1967, **64**, 1818.
[38] E. Schreiner, *Z. Phys. Chem. (Leipzig)*, 1924, **111**, 419.
[39] L. J. Nunez and M. C. Day, *J. Phys. Chem.*, 1961, **65**, 1993.
[40] I. M. Kolthoff and A. Willman, *J. Am. Chem. Soc.*, 1934, **56**, 1007.
[41] T. L. Smith and T. H. Elliot, *J. Am. Chem. Soc.*, 1953, **75**, 3566.

Table 5.4

Protonation constants of organic substances for mixed solvents and non-aqueous media† (temperature 25°)

Compound Emp. form. Name	Symbol	Medium ($p\%$) [I, temp.]	Protonation constants for Brønsted base L The constant, value of log K_{Hn}	Ref.
CH_2O_2 formic acid	HL	dioxane (50%)	K_{H1} 4.75 [I 0.1]	[1]
$C_2H_4O_2$ acetic acid	HL	CH_3OH	K_{H1} 5.660 (50%) [$I \to 0$]; 6.97 (95%) [I 0.5]	[2]
		dioxane (50%)	K_{H1} 6.01 [I 0.1]; 5.62 [0.6]	[1, 3, 4]
$C_2H_4O_3$ glycollic acid	HL	dioxane (50%)	K_{H1} 4.88 [I 0.1]	[1]
$C_2H_3O_2Cl$ monochloroacetic acid	HL	dioxane (50%) C_2H_5OH (20%)	K_{H1} 4.08 [I 0.1] K_{H1} 3.19	[3] [5]
$C_2H_5NO_2$ glycine	HL	dioxane [I 0.001] dioxane (75%) [30°C]	K_{H1} 10.13 (45%); 11.18 (70%) K_{H1} 10.7; K_{H2} 4.2	[6] [7]
C_2H_7NO 2-hydroxyethylamine	L	C_2H_5OH (50 mole %)	K_{H1} 9.34	[8]
$C_2H_8N_2$ ethylenediamine	L	dioxane (75%) [30°C]	K_{H1} 9.2; K_{H2} 6.2	[7]
$C_3H_6O_2$ propionic acid	HL	C_2H_5OH (20%) dioxane (50%) [I 0.1]	K_{H1} 5.29 K_{H1} 6.29	[5] [1]
$C_4H_6O_6$ tartaric acid	H_2L	dioxane (75%) [30°C]	K_{H1} 16.5; K_{H2} 14.3	[7]
$C_4H_8NO_2$ butane-2,3-dione dioxime	HL	dioxane (50%)	K_{H1} 12.84; $K_{H2} \sim 2$	[9]
		dioxane (75%) [I 0.1]	K_{H1} 13.53	[10]
$C_4H_8N_2O_3$ glycylglycine	HL	CH_3OH ($p\%$)	K_{H1} 8.03 (39.1%); 7.93 (70%); K_{H2} 3.69 (39.1%); 4.28 (70%)	[11]
		dioxane ($p\%$)	K_{H1} 8.36 (45%); 8.73 (60%); K_{H2} 4.06 (45%); 4.74 (60%)	[11]
$C_4H_{11}NO_2$ 2,2'-iminodiethanol	L	C_2H_5OH (50 mole %)	K_{H1} 8.73	[8]
$C_4H_{11}NO_3$ 2-amino-2-(hydroxymethyl)propan-1,3-diol (tris)	L	CH_3OH (50%)	K_{H1} 7.962	[12]
$C_5H_2F_6O_2$ hexafluoroacetylacetone	HL	dioxane (75%) [30°C]	K_{H1} 6.0	[14, 15]
$C_5H_3ClN_4$ 6-chloropurine	L	dioxane (50% v/v) [I 0.01]	K_{H1} 8.21; $K_{H2} < 2$	[13]
$C_5H_4N_4O$ 6-hydroxypurine	L	dioxane (50% v/v) [I 0.01]	K_{H1} 8.21; $K_{H2} < 2$	[13]

† The solvent is named or given by formula; the composition of a mixed solvent (usually with water) may be given as percentage by weight (%), mole percentage (mole %), or mole fraction (x).

Table 5.4 (Continued)

Compound Emp. form. Name	Symbol	Medium $(p\%)$ $[I$, temp.$]$	Protonation constants for Brønsted base L The constant, value of log K_{Hn}	Ref.
$C_5H_4N_4S$ 6-mercaptopurine	L	dioxane (50% v/v) $[I\ 0.01]$	K_{H1} 11.9; K_{H2} 8.67; $K_{H3} < 2$	[13]
$C_5H_4O_2S$ thiophene-2-carboxylic acid	HL	dioxane (50%) $[I\ 0.1]$	K_{H1} 5.03	[3]
$C_5H_5F_3O_2$ 1,1,1-trifluoroacetylacetone	HL	dioxane $(p\%)$	K_{H1} 6.7 (50%) $[20°C]$; 8.70 (75%) $[30°C]$	[16] [17]
C_5H_5N pyridine	L	C_2H_5OH (50 mole %)	K_{H1} 4.34	[8]
$C_5H_5NO_2$ 3-furancarboxaldehyde oxime	HL	dioxane (75%) $[I\ 0.1]$	K_{H1} 12.10	[18]
$C_5H_5N_5$ adenine	L	dioxane (50% v/v)	K_{H1} 10.65; K_{H2} 3.54	[13]
$C_5H_8O_2$ acetylacetone	HL	CH_3OH $[I\ 0.1]$	K_{H1} 11.8	[19, 20]
		CH_3OH $[I\ 0.0172]$ (x)	K_{H1} 8.98 (0); 9.14 (0.10); 9.45 (0.30); 10.03 (0.61)	[21]
		C_2H_5OH $[I]$	K_{H1} 11.8 [0.1]; 11.5 [0.025]	[20, 22]
		C_2H_5OH (x) $[I\ 0.0172]$	K_{H1} 9.12 (0.07); 9.94 (0.39); 10.31 (0.52)	[21]
		C_3H_7OH (x) $[I\ 0.0172]$	K_{H1} 9.16 (0.06); 9.91 (0.26); 10.66 (0.47)	[21]
		$(CH_3)_2CHOH$ (x)	K_{H1} 9.21 (0.06); 9.71 (0.20); 10.01 (0.27)	[23, 24]
		acetone (75%)	K_{H1} 9.80	[27]
		dioxane $(p\%)$ $[30°C]$	K_{H1} 8.94 (20%); 9.70 (50%); 12.70 (75%)	[17, 23, 24, 36]
		dioxane (x)	K_{H1} 9.25 (0.05); 9.85 (0.13); 10.31 (0.18); 10.92 (0.25)	[23, 24]
$C_6H_5NO_2$ pyridine-2-carboxylic acid	HL	dioxane (50%) $[I\ 0.1]$	K_{H1} 5.86; K_{H2} 1.31	[31]
$C_6H_6O_2$ pyrocatechol	H_2L	CH_3OH $[I\ 1.0]$	K_{H1} 15.5; K_{H2} 13.15	[25]
		CH_3OH (50%) $[I]$	K_{H1} 14.1 [0.1]; 13,05 [1]; K_{H2} 10.012 [0.1]; 9.229 [1]	[26]
		dioxane (75%) $[30°C]$	K_{H1} 14.2; K_{H2} 13.7	[7]
$C_6H_6O_4$ kojic acid	HL	dioxane $(p\%)$ $[I\ 0.1, 30°C]$	K_{H1} 9.66 (50%); 12.29 (75%)	[28, 29]
C_6H_7N aniline	L	C_2H_5OH (50 mole %)	K_{H1} 4.42	[8]
C_6H_7N 4-methylpyridine	L	dioxane (50%) $[I\ 0.1]$	K_{H1} 5.11	[30]

Table 5.4 (Continued)

Compound Emp. form. Name	Symbol	Medium ($p\%$) [I, temp.]	Protonation constants for Brønsted base L The constant, value of log K_{Hn}	Ref.
C_6H_7NO 2-aminophenol	HL	dioxane (50%) [I 0.1]	K_{H1} 11.57; K_{H2} 3.75	[32, 33]
$C_6H_{10}N_2O_2$ cyclohexane-1,2-dione dioxime	HL	dioxane ($p\%$) [I 0.1]	K_{H1} 12.0 (50%); 13.11 (75%)	[10, 35]
$C_6H_9NO_6$ nitrilotriacetic acid (NTA)	H_3L	dioxane (50%) [I 0.1]	K_{H1} 10.11; K_{H2} 4.18	[34]
$C_6H_{15}NO_3$ 2,2',2''-trihydroxytriethylamine	HL	C_2H_5OH (50 mole %)	K_{H1} 7.40	[8]
$C_7H_6O_2$ benzoic acid	HL	C_2H_5OH (60.1%) dioxane (50%) [I 0.1]	K_{H1} 5.98 K_{H1} 5.79	[37] [1, 3]
$C_7H_6O_2$ 2-hydroxybenzaldehyde	HL	C_2H_5OH (50%) dioxane ($p\%$) [I 0.3]	K_{H1} 8.7 K_{H1} 9.50 (50%); 10.35 (75%)	[38] [39]
$C_7H_6O_3$ salicylic acid	H_2L	C_2H_5OH (50%) dioxane (75%) [I 0.1]	K_{H1} 13.13; K_{H2} 3.91 K_{H1} 13.27; K_{H2} 4.73	[40] [41]
$C_7H_8O_2$ saligenine	HL	dioxane (75%) [I 0.1]	K_{H1} 11.93 [$30°C$]	[41]
$C_7H_6O_2S$ thiosalicylic acid	H_2L	dioxane (50%) [I 0.1]	K_{H1} 9.96; K_{H2} 4.92	[42]
$C_7H_6O_6S$ 5-sulphosalicylic acid	H_3L	C_2H_5OH (50%)	K_{H1} 12.88; K_{H2} 3.41; K_{H3} 1.60	[40]
$C_7H_7NO_2$ anthranilic acid	HL	dioxane (50%)	K_{H1} 6.6	[44]
$C_7H_7NO_2$ 2-hydroxybenzaldehyde oxime	H_2L	dioxane (75%) [I 0.1, $30°C$] C_2H_5OH (50%) [I 0.0045]	K_{H1} 12.08; K_{H2} 11.72; K_{H3} 1.72 (?) K_{H2} 10.23; K_{H3} < 3	[41] [45]
$C_8H_5F_3O_2S$ 2-thenoyl- trifluoroacetone	HL	CH_3OH [I 0.1] C_2H_5OH [I 0.1] dioxane (75%) [$I \to 0$, $30°C$]	K_{H1} 8.59 K_{H1} 8.20 K_{H1} 8.63	[20] [20] [46]
$C_8H_8O_2S$ 3-thenoylacetone	HL	dioxane (75%) [$I \to 0$, $30°C$]	K_{H1} 12.41	[46]
C_9H_7NO 8-hydroxyquinoline (oxine)	HL	C_2H_5OH (75%) [I 0.01] dioxane (50%) [I] dioxane (75%) [I 0.3]	K_{H1} 11.19; K_{H2} 3.76 K_{H1} 10.95 [0.1]; 10.82 [0.3]; K_{H2} 4.13 [0.1]; 4.08 [0.3] K_{H1} 11.22; K_{H2} 3.14	[47] [48–50] [39, 49]
C_9H_7NS 8-mercaptoquinoline	HL	dioxane (50%) [I 0.1]	K_{H1} 9.22; K_{H2} 1.79	[51]

Table 5.4 (Continued)

Compound Emp. form. Name	Symbol	Medium ($p\%$) [I, temp.]	Protonation constants for Brønsted base L The constant, value of log K_{Hn}	Ref.
$C_9H_7NO_4S$ 8-hydroxyquinoline-5-sulphonic acid	H_2L	C_2H_5OH ($p\%$) [I 0.01]	K_{H1} 9.25 (26.3%); 9.79 (51.7%); 10.30 (75%); K_{H2} 3.82 (26.3%); 3.54 (51.7%); 3.39 (75%)	[47]
		dioxane (50%) [I]	K_{H1} 9.85 [0.1]; 9.68 [0.3]; K_{H2} 3.56 [0.1]; 3.40 [0.3]	[48, 49]
$C_9H_6INO_4S$ 7-iodo-8-hydroxyquinoline-5-sulphonic acid	H_2L	dioxane (50%) [I]	K_{H1} 8.83 [0.1]; 8.13 [0.3]; K_{H2} 2.18 [0.1]; 1.75 [0.3]	[39, 49, 52]
$C_9H_7N_3O_2S$ 4-(2'-thiazolylazo)-1,3-dihydroxybenzene (TAR)	H_2L	CH_3OH (50%) [I 0.1]	K_{H1} 10.76; K_{H2} 6.53; K_{H3} 0.9	[53]
		dioxane (50%)	K_{H1} 12.80; K_{H2} 7.37; K_{H3} 1.65	[54]
$C_{10}H_8N_2$ 2,2'-bipyridyl	L	C_2H_5OH (75%) [20°C]	K_{H1} 3.94	[55]
		dioxane (50%) [I 0.3]	K_{H1} 3.62	[56]
$C_{10}H_9NO$ 2-methyl-8-hydroxyquinoline	HL	dioxane (50%) [I 0.1]	K_{H1} 11.30; K_{H2} 4.68	[48, 57]
$C_{10}H_9NO$ 4-methyl-8-hydroxyquinoline	HL	dioxane (50%) [I 0.1]	K_{H1} 11.10; K_{H2} 4.83	[48]
$C_{10}H_{10}O_2$ benzoylacetone	HL	CH_3OH or C_2H_5OH	K_{H1} 12.02 [$(C_4H_9)_4NClO_4$, I 0.1]	[20]
		CH_3OH ($p\%$) [NaCl, I 0.1]	K_{H1} 9.53 (80%); 11.95 (100%)	[58–60]
		dioxane ($p\%$)	K_{H1} 9.8 (50%) [I 0.1, 20°C]; 12.85 (75%) [30°C]	[61–64]
$C_{10}H_7NO_2$ 1-nitroso-2-naphthol	HL	dioxane ($p\%$) [30°C]	K_{H1} 9.47 (50%); 11.61 (75%)	[65, 66]
$C_{10}H_7NO_2$ 2-nitroso-1-naphthol	HL	dioxane ($p\%$) [30°C]	K_{H1} 8.90 (50%); 11.14 (75%)	[65, 66]
$C_{11}H_9N_3O$ 2-(2'-pyridylazo)phenol	HL	CH_3OH (50%) [I 0.1, 20°C]	K_{H1} 9.42; K_{H2} 1.85	[67]
$C_{11}H_9N_3O$ 4-(2'-pyridylazo)phenol	HL	CH_3OH (50%) [I 0.1, 20°C]	K_{H1} 8.29; K_{H2} 2.47	[67]
$C_{11}H_9N_3O_2$ 1,3-dihydroxy-4-(2'-pyridylazo)benzene	H_2L	dioxane (50%) [I 0.01]	K_{H1} 13.42; K_{H2} 6.87; K_{H3} 2.31	[54]
$C_{12}H_6Cl_2N_2$ 4,7-dichloro-1,10-phenanthroline	L	dioxane (50%) [I 0.3]	K_{H1} 2.65	[56]
$C_{12}H_7N_3O_2$ 5-nitro-1,10-phenanthroline	L	dioxane (50%) [I 0.3]	K_{H1} 2.80	[56]
$C_{12}H_8N_2$ 1,10-phenanthroline	L	C_2H_5OH (50%) [I 0.1]	K_{H1} 4.27	[68]
		dioxane (50%) [I 0.3]	K_{H1} 4.63	[56]
		CH_3CN [20°C]	K_{H1} 4.54	[55]
$C_{12}H_{11}NOS$ 2-mercapto-N-(2'-naphthyl)acetamide	HL	dioxane (75%) [I 0.1, 20°C]	K_{H1} 10.20	[69]

Table 5.4 (Continued)

Compound Emp. form. Name	Symbol	Medium ($p\%$) [I, temp.]	Protonation constants for Brønsted base L The constant, value of log K_{Hn}	Ref.
$C_{13}H_{12}N_4O$ 1,5-diphenylcarbazone	L	dioxane [I 0.1]	K_{H1} 9.26	[70]
$C_{13}H_{12}N_4S$ dithizone	HL	dioxane (50%) [I 0.1] extn. H_2O [I 0.1] –$CHCl_3$ –CCl_4 –cyclohexane	K_{H1} 5.80 K_{H1} 5.7 K_{H1} 4.6 K_{H1} 4.2	[70] [71] [71] [71]
$C_{14}H_{12}N_2$ 4,7-dimethyl-1,10-phenanthroline	L	dioxane (50%) [I 0.3]	K_{H1} 5.40	[56]
$C_{14}H_{12}N_2$ 5,6-dimethyl-1,10-phenanthroline	L	dioxane (50%) [I 0.3]	K_{H1} 5.00	[56]
$C_{15}H_{11}N_2O$ 1-(2'-pyridylazo)-2-naphthol (PAN)	HL	dioxane ($p\%$) [I]	K_{H1} 12.2 (20%) [0.08]; 12.3 (50%); K_{H2} 1.90 (20%) [0.08]; <2 (50%)	[72, 73]
$C_{15}H_{11}N_2O$ 4-(2'-pyridylazo)-1-naphthol	HL	dioxane (50%) [33°C]	K_{H1} 10.74; K_{H2} 2.54	[74]
$C_{15}H_{12}O_2$ dibenzoylmethane	HL	CH_3OH [I 0.1] C_2H_5OH [I 0.1] dioxane (75%)	K_{H1} 13.03 K_{H1} 13.4 K_{H1} 13.75	[20] [20] [17, 75, 76]

REFERENCES

[1] R. Griesser, B. Prijs and H. Sigel, *Inorg. Nucl. Chem. Lett.*, 1968, **4**, 443.
[2] M. Paabo, R. A. Robinson and R. G. Bates, *J. Am. Chem. Soc.*, 1965, **87**, 415.
[3] H. Erlenmeyer, R. Griesser, B. Prijs and H. Sigel, *Helv. Chim. Acta*, 1968, **51**, 339.
[4] D. L. Martin, F. J. C. Rossotti and K. Schlyter, unpublished results; cf. *Stability Constants*, L. G. Sillén and A. E. Martell, (eds.), p. 364, Chemical Society, London, 1964.
[5] W. P. Bale and C. B. Monk, *Trans. Faraday Soc.*, 1957, **53**, 450.
[6] N. C. Li, J. M. White and R. L. Yoest, *J. Am. Chem. Soc.*, 1956, **78**, 5218.
[7] L. G. Van Uitert and W. C. Fernelius, *J. Am. Chem. Soc.*, 1954, **76**, 375.
[8] C. T. Anderson, *Thesis*, Ohio State Univ., 1955.
[9] R. G. Charles and H. Freiser, *Anal. Chim. Acta*, 1954, **11**, 101.
[10] C. V. Banks and S. Anderson, *Inorg. Chem.*, 1963, **2**, 112.
[11] J. Vaissermann and M. Quintin, *J. Chim. Phys. Physicochim. Biol.*, 1966, **63**, 731.
[12] M. Woodhead, M. Paabo, R. A. Robinson and R. G. Bates, *J. Res. Natl. Bur. Stds.*, 1965, **A69**, 263.
[13] G. E. Cheney, H. Freiser and Q. Fernando, *J. Am. Chem. Soc.*, 1959, **81**, 2611.
[14] L. G. Van Uitert, W. C. Fernelius and B. E. Douglas, *J. Am. Chem. Soc.*, 1953, **75**, 457.
[15] L. G. Van Uitert, *Thesis*, Pennsylvania State College, 1951.

[16] M. Calvin and K. W. Wilson, *J. Am. Chem. Soc.*, 1945, **67**, 2003.
[17] L. G. Van Uitert, W. C. Fernelius and B. E. Douglas, *J. Am. Chem. Soc.*, 1953, **75**, 457, 2736, 2739.
[18] K. M. J. Al-Komser and B. Sen, *Inorg. Chem.*, 1963, **2**, 1219.
[19] R. Gut, H. Buser and E. Schmid, *Helv. Chim. Acta*, 1965, **48**, 878.
[20] D. C. Luehrs, R. T. Iwamoto and J. Kleinberg, *Inorg. Chem.*, 1965, **4**, 1739.
[21] P. S. Gentile and A. Dadgar, *J. Chem. Eng. Data*, 1968, **13**, 236.
[22] M. C. Day and G. M. Rouayheb, *J. Chem. Eng. Data*, 1960, **5**, 508.
[23] P. S. Gentile, M. Cefola and A. V. Celiano, *J. Phys. Chem.*, 1963, **67**, 1083.
[24] R. W. Green and P. W. Alexander, *J. Phys. Chem.*, 1963, **67**, 905.
[25] R. Gut, *Helv. Chim. Acta*, 1964, **47**, 2262.
[26] C. A. Tyson and A. E. Martell, *J. Am. Chem. Soc.*, 1968, **90**, 3379.
[27] N. K. Dutt and P. Bandhyopandhyay, *Sci. Culture (India)*, 1956, **22**, 690; 1957, **23**, 105.
[28] N. J. Clark and B. R. Willford, Jr., *J. Am. Chem. Soc.*, 1957, **79**, 1296.
[29] H. Kido and W. C. Fernelius, *Pennsylvania State Univ. Contract* AT (30-1)-907, 1960.
[30] H. Sigel, *Chimia*, 1967, **21**, 489.
[31] J. R. Walter and S. M. Rosalie, *J. Inorg. Nucl. Chem.*, 1966, **28**, 2969.
[32] R. G. Charles and H. Freiser, *J. Am. Chem. Soc.*, 1952, **74**, 1385.
[33] H. Freiser, R. G. Charles and W. D. Johnston, *J. Am. Chem. Soc.*, 1952, **74**, 1383.
[34] H. Berge, *J. Prakt. Chem.*, 1966, **34**, 15.
[35] V. M. Peshkova and V. M. Bochkova, *Nauk. Dokl. Vyssh. Shkoly*, **1958**, 62.
[36] L. E. Maley and D. P. Mellor, *Aust. J. Sci.*, 1949, **2A**, 92.
[37] J. Chatt and A. A. Williams, *J. Chem. Soc.*, **1954**, 4403.
[38] C. Hertnee and J. Shamir, *Bull. Soc. Chim. France*, **1957**, 1334.
[39] J. G. Jones, J. B. Poole, J. C. Tomkinson and R. J. P. Williams, *J. Chem. Soc.*, **1958**, 2001.
[40] R. P. Agarwal and R. C. Mehrotra, *J. Less-Common Metals*, 1961, **3**, 398.
[41] K. E. Jabalpurwala, K. A. Venkatachalam and M. B. Kabadi, *J. Inorg. Nucl. Chem.*, 1964, **26**, 1011, 1027.
[42] S. V. Larionov, V. M. Shul'man and L. A. Podolskaya, *Zh. Neorgan. Khim.*, 1964, **9**, 2333.
[43] K. Burger and I. Egyed, *J. Inorg. Nucl. Chem.*, 1965, **27**, 2361.
[44] A. Young and T. R. Sweet, *J. Am. Chem. Soc.*, 1958, **80**, 800.
[45] Z. Holzbecher, *Collection Czech. Chem. Commun.*, 1959, **24**, 3915.
[46] J. L. Rosenstreich and D. E. Goldberg, *Inorg. Chem.*, 1965, **4**, 909.
[47] M. S. Janssen, *Rec. Trav. Chim.*, 1956, **75**, 1397.
[48] G. Gutnikov and H. Freiser, *Anal. Chem.*, 1968, **40**, 39.
[49] J. C. Tomkinson and R. J. P. Williams, *J. Chem. Soc.*, **1958**, 2010.
[50] D. Fleischer and H. Freiser, *J. Phys. Chem.*, 1959, **63**, 260.
[51] D. Kealey and H. Freiser, *Anal. Chem.*, 1966, **38**, 1577.
[52] J. C. Tomkinson and R. J. P. Williams, *J. Chem. Soc.*, **1958**, 1153.
[53] G. Nickless, F. H. Pollard and T. J. Samuelson, *Anal. Chim. Acta*, 1967, **39**, 37.
[54] L. Sommer and V. M. Ivanov, *Talanta*, 1967, **14**, 171.
[55] W. J. Peard and R. T. Pflaum, *J. Am. Chem. Soc.*, 1958, **80**, 1593.
[56] B. R. James and R. J. P. Williams, *J. Chem. Soc.*, **1961**, 2007.
[57] R. J. Stevenson and H. Freiser, *Anal. Chem.*, 1967, **39**, 1354.
[58] K. B. Yatsimirskii, N. K. Davidenko and L. N. Lugina, *Dokl. Akad. Nauk SSSR*, 1966 **170**, 864.
[59] N. K. Davidenko and A. A. Zholdakov, *Zh. Neorgan. Khim.*, 1967, **12**, 1195.
[60] A. A. Zholdakov and N. K. Davidenko, *Zh. Neorgan. Khim.*, 1967, **12**, 3066.

[61] N. K. Dutt and P. Bandyopandhyay, *Sci. Culture (India)*, 1958, **23**, 365.
[62] L. G. Van Uitert, W. C. Fernelius and B. E. Douglas, *J. Am. Chem. Soc.*, 1953, **75**, 457, 2736.
[63] E. H. Holst, *Thesis*, Pennsylvania Univ., 1955.
[64] M. Calvin and K. W. Wilson, *J. Am. Chem. Soc.*, 1945, **67**, 2003.
[65] C. M. Callahan, W. C. Fernelius and P. B. Bloch, *Anal. Chim. Acta*, 1957, **16**, 101.
[66] L. G. Van Uitert and W. C. Fernelius, *J. Am. Chem. Soc.*, 1954, **76**, 375.
[67] R. G. Anderson and G. Nickless, *Anal. Chim. Acta*, 1967, **39**, 469.
[68] C. J. Hawkins, H. Duewell and W. F. Pickering, *Anal. Chim. Acta*, 1961, **25**, 257.
[69] K. Burger, L. Korecz and A. Tóth, *Acta Chim. Acad. Sci. Hung.*, 1968, **55**, 1.
[70] K. S. Math, Q. Fernando and H. Freiser, *Anal. Chem.*, 1964, **36**, 1762.
[71] H. Irving and C. F. Bell, *J. Chem. Soc.*, **1952**, 1216.
[72] B. F. Pease and M. B. Williams, *Anal. Chem.*, 1959, **31**, 1044.
[73] O. Navrátil, *Collection Czech. Chem. Commun.*, 1964, **29**, 2490.
[74] D. Betteridge, P. K. Todd, Q. Fernando and H. Freiser, *Anal. Chem.*, 1963, **35**, 729.
[75] L. G. Van Uitert and W. C. Fernelius, *J. Am. Chem. Soc.*, 1953, **75**, 3862.
[76] W. C. Fernelius and L. G. Van Uitert, *Acta Chem. Scand.*, 1954, **8**, 1726.

CHAPTER 6

Stability constants of complexes with inorganic ligands

Experimental approaches to the investigation of complexation equilibria and to the determination of stability constants of complexes are analogous to those used in the study of protonation reactions. The differences, and many times the difficulties, arise from the fact that complexation reactions are intrinsically more complicated: in an amphiprotic or protophilic solvent the protonation and complexation equilibria form a competitive system. Besides the molecules of the solvent and its autoprotolysis species other kinds of ligands present in solution may also take part in formation of complex species. For example, in an aqueous medium it is always necessary to consider the acid–base equilibria of the ligand, the formation of hydroxo-complexes of the metal ion, the existence of various hydrated species, and last but not least the contributions of various hydrogen or hydroxo and other mixed complexes.

The interpretation of complex-formation reactions is thus more difficult, but the presence of the metal ion and its compounds allows various other experimental techniques to be used. Experimental approaches to the study of complexation equilibria and interpretations of experimental results are dealt with in many monographs, e.g. [1–12], where references to original papers can be found. When pH measurement is included in the experiment, the rules outlined in the introduction to Chapter 5 (cf. p. 87–88) should be observed.

Tables 6.1–6.20 summarize the stability constants of some of the more important complexes with inorganic ligands, namely those with the hydroxide ion, halides or pseudo-halides, various anionic ligands containing carbon, phosphorus or sulphur, and finally with ammonia, which represents an uncharged ligand. Selected values of logarithms of stability constants refer to a temperature of 25°C if not otherwise stated. The ionic strength I is given after the relevant value of log K. For ionic strength approaching zero the thermodynamic constants are defined by the expressions in Eqs. (2.43) and (2.45). For $I > 0$ the concentration constants are defined as indicated by the expression $\beta_{nm} = [M_m L_n]/([M]^m [L]^n)$.

The complexes of inorganic ligands are arranged separately from those of organic ligands only for systematic reasons. The ligands are listed in the same order as in Chapter 5. The metals follow in sequence of subgroups of the periodic system. The reliability of the values of the stability constants is indicated in the way already used in Chapter 5. Data listed without a reference to the original literature are quoted from reference [13].

REFERENCES

[1] J. Bjerrum, *Metal Ammine Formation in Aqueous Solution*, Haase, Copenhagen, 1941; reprinted 1957.

[2] A. K. Babko, *Fiziko-khimicheskii analiz kompleksnykh soedinenii v rastvorakh, opticheskii metod* (*Physicochemical Analysis of Complex Compounds in Solution. Optical Methods*), Izd. Akad. Nauk USSR, Kiev, 1955.

[3] K. B. Yatsimirskii and V. P. Vasil'ev, *Konstanty nestoikosti kompleksnykh soedinenii*, Izd. Akad. Nauk USSR, Moscow, 1959; *Instability Constants of Complex Compounds*, Pergamon Press, Oxford, 1960.

[4] H. L. Schläfer, *Komplexbildung in Lösung*, Springer, Berlin, 1961.

[5] F. J. C. Rossotti and H. S. Rossotti, *The Determination of Stability Constants*, McGraw-Hill, New York, 1961.

[6] C. W. Davies, *Ion Association*, Butterworths, London, 1962.

[7] J. N. Butler, *Ionic Equilibrium*, Addison-Wesley, Reading, Massachusetts, 1964.

[8] G. H. Nancollas, *Interactions in Electrolyte Solutions*, Elsevier, Amsterdam, 1966.

[9] D. Dyrssen, D. Jagner and F. Wengelin, *Computer Calculation of Ionic Equilibria and Titration Procedures*, Almqvist and Wiksell, Stockholm; Wiley, New York, 1968.

[10] J. Inczédy, *Analytical Applications of Complex Equilibria*, Horwood, Chichester, 1976.

[11] M. T. Beck, *Chemistry of Complex Equilibria*, Van Nostrand Reinhold, London, 1970.

[12] F. C. Baes, Jr. and R. E. Mesmer, *The Hydrolysis of Cations*, Wiley, New York, 1976.

[13] R. M. Smith and A. E. Martell, *Critical Stability Constants*, Vol. 4, *Inorganic Complexes*, Plenum Press, New York, 1976.

Ch. 6] **Stability Constants of Complexes with Inorganic Ligands** 111

Table 6.1
Stability constants of borato-complexes [L = B(OH)$_4^-$]

Metal ion	Equilibrium constants (Symbol for the constant, value of log β, [I])	Ref.
Na$^+$	β_1 1.87 [var][a]	[1]
Fe^{3+}	β_1 8.5, β_2 15.6, β_3 20.6, β_4 20.3 [var][b]	[2, 3]
Co^{2+}	β_4 10.03 [0][c]	[4]
Ni^{2+}	β_3 8.44 [0][c]	[4]
Cu^{2+}	β_1 7.13, β_2 12.45, β_3 15.17 [var][c]	[5]
Ag$^+$	β_1 0.45 [3.0]	
Zn^{2+}	β_4 (11.8) [var][c]	[6]
Cd^{2+}	β_4 (10.6) [var][c]	[6]
Pb^{2+}	β_1 5.21, β_3 11.17 [0][c]	[7]

[a] 20°C; [b] temperature not given; [c] 22°C, influence of polyborates neglected

REFERENCES

[1] V. Frei and A. Ustyanovichova, *Zh. Fiz. Khim.*, 1963, **37**, 1153.
[2] M. B. Shchigol, *Zh. Neorgan. Khim.*, 1961, **6**, 337.
[3] N. B. Burchinskaya, *Ukr. Khim. Zh.*, 1964, **30**, 177.
[4] M. B. Shchigol, *Zh. Neorgan. Khim.*, 1961, **6**, 2693.
[5] M. B. Shchigol, *Zh. Neorgan. Khim.*, 1965, **10**, 2097.
[6] M. B. Shchigol, *Zh. Neorgan. Khim.*, 1959, **4**, 2014.
[7] M. B. Shchigol, *Zh. Neorgan. Khim.*, 1963, **8**, 1361.

Table 6.2
Stability constants of carbonato-complexes (L = CO$_3^{2-}$)

Metal ion	Equilibrium constants (Symbol for the constant or scheme of the equilibrium, value of log K or log β, [I])	Ref.
Mg^{2+}	β_1 2.88 [0]; (2.37) [0.15][a]; MHL/M.HL 0.95 [0]; 0.77 [0.15][a] MHL/ML.H·8.00 [0][a]	[1]
Ca^{2+}	β_1 3.15 [0]; 3.00 [0.15][a]; MHL/M.HL 1.0 [0]; 0.81 [0.15][a]	
Ba^{2+}	β_1 2.78 [0]	
La^{3+}	ML.H^2/M.H$_2$L < −9	[2]
M^{3+}(4f)	β_3 10.6 (Er), β_4 11.2 (Pr, Nd) [var]	[3]
UO$_2^{2+}$	β_2 14.6 [0]; 15.6 [0.2][b]; β_3 18.3 [0]; 20.7 [0.2][b]; 22.8 [1.0]	[4–6]
PuO$_2^{2+}$	β_2 15.1, MOHL/M.OH.L 23.85 [0][c]	[7]
Mn^{2+}	MHL/M.HL 1.8 [0]; 0.45 [3.0]	
Cu^{2+}	β_1 6.75, β_2 9.92 [0]; 8.6 [1.7][d]	
Cd^{2+}	β_3 6.24 [0]	[8]
Pb^{2+}	β_2 9.09 [1]; 8.2 [1.7][c]; M(HL)$_2$/M.(HL)^2 4.77, M(HL)$_3$/M.(HL)^3 5.19 [var][c]; 5.19 [1][d]	[9, 10]

[a] 22°C; [b] laboratory temperature; [c] 20°C; [d] 18°C

REFERENCES

[1] P. B. Hostetler, *J. Phys. Chem.*, 1963, **67**, 720.
[2] L. G. Sillén and A. E. Martell, *Stability Constants of Metal-Ion Complexes*, p. 138, The Chemical Society, London, 1964.
[3] Ya. D. Fridman and N. V. Dolgashova, *Zh. Neorgan. Khim.*, 1964, **9**, 623.
[4] L. G. Sillén and A. E. Martell, *op. cit.*, p. 139.
[5] A. E. Klygin and I. D. Smirnova, *Zh. Neorgan. Khim.*, 1959, **4**, 42.
[6] A. K. Babko and V. S. Kodenskaya, *Zh. Neorgan. Khim.*, 1960, **5**, 2568.
[7] A. D. Gel'man, A. I. Moskvin and V. P. Zaitseva, *Radiokhimiya*, 1962, **4**, 154.
[8] P. E. Lake and J. M. Goodings, *Can. J. Chem.*, 1958, **36**, 1089.
[9] N. N. Baranova and V. L. Barsukov, *Geokhimiya*, **1965**, 1093; **1968**, 17.
[10] N. N. Baranova, *Zh. Neorgan. Khim.*, 1967, **12**, 1438.

Table 6.3
Stability constants of cyano-complexes (L = CN^-)

Metal ion	Equilibrium constants (Symbol for the constant or scheme of the equilibrium, value of log K or log β, $[I]$)	Ref.
Fe^{2+}	β_6 35.4 [0]	
Fe^{3+}	β_6 43.6 [0]	
Ni^{2+}	β_1 7.03 [3.0]; β_4 30.22 [0]; 30.5 [0.1]; 31.06 [3.0]; $MHL_4/ML_4 \cdot H$ 5.4, $MH_2L_4/MHL_4 \cdot H$ 4.5, $MH_3L_4/MH_2L_4 \cdot H$ 2.6 [0.1]	
Pd^{2+}	β_4 42.4, β_5 45.3 [0]	
Pt^{2+}	β_4 (41) [1][a]	[1]
Cu^+	β_2 16.26, β_3 21.6, β_4 23.1 [0]	
Cu^{2+}	β_4 (25) [?]; 26.7 [60% CH_3OH][d]	[2, 3]
Ag^+	β_2 20.48 [0]; 20.0 [1.0]; β_3 21.4 [0]; 20.3, β_4 20.8 [1.0][b] MOHL/M.OH.L 13.2 [0]; 12.7 [1.0]	
Au^+	β_2 38.3 [0]	[4]
Au^{3+}	β_4 (56) [?]	[5]
Zn^{2+}	β_1 5.3 [3.0]; β_2 11.07 [0]; 11.7 [3.0]; β_3 16.05 [0]; 16.7 [3.0]; β_4 19.62 [0]; 21.6 [3.0]	
Cd^{2+}	β_1 6.01 [0]; 5.55 [3.0]; β_2 11.12 [0]; 10.7 [3.0]; β_3 15.65 [0]; 15.5 [3.0]; β_4 17.92 [0]; 19.0 [3.0]	
Hg^{2+}	β_1 17.00 [0]; 18.00 [0.1][c]; β_2 32.75 [0]; 34.71 [0.1][c]; 33.9 [2.0][b]; β_3 36.31 [0]; 38.54 [0.1][c]; 38.1 [2.0][b]; β_4 38.97 [0]; 41.5 [0.1][c]; 40.6 [2.0][b]; MOHL/M.OH.L 28.9 [2.0][b]	
Tl^{3+}	β_4 (35) [var]	[2]
Pb^{2+}	β_4 (10.3) [1]	[6]

[a] 18°C; [b] 30°C; [c] 20°C; [d] −45°C

REFERENCES

[1] A. A. Grinberg and M. I. Gel'fman, *Dokl. Akad. Nauk SSSR*, 1960, **133**, 1081.
[2] J. Bjerrum, *Chem. Rev.*, 1950, **46**, 381.
[3] R. Paterson and J. Bjerrum, *Acta Chem. Scand.*, 1965, **19**, 729.
[4] W. M. Latimer, *Oxidation Potentials*, 2nd Ed., p. 186. Prentice-Hall, New York, 1952.
[5] H. T. S. Britton and T. N. Dodd, *J. Chem. Soc.*, **1935**, 100.
[6] I. M. Kolthoff and J. J. Lingane, *Polarography*, p. 268. Interscience, New York, 1941.

Table 6.4
Stability constants of thiocyanato-complexes (L = SCN^-)

Metal ion	Equilibrium constants (Symbol of the constant or scheme of the equilibrium, value of log K or log β, [I])	Ref.
Sc^{3+}	β_1 0.20 [0.6]^a; 0.8 [4.0]^a	
La^{3+}	β_1 0.12 [1.0]^b; 0.24 [4.0]	
$M^{3+}(4f)$	β_1 (0.15) [0.6]^a; (0.2) [1.0]^b; (0.5) [4.0]	
Ti^{4+}	MOHL/MOH.L 1.7 [1]^c	[1]
Zr^{4+}	MOOHL/MOOH.L 1.15 [var]^a	[2]
Th^{4+}	β_1 1.08, β_2 1.78 [1.0]	
V^{2+}	β_1 1.43 [1.0]	
V^{3+}	β_1 2.07 [1]	[3]
VO^{2+}	β_1 2.32, β_2 3.68 [0]	
Cr^{2+}	β_1 1.09, β_2 0.77 [0]	
Cr^{3+}	β_1 3.08 [0]; (1.87) [1.0]; β_2 (2.98) [1.0]	
U^{4+}	β_1 1.49 [1.0]^a; 1.49 [2.0]; β_2 1.95 [1.0]^a; 2.11 [2.0]; β_3 2.2 [1.0]^a	
UO_2^{2+}	β_1 0.93 [0]; 0.75, β_2 0.72, β_3 0.18 [1.0]	
$M^{3+}(5f)$	β_1 (0.5) [1.0]	
Mn^{2+}	β_1 1.23 [0]; 0.80 [0.5]; 0.65 [1.0]	
Fe^{2+}	β_1 1.31 [0]; 0.81 [3.0]	
Fe^{3+}	β_1 3.02 [0]; 2.14 [0.5]; 2.10 [1.0]; 2.21 [2.0]; β_2 4.64 [0]^d; 3.3 [0.5]; 3.2 [1.0]; 3.64 [3.0]; β_3 5.0, β_4 6.3, β_5 6.2, β_6 6.1 [3.0]	
Co^{2+}	β_1 1.72 [0]; 1.11 [0.5]; 0.98 [1.0]; 1.27 [3.0]; β_2 1.32 [1.0]; β_3 −0.38 [3.0]	[4]
Ni^{2+}	β_1 1.76 [0]; 1.23 [0.5]; 1.13 [1.0]; 1.34 [3.0]; β_2 1.58 [1.0]; β_3 1.5 [1.0]	
Pd^{2+}	β_4 27.2 [1.0]	
Cu^+	β_2 11.00, β_3 10.9, β_4 10.4 [5.0]	
Cu^{2+}	β_1 2.33 [0]; 1.90 [0.1]; 1.74 [1.0]; 1.91 [3.0]; β_2 3.65 [0]; 3.00 [0.1]; 2.74 [1.0]	
Ag^+	β_1 4.8 [0]; 4.6 [0.1]; 4.6 [4.0]; β_2 8.23 [0]; 8.06 [0.1]; 8.18 [2.0]; 8.29 [4.0]; β_3 9.5 [0]; 9.6 [0.1]; 9.3 [2.0]; 10.0 [4.0]; β_4 9.7 [0]; 10.5 [0.1]; 10.0 [2.0]; 11.3 [4.0]	
Au^+	β_1 15.27; β_2 16.98 [3.0]	
Au^{3+}	β_4 42 [0]	[5]

Table 6.4 (Continued)

Metal ion	Equilibrium constants (Symbol of the constant or scheme of the equilibrium, value of log K or log β, [I])	Ref.
Zn^{2+}	β_1 1.33 [0]; 0.71 [1.0]; 0.74 [2.0]; 1.11 [4.0]; β_2 1.91 [0]; 1.04 [1.0]; 1.15 [2.0]; 1.81 [4.0]; β_3 2.0 [0]; 1.2 [1.0]; 1.3 [2.0]; 2.8 [4.0]; β_4 1.6 [0]; 1.5 [1.0]; 1.7 [2.0]; 2.8 [4.0]	
Cd^{2+}	β_1 1.89 [0]; 1.53 [0.1]; 1.35 [0.5]; 1.32 [1.0]; 1.34 [2.0]; 1.41 [3.0]; β_2 2.78 [0]; 2.04 [0.5]; 1.99 [1.0]; 2.05 [2.0]; 2.24 [3.0]; β_3 2.8 [0]; 2.1 [0.5]; 2.0 [1.0]; 2.2 [2.0]; 2.5 [3.0]; β_4 2.3 [0]; 2.0 [0.5]; 1.9 [1.0]; 2.0 [2.0]; 2.5 [3.0]	
Hg^{2+}	β_1 9.08 [1.0]; β_2 17.26 [0]; 16.43 [0.1]; 16.86 [1.0]; β_3 19.97 [0]; 19.14 [0.1]; 19.70 [1.0]; β_4 21.8 [0]; 21.2 [0.1]; 21.7 [1.0]	
Al^{3+}	β_1 0.42 [0][f]	
Ga^{3+}	β_1 2.15 [0][b]; 1.18 [0.6][a]	
In^{3+}	β_1 3.15 [0][b]; 2.34 [0.6][a]; 2.56 [2.0]; β_2 3.53; β_3 4.6 [2.0]	
Tl^{+}	β_1 0.58 [0]; 0.24 [0.5]; 0.17 [1.0]; 0.12 [2.0]; 0.15 [3.0]; β_2 0.00 [0.5]; -0.05 [1.0]; -0.12 [2.0]; -0.12 [3.0]; β_3 -0.4 [1.0]; -0.5 [2.0]; -0.5 [3.0]	
Bi^{3+}	β_1 2.21 [0]; 1.67 [0.1]; 1.32 [1.0]; (1.28) [3.0]; 2.02 [4.0]; β_2 2.7 [0]; 3.0 [0.1]; 2.1 [1.0]; 2.7 [3.0]; 3.5 [4.0]; β_3 4.4 [0]; 4.0 [0.1]; 3.0 [1.0]; 3.8 [3.0]; 4.8 [4.0]; β_4 5.2 [0]; 4.8 [0.1]; (2.9) [1.0]; 5.3 [3.0]; 6.3 [4.0]; β_5 5.8 [0]; 5.5 [0.1]; 3.6 [1.0]; 6.0 [3.0]; 6.8 [4.0]; β_6 5.4 [0]; 6.1 [0.1]; 4.0 [1.0]; 6.9 [3.0]; 8.3 [4.0]	

[a] 20°C; [b] 30°C; [c] temperature not given; [d] 18°C; [e] 23°C; [f] 22–23°C

REFERENCES

[1] D. Delafosse, *Compt. Rend.*, 1953, **236**, 2313.
[2] J. Selbine and L. H. Holmes, Jr., *J. Inorg. Nucl. Chem.*, 1962, **24**, 1111.
[3] B. R. Baker, N. Sutin and T. J. Welch, *Inorg. Chem.*, 1967, **6**, 1948.
[4] V. N. Kumok, *Zh. Neorgan. Khim.* 1964, **9**, 362.
[5] W. M. Latimer, *Oxidation Potentials*, 2nd Ed., p. 194, Prentice-Hall, New York, 1952.

Table 6.5
Stability constants of ammine complexes (L = NH_3)

Metal ion	Equilibrium constants (Symbol for the constant or scheme of the equilibrium, value of log K or log β, [I])	Ref.
Li^+	β_1 −0.3, β_2 −1.1, β_3 −2.4 [2][a]	[1]
Mg^{2+}	β_1 0.23, β_2 0.08, β_3 −0.3 [2][a]	
Ca^{2+}	β_1 −0.2, β_2 −0.8, β_3 −1.6, β_4 −2.7, β_5 −4.0, β_6 −5.7 [2][a]	[1]
Cr^{3+}	β_6 (13), $M_2(OH)_3L_6/M^2 \cdot (OH)^3 \cdot L^6$ 46.42 [4.5]	[2]
Mn^{2+}	β_1 1.00, β_2 1.54, β_3 1.70, β_4 1.3 [2.0][b]	
Fe^{2+}	β_1 (1.4), β_2 (2.2) [0.5−5.0]; β_4 (3.7) [0]	[3, 4]
Co^{2+}	β_1 1.99 [0][b]; 2.10 [2][c]; β_2 3.50 [0][b]; 3.67 [2][c]; β_3 4.43 [0][b]; 4.78 [2][c]; β_4 5.07 [0][b]; 5.53 [2][c]; β_5 5.13 [0][b]; 5.75 [2][c]; β_6 4.39 [0][b]; 5.14 [2][c]	
Co^{3+}	β_1 7.3, β_2 14.0, β_3 20.1, β_4 25.7, β_5 30.8, β_6 35.2 [2]; K_5 4.15 [0]; 4.33 [1.0]; 4.50 [2]	[1]
	$ML_5OH \cdot LH/ML_6$ −1.0, $ML_4(OH)_2 \cdot LH/ML_5OH$ −2.6 [0.12][d]	[5]
Ni^{2+}	β_1 2.72 [0]; 2.81 [2.0]; β_2 4.89 [0]; 5.08 [2.0]; β_3 6.55 [0]; 6.85 [2.0]; β_4 7.67 [0]; 8.12 [2.0]; β_5 8.34 [0]; 8.93 [2.0]; β_6 8.31 [0]; 9.08 [2.0]	
Pd^{2+}	β_1 9.6, β_2 18.5, β_3 26.0, β_4 32.8 [1.0]	
Pt^{2+}	β_4 35.3 [1][e]	[6]
Cu^+	β_1 5.93, β_2 10.58 [2.0][e]	
Cu^{2+}	β_1 4.04 [0]; 4.12 [1.0]; 4.24 [2.0]; β_2 7.47 [0]; 7.63 [1.0]; 7.83 [2.0]; β_3 10.27 [0]; 10.51 [1.0]; 10.80 [2.0]; β_4 11.75 [0]; 12.6 [1.0]; 13.00 [2.0]; β_5 12.43 [2.0]	
Ag^+	β_1 3.31 [0]; 3.30 [0.5]; 3.26 [2.0]; 3.20 [5.0]; β_2 7.22 [0]; 7.21 [1.0]; 7.20 [2.0]; 7.13 [5.0]	
Au^+	β_2 27 [var][f]	[7]
Au^{3+}	β_4 30 [?][d]; K_4 10.3, $MH_{-1}L_4 \cdot H/ML_4$ −7.48 [1.0]	[8]
Zn^{2+}	β_1 2.21 [0]; 2.32 [1.0]; 2.38 [2.0]; β_2 4.50 [0]; 4.81 [1.0]; 4.88 [2.0]; β_3 6.86 [0]; 7.11 [1.0]; 7.43 [2.0]; β_4 8.89 [0]; 9.32 [1.0]; 9.65 [2.0]	
Cd^{2+}	β_1 2.55 [0]; 2.62 [1.0]; 2.72 [2.0]; β_2 4.56 [0]; 4.79 [1.0]; 4.90 [2.0]; β_3 5.90 [0]; 6.16 [1.0]; 6.32 [2.0]; β_4 6.74 [2]; 7.1 [1.0]; 7.38 [2.0]; β_5 6.9 [1.0]; 7.02 [2.0]; β_6 5.41 [2.0]	
Hg^{2+}	β_1 8.8 [2.0][a]; β_2 (17.8) [1.0]; 17.4 [2.0]; β_3 18.4 [2.0]; β_4 (19.3) [0.1]; 19.1 [2.0]	
Tl^+	β_1 −0.9 [2.0][a]	
Tl^{3+}	β_4 (17) [?][f]; $M(OH)_2L/M(OH)_2 \cdot L$ 4.6, $M(OH)_2L_2/M(OH)_2L \cdot L$ 4.7, $M(OH)_2L_3/M(OH)_2L_2 \cdot L$ 2.3, $M(OH)_2L_4/M(OH)_2L_3 \cdot L$ 1.5 [1.0]	[7, 9]

[a] 23°C; [b] 20°C; [c] 30°C; [d] temperature not given; [e] 18°C; [f] laboratory temperature

REFERENCES

[1] J. Bjerrum, *Metal Ammine Formation in Aqueous Solution*, pp. 149, 251, 280, 285. Haase, Copenhagen, 1957.

[2] C. E. Schäffer and P. Andersen, in *Theory and Structure of Complex Compounds*, p. 571. Pergamon Press, Oxford, 1964.
[3] D. L. Leussing and I. M. Kolthoff, *J. Am. Chem. Soc.*, 1953, **75**, 2476.
[4] K. B. Yatsimirskii and V. P. Vasil'ev, *Konstanty nestoikosti kompleksnykh soedinenii (Instability Constants of Complex Compounds)*, p. 88. Izdat. Akad. Nauk SSSR, Moscow, 1959.
[5] A. Jensen, J. Bjerrum and F. Wolbye, *Acta Chem. Scand.*, 1958, **12**, 1202.
[6] A. A. Grinberg and M. I. Gel'fman, *Dokl. Akad. Nauk SSSR*, 1961, **137**, 87.
[7] J. Bjerrum, *Chem. Rev.*, 1950, **46**, 381.
[8] L. G. Sillén and A. E. Martell, *Stability Constants of Metal-Ion Complexes*, p. 154. The Chemical Society, London, 1964.
[9] V. I. Lobov, F. Ya. Kul'ba and V. E. Mironov, *Zh. Neorgan. Khim.*, 1967, **12**, 334.

Table 6.6

Stability constants of nitrato-complexes (L = NO_3^-)

Metal ion	Equilibrium constants (Symbol for the constant or scheme of the equilibrium, value of log K or log β, [I])	Ref.
Na^+	β_1 (−0.6) [0]^a	
K^+	β_1 (−0.15) [0]; −0.37 [0.1]	
Be^{2+}	β_1 −0.6 [4.0]	
Ca^{2+}	β_1 0.7 [0]; 0.06 [0.5]; −0.06 [1.0]; −0.02 [2.0]; 0.04 [3.0]	
	β_2 0.6 [0]; −0.3 [0.5]; −0.5 [1.0]; −0.4 [2.0]; −0.4 [3.0]	
Sr^{2+}	β_1 0.8 [0]; (0.06) [0.5]; 0.05 [1.0]; 0.06 [2.0]; 0.08 [3.0];	
	β_2 0.8 [0]; (−0.5) [0.5]; −0.3 [1.0]; −0.2 [2.0]; (−0.4) [3.0]	
Ba^{2+}	β_1 0.9 [0]; 0.21 [0.5]; 0.16 [1.0]; 0.14 [2.0]; 0.20 [3.0];	
	β_2 1.0 [0]; 0.1 [0.5]; 0.0 [1.0]; 0.0 [2.0]; (−0.1) [3.0]	
Sc^{3+}	β_1 0.55 [2.0]; 0.28 [4.0]; β_2 0.08 [2.0]; −0.3 [4.0]	[1]
La^{3+}	β_1 0.1 [1.0]	
$M^{3+}(4f)$	β_1 (0.3) [1.0] (Ce–Eu); β_1 0.0 [1.0] (Gd); β_1 −0.2 [1.0] (Tb–Lu)	
Ce^{4+}	β_1 0.78 [2.0]^b; 0.33 [3.5]^c; β_2 1.20, β_3 1.34, β_4 1.24, β_5 1.07 [2.0]^b	[2]
Zr^{4+}	β_1 0.3 [2.0]; 0.34 [4.0]^d; β_2 0.1, β_3 −0.3, β_4 −0.8 [4.0]^d	
Hf^{4+}	β_1 0.34 [2.0]; 0.40 [4.0]^d; 0.0 [2.0]; 0.1 [4.0]^d; β_3 −0.7 [2.0]	
Th^{4+}	β_1 0.67 [0.5]; 0.45 [6.0]; β_2 0.15 [6.0]	
VO_2^+	β_1 −0.5 [1.0]^d	
Cr^{3+}	β_1 −2.01 [1.0]	[3]
Mo(VI)	$MO_2L/MO_2 \cdot L$ 0.14 [0.5]^d; 0.10 [1.0]^d; 0.14 [2.0]^d;	
	$MO_2L_2/MO_2 \cdot L^2$ −0.08 [0.5]^d; −0.26 [1.0]^d; −0.17 [2.0]^d	[4]
U^{4+}	β_1 0.20 [2.0]; 0.28 [3.0]; β_2 0.2 [2.0]; 0.3 [3.0];	
	β_3 0.0 [2.0]; 0.2 [3.0]; β_4 −0.5 [2.0]; −0.2 [3.0]	
UO_2^{2+}	β_1 −0.3 [1.0]^d; −0.6 [2.0]	
Np^{4+}	β_1 1.7 [0]^d; 0.38 [1.0]; 0.34 [2.0]; β_2 0.1 [1.0]^d; 0.2 [2.0];	
	β_3 −0.3 [1.0]^d	
NpO_2^+	β_1 −0.25 [2.0]	

Stability Constants of Complexes with Inorganic Ligands

Table 6.6 (Continued)

Metal ion	Equilibrium constants (Symbol for the constant or scheme of the equilibrium, value of $\log K$ or $\log \beta$, $[I]$)	Ref.
NpO_2^{2+}	β_1 −0.9 [0.5]; −0.4 [2.0]	
Am^{3+}	β_1 0.26 [1.0]; 0.20 [2.0]	
Mn^{2+}	β_1 0.2 [0]; −0.38 [0.5]; −0.43 [1.0]; −0.41 [2.0]; −0.24 [3.0]; β_2 0.6 [0]; −0.3 [0.5]; −0.6 [1.0]; −0.9 [2.0]; −0.8 [3.0]	
Fe^{3+}	β_1 1.0 [0]; −0.22 [0.6][d]; −0.5 [1.0][e]	
Co^{2+}	β_1 0.2 [0]; (−0.46) [0.5]; −0.46 [1.0]; −0.48 [2.0]; −0.60 [3.0]; β_2 −0.3 [0.5]; −0.4 [1.0]; −0.6 [2.0]; −0.6 [3.0]	
Ni^{2+}	β_1 0.4 [0]; −0.22 [1.0]; −0.44 [2.0]; −0.55 [3.0]; β_2 −0.5 [2.0]; −0.9 [3.0]	
Cu^{2+}	β_1 0.5 [0]; (−0.13) [0.5]; −0.01 [1.0]; −0.06 [2.0]; −0.02 [3.0]; β_2 −0.4 [0]; −0.6 [1.0]; −0.6 [2.0]; −0.5 [3.0]	
Ag^+	β_1 (−0.2) [0]; −0.34 [2.0]	
Zn^{2+}	β_1 0.4 [0]; −0.18 [0.5]; −0.19 [1.0]; −0.14 [2.0]; 0.01 [3.0]; β_2 −0.3 [0]; −0.6 [1.0]; −0.8 [2.0]; −1.1 [3.0]	
Cd^{2+}	β_1 0.5 [0]; −0.11 [0.5]; −0.05 [1.0]; 0.02 [2.0]; 0.04 [3.0]; β_2 0.2 [0]; −0.8 [1.0]; −0.4 [2.0]; (−0.6) [3.0]	
Hg_2^{2+}	β_1 0.08 [0.5]; 0.02 [3.0]; β_2 −0.3 [3.0]	
Hg^{2+}	β_1 0.11, β_2 0.0 [3.0]	
In^{3+}	β_1 0.18, β_2 −0.3 [0.7][d]	
Tl^+	β_1 0.33 [0]; −0.48 [3.0]	
Tl^{3+}	β_1 0.90, β_2 0.1, β_3 1.1 [3.0]	
Pb^{2+}	β_1 1.17 [0]; 0.25 [0.5]; 0.33 [1.0]; 0.40 [2.0]; 0.51 [3.0]; β_2 1.4 [0]; 0.4 [0.5]; 0.4 [1.0]; 0.4 [2.0]; 0.4 [3.0]; β_3 0.1 [2.0]; 0.2 [3.0]; β_4 −0.3 [3.0]	
Bi^{3+}	β_1 1.7 [0]; 0.72 [0.5]; 0.81 [1.0]; (0.72) [2.0]; (0.72) [3.0]; β_2 2.5 [0]; (0.94) [0.5]; 0.90 [1.0]; 0.98 [2.0]; 0.96 [3.0]; β_3 0.7 [1.0]; (0.2) [2.0]; (0.1) [3.0]; β_4 (0.6) [2.0]; (−0.2) [3.0]	

[a] 18°C; [b] temperature not given; [c] 23°C; [d] 20°C; [e] 26°C;

REFERENCES

[1] A. P. Samodelov, *Radiokhimiya*, 1964, **6**, 568.
[2] V. S. Smelov and Yu. I. Vereshchagin, *Zh. Neorgan. Khim.*, 1964, **9**, 2775.
[3] M. Ardon and N. Sutin, *Inorg. Chem.*, 1967, **6**, 2268.
[4] S. P. Vorobev, I. P. Davidov and I. V. Shilin, *Zh. Neorgan. Khim.*, 1967, **12**, 2142.

Table 6.7

Stability constants of phosphato-complexes (L = PO_4^{3-})

Metal ion	Equilibrium constants (Symbol for the constant or scheme of the equilibrium, value of log K or log β, [I])	Ref.
Li^+	MHL/M.HL 0.72 [0.2][a]	
Na^+	MHL/M.HL 0.60 [0.2][a]	
K^+	MHL/M.HL 0.49 [0.2][a]	
Mg^{2+}	β_1 3.4 [0.15][b]; MHL/M.HL 2.91 [0]; 1.7 [0.1]; 1.8 [0.15][b]; 1.88 [0.2][a]; 1.42 [3.0]; $MH_2L/M.H_2L$ 0.7 [0.15][b]; 0.16 [3.0]	
Ca^{2+}	β_1 6.46 [0]; MHL/M.HL 2.74 [0]; 1.50 [0.2]; 1.3 [0.15][b]; 1.70 [0.2][a]; $MH_2L/M.H_2L$ 1.4 [0]; 0.6 [0.15][b]	
Sr^{2+}	β_1 (4.2) [0.1][c]; MHL/M.HL 1.2 [0.1][c]; 1.52 [0.2][a]; $MH_2L/M.H_2L$ 0.3 [0.1][c]	
Y^{3+}	$MH_2L/M.H_2L$ 2.65 [0]; 1.84 [0.2][c]	
La^{3+}	$MH_2L/M.H_2L$ 1.61 [0.5]	
Ce^{3+}	β_1 (18.52) [0]; $MH_2L/M.H_2L$ 2.33 [0]; 1.52 [0.2][c]	
Th^{4+}	MHL/M.HL 10.8, $M(HL)_2/M.(HL)^2$ 22.8, $M(HL)_3/M(HL)^3$ 31.3 [0.35]; $MH_2L/M.H_2L$ 3.96, $M(H_2L)_2/M.(H_2L)^2$ 7.5, $MH(H_2L)_2/M.H_2L.H_3L$ 6.0, $MH_3L/M.H_3L$ 1.89 [2.0]	[1, 2]
Cr^{3+}	MHL/M.HL 9.4 [0]	[3]
Cr(VI)	$HMO_3L/HMO_4.H_2L$ 0.48 [0.25]; 0.8 [3.0]; $H_2MO_3L/HMO_4.H_3L$ 0.95 [0.25]	[4, 5]
Mo(VI)	$MO_2L/H_2MO_4.H_2L$ 2.7, $MO_2L_2/H_2MO_4.(HL)^2$ 19.2 [0.48][d]; $MO_2HL/H_2MO_4.H_3L$ 3.16, $MO_2(HL)_2/H_2MO_4.(H_3L)^2$ 3.35 [var]	[6, 7]
U^{4+}	MHL/M.HL 12.0, $M(HL)_2/M.(HL)^2$ 22.0, $M(HL)_3/M.(HL)^3$ 30.6, $M(HL)_4/M.(HL)^4$ 38.6 [var][e]	[2, 8]
UO_2^{2+}	MHL/M.HL 8.43 [0]; 7.18 [0.5]; $M(HL)_2/M.(HL)^2$ 18.57 [0]; 17.30 [0.5]; $MH_2L/M.H_2L$ 3.0 [0][c]; 2.9 [1.07]; $M(H_2L)_2/M.(H_2L)^2$ 5.5 [0][c]; 4.7 [1.07]; $M(H_2L)_3/M.(H_2L)^3$ 7.4 [0]; $MH_3L/M.H_3L$ < 1.8 [0]; 0.76 [1.0]; $M(H_3L)_2/M.(H_3L)^2$ 3.9, $M(H_3L)_3/M.(H_3L)^3$ 5.3 [0]; $MH(H_2L)_2/M.H_2L.H_3L$ 3.0 [1.0]	[9–12]
Np^{4+}	MHL/M.HL 12.4; $M(HL)_2/M.(HL)^2$ 23.1; $M(HL)_3/M.(HL)^3$ 32.0; $M(HL)_4/M.(HL)^4$ 41.0 [?]	[2]
NpO_2^+	MHL/M.HL 3.38 [0][c]; 2.85, $MH_2L/M.H_2L$ 0.81 [0.2][c]	[13]
Pu^{4+}	MHL/M.HL 12.92, $M(HL)_2/M.(HL)^2$ 23.74, $M(HL)_3/M.(HL)^3$ 33.42, $M(HL)_4/M.(HL)^4$ 43.22, $M(HL)_5/M.(HL)^5$ 52.02 [2.0]	[14]
PuO_2^{2+}	MHL/M.HL (8.2) [var][e]; $MH_2L/M.H_2L$ 3.93 [var]	[15]
Am^{3+}	$MH_2L/M.H_2L$ 2.51 [0]; 1.69 [0.2][c]	
Mn^{2+}	MHL/M.HL 2.58 [0.2][a]	[16]
Fe^{2+}	MHL/M.HL 3.6, $MH_2L/M.H_2L$ 2.7 [0]	
Fe^{3+}	MHL/M.HL 8.30; $MH_2L/M.H_2L$ 3.47 [0.5]; $M_2HL/M^2.HL$ 11.14 [var][e]	[17]
Co^{2+}	MHL/M.HL 2.18 [0.1]	
Ni^{2+}	MHL/M.HL 2.08 [0.1]	

Table 6.7 (Continued)

Metal ion	Equilibrium constants (Symbol for the constant or scheme of the equilibrium, value of log K or log β, $[I]$)	Ref.
Cu^{2+}	MHL/M.HL 3.2 [0.1]; 3.3 [0.15][b]; $MH_2L/M.H_2L$ 1.3 [0.15][b]	
Zn^{2+}	MHL/M.HL 2.4 [0.1]; 2.4 [0.15][b]; $MH_2L/M.H_2L$ 1.2 [0.15][b]	
Al^{3+}	$MH_2L/M.H_2L$ 3, $M(H_2L)_2/M.(H_2L)^2$ 5.3, $M(H_2L)_3/M.(H_2L)^3$ 7.6 [0.1][f]	[18]
In^{3+}	$MH_2L/M.H_2L$ 1.43 [0.9][c]	
Pb^{2+}	MHL/M.HL 3.1, $MH_2L/M.H_2L$ 1.5 [0]	

[a] $(C_3H_7)_4NCl$ used as electrolyte to adjust ionic strength; [b] 37°C; [c] 20°C; [d] 22–23°C; [e] temperature not given; [f] 18°C

REFERENCES

[1] E. L. Zebroski, H. W. Alter and F. K. Heumann, *J. Am. Chem. Soc.*, 1951, **73**, 5646.
[2] A. I. Moskvin, L. N. Essen and T. N. Bukhtyarova, *Zh. Neorgan. Khim.*, 1967, **12**, 3390.
[3] S. C. Lahiri and S. Aditya, *J. Indian Chem. Soc.*, 1966, **43**, 513.
[4] F. Holloway, *J. Am. Chem. Soc.*, 1952, **74**, 224.
[5] S.-Å. Frennesson, J. K. Beattie and G. P. Haight, Jr., *J. Am. Chem. Soc.*, 1968, **90**, 6018.
[6] K. B. Yatsimirskii and I. I. Alekseeva, *Zh. Neorgan. Khim.*, 1965, **1**, 952.
[7] K. B. Yatsimirskii and L. I. Budarin, *Collection Czech. Chem. Commun.*, 1961, **26**, 215.
[8] I. V. Moiseev, in *Uran, metody ego opredeleniya* (*Uranium, Methods of its Determination*), V. K. Markov (ed.), 2nd Ed., p. 92. Atomizdat, Moscow, 1964.
[9] L. G. Sillén and A. E. Martell, *Stability Constants of Metal-Ion Complexes*, p. 184, The Chemical Society, London, 1964.
[10] C. F. Baes, Jr., *J. Phys. Chem.*, 1956, **60**, 878.
[11] J. Thamer, *J. Am. Chem. Soc.*, 1957, **79**, 4298.
[12] A. I. Moskvin, A. M. Shelyakina and P. S. Perminov, *Zh. Neorgan. Khim.*, 1967, **12**, 3319.
[13] A. I. Moskvin and V. F. Peretrukhin, *Radiokhimiya*, 1964, **6**, 206.
[14] R. G. Denotkina, A. I. Moskvin and V. B. Shevchenko, *Zh. Neorgan. Khim.*, 1960, **5**, 1509.
[15] R. G. Denotkina and V. B. Shevchenko, *Zh. Neorgan. Khim.*, 1967, **12**, 2345.
[16] R. M. Smith and R. A. Alberty, *J. Am. Chem. Soc.*, 1956, **78**, 2376.
[17] L. N. Filatova and M. I. Shepelevetskii, *Dokl. Akad. Nauk SSSR*, 1966, **166**, 140; *Zh. Neorgan. Khim.*, 1966, **11**, 1662.
[18] N. Bjerrum and C. R. Dahm, *Z. Phys. Chem.* (*Leipzig*) (*Bodenstein-Festband*),**1931**, 627.

Table 6.8

Stability constants of diphosphato-complexes (L = $P_2O_7^{4-}$)

Metal ion	Equilibrium constants (Symbol for the constant or scheme of the equilibrium, value of log K or log β, [I])	Ref.
Li^+	β_1 3.4 [0]; 2.39 [1.0][a]; MHL/M.HL 2.0 [0]; 1.03 [1.0][a]	
Na^+	β_1 2.29 [0]; 1.00 [1.0][a]; 0.21 [2.0]; $M_2L/ML.M$ 1.9 [0]; −0.8 [2.0]; MHL/M.HL 1.4 [0]; −0.5 [2.0]	
K^+	β_1 2.1 [0]; 0.80 [1.0][a]	
Mg^{2+}	β_1 7.2 [0]; 5.45 [0.1]; 5.42, β_2 7.80 [1.0][a]; MHL/M.HL 3.18 [0.1][b]; 3.06 [1.0][a]; MOHL/ML.OH 2.1 [0]	
Ca^{2+}	β_1 6.8 [0]; 5.4 [0.1][a]; 4.9 [1.0][a]; MHL/M.HL 3.3 [0.1][a]; 2.3 [1.0][a]; MOHL/ML.OH 2.1 [0]	
Sr^{2+}	β_1 5.4, MOHL/ML.OH 2.3 [0]	
La^{3+}	β_1 16.72, β_2 18.57, β_{21} 19.59 [0]	
Ce^{3+}	β_1 17.15 [0]	
Ce^{4+}	β_1 (18.41) [0.1]	
$M^{3+}(4f)$	β_{21} (20.5) [0] (Nd–Er)	
Yb^{3+}	β_1 17.5, β_2 19.4, β_{21} 21.88 [0]	
Lu^{3+}	β_{21} 22.23 [0]	
Mn^{3+}	β_1 16.68, β_2 31.85, $MH_2L/M.H_2L$ 5.11, $M(H_2L)_2/M.(H_2L)^2$ 8.41, $M(H_2L)_3/M.(H_2L)^3$ 11.24 [0.3]	
Fe^{3+}	β_2 (5.55) [var][c]; β_{21} 23.4 [0]; $M(OH)_2L/M.(OH)^2.L$ 31 [0.225][d]; $M(HL)_2/M.(HL)^2$ 22.19 [var][c]; $MH_2L/M.H_2L$ 6.97, $MH_3L/M.H_3L$ 6.43 [0]	[1–4]
Co^{2+}	β_1 6.1, 7.36[a], MHL/M.HL 3.4, 4.07[a] [0.1]	
Ni^{2+}	β_1 5.94, 7.01[a], MHL/M.HL (3.71), 3.81[a] [0.1]; $ML_2/M.L^2$ 2.0 [var]	
Cu^{2+}	β_1 7.6, 9.07[a], β_2 12.45, 16.65[a], MHL/M.HL 4.45, 5.37[a], $MHL_2/ML_2.H$ 4.9, 6.61[a], $MH_2L/M.H_2L$ 1.99, 2.55[a], $MH_2L_2/MHL_2.H$ 4.7, 5.63[a], $MH_3L_2/MH_2L_2.H$ 3.7, 4.25[a], $MH_4L_2/MH_3L_2.H$ 2.7, 3.06[a], $MH_5L_2/MH_4L_2.H$ 1.7 [1.0]	
Zn^{2+}	β_1 8.7, β_2 11.0, MOHL/ML.OH 4.4 [0]	
Cd^{2+}	β_1 (8.7), MOHL/M.OH.L 11.8 [0]	
Hg_2^{2+}	β_2 12.38, MOHL/M.OH.L 15.64 [0.75][e]	
Hg^{2+}	MOHL/M.OH.L 17.45 [0.75][e]	
Tl^+	β_1 1.69, β_2 1.9 [2.0][f]	
Sn^{2+}	β_1 (14) [var][g]; β_2 16.42 [1.0][c]; $MH_2L/M.H_2L$ 4.48, $M(H_2L)_2/M.(H_2L)^2$ 6.08, $MOH(H_2L)/MOH.H_2L$ 5.48, $MOH(H_2L)_2/MOH.(H_2L)^2$ 7.30 [2.0]	[5, 6]
Pb^{2+}	β_1 7.3, β_2 10.15 [1.0]	

[a] Medium $(CH_3)_4N^+$; [b] 15°C; [c] temperature not given; [d] 20°C; [e] 27°C; [f] 35°C; [g] laboratory temperature

REFERENCES

[1] S. Banerjee and S. K. Mitra, *Sci. Culture (India)*, 1951, **16**, 530.
[2] Z. A. Sheka, L. P. Andrusenko and I. A. Sheka, *Zh. Neorgan. Khim.*, 1967, **12**, 74.
[3] E. Eriksson, *Kgl. Lantbruks-Högskol Annaler*, 1949, **16**, 39.
[4] P. I. Yakshova, *Tr. Voronezh. Univ.*, 1956, **42**, No. 2, 63.
[5] L. G. Sillén and A. E. Martell, *Stability Constants of Metal-Ion Complexes*, p. 193; The Chemical Society, London, 1964.
[6] B. A. Purin and I. A. Vitina, *Izvest. Akad. Nauk Latv. SSR, Ser. Khim.*, 1968, **277**, 372.

Table 6.9
Stability constants of triphosphato-complexes (L = $P_3O_{10}^{5-}$)

Metal ion	Equilibrium constants (Symbol for the constant or scheme of the equilibrium, value of log K or log β, [I])	Ref.
Li$^+$	β_1 3.9 [0]; 2.87, MHL/M.HL 1.88 [1.0]^a	
Na$^+$	β_1 2.7 [0]; 1.64, MHL/M.HL 0.77 [1.0]^a	
K$^+$	β_1 2.8 [0]; 1.39 [1.0]^a	[1]
Cs$^+$	β_1 2.8 [0]	[1]
Be^{2+}	MHL/ML.H 5.35 [0.1]^b	
Mg^{2+}	β_1 8.6 [0]; 5.76, 7.11^a [0.1]; 5.82 [1.0]^a; MHL/M.HL 3.5, 4.45^{ab} [0.1]; 3.35 [1.0]^a; MOHL/ML.OH 2.4 [0]	
Ca^{2+}	β_1 8.1 [0]; 5.20, 6.38^a [0.1]; 5.40 [1.0]^a; MHL/M.HL 3.04, 4.02^{ab} [0.1]; 2.9 [1.0]^a; MOHL/ML.OH 2.3 [0]	
Sr^{2+}	β_1 7.2 [0]; 4.10, 5.50^a [0.1]; MHL/M.HL 2.53, 3.56^{ab} [0.1]; MOHL/ML.OH 2.1 [0]	
Ba^{2+}	β_1 6.3 [0]; 3.3, MHL/M.HL 1.8 [0.1]	
Y^{3+}	β_2 17.21, MHL/M.HL 4.97, M(HL)$_2$/M.(HL)^2 8.87 [0.1]^c	[2]
La^{3+}	β_1 6.56 [0.1]^a; β_2 15.8, MHL/M.HL 4.6, M(HL)$_2$/M.(HL)^2 8.3 [0.1]^c	[2, 3]
M^{3+}(4f)	β_2 (17.2), MHL/M.HL 5.0, M(HL)$_2$/M.(HL)^2 (8.8) [0.1]^c (Pr, Sm, Eu, Gd, Dy, Er, Yb)	[2]
Mn^{2+}	β_1 7.15, 8.08^a, MHL/M.HL 3.77, 5.08^{ab} [0.1]	
Fe^{2+}	MH$_2$L/M.H$_2$L 2.38, MH$_3$L/M.H$_3$L 2.12 [1.0]	[4]
Fe^{3+}	M(HL)$_2$/M.(HL)^2 20.63 [0]; 18.85 [0.1]; MH$_2$L/M.H$_2$L 7.03 [0]; 5.03 [0.1]; MH$_3$L/M.H$_3$L 6.37 [0]; 5.04 [0.1]	[5, 6]
Co^{2+}	β_1 6.94, 8.01^a [0.1]; MHL/M.HL 3.81, 4.93^{ab} [0.1]	
Ni^{2+}	β_1 6.75, 7.86^a, MHL/M.HL 3.65, 4.9^{ab} [0.1]	
Cu^{2+}	β_1 8.3, 9.36^a, MHL/M.HL 4.34, 6.1^{ab} [0.1]	
Zn^{2+}	β_1 (9.7) [0]; 7.5, 8.43^a, MHL/M.HL 3.92, 5.13^{ab} [0.1]; MOHL/ML.OH 3.3 [0]	
Cd^{2+}	β_1 9.8 [0]; 6.58, 8.13^a, MHL/M.HL 3.60, 4.97^{ab} [0.1]; MOHL/ML.OH 2.8 [0]	
Hg$_2^{2+}$	β_2 11.23, MOHL/M.OH.L 15.00 [0.75]^d	
In^{3+}	M(H$_2$L)$_2$/M(H$_2$L)^2 14.16 [0]; 12.18 [0.1]^b	[7]
Tl$^+$	β_1 1.34, β_2 2.26 [2.4]^b	
Pb^{2+}	MHL/M.HL 6.32 [1.0]^e	[8]

^{a} Medium (CH$_3$)$_4$N$^+$; ^{b} 20°C; ^{c} temperature not given; ^{d} 27°C; ^{e} 30°C

REFERENCES

[1] J. A. Wollhoff and J. T. G. Overbeek, *Rec. Trav. Chim.*, 1959, **78**, 759.
[2] A. Roppongi and T. Kato, *Bull. Chem. Soc. Japan*, 1962, **35**, 1092.
[3] W. M. McNabb, J. F. Hazel and R. A. Baxter, *J. Inorg. Nucl. Chem.*, 1968, **30**, 1585.
[4] C. F. Wells and M. A. Salam, *J. Chem. Soc. A*, **1968**, 308.
[5] L. P. Andrusenko and I. A. Sheka, *Zh. Neorgan. Khim.*, 1968, **13**, 2645.
[6] I. A. Sheka and L. P. Andrusenko, *Zh. Neorgan. Khim.*, 1968, **13**, 347.
[7] L. P. Andrusenko and I. A. Sheka, *Zh. Neorgan. Khim.*, 1967, **12**, 638.
[8] P. S. Shetty, P. R. Subbaraman and J. Gupta, *Indian J. Chem.*, 1964, **2**, 8.

Table 6.10
Stability constants of polyphosphato-complexes ($L = P_nO_{3n+1}^{(n+2)-}$; $n = 4, 5$)

Metal ion	Equilibrium constants (Symbol for the constant or scheme of the equilibrium, value of $\log K$ or $\log \beta$, $[I]$)	Ref.
Li^+	$n = 4$: β_1 (2.64), MHL/M.HL (1.59) $[1.0]^a$	
	$n \sim 5$: β_1 1.5 $[^b]$	[1]
Na^+	$n = 4$: β_1 1.79, MHL/M.HL 1.10 $[1.0]^a$	
	$n \sim 5$: β_1 1.3 $[^b]$	[1]
K^+	$n = 4$: β_1 1.71, MHL/M.HL (1.11) $[1.0]^a$	
	$n \sim 5$: β_1 0.7 $[^b]$	[1]
Mg^{2+}	$n = 4$: β_1 6.04, MHL/M.HL 3.74, $M_2L/ML.M$ 2.19 $[1.0]^a$	
	$n \sim 5$: β_1 3.2 $[^b]$	[1]
Ca^{2+}	$n = 4$: β_1 5.46, MHL/M.HL 3.54, $M_2L/ML.M$ 3.07 $[1.0]^a$	
	$n \sim 5$: β_1 3.0 $[^b]$	[1]
Sr^{2+}	$n = 4$: β_1 4.82, MHL/M.HL 3.49, $M_2L/ML.M$ 3.42 $[1.0]^a$	
	$n \sim 5$: β_1 2.8 $[^b]$	[1]
Ba^{2+}	$n = 4$: β_1 4.99 $[0]$	[2]
	$n \sim 5$: β_2 5.5 $[^b]$; 9.5 $[0.1]^a$	[1]
La^{3+}	$n = 4$: β_1 6.59, MHL/M.HL 3.29 $[0.1]^a$	[3]
UO_2^{2+}	$n \sim 5$: β_1 3.0 $[^b]$	[1]
Mn^{2+}	$n \sim 5$: β_2 5.5 $[^b]$	[1]
Fe^{2+}	$n \sim 5$: β_1 3.0 $[^b]$	[1]
Fe^{3+}	$n \sim 5$: β_2 6.5 $[^b]$	[1]
Co^{2+}	$n \sim 5$: β_1 3.0 $[^b]$	[1]
Ni^{2+}	$n \sim 5$: β_1 3.0 $[^b]$	[1]
Cu^{2+}	$n = 4$: β_1 9.44, β_2 10.60, MHL/M.HL 6.66, $MH_2L/M.H_2L$ 3.48, MOHL/ML.OH 3.86, $MHL_2/ML_2.H$ 8.40, $MH_2L_2/MHL_2.H$ 7.28, $MH_3L_2/MH_2L_2.H$ 4.52, $MH_4L_2/MH_3L_2.H$ 3.55 $[1.0]^a$	
Ag^+	$n \sim 5$: β_1 3.5 $[^b]$	[1]
Zn^{2+}	$n \sim 5$: β_1 2.5 $[I]^b$	[1]
Hg_2^{2+}	$n = 4$: β_1 7.32^c, 6.98^d $[1.0]$; β_2 9.88^c, 9.42^d $[0.15]^e$; MOHL/M.OH.L 15.26^c, 15.91^d, $M(OH)_2L$ 22.45^c, 22.64^d $[0.15]^e$	[4]
Pb^{2+}	$n \sim 5$: β_2 5.5 $[I]^b$	[1]

Table 6.10 (Notes)

[a] $(CH_3)_4NCl$ used as electrolyte to adjust ionic strength; [b] dilute solution, ligand concentration refers to PO_3; [c] in presence of K^+; [d] in presence of Na^+; [e] medium $C(NH_2)_3^+$.
Note: for the guanidinium ion the stability of the tetraphosphato-complexes is given by β_1 1.84, β_{12} 2.76, MHL/M.HL 1.16 [1.0][a] (ref. [5]).

REFERENCES

[1] J. R. Van Wazer and D. A. Campanella, *J. Am. Chem. Soc.*, 1950, **72**, 655.
[2] H. W. Jones and C. B. Monk, *J. Chem. Soc.*, **1950**, 3475.
[3] W. M. McNabb, J. F. Hazel and R. A. Baxter, *J. Inorg. Nucl. Chem.*, 1968, **30**, 1585.
[4] L. G. Sillén and A. E. Martell, *Stability Constants of Metal-Ion Complexes*, p. 201. The Chemical Society, London, 1964.
[5] J. I. Watters and S. Matsumoto, *J. Am. Chem. Soc.*, 1964, **86**, 3961.

Table 6.11
Stability constants of polymetaphosphato-complexes (L = $P_nO_{3n}^{n-}$)

Metal ion	Equilibrium constants (Symbol for the constant or scheme for the equilibrium, value of log K or log β, [I])
Na^+	$n = 3$: β_1 1.40 [0]
	$n = 4$: β_1 2.10 [0]; 0.81 [1.0][ab]
	$n = 6$: β_1 4.3 [0]
	$n = 8$: β_1 4.6 [0]
Mg^{2+}	$n = 3$: β_1 3.31 [0]; 1.11 [0.1][c]
	$n = 4$: β_1 5.17 [0]
Ca^{2+}	$n = 3$: β_1 3.47 [0]; 2.06 [0.1][c]; 1.64 [1.0][a]
	$n = 4$: β_1 5.37 [0]; 3.28 [0.1][c]; 2.90 [0.2][c]; 3.1 [1.0][a]; β_2 8.02 [0]
	$n = 6$: β_1 6.9 [0][c]; 4.59 [0.1][c]; 4.11 [0.2]
	$n = 8$: β_1 8.1 [0][c]; 5.18 [0.1][c]; 4.62 [0.2]
Sr^{2+}	$n = 3$: β_1 3.35 [0]; 1.99 [0.2][c]
	$n = 4$: β_1 5.12 [0]; 2.80 [0.15][c]; β_2 7.54 [0]
	$n = 6$: β_1 3.73 [0.2]
	$n = 8$: β_1 4.30 [0.2]
Ba^{2+}	$n = 3$: β_1 3.35 [0]
	$n = 4$: β_1 4.99 [0]
La^{3+}	$n = 3$: β_1 5.70 [0]
	$n = 4$: β_1 6.66 [0]
Pm^{3+}	$n = 3$: β_1 5.74 [0]; 3.80 [0.2][c]
Am^{3+}	$n = 3$: β_1 6.06 [0]; 3.48 [0.2][c]
Cm^{3+}	$n = 3$: β_1 5.92 [0]; 3.64 [0.2][c]
Mn^{2+}	$n = 3$: β_1 3.57 [0]
	$n = 4$: β_1 5.74 [0]
Co^{2+}	$n = 4$: β_1 2.62 [0.2][c]
	$n = 6$: β_1 3.65 [0.2]
	$n = 8$: β_1 4.80 [0.2]

Table 6.11 (Continued)

Metal ion	Equilibrium constants (Symbol for the constant or scheme of the equilibrium, value of log K or log β, $[I]$)
Ni^{2+}	$n = 3$: β_1 3.22 [0]
	$n = 4$: β_1 4.95 [0]; 2.63 [1.0][ab]; β_2 3.48 [1.0][ab]
Cu^{2+}	$n = 3$: β_1 1.58, β_2 2.2 [1.0][a]
	$n = 4$: β_1 3.04, 3.18[b], β_2 4.28, 4.64[b] [1.0][a]
Zn^{2+}	$n = 3$: β_1 1.94 [0.2][c]
	$n = 4$: β_1 2.86 [0.2][c]
	$n = 6$: β_1 3.95 [0.2]
	$n = 8$: β_1 5.02 [0.2]

[a] $(CH_3)_4N.NO_3$ used to adjust ionic strength; [b] $30°C$; [c] $20°C$

Table 6.12
Stability of hydroxo-complexes (L = OH^-)

Metal ion	Equilibrium constants* (Symbol for the constant or scheme of the equilibrium, value of log K or log β, $[I]$)	Ref.
H^+	β_1 13.997, *13.997* [0]; 13.78, *13.78*, 13.95[a] [0.1]; 13.74, *13.74* [0.5] 13.79, *13.80* [1.0]; 13.96, *13.97* [2.0]; 14.18, *14.20* [3.0]; 14.42 [4.0]	
Li^+	β_1 0.36 [0]; -0.18 [3.0]	
Na^+	β_1 -0.2 [0]	
K^+	β_1 -0.5 [0]	
Be^{2+}	β_1 8.6, *8.6* [0]; 8.3 [0.1][a]; *8.1* [1.0]; β_2 (14.4), *14.3* [0]; 16.7 [0.1][a]; (16.5) [0.5]; *13.7* [1.0]; (17.5) [3.0]; β_3 18.8, *18.7* [0]; *18.1* [1.0]; β_4 18.6, *18.6* [0]; *18.5* [1.0]; β_{12} (10.0) *10.0* [0]; 10.54 [0.5]; *10.4* [1.0]; 10.68 [2.0]; 10.95 [3.0]; β_{33} *33.1*, 33.1 [0]; 32.41 [0.5]; *32.6* [1.0]; 32.98 [2.0]; 33.88 [3.0]; β_{86} (85); *84.8* [0]; *83.0* [1.0]	
Mg^{2+}	β_1 2.58, *2.55* [0]; *1.81* [1.0]; 1.85 [3.0]; β_{44} 16.3; *16.3* [0]; *16.1* [1.0]; 16.93 [3.0]	
Ca^{2+}	β_1 1.3, *1.15* [0]; *0.50* [1.0]; 0.64 [3.0]	
Sr^{2+}	β_1 0.8, *0.7* [0]; *0.05* [1.0]; 0.23 [3.0]	
Ba^{2+}	β_1 0.6, *0.5* [0]; 0.4 [0.1]; (-0.2) [1.0]	
Sc^{3+}	β_1 9.7; *9.7* [0]; 9.31, 9.1[a] [0.1]; 9.06 [0.5] 8.63, *8.6* [1.0]; β_2 18.3, *18.3* [0]; 17.4 [0.1][a]; *17.4*, *16.9* [1.0]; β_3 25.9, *25.9* [0]; 24.9 [0.1][a]; *24.2* [1.0]; β_4 30, (*30*) [0]; (*28.4*) [1.0]; β_{22} 22.0, *22.0* [0]; 21.53 [0.1]; *21.50* [0.5]; 21.49, *21.5* [1.0]; β_{53} 53.8, *53.6* [0]; 51.88 [0.1]; 51.69 [0.5]; 51.55, *51.6* [1.0]	

* The values of log K printed in italics were calculated from the data of Baes and Mesmer [1].

Stability Constants of Complexes with Inorganic Ligands

Table 6.12 (Continued)

Metal ion	Equilibrium constants* (Symbol for the constant or scheme of the equilibrium, value of log K or log β, [l])	Ref.
Y^{3+}	β_1 6.3, *6.3* [0]; 5.39 [0.3]; *5.4* [1.0]; 5.1 [3.0]; β_2 *(11.6)* [0]; (10.2) [1.0]; β_3 *(16.0)* [0]; *(14.3)* [1.0]; β_4 (19.5) [0]; *(17.8)* [1.0]; β_{22} 13.8; *13.8* [0]; *13.4* [1.0]; 14.06 [3.0]; β_{53} 38.4; *38.4* [0]; 35.9 [1.0]; 37.1 [3.0]	
La^{3+}	β_1 5.5, *5.5* [0]; 4.67 [0.3]; *4.6* [1.0]; 4.1 [3.0]; β_2 ~ *10* [1.0]; β_3 ~ *15* [1.0]; β_4 ~ *18* [1.0]; β_{12} (4.2) [3.0]; β_{22} *10.5* [0]; 10.1 [1.0]; β_{53} *31.7* [0]; 29.2 [1.0]; β_{95} 54.8; *54.8* [0]; *53.0* [1.0] 56.2 [3.0]	
Ce^{3+}	β_1 *5.7* [0]; *4.8* [1.0]; β_{22} *(12.5)* [0]; *(12.1)* [1.0]; β_{53} 36.5, *36.5* [0]; 33.9 [1.0]; 35.1 [3.0]	
Ce^{4+}	β_1 *14.9* [1.0]; 13.17 [1.5]; *15.3* [3.0]; β_2 27.9 [1.0]; *28.7* [3.0]; β_{22} 31.2 [1.0]; 32.0 [3.0]; β_{32} 45.5 [1.0]; 46.7, 40.9^b [3.0]; β_{42} 58.7 [1.0]; (60.3) 54.5^b [3.0]; $\beta_{12,6}$ *186*, 168.4^b [3.0]	
$M^{3+}(4f)$ (Ln, Pr–Lu)	β_1 6 [0]; 5.6 [0.3]; *5* [1.0]; β_2 *12* [0]; *11* [1.0]; β_3 *(17)* [0]; *(15)* [1.0]; β_4 *(22)* [0]; *(20)* [1.0]; β_{22} *(14)* [0]; *(14)* [1.0]; β_{53} *(41)* [0]; *(38)* [1.0]	
Ti^{3+}	β_1 12.7, *11.8* [0]; 11.8 [0.5]; *11.0* [1.0]; 11.5 [3.0]; β_{22} 24.4 [0]; 24.0 [1.0]; 24.8 [3.0]	
$Ti(OH)_2^{2+}$	β_1 11.7 [0]; *11.1* [1.0]; β_2 23.2 [0]; 22.6 [0.1]; *22.3* [1.0]	
Ti^{4+}	β_1 18.0, β_2 35.2, β_3 47.7, β_4 58.7 [0]^c	[2]
Zr^{4+}	β_1 14.3, *14.3* [0]; *13.2* [1.0]; 13.9 [4.0]; β_2 *(26.3)* [0]; *(25.3)* [1.0]; β_3 *(36.9)* [0]; *(35.0)* [1.0]; β_4 46.3 [0]; 44.2 [1.0]; β_5 54.0; *54.0* [0]; 51.9 [1.0]; β_{43} 55.4, *55.4* [0]; *59.7* [1.0]; 50.5 [2.0]; β_{53} *73.7* [0]; 74.2 [1.0]; β_{84} 106.0; *118.0* [0]; 118.4 [1.0]	
Hf^{4+}	β_1 13.7, *13.7* [0]; *12.6* [1.0]; 13.3 [4.0]; β_2 25.6 [0]; *24.6* [1.0]; β_3 *(36.0)* [0]; *(34.1)* [1.0]; β_4 45.3 [0]; *43.2* [1.0]; β_5 52.8, *52.8* [0]; 50.7 [1.0]	
Th^{4+}	β_1 10.8, *10.8* [0]; 9.6, *9.7* [1.0]; 9.1 [3.0]; β_2 21.1, *21.1* [0]; 19.89, *20.1* [1.0]; β_3 *(30.3)* [0]; *(28.4)* [1.0]; β_4 40.1 [0]; 38.0 [1.0]; β_{22} *(21.9)*, 21.9 [0]; 22.97, *23.0* [1.0]; 23.58 [3.0]^d; β_{32} 33.8 [3.0]^d; β_{52} (53.7) [3.0]^d; β_{13} (12.7) [3.0]^d; β_{33} (35.7) [3.0]^d β_{84} (90.9), 90.9 [0]; 91.2, *91.3* [1.0]; $\beta_{15,6}$ 173.2, *173.2* [0]; 169.8, *170.2* [1.0]	
V^{3+}	β_1 11.7, *11.74* [0]; 11.01^a, *10.96* [1.0]; 11.11 [3.0]; β_{22} 24.2, *24.2* [0]; 23.8^a, *23.7* [1.0]; 23.43 [3.0]; β_{32} 34.5 [3.0]	
VO^{2+}	β_1 8.3, *8.33* [0]; *7.76* [1.0]; 7.9 [3.0]; β_2 18.31 [3.0]; β_{22} 21.32, 21.3 [0]; *20.56* [1.0]; 28.35 [3.0]	
VO_2^+	β_1 10.7 [0]; *10.5* [1.0]; β_2 20.7 [0]; *20.6* [1.0]; β_3 26.1 [0]; *26.7* [1.0]; β_4 25.8 [0]; *27.3* [1.0]; β_{52} 48.9 [0]; *50.2* [1.0]; β_{62} 52.8 [0]; 54.7 [1.0]; β_{63} 68.2 [0]; *69.2* [1.0]; $\beta_{14,10}$ 185.3 [0]; *186.5* [1.0]; $\beta_{15,10}$ 195.7 [0]; *196.3* [1.0]; $\beta_{16,10}$ 203.6 [0]; *204.3* [1.0]	

Table 6.12 (Continued)

Metal ion	Equilibrium constants* (Symbol for the constant or scheme of the equilibrium, value of log K or log β, [I])	Ref.
Nb(V)	$ML_5/ML_4.L$ (14.4); $ML_6/ML_5.L$ (6.4) [1.0][e]	
Ta(V)	$ML_5/ML_4.L$ (12.8); $ML_6/ML_5.L$ (4.2) [1.0][e]	
Pa^{4+}	β_1 14.83 [0]; 13.72 [1.0]; β_2 27.99 [0]; 27.04 [1.0]; β_3 40.5 [0]; 38.63 [1.0]	
Pa(V)	$ML_4/ML_3.L$ 14.5, 14.5 [0]; 13.6 [1.0]; 13.13 [3.0]; $ML_5/ML_4.L$ 9.5, 9.5 [0]; 9.3 [1.0]; 9.68 [3.0]	
Cr^{3+}	β_1 10.07, 10.0 [0]; 9.77 [0.1][a]; 9.41 [0.5][a]; 9.41, 9.2 [1.0]; β_2 18.3 [0]; 17.3 [0.1][a]; 16.8 [1.0]; β_3 24 [0]; 22 [1.0]; β_4 28.6 [0]; 26.9 [1.0]; β_{22} 22.93 [0]; 24.1, 22.68 [1.0]; 24.6 [2.0]; β_{43} 47.84 [0]; 47.3 [1.0]; β_{44} 50.7 [2.0]; β_{64} 72.8 [2.0]	
U^{4+}	β_1 13.34, 13.3 [0]; 12.50 [0.1]; 12.24 [0.5]; 12.23, 12.23 [1.0]; 12.31 [2.0]; (12.1) [3.0]; β_2 (25.4) [0]; (24.4) [1.0]; β_3 (36.2) [0]; (34.3) [1.0]; β_4 (45.7) [0]; (43.6) [1.0]; β_5 54.0, 54.0 [0]; 51.9 [1.0]; $\beta_{15,6}$ 192.8, 192.8 [0]; 189.8 [1.0]; 196.1 [3.0]	
UO$_2^{2+}$	β_1 8.19 [0]; (7.7) [0.1]; 8.0 [0.5]; 8.0 [1.0]; 8.1 [3.0]; β_{12} 9.6 [1.0]; β_{22} 22.4, 22.4 [0]; 21.73 [0.1]; 21.55 [0.5]; 21.64, 21.7 [1.0]; 22.32 [3.0]; β_{43} (44.24) [0]; 42.4, 42.89 [1.0]; 43.5 [3.0]; β_{53} 54.4; 54.35 [0]; 52.4 [0.5]; 52.6, 52.64 [1.0]; 54.4 [3.0]	
Np^{4+}	β_1 12.5, 12.50 [0]; 11.4 [1.0]; 11.7 [2.0]	
NpO$_2^+$	β_1 5.14 [0]; 4.95 [1.0]	
NpO$_2^{2+}$	β_1 8.9, 8.84 [0]; 8.6, 8.64 [1.0]; β_{22} 21.6, 21.60 [0]; 20.9, 20.94 [1.0]; β_{53} 52.5, 52.50 [0]; 50.7, 50.79 [1.0]	
Pu^{3+}	β_1 7.0, 6.0 [0]; 6.7 [0.1]; 5.8 [1.0]	
Pu^{4+}	β_1 13.5 [0]; 12.14 [0.5]; 12.1, 12.3 [1.0]; 21.23 [2.0]; β_2 (25.7) [0]; (24.7) [1.0]; β_3 (36.7) [0]; (34.8) [1.0]; β_4 46.5 [0]; 44.4 [1.0]; β_5 55.0, 52.9 [1.0]	
PuO$_2^+$	β_1 4.3 [0]; 3.9 [1.0]	
PuO$_2^{2+}$	β_1 8.4, 8.4 [0]; 7.8, 7.8 [1.0]; 7.9 [3.0]; β_{22} 19.6, 19.63 [0]; 19.1, 19.09 [1.0]; 20.1 [3.0]; β_{53} 48.4, 48.33 [0]; 46.8, 46.90 [1.0]; 49.3 [3.0]	
M^{3+}(5f) (Am–Cf)	β_1 (8.0) [0.1]	
Mn^{2+}	β_1 3.4, 3.40 [0]; 2.9 [0.1]; 3.0; 2.85 [1.0]; 3.5 [2.0]; β_2 (5.8) [0]; (4.9) [1.0]; β_3 (7.2) [0]; (6.4) [1.0]; β_4 7.7, 7.7 [0]; 7.6 [1.0]; β_{12} (3.4), 3.44 [0]; 3.47 [0.1]; 3.75 [1.0]; β_{32} 18.1, 18.09 [0]; 16.48 [1.0]; 16.4 [2.0]	
Mn^{3+}	β_1 14.4 [4.0]	
Fe^{2+}	β_1 4.5, 4.5 [0]; 4.3, 4.0 [1.0]; β_2 (7.4), 7.4 [0]; 6.5 [1.0]; β_3 10.0, 11 [0]; 10.2 [1.0]; β_4 9.6, 10 [0]; 9.9 [1.0]	
Fe^{3+}	β_1 11.81, 11.81 [0]; 11.17 [0.1]; 11.01 [0.5]; 11.09, 11.03 [1.0]; 11.14 [2.0]; 11.21 [3.0]; β_2 22.3, 22.32 [0]; 21.9[a], 21.18 [1.0];	

Stability Constants of Complexes with Inorganic Ligands

Table 6.12 (Continued)

Metal ion	Equilibrium constants* (Symbol for the constant or scheme of the equilibrium, value of log K or log β, $[I]$)	Ref.
	Fe^{3+} (continued) 22.1 [3.0]; β_3 30 [0]; 28.7 [1.0]; β_4 34.4, *34.4* [0]; *33.3* [1.0]; β_{22} 25.1, *25.04* [0]; 24.7 [0.1]; 24.7 [0.5]; 24.9, *24.97* [1.0]; 25.3 [2.0]; 25.6 [3.0]; β_{43} 49.7, *49.7* [0]; *49.4* [1.0]; 51.0 [3.0]	
Co^{2+}	β_1 4.3, *4.35* [0]; (3.9) [0.5]; (3.9), *3.78* [1.0]; 4.2 [3.0]; β_2 8.4, *9.19* [0]; 8.5, *8.32* [1.0]; β_3 9.7, *10.49* [0]; 9.7, *9.66* [1.0]; β_4 10.2, *9.68* [0]; *9.54* [1.0]; β_{12} (2.7), *2.80* [0]; 3.10 [1.0]; (3.7), [3.0]; β_{44} (25.6), *25.46* [0]; *25.65* [1.0]; 27.5 [3.0]	
Co^{3+}	β_1 13.52 [3.0]	
Ni^{2+}	β_1 4.1, *4.14* [0]; 3.7 [0.1]; 3.8, *3.58* [1.0]; β_2 8, 9 [0]; *8.1* [1.0]; β_3 11, *12* [0]; 11.2 [1.0]; β_4 12 [0]; *11.9* [1.0]; β_{12} (3.3), *3.3* [0]; 3.6 [1.0]; 4.2 [3.0]; β_{44} 28.3, *28.25* [0]; 28.22 [1.0]; 29.37 [3.0]	
Rh^{3+}	β_1 10.67 [2.5]	
Pd^{2+}	β_1 13.0 [0]; 12.4 [0.1]; *11.8* [1.0]; β_2 25.8 [0]; 25.2 [0.1]; *23.6* [1.0]; β_3 (29.4), β_4 (42.2) [1.0]	
51 RuO_4	β_1 2.1 [0]	[3]
OsO_4	β_1 *1.9* [0]; 1.8 [1.0]; K_2 −0.6, β_2 1.1 [1.0]	[4]
Cu^{2+}	β_1 6.3, (6) [0]; (5.5) [1.0]; β_2 (10.7) [0]; 12.8, *10.0* [1.0]; β_3 (14.2) [0]; 14.5, *13.6* [1.0]; β_4 16.4, *16.4* [0]; 15.6, *16.5* [1.0]; β_{22} 17.7, *17.63* [0]; 17.02 [0.1][a]; 17.28[a], 16.81 [1.0]; 17.8 [3.0]	
Ag^+	β_1 2.0, *2.0* [0]; *1.6* [1.0]; β_2 3.99, *4.0* [0]; *3.7* [1.0]; 3.55 [3.0]	
Au^{3+}	β_1 15.51 [0]; *14.80* [1.0]; β_2 29 [0]; 28.7 [1.0]; β_3 42 [0]; 40.5 [1.0]; K_4 2.3 [0]; β_4 44.22 [0]; *43.64* [1.0]; K_5 0.6, β_5 44.86 [0]; *44.90* [1.0]; β_6 42.9 [0]; *43.96* [1.0]	
Zn^{2+}	β_1 5.0, *5.04* [0]; 4.64 [1.0]; β_2 (11.1), *11.09* [0]; 10.4 [1.0]; 8.3 [3.0]; β_3 13.6, *13.52* [0]; 12.93 [1.0]; 13.7 [3.0]; β_4 *14.8*, (14.8) [0]; 14.82 [1.0]; 18.0 [3.0]; β_{12} 5.0, *5.0* [0]; 5.5, 4.9 [1.0]; β_{62} 26.2 [0]; 25.5 [1.0]	
Cd^{2+}	β_1 3.9, *3.92* [0]; *3.37* [1.0]; 4.1 [3.0]; β_2 7.7, *7.64* [0]; *6.81* [1.0]; 7.7 [3.0]; β_3 (8.69) [0]; 8.07 [1.0]; 10.3 [3.0]; β_4 (8.7) 8.64 [0]; *8.34* [1.0]; 12.0 [3.0]; β_{12} 4.6, *4.61* [0]; *4.80* [1.0]; 5.08 [3.0]; β_{44} 23.2, *23.13* [0]; *23.27* [1.0]; 24.9 [3.0]	
Hg^{2+}	β_1 10.6, *10.59* [0]; 10.0 [0.5]; 10.1, *10.13* [1.0]; 10.8 [3.0]; β_2 21.8, *21.82* [0]; 21.2 [0.1]; 21.0 [0.5]; 21.1, *21.28* [1.0]; 21.6 [2.0]; 22.2 [3.0]; β_3 20.9, *20.89* [0]; *20.30* [1.0]; β_{12} 10.7, *10.67* [0]; 10.98 [1.0]; 11.5 [3.0]; β_{33} 35.6, *35.57* [0]; 34.98 [1.0]; 36.1 [3.0]	
Hg_2^{2+}	β_1 8.7, *8.74* [0.5]	
Al^{3+}	β_1 9.01, *9.03* [0]; 8.48 [0.1]; 8.31, *8.32* [1.0]; β_2 (18.7), *18.69* [0]; (17.6) [0.1]; *17.31* [1.0];	

Table 6.12 (Continued)

Metal ion	Equilibrium constants* (Symbol for the constant or scheme of the equilibrium, value of log K or log β, [I])	Ref.
	Al^{3+} (continued)	
	β_3 (27.0), 26.99 [0]; (25.7) [0.1]; 25.32 [1.0];	
	β_4 33.0, 32.99 [0]; 31.53 [1.0]; β_{22} 20.3, 20.30 [0]; 20.0, 19.9 [1.0];	
	β_{43} 42.1, 42.05 [0]; 42.5, 41.77 [1.0];	
	$\beta_{32,13}$ 349.2, 349.17 [0]; 336.5, 336.90 [1.0]	
Ga^{3+}	β_1 11.4; 11.4 [0]; 10.9 [0.1]; 10.8 [0.5]; 10.6 [1.0];	
	β_2 22.1; 22.1 [0]; 21.5 [0.1]; 20.6 [1.0];	
	β_3 31.7; 31.7 [0]; 30.9 [0.1]; 29.8 [1.0];	
	β_4 39.4; 39.4 [0]; 34.7 [1.0]; $\beta_{65,26}$ 770.7 [0]; 757.9 [1.0]	
In^{3+}	β_1 10.0; 10.00 [0]; (10.5) [0.1]; 9.49 [1.0]; 9.76 [3.0];	
	β_2 20.2; 20.17 [0]; (20.3) [0.1]; 19.12 [1.0]; 19.6 [3.0];	
	β_3 29.6; 29.59 [0]; (29.3) [0.1]; 28.21 [1.0];	
	β_4 33.9; 33.92 [0]; 32.61 [1.0];	
	β_{43} 50.2; 50.17 [0]; 49.89 [1.0]; 47.4 [3.0]	
Tl^+	β_1 0.79 [0]; 0.30 [0.5]; 0.26 [1.0]; 0.08 [3.0]; -0.10 [5.0];	
	β_2 (-0.8) [3.0]	
Tl^{3+}	β_1 13.4; 13.37 [0]; (12.8) [0.1]; 12.79 [1.0]; 13.02 [3.0];	
	β_2 26.4; 26.42 [0]; (25.3) [0.1]; 25.28 [1.0]; 25.75 [3.0];	
	β_3 (38.7); 38.69 [0]; (37.6) [0.1]; 37.2 [1.0];	
	β_4 41.0; 41.0 [0]; 39.6 [1.0]; 40.0 [3.0]	
$Ge(OH)_4$	β_1 4.69 [0]; 4.80 [1.0]; β_2 6.1 [0]; 6.8 [1.0];	
	$\beta_{35,8}$ 27.75 [0]; 30.42 [1.0]	
Sn^{2+}	β_1 10.60 [0]; 9.95 [1.0]; 10.4 [3.0];	
	β_2 20.93 [0]; 20.03 [1.0]; β_3 25.38 [0]; 24.68 [1.0];	
	β_{22} 23.22 [0]; 22.62 [1.0]; 23.9 [3.0];	
	β_{43} 49.11 [0]; 47.75 [1.0]; 49.93 [3.0]	
Sn^{4+}	β_1 (14); β_2 (28); β_3 (42); β_4 (54) [1.0][f]	[5]
Pb^{2+}	β_1 6.3, 6.29 [0]; 6.0 [0.3]; 6.05 [1.0]; 6.3 [3.0];	
	β_2 10.9, 10.87 [0]; 10.3 [0.3]; 10.37 [1.0]; 10.9 [3.0];	
	β_3 13.9, 13.93 [0]; 13.3 [0.3]; 13.35 [1.0]; 13.7 [3.0];	
	β_{12} 7.6, 7.64 [0]; 7.62 [1.0]; 7.9 [3.0];	
	β_{43} 32.1, 32.10 [0]; 31.7 [0.3]; 32.43 [1.0]; 33.8 [3.0];	
	β_{44} (35.1), 35.11 [0]; 35.1 [0.3]; 35.75 [1.0]; 37.5 [3.0];	
	β_{86} (68.4), 68.36 [0]; 67.4 [0.3]; 68.38 [1.0]; 71.3 [3.0]	
Sb^{3+}	β_2 21.5, β_3 32.9, β_4 35.1 [1.0]	
Bi^{3+}	β_1 12.9, 12.91 [0]; 12.36 [0.1]; 12.29 [1.0]; 12.60 [3.0];	
	β_2 23.99 [0]; 22.54 [1.0]; β_3 33.1, 33.13 [0]; 31.9 [0.1]; 31.43 [1.0];	
	β_4 34.2, 34.19 [0]; 32.57 [1.0];	
	$\beta_{12,6}$ 164.95, 165 [1.0]; 170.49 [3.0]; $\beta_{20,9}$ 272 [0.1–1.0];	
	$\beta_{21,9}$ 282 [0.1–1.0]; $\beta_{22,9}$ 294 [0.1–1.0]	

[a] 20°C; [b] in nitrate solutions; [c] 18°C; [d] in chloride solutions; [e] 19°C; [f] in 3% aqueous ethanol

REFERENCES

[1] C. F. Baes, Jr. and R. E. Mesmer, *The Hydrolysis of Cations*, Wiley, New York, 1976.
[2] A. I. Zhukov and A. S. Nasarov, *Zh. Neorgan. Khim.*, 1964, **9**, 1465.
[3] M. D. Silverman and H. A. Levy, *Thesis*, University of Tennessee, 1950.
[4] R. D. Sauerbrunn and E. B. Sandell, *J. Am. Chem. Soc.*, 1953, **75**, 4170.
[5] V. A. Nazarenko, V. P. Antonovich and E. M. Nevskaya, *Zh. Neorg. Khim.*, 1971, **16**, 980.

Table 6.13
Stability constants of sulphido-complexes (L = S^{2-})

Metal ion	Equilibrium constants (Symbol for the constant or scheme of the equilibrium, value of log K or log β, [I])	Ref.
Cu^+	$ML_5/M \cdot L_5$ (21) [var] (L = S_5^{2-})	[1]
Ag^+	$MHL/M \cdot HL$ 16.2 [0]; 13.6 $[0.1]^a$; 13.30 $[1.0]^a$; $M(HL)_2/M \cdot (HL)^2$ 18.0 [0]; 17.7 $[0.1]^a$; $MHL/ML \cdot H$ 8.3, $M(HL)_2/MHL_2 \cdot H$ 9.5 $[0.1]^a$; $M_2H_2L_3 \cdot H_2L/(M(HL)_2)^2$ 3.2 $[1.0]^a$; $(MHL)^2/M_2L_{(s)} \cdot H_2L$ -15.78, $(M(HL)_2)^2/M_2L_{(s)} \cdot H_2L \cdot (HL)^2$ -8.04, $M_2H_2L_3/M_2L_{(s)} \cdot (HL)^2$ -4.82 $[1.0]^a$	[2, 3]
Au^{3+}	$(ML_2)^2/M_2L_{3(s)} \cdot HL \cdot OH$ -2.81, $(ML_2)^2/M_2L_{3(s)} \cdot L$ -1.89 $[var]^b$	[4]
Zn^{2+}	$ML/M \cdot OH \cdot HL$ 19.0, $MHLOH/ML_{(s)}$ -5.87 [1.0]	[5]
Cd^{2+}	$MHL/M \cdot HL$ 7.6, $M(HL)_2/M \cdot (HL)^2$ 14.6, $M(HL)_3/M \cdot (HL)^3$ 16.5, $M(HL)_4/M \cdot (HL)^4$ 18.9 [1.0]	
Hg^{2+}	$M(HL)_2/M \cdot (HL)^2$ 37.71, $M(HL)_2/MHL_2 \cdot H$ 6.19, $MHL_2/ML_2 \cdot H$ 8.30 $[1.0]^a$; $M(HL)_2/ML_{(s)} \cdot H_2L$ -5.97, $ML_2/ML_{(s)} \cdot L$ 0.57 $[1.0]^a$	[6]
Tl^+	$MHL/M \cdot HL$ 2.27, $M_2HL/MHL \cdot M$ 5.77, $M_2OH(HL)_3/M^2 \cdot OH \cdot (HL)^3$ 14.96, $M_2(OH)_2(HL)_2/M^2 \cdot (OH)^2 \cdot (HL)^2$ 16.70 [1.0]	
As(III)	$(ML_2)^2/M_2L_{3(s)} \cdot L$ 2.0, $(HML_2)^2/M_2L_{3(s)} \cdot H_2L$ -10.6 $[var]^c$; $(M_3L_6)^2 \cdot H^6/(M_2L_{3(s)})^3 \cdot (H_2L)^3$ -33.19 $[1]^d$	[7, 8]
Sb(III)	$(ML_2)^2/M_2L_{3(s)} \cdot L$ 0.9 $[var]^a$; $(ML_3)^2/M_2L_{3(s)} \cdot L^3$ 1.78 [0]	[9, 10]
	$M_4L_7/(M_2L_{3(s)})^2 \cdot HL \cdot OH$ 0.7 [var]	[11]
Te(IV)	$ML_3/ML_{2(s)} \cdot L$ 7.61 $[var]^d$	[12]

^{a} 20°C; ^{b} 30°C; ^{c} 0°C; ^{d} temperature not given

REFERENCES

[1] L. G. Sillén and A. E. Martell, *Stability Constants of Metal-Ion Complexes*, p. 218. The Chemical Society, London, 1964.
[2] L. G. Sillén and A. E. Martell, *op. cit.*, Supplement No. 1, p. 129. The Chemical Society, London, 1971.

[3] G. Schwarzenbach and M. Widmer, *Helv. Chim. Acta*, 1966, **49**, 111.
[4] K. P. Dubey, *Z. Anorg. Chem.*, 1965, **337**, 309.
[5] A. O. Gubeli and J. Ste-Marie, *Can. J. Chem.*, 1967, **45**, 2101.
[6] G. Schwarzenbach and M. Widmer, *Helv. Chim. Acta*, 1963, **46**, 2613.
[7] A. Ringbom, *Solubilities of Sulphides; Report to Analytical Section*, IUPAC, July 1953.
[8] J. Angeli and P. Souchay, *Compt. Rend.*, 1960, **250**, 713.
[9] Ref. [1], p. 221.
[10] A. K. Babko and G. S. Lisetskaya, *Zh. Neorgan. Khim.*, 1956, **1**, 969.
[11] R. H. Arntson, F. W. Dickson and G. Tunell, *Science*, 1966, **153**, 1673.
[12] K. P. Dubey and S. Ghosh, *J. Indian Chem. Soc.*, 1963, **40**, 479.

Table 6.14
Stability constants of sulphito-complexes (L = SO_3^{2-})

Metal ion	Equilibrium constants (Symbol for the constant or scheme of the equilibrium, value of log K or log β, [I])	Ref.
Ce^{3+}	β_1 8.04 [0]	
UO_2^{2+}	β_1 6.0 [0.1][a], β_2 7.1 [var]; 9.2 [1.0][a]	[1, 2]
NpO_2^+	β_1 2.15, β_2 3.00 [0][b]	[3]
Cu^+	β_1 7.85, β_2 8.7, β_3 9.4 [1.0]	
Ag^+	β_1 5.60 [0]; 5.4 [2.0]; β_2 8.68 [0]; 7.8 [2.0]; β_3 9.00 [0]	
Cd^{2+}	β_2 4.2 [1.0]	
Hg^{2+}	β_2 24.07 [0][c]; 22.85 [3.0][c]; β_3 25.96 [0][c]	

[a] 23°C; [b] temperature not given; [c] 18°C

REFERENCES

[1] A. E. Klygin and N. S. Kolyada, *Zh. Neorgan. Khim.*, 1959, **4**, 239.
[2] F. A. Zakharova and M. M. Orlova, *Zh. Neorgan. Khim.*, 1967, **12**, 3016.
[3] A. I. Moskvin and M. P. Mefodeva, *Radiokhimiya*, 1965, **7**, 410.

Table 6.15
Stability constants of sulphato-complexes (L = SO_4^{2-})

Metal ion	Equilibrium constants (Symbol for the constant or scheme of the equilibrium, value of log K or log β, [I])
Li^+	β_1 0.64 [0]
Na^+	β_1 0.70 [0]
K^+	β_1 0.85 [0]; 0.4 [0.1]
NH_4^+	β_1 1.11 [0][a]
Be^{2+}	β_1 1.95 [0]; 0.72 [0.5][a]; β_2 1.78, β_3 2.08 [1.0]
Mg^{2+}	β_1 2.23 [0]; 1.01 [0.7]
Ca^{2+}	β_1 2.31 [0]; 1.03 [0.7]
Sr^{2+}	β_1 2.55 [0]; 1.14 [0.5]

Table 6.15 (Continued)

Metal ion	Equilibrium constants (Symbol for the constant or scheme of the equilibrium, value of log K or log β, $[I]$)
Ba^{2+}	β_1 2.7 [0]; 0.66 [1.0]; β_2 1.42 [1.0]
Sc^{3+}	β_1 4.20 [0]; 2.59 [0.5]; β_2 5.7 [0]; 3.96 [0.5]
Y^{3+}	β_1 3.47 [0]; 1.24 [2.0]; β_2 5.3 [0]; 1.68 [2.0]
La^{3+}	β_1 3.64 [0]; 1.43 [1.0]; 1.29 [2.0]; β_2 5.29 [0]; 2.46 [1.0]
Ce^{3+}	β_1 3.59 [0]; 1.77 [0.5]; 1.24 [2.0]; β_2 5.2 [0]; 2.90 [0.5]
$M^{3+}(4f)$	β_1 (3.6) [0]; (1.2) [2.0]; β_2 (5.1) [0]; (1.8) [2.0]
TiO^{2+}	β_1 2.5 [0]; 2.15 [3.0]; 2.26 [4.0]; 2.47 [5.0]
Zr^{4+}	β_1 3.67, β_2 6.40, β_3 7.4 [2.0]
Hf^{4+}	β_1 3.04, β_2 5.44 [2.0]
Th^{4+}	β_1 3.22, β_2 5.53 [2.0]
V^{3+}	β_1 1.45 [1.0]
VO^{2+}	β_1 2.44 [0]
VO_2^+	β_1 0.97 [1.0][b]
Cr^{3+}	β_1 2.60 [1.0][c]
U^{4+}	β_1 3.42, β_2 5.82 [2.0]
UO_2^{2+}	β_1 2.95 [0]; 1.81 [1.0]; β_2 4.0 [0]; 2.5 [1.0]; β_3 3.7 [1.0][b]
Np^{4+}	β_1 3.51 [2.0]; 3.41 [3.0][d]; β_2 5.42 [3.0][d]
NpO_2^{2+}	β_1 3.27 [0]; 2.20 [0.1]; 2.07 [0.5]; 1.82 [1.0][b]; β_2 3.8 [0.1]; 3.4 [0.5]; 2.62 [1.0][b]
Pu^{3+}	β_1 1.26, $M(HL)_2/M.(HL)^2$ 1.00 [1.0]
Pu^{4+}	β_1 3.66 [1.0]
Am^{3+}	β_1 1.86 [0.5]; 1.57 [1.0]; 1.43 [2.0]; β_2 2.82 [0.5]; 2.66 [1.0]; 1.85 [2.0]
Cm^{3+}	β_1 1.86, β_2 2.7 [0.5]; 1.86 [2.0]
Cf^{3+}	β_1 1.36, β_2 2.07 [2.0]
Mn^{2+}	β_1 2.26 [0]
Mn^{3+}	β_1 1.20 [2.7][d]; 1.63 [4.0]; 1.90 [5.0]
Fe^{2+}	β_1 2.2 [0]
Fe^{3+}	β_1 4.04 [0]; 2.24 [0.5]; 2.02 [1.0]; 1.93 [3.0]; β_2 5.38 [0]; 2.11 [3.0]
Co^{2+}	β_1 2.36 [0]; 0.23 [3.0]
Ni^{2+}	β_1 2.32 [0]; 0.57 [1.0]; 0.26 [3.0]; β_2 1.42 [1.0]
Ru^{2+}	β_1 2.72 [0][b]; 1.88 [0.5][b]; 1.70 [1.0][b]; 1.30 [2.0][b]
Ru^{3+}	β_1 2.04, β_2 3.57 [2.0]
Cu^{2+}	β_1 2.36 [0], 0.95 [1.0]; 0.70 [3.0]
Ag^+	β_1 1.3 [0]; 0.31 [2.0]; 0.23 [3.0]; β_2 0.19 [2.0]; 0.00 [3.0]; β_3 0.40 [2.0]
Zn^{2+}	β_1 2.38 [0]; 0.93 [0.5]; 0.89 [1.0]; 0.76 [2.0]; 0.70 [3.0]; β_2 1.9 [0.5]; 1.2 [1.0]; 1.2 [2.0]; 0.7 [3.0]; β_3 1.7 [1.0]; (1.1) [2.0]; 0.9 [3.0]; β_4 1.7 [1.0]; 1.4 [2.0]; 0.9 [3.0]
Cd^{2+}	β_1 2.46 [0]; 1.08 [0.5]; 0.95 [1.0]; 0.86 [2.0]; 0.71 [3.0]; β_2 2.0 [0.5]; 1.6 [1.0]; 1.3 [2.0]; 0.9 [3.0]; β_3 2.7 [0.5]; 1.8 [1.0]; 1.6 [2.0]; 1.0 [3.0]; β_4 (2.3) [1.0]; 1.5 [2.0]; 1.0 [3.0]
Hg_2^{2+}	β_1 1.30, β_2 3.54 [0.5]
Hg^{2+}	β_1 1.34, β_2 2.4 [0.5]
In^{3+}	β_1 1.85 [1.0][b]; 1.78 [2.0][b]; β_2 2.60 [1.0][b]; 1.88 [2.0][b]; β_3 3.00 [1.0][b]; 2.36 [2.0][b]

Table 6.6 (Continued)

Metal ion	Equilibrium constants (Symbol for the constant or scheme of the equilibrium, value of log K or log β, [I])
Tl^+	β_1 1.37 [0]
Tl^{3+}	β_1 2.27 [3.0]
Pb^{2+}	β_1 2.75 [0]; (0.74), β_2 1.99 [3.0]
Bi^{3+}	β_1 1.98, β_2 3.41, β_3 4.08, β_4 4.34, β_5 4.60 [3.0]

[a] 18°C; [b] 20°C; [c] 48°C; [d] 23°C

Table 6.16
Stability constants of thiosulphato-complexes (L = $S_2O_3^{2-}$)

Metal ion	Equilibrium constants (Symbol for the constant or scheme of the equilibrium, value of log K or log β, [I])	Ref.
Na^+	β_1 0.53 [0]; 0.04 [0.5]	
K^+	β_1 0.96 [0]; 0.28 [0.5]	
Mg^{2+}	β_1 1.82 [0]; 0.56 [0.5]	
Ca^{2+}	β_1 1.98 [0]; 0.70 [0.5]	
Sr^{2+}	β_1 2.04 [0]	
Ba^{2+}	β_1 2.27 [0]	
La^{3+}	β_1 2.99 [0]; 0.8 [1.0]	
$M^{3+}(4f)$	β_1 2.82 [0][a]; β_{32} 10.5 (Ce, Pr, Nd, Sm) [b][c]	[1, 2]
Cr(VI)	$H_2MO_4L/HMO_4 \cdot HL$ 4.09 [0.11][d]	[3]
UO_2^{2+}	β_1 2.04 [b]	[4]
Mn^{2+}	β_1 1.95 [0]; 0.67 [0.5]	
Fe^{2+}	β_1 2.0 [0]; 2.17 [0][e]; 0.92 [0.48][e]	[5, 6]
Fe^{3+}	β_1 1.98 [0.1]; 1.18 [1.0][f]	
Co^{2+}	β_1 2.05 [0]; 0.77 [0.5]	
Ni^{2+}	β_1 2.06 [0]; 0.78 [0.5]	
Cu^+	β_1 10.35, β_2 12.27, β_3 13.71 [1.6]	
Cu^{2+}	β_2 12.29 [b]	[7]
Ag^+	β_1 8.82 [0][d]; 7.36 [4.0]; β_2 13.67 [0][d]; 12.63 [1.0]; 12.78 [2.0]; 12.72 [4.0]; β_3 14.2 [0][d]; 12.8 [1.0]; 13.1 [2.0]; 13.5 [4.0]; β_{42} 26.3, β_{53} 39.8, β_{86} 78.6 [4.0]	
Zn^{2+}	β_1 2.35 [0]; (1.12) [0.5]; (0.62) [1.0]; 0.96 [3.0]; β_2 1.94, β_3 3.3, β_{22} 5.84 [3.0]	
Cd^{2+}	β_1 3.92 [0]; 3.2 [0.1][g]; 2.82 [1.0]; 2.74 [3.0]; β_2 6.3 [0]; 4.57 [1.0]; 4.70 [3.0]; β_3 6.4 [1.0]; 6.9, β_4 7.1, β_{22} 11.18 [3.0]	
Hg^{2+}	β_2 29.23 [0]; (29.93) [1.0]; β_3 30.6 [0]; (33.26) [1.0]	
Tl^+	β_1 0,86, β_2 0.72, β_3 0.2, β_{42} 2.54 [4.0]	
Pb^{2+}	β_1 2.42, β_2 4.86, β_3 6.2, β_4 6.2 [3.0]	

[a] Temperature not given; [b] varied I; [c] 32°C; [d] 20°C; [e] 6°C; [f] 18°C; [g] 30°C

REFERENCES

[1] L. G. Sillén and A. E. Martell, *Stability Constants of Metal-Ion Complexes*, Supplement No. 1, p. 131, The Chemical Society, London, 1971.
[2] N. K. Dutt, *J. Indian Chem. Soc.*, 1950, **27**, 191.
[3] I. Baldea and G. Niac, *Inorg. Chem.*, 1968, **7**, 1232.
[4] S. L. Melton and E. S. Amis, *Anal. Chem.*, 1963, **35**, 1626.
[5] F. M. Page, *Trans. Faraday Soc.*, 1954, **50**, 120.
[6] K. B. Yatsimirskii, *Zh. Obshch. Khim.*, 1954, **24**, 1498.
[7] A. I. Levin and G. A. Ukshe, *Sb. Stat. Obshch. Khim.*, 1953, **2**, 798.

Table 6.17
Stability constants of fluoro-complexes (L = F^-)

Metal ion	Equilibrium constants (Symbol for the constant or scheme of the equilibrium, value of $\log K$ or $\log \beta$, $[I]$)	Ref.
Be^{2+}	β_1 5.99 [0]; 4.71 [0.5]; 4.99 [1.0]; 4.99 [2.0]^a; β_2 8.32 [0.5]; 8.80 [1.0]; 8.78 [2.0]^a; β_3 11.1 [0.5]; 11.7 [2.0]^a; β_4 13.1 [1.0]; 13.4 [2.0]^a	[1]
Mg^{2+}	β_1 1.8 [0]; 1.46 [0.1]; 1.31 [0.5]; 1.32 [1.0]	
Ca^{2+}	β_1 1.1 [0]; 0.6 [0.5]; 0.58 [1.0]	
Sr^{2+}	β_1 0.1 [1.0]	
Ba^{2+}	β_1 −0.3 [1.0]	
Sc^{3+}	β_1 7.1 [0]; 6.18 [0.5]; β_2 12.9 [0]; 11.46 [0.5]; β_3 17.4 [0]; 15.5 [0.5]; β_4 20.3 [0]; 18.4 [0.5]	
Y^{3+}	β_1 4.8 [0]; 3.90 [0.5]; 3.60 [1.0]; β_2 8.5 [0]; 7.13 [0.5]; β_3 12.1 [0]; 10.3 [0.5]	
La^{3+}	β_1 3.6 [0]; 2.69 [0.5]; 2.67 [1.0]	
Ce^{3+}	β_1 4.0 [0]; 3.13 [0.5]; 2.81 [1.0]	
$M^{3+}(4f)$	β_1 (3.4) [1.0]	
TiO^{2+}	β_1 6.65 [0.2]^b; 5.4 [3.0]; β_2 11.74 [0.2]^b; 9.7 [3.0]; β_3 16.32 [0.2]^b; 13.7 [3.0]; β_4 20.38 [0.2]; 17.4 [3.0]	[2, 3]
Zr^{4+}	β_1 9.8 [0]; 8.94 [2.0]; 9.4 [4.0]^a; β_2 16.4 [2.0]; 17.2 [4.0]^a; β_3 22.4 [2.0]; 23.7, β_4 29.5, β_5 23.5, β_6 28.3 [4.0]^a	
Hf^{4+}	β_1 9.0, β_2 16.5, β_3 23.1, β_4 28.8, β_5 34.0, β_6 38.0 [4.0]^a	
Th^{4+}	β_1 8.44 [0]; 7.59 [0.5]; 7.80 [3.0]; 8.2 [4.0]^a; β_2 15.08 [0]; 13.44 [0.5]; 13.82 [3.0]; 14.7 [4.0]^a; β_3 19.8 [0]; 17.9 [0.5]; 18.8 [3.0]; β_4 23.2 [0]	
VO^{2+}	β_1 3.37, β_2 5.74, β_3 7.29, β_4 8.1 [1.0]	
VO_2^+	β_1 3.04, β_2 5.60, β_3 6.9, β_4 7.0 [1.0]^a	
Nb(V)	$M(OH)_3L_2/M(OH)_3L_{2(s)}$ −4.82 [3.0]^c; $M(OH)_4L/M(OH)_4L_{(s)}$ −5.22 [0.5]^c; $M(OH)_2L_3/M(OH)_3L_2 \cdot HL$ 4.2 [3.0]^c; $M(OH)_4L_2/M(OH)_4L \cdot L$ 6.81 [0.5]^c; $M(OH)_2L_5/M(OH)_2L_4$ 2.51 [3.0]	[4]

Table 6.17 (Continued)

Metal ion	Equilibrium constants (Symbol for the constant or scheme of the equilibrium, value of log K or log β, [I])	Ref.
Ta(V)	$ML_4/ML_3 \cdot L$ 5.86 [3.0]; $ML_5/ML_4 \cdot L$ 4.8 [1.0]; 4.9 [3.0]; $ML_6/ML_5 \cdot L$ 3.7 [1.0]; 4.9 [3.0]; $ML_7/ML_6 \cdot L$ 3.2 [1.0]; 4.5 [3.0]; $ML_8/ML_7 \cdot L$ 3 [1.0]	[5–7]
Cr^{3+}	β_1 5.2 [0]; 4.36 [0.5]; β_2 7.70, β_3 10.2 [0.5]	
U^{4+}	β_1 9.0, β_2 15.7, β_3 21.2 [4.0]^a	
UO_2^{2+}	β_1 4.3 [0.5]; 4.54 [1.0]; 4.55 [2.0]; 5.0 [4.0]; β_2 7.97, β_3 10.55, β_4 12.0 [1.0]	
Np^{4+}	β_1 8.3, β_2 14.5, β_3 20.3, β_4 25.1 [4.0]^a	
NpO_2^{2+}	β_1 4.6 [0]; 4.12 [0.1]; 4.04 [0.5]; 3.85 [1.0]^a; β_2 7.01 [0.1]; 7.00 [0.5]; 6.97 [1.0]^a	
Pu^{4+}	β_1 6.77 [1.0]	
PuO_2^{2+}	β_1 5.07 [1.0]; 5.13 [2.0]; β_2 10.07 [1.0]; 10.08 [2.0]; β_3 14.96 [1.0]; 14.91 [2.0]; β_4 18.14 [1.0]; 19.2 [2.0]	[8]
Mn^{2+}	β_1 0.7 [1.0]	
Mn^{3+}	β_1 5.65 [2.0]	
Fe^{2+}	β_1 0.8 [1.0]	
Fe^{3+}	β_1 6.0 [0]; 5.18 [0.5]; 5.18 [1.0]; β_2 9.13 [0.5]; 9.07 [1.0]; β_3 11.9 [0.5]; 12.1 [1.0]	
Co^{2+}	β_1 0.4 [1.0]	
Ni^{2+}	β_1 0.5 [1.0]	
Cu^{2+}	β_1 1.2 [0]; 0.7 [0.5]; 0.9 [1.0]	
Ag^+	β_1 0.4 [0]; −0.17 [0.5]; −0.32 [1.0]	
Zn^{2+}	β_1 1.15 [0]; 0.73 [0.5]; 0.78 [1.0]; 0.8 [3.0]	
Cd^{2+}	β_1 0.46 [1.0]; 0.57 [3.0]; β_2 0.53 [1.0]; 0.85 [3.0]	
Hg^{2+}	β_1 1.6 [0]; 1.03 [0.5]	
B(III)	$M(OH)_3L/M(OH)_3 \cdot L$ −0.30, $M(OH)_2L_2/M(OH)_3 \cdot L^2 \cdot H$ 7.52, $MOHL_3/M(OH)_3 \cdot L^3 \cdot H^2$ 13.38, $ML_4/M(OH)_3 \cdot L^4 \cdot H^3$ 19.77 [1.0]	
Al^{3+}	β_1 7.0 [0]; 6.43 [0.1]; 6.11 [0.5]; 6.09 [1.0]; β_2 12.6 [0]; 11.63 [0.1]; 11.12 [0.5]; β_3 16.7 [0]; 15.5 [0.1]; 15.0 [0.5]; β_4 19.1 [0]; 18.3 [0.1]; 18.0 [0.5]; β_5 19.4; β_6 19.8 [0.5]	
Ga^{3+}	β_1 5.9 [0]; 4.49 [0.5]; 4.38 [1.0]; β_2 8.00, β_3 10.5 [0.5]	
In^{3+}	β_1 4.6 [0]; 3.75 [0.5]; 3.70 [1.0]; 3.74 [2.0]; β_2 8.1 [0]; 6.5 [0.5]; 6.4 [1.0]; 6.6 [2.0]; β_3 10.3 [0]; 8.6 [0.5]; 8.6 [1.0]; 9.0 [2.0]; β_4 11.5 [0]; 9.9 [0.5]; 9.8 [1.0]; 10.3 [2.0]	
Tl^+	β_1 0.10 [0]	
Ge(IV)	$ML_4/M(OH)_4 \cdot (HL)^4$ 7.30, $MHL_5/M(OH)_4 \cdot (HL)^5$ 8.94 [0.5], $ML_6/MO_{2(s)} \cdot L^6 \cdot H^4$ −25.87 [^{d}]	[9]
Sn^{2+}	β_1 4.08, β_2 6.68, β_3 9.5 [1.0]	
Sn^{4+}	β_6 (25) [^{d}]	[10]

Table 6.17 (Continued)

Metal ion	Equilibrium constants (Symbol for the constant or scheme of the equilibrium, value of log K or log β, [I])	Ref.
Pb^{2+}	β_1 1.44 [1.0]; 1.26 [2.0]; β_2 2.54 [1.0]; 2.55 [2.0]	
As(V)	$MO_3L/H_2MO_4 \cdot L$ -0.75 [1.0]^b	[11]
Sb^{3+}	β_1 3.0, β_2 5.7, β_3 8.3, β_4 10.9 [2.0]^e	
Bi^{3+}	β_1 1.42 [2.0]^e	

^{a} 20°C; ^{b} temperature not given; ^{c} 18°C; ^{d} varied I; ^{e} 30°C

REFERENCES

[1] K. P. Anderson, W. O. Greenhalgh and E. A. Butler, *Inorg. Chem.*, 1967, **6**, 1056.
[2] B. I. Nabivanets, *Ukr. Khim. Zh.*, 1966, **32**, 886.
[3] V. Caglioti, L. Ciavatta and A. Liberti, *J. Inorg. Nucl. Chem.*, 1960, **15**, 115.
[4] A. K. Babko, B. I. Nabivanets and V. V. Lukachina, *Zh. Neorgan. Khim.*, 1967, **12**, 2965.
[5] L. P. Varga, W. D. Wakley, L. S. Nicolson, M. L. Madden and J. Patterson, *Anal. Chem.*, 1965, **37**, 1003.
[6] M. N. Bukhsh, J. Flegenheimer, F. M. Hall, A. G. Maddock and C. Ferreira de Miranda, *J. Inorg. Nucl. Chem.*, 1966, **28**, 421.
[7] L. P. Varga and H. Ferund, *J. Phys. Chem.*, 1962, **66**, 21.
[8] V. N. Krylov, E. V. Komarov, M. F. Pushlenkov, *Radiokhimiya*, 1968, **10**, 717, 719, 723.
[9] I. G. Ryss and N. F. Kulish, *Zh. Neorgan. Khim.*, 1965, **10**, 1827.
[10] W. B. Schaap, J. A. Davies and W. H. Nebergall, *J. Am. Chem. Soc.*, 1954, **76**, 5226.
[11] N. K. Dutt and A. Gupta, *J. Indian Chem. Soc.*, 1961, **38**, 249.

Table 6.18
Stability constants of chloro-complexes (L = Cl^-)

Metal ion	Equilibrium constants (Symbol for the constant or scheme of the equilibrium, value of log K or log β, [I])	Ref.
Na^+	β_1 -1.85 [0]	[1]
M^+(IA)	β_1 (-0.5) [0] (K, Rb, Cs)	
Be^{2+}	β_1 -0.3 [0.7]^a; (-0.8); β_2 -0.7 [4.0]	
Mg^{2+}	β_1 -1.0 [3.0]	
Ba^{2+}	β_1 -0.13 [0]^b	
Sc^{3+}	β_1 0.04 [0.7]^a; -0.12 [4.0]; β_2 -0.1 [0.7]^a; -0.8 [4.0]	
La^{3+}	β_1 -0.1 [1.0]; -0.2, β_2 -0.6 [4.0]	
$M^{3+}(4f)$	β_1 (-0.2), β_2 (-0.7) [1.0] (Ce, Pr, Eu, Tm, Yb, Lu, Ac)	

Table 6.18 (Continued)

Metal ion	Equilibrium constants (Symbol for the constant or scheme of the equilibrium, value of log K or log β, [I])	Ref.
Ti^{3+}	β_1 0.34 [0.5]^c	
TiO^{2+}	β_1 0.55, β_2 0.15 [3.0]^d	[2]
Zr^{4+}	β_1 0.30 [2.0]; 0.92, β_2 1.32, β_3 1.51, β_4 1.2 [6.5 H^+]^a	
Hf^{4+}	β_1 0.38 [2.0]; 0.34 [3.0 H^+]^a; β_2 0.07 [2.0]; -0.02 [3.0 H^+]^a; β_3 -0.6 [2.0]	
Th^{4+}	β_1 1.38 [0]; 0.30 [0.5]; 0.18 [1.0]; 0.08 [2.0]; 0.17 [4.0]; β_2 -1.0 [2.0]; -0.9, β_3 -0.9, β_4 -1.5, β_5 -2.5 [4.0]	[3]
VO^{2+}	β_1 0.04 [1.0]^a	
Cr^{3+}	β_1 -0.5 [1.0]; -0.4 [2.0]; 0.05 [4.0]	
Cr(VI)	$MO_3L/MO_4 \cdot L \cdot H$ 1.1 [1.0]	[4]
U^{3+}	β_1 -2.89 [0]	[5]
U^{4+}	β_1 0.26 [2.0]; 0.30 [1.0]^a	
UO_2^{2+}	β_1 0.21 [0]; -0.10 [1.0]^a; -0.06 [2.0]	
Np^{4+}	β_1 0.15 [0.5]^a; -0.04 [1.0]^a; 0.04 [2.0]^a; β_2 -0.15 [0.5]^a; -0.24 [1.0]^a; β_3 -0.5 [1.0]^a	
NpO_2^{2+}	β_1 -0.3 [0.5]; -0.09, β_2 -0.8 [3.0]^e	
Pu^{4+}	β_1 0.14 [1.0]; 0.15 [4.0]^a; β_2 -0.17 [1.0]; 0.08, β_3 -1.0 [4.0]^a	
PuO_2^{2+}	β_1 0.10, β_2 -0.35 [2.0]	
Pu^{3+}	β_1 -0.1 [1.0]	
Am^{3+}	β_1 -0.60 [1.0]^f; -0.15 [4.0]; β_2 -0.42 [1.0]^f; -0.69 [4.0]	[6–8]
Bk^{3+}, Es^{3+}	β_1 -0.02 [1.0]	
Mn^{2+}	β_1 0.04 [1.0]	
Mn^{3+}	β_1 0.9 [2.0]; 1.12 [3.0]	
Co^{2+}	β_1 -0.05 [1.0]; -0.14 [2.0]; -0.26 [3.0]	
Ni^{2+}	β_1 0.00 [1.0]; -0.21 [2.0]; -0.57 [3.0]	
Fe^{2+}	β_1 -0.3 [1.0]; 0.36, β_2 0.4 [2.0]^a	[9, 10]
Fe^{3+}	β_1 1.48 [0]; 0.64 [0.5]; 0.63 [1.0]; 0.7 [2.0]; 0.8 [3.0]; 1.0 [4.0]; β_2 2.13 [0]; 0.75 [1.0]; β_3 1.1 [0]; -0.7 [1.0]; β_4 -1.3 [0]	[11, 12]
Ru^{3+}	K_2 1.4, K_3 0.4 [0.1]	[13]
Ru^{4+}	$MOHL_5/MO_{2(s)} \cdot L^5 \cdot H^3$ 12.65	[14]
Rh^{3+}	β_1 2.45, β_2 4.54, β_3 5.92, β_4 7.08, β_5 8.75, β_6 8.43 [1.0]	[15]
Pd^{2+}	β_1 6.1 [0]; 4.47 [1.0]; β_2 10.7 [0]; 7.74 [1.0]; β_3 13.1 [0]; 10.2 [1.0]; β_4 15.4 [0]; 11.5 [1.0]	
Pt^{2+}	K_1 (7), K_2 4.0, K_3 2.96, K_4 1.90 [0.5]; β_4 (16) [1.0]; ML_2 (cis)/ML_2 (trans) 0.08 [0.5]; $M(OH)_3L/M(OH)_4 \cdot L \cdot H$ 10.5, $M(OH)_2L_2/M(OH)_3L \cdot L \cdot H$ 10.0, $MOHL/M(OH)_2L_2 \cdot L \cdot H$ 9.5, $ML_4/MOHL_3 \cdot L \cdot H$ 8.7 [1.0]	[16, 17]
Cu^+	β_1 2.70 [5.0]; β_2 5.5 [0]^a; 5.19 [0.1]^a; 5.84 [10.0]^g, 6.00 [5.0]; β_3 5.7 [0]^a; 6.0 [5.0]; 5.45 [10.0]^g; β_{32} 12.08 [10.0]^g; β_{42} 13.1 [5.0]; 11.84 [10.0]^g; β_{52} 11.60, β_{62} 11.30, β_{43} 18.18, β_{53} 18.04, β_{63} 17.70, β_{54} 24.48, β_{64} 24.25 [10.0]^g	[18]

Stability Constants of Complexes with Inorganic Ligands

Table 6.18 (Continued)

Metal ion	Equilibrium constants (Symbol for the constant or scheme of the equilibrium, value of $\log K$ or $\log \beta$, $[I]$)	Ref.
Cu^{2+}	β_1 0.40 [0]; 0.09 [2.0]; -0.06 [3.0]; β_2 -0.4 [0]; -0.5 [1.0]	[19, 20]
Ag^+	β_1 3.31 [0]; 3.08 [0.1]; 3.36 [1.0]; 3.45 [4.0][a]; 3.70 [5.0]; β_2 5.25 [0]; 5.08 [0.1]; 5.20 [1.0]; 5.67 [4.0][a]; 5.62 [5.0]; β_3 6.0 [4.0][a], 6.1 [5.0]	
Au^{3+}	$M(OH)_3L/M(OH)_4 . H . L$ 8.51, $M(OH)_2L_2/M(OH)_4 . H^2 . L^2$ 16.57, $MOHL_3/M(OH)_4 . H^3 . L^3$ 23.6, $ML_4/M(OH)_4 . H^4 . L^4$ 29.6 [0][a]	
Zn^{2+}	β_1 0.43 [0]; 0.11 [1.0]; -0.49 [2.0]; -0.19 [3.0]; 0.30 [4.0]; β_2 0.61 [0]; 0.02 [2.0]; -0.6 [3.0]; 0.0 [4.0]; β_3 0.5 [0]; -0.1 [2.0]; 0.1 [3.0]; 1.0 [4.0]; β_4 0.2 [0]; -1 [4.0]	
Cd^{2+}	β_1 1.98 [0]; 1.35 [0.5]; 1.35 [1.0]; 1.44 [2.0]; 1.54 [3.0]; 1.66 [4.0]; β_2 2.6 [0]; 1.7 [0.5]; 1.7 [1.0]; 1.9 [2.0]; 2.2 [3.0]; 2.4 [4.0]; β_3 2.4 [0]; 1.5 [1.0]; 1.9 [2.0]; 2.3 [3.0]; 2.8 [4.0]; β_4 1.7 [0]; 1.6 [3.0]; 2.2 [4.0]	
Hg^{2+}	β_1 6.74 [0.5]; 6.72 [1.0]; 7.07 [3.0]; β_2 13.22 [0.5]; 13.23 [1.0]; 13.98 [3.0]; β_3 14.1 [0.5]; 14.2 [1.0]; 14.7 [3.0]; β_4 15.1 [0.5]; 15.3 [1.0]; 16.2 [3.0]; $MOHL/ML . OH$ 10.7 [1.0]; $MOHL . L/ML_2 . OH$ 4.09, $MOHL . OH/M(OH)_2 . L$ -3.77 [0]; -3.8 [1.0]	
Ga^{3+}	β_1 0.01 [0.7][a]	
In^{3+}	β_1 2.32 [0.7][a]; 2.20 [1.0][a]; 2.45 [2.0]; 2.58 [3.0]; β_2 3.62 [0.7][a]; 3.56 [1.0][a]; 3.4 [2.0]; 3.84 [3.0]; β_3 4.0 [0.7][a]; 3.7 [2.0]; 4.2 [3.0]; $MOHL/ML . OH$ 10.3; $M_2OHL/MOHL . M$ 1.6 [3.0]	
Tl^+	β_1 0.49 [0]; 0.11 [0.5]; 0.04 [1.0]; -0.10 [2.0]; -0.1 [3.0]; 0.0 [4.0]; β_2 0.0 [0]; -0.1 [1.0]; -0.6 [2.0]; -1.0 [3.0]; -0.8 [4.0]	
Tl^{3+}	β_1 7.72 [0]; 6.72 [0.5]; 7.10 [3.0]; 7.46 [4.0]; β_2 13.48 [0]; 11.76 [0.5]; 12.46 [3.0]; 13.25 [4.0]; β_3 16.5 [0]; 14.4 [0.5]; 15.8 [3.0]; 16.7 [4.0]; β_4 18.3 [0]; 16.3 [0.5]; 18.0 [3.0]; 19.4 [4.0]; $MOHL/ML . OH$ 12.31 [3.0]	
Sn^{2+}	β_1 1.51 [0]; 1.08 [0.5]; 1.17 [3.0]; 14.5 [4.0]; β_2 2.25 [0]; 1.72 [2.0]; 1.72 [3.0]; 2.35 [4.0]; β_3 2.0 [0]; 1.5 [2.0]; 1.7 [3.0]; 2.5 [4.0]; β_4 1.5 [0]; 2.3 [4.0]	
Pb^{2+}	β_1 1.59 [0]; 0.9 [0.5]; 0.90 [1.0]; 1.02 [2.0]; 1.17 [3.0]; 1.29 [4.0]; β_2 1.8 [0]; 1.3 [0.5]; 1.3 [1.0]; 1.4 [2.0]; 1.7 [3.0]; 2.0 [4.0]; β_3 1.7 [0]; 1.4 [1.0]; 1.5 [2.0]; 1.9 [3.0]; 2.3 [4.0]; β_4 1.4 [0]; 0.7 [2.0]; 1.2 [3.0]; 1.7 [4.0]	
As(III)	$M(OH)_2L/M(OH)_3 . H . L$ -1.07, $MOHL_2/M(OH)_3 . H^2 . L^2$ -4.54, $ML_3/M(OH)_3 . H^3 . L^3$ -8.7 [0]	
Sb^{3+}	β_1 2.3, β_2 3.5, β_3 4.2, β_4 4.7, β_5 4.7, β_6 4.1 [4.0]	

Table 6.18 (Continued)

Metal ion	Equilibrium constants (Symbol for the constant or scheme of the equilibrium, value of log K or log β, $[I]$)	Ref.
Bi^{3+}	β_1 2.36 $[2.0]^a$; 2.2 $[3.0]$; β_2 3.5 $[2.0]^a$; 3.5 $[3.0]$; β_3 5.4 $[2.0]^a$; 5.8 $[3.0]$; β_4 6.1 $[2.0]^a$; 6.8 $[3.0]$; β_5 6.7 $[2.0]^a$; 7.3 $[3.0]$; β_6 6.6 $[2.0]^a$; 7.4 $[3.0]$	

[a] 20°C; [b] 18°C; [c] 40°C; [d] temperature not given; [e] 10°C; [f] 26°C; [g] 50°C

REFERENCES

[1] W. N. Latimer, P. W. Shutz and J. F. G. Hicks, *J. Chem. Phys.*, 1934, **2**, 82.
[2] B. I. Nabivanets and L. N. Kudritskaya, *Zh. Neorgan. Khim.*, 1967, **12**, 1500.
[3] B. I. Nabivanets and L. N. Kudritskaya, *Ukr. Khim. Zh.*, 1964, **30**, 1007.
[4] J. Y. Tong and R. L. Johnson, *Inorg. Chem.*, 1965, **5**, 1902.
[5] M. Shiloh and Y. Marcus, *Isr. J. Chem.*, 1965, **3**, 123.
[6] B. M. L. Bansal, S. K. Patil and H. D. Sharma, *J. Inorg. Nucl. Chem.*, 1964, **26**, 993.
[7] L. G. Sillén and A. E. Martell, *Stability Constants of Metal-Ion Complexes*, Supplement No. 1, p. 169, The Chemical Society, London, 1971.
[8] T. Sekine, *Acta Chem. Scand.*, 1965, **19**, 1435.
[9] L. G. Sillén and A. E. Martell, *Stability Constants of Metal-Ion Complexes*, p. 280, The Chemical Society, London, 1964.
[10] H. N. Po and N. Sutin, *Inorg. Chem.*, 1968, **7**, 621.
[11] E. Rabinowitch and W. H. Stockmayer, *J. Am. Chem. Soc.*, 1942, **64**, 335.
[12] C. Dragulescu and R. Pomoje, *Rev. Roum. Chim.*, 1967, **12**, 37.
[13] R. E. Conninck and D. A. Fine, *J. Am. Chem. Soc.*, 1961, **83**, 3414.
[14] Cf. ref. [9], p. 283.
[15] W. C. Wolsey, C. A. Reynolds and J. Kleinberg, *Inorg. Chem.*, 1963, **2**, 463.
[16] A. A. Grinberg and M. I. Gel'fman, *Dokl. Akad. Nauk SSSR*, 1960, **133**, 1081.
[17] B. I. Peshkevitskii, B. V. Ptitsyn and N. M. Leskova, *Izv. Sib. Otd. Akad. Nauk SSSR, Ser. Khim. Nauk*, **1962**, No. 11, 143.
[18] T. G. Sukhova, O. N. Temkin and R. M. Flid, *Zh. Neorgan. Khim.*, 1969, **14**, 928.
[19] H. McConnell and N. Davidson, *J. Am. Chem. Soc.*, 1950, **72**, 3164, 3168.
[20] J. Bjerrum, *Kgl. Danske Videnskab. Selskab., Mat.-Fys. Medd.*, 1946, **22**, No. 18.

Table 6.19
Stability constants of bromo-complexes (L = Br$^-$)

Metal ion	Equilibrium constants (Symbol for the constant or scheme of the equilibrium, value of log K or log β, [I])	Ref.
Na$^+$	β_1 −3.1 [0]	[1]
M$^+$(IA)	β_1 (−0.6) (K, Rb, Cs) [4.0]	[2]
Be^{2+}	β_1 −0.4 [0.7]^a; −0.7, β_2 −0.8 [4.0]	
Mg^{2+}	β_1 −1.4 [3.0]	
Sc^{3+}	β_1 −0.07, β_2 −0.3 [0.7]^a	
Y^{3+}	β_1 −0.15 [1.0]	
M^{3+}(4f)	β_1 (−0.3) (Ce, Pr, Sm, Eu, Ho, Er) [1.0–3.0]	
Hf^{4+}	β_1 −0.1 [3.0]^a	
U^{3+}	β_1 −3.95 [0]	[3]
U^{4+}	β_1 0.18 [1.0]^a	
UO$_2^{2+}$	β_1 0.2 [0]; −0.3 [1.0]^a	
M^{3+}(5f)	β_1 (−3.4), β_2 (−6.5) (Np, Pu, Am) [^{b}]^c	[4]
Mn^{2+}	β_1 0.27, β_2 0.01 [0.7]^a	[5]
Fe^{3+}	β_1 0.6 [0]; −0.2 [1.0]; β_2 −0.5 [1.0]	
Co^{2+}	β_1 −0.13 [0.7]; −0.11 [2.0]; −0.7 [3.0]; β_2 −0.4 [0.7]	
Ni^{2+}	β_1 −0.12 [2.0]; −0.8 [3.0]	
Pd^{2+}	β_1 5.17, β_2 9.42, β_3 12.7, β_4 14.9 [1.0]	
Pt^{2+}	M(OH)$_3$L/M(OH)$_4$.L.H 10.75, M(OH)$_2$L$_2$/M(OH)$_3$L.L.H 10.3, MOHL$_3$/M(OH)$_2$L$_2$.L.H 9.6, ML$_4$/MOHL$_3$.L.H 7.75 [0.1]^d	[6]
Pt^{4+}	M(OH)$_5$L/M(OH)$_6$.L.H 9.55, M(OH)$_4$L$_2$/M(OH)$_5$L.L.H 9.5, M(OH)$_3$L$_3$/M(OH)$_4$L$_2$.L.H 9.3, M(OH)$_2$L$_4$/M(OH)$_3$L$_3$.L.H 9.0, MOHL$_5$/M(OH)$_2$L$_4$.L.H 8.9, ML$_6$/MOHL$_5$.L.H 8.5 [0.1]^d	[7]
Cu$^+$	β_2 5.9 [0]	
Cu^{2+}	β_1 −0.03 [0]; −0.07 [2.0]; −0.5 [3.0]	
Ag$^+$	β_1 4.68 [0]^e; 4.30 [0.1]; β_2 7.7 [0]^e; 6.64 [0.1]; 7.23 [5.0]; β_3 8.7 [0]; 8.1 [0.1]; 8.3 [1.0]; 9.2 [5.0]; β_4 9.0 [0]; 8.9 [0.1]; 9.5 [5.0]	
Zn^{2+}	β_1 −0.19, β_2 −1.15, β_3 −0.55, β_4 −1.5 [3.0]	[8]
Cd^{2+}	β_1 2.14 [0]; 1.55 [0.5]; 1.57 [1.0]; 1.63 [2.0]; 1.74 [3.0]; β_2 3.0 [0]; 2.1 [1.0]; 2.2 [2.0]; 2.4 [3.0]; β_3 3.0 [0]; 2.6 [1.0]; 2.8 [2.0]; 3.3 [3.0]; β_4 2.9 [0]; 2.6 [1.0]; 3.2 [2.0]; 3.8 [3.0]	
Hg^{2+}	β_1 9.00 [0.5]; 9.40 [3.0]; β_2 17.1 [0.5]; 17.98 [3.0]; β_3 19.4 [0.5]; 20.7 [3.0]; β_4 21.0 [0.5]; 22.23 [3.0]	
Ga^{3+}	β_1 −0.10 [0.7]^a	
In^{3+}	β_1 1.93 [1.0]; 2.04 [0.7]^a; 1.99 [2.0]; 2.08 [4.0]; β_2 3.1 [0.7]; 2.6 [1.0]; 2.6 [2.0]; 3.4 [4.0]; β_3 3.4 [0.7]; 4.0 [4.0]; β_3 4.8 [4.0]	
Tl$^+$	β_1 0.91 [0]; 0.48 [0.5]; 0.41 [1.0]; 0.38 [2.0]; 0.34 [3.0]; 0.33, β_2 0.17, β_3 −0.1, β_4 −0.6 [4.0]	
Tl^{3+}	β_1 9.7 [0]^e; 8.3 [0.4]^a; 8.9 [1.0]; 9.28 [3.0]; 9.51 [4.0];	

Table 6.19 (Continued)

Metal ion	Equilibrium constants (Symbol for the constant or scheme of the equilibrium, value of log K or log β, $[I]$)	Ref.
	Tl^{3+} (continued)	
	β_2 16.6 [0]^a; 14.6 [0.4]^a; 16.4 [1.0]; 16.70 [3.0]; 16.88 [4.0];	
	β_3 21.2 [0]^a; 19.2 [0.4]^a; 22.1 [3.0]; 22.3 [4.0];	
	β_4 23.9 [0]^a; 22.3 [0.4]^a; 25.7 [3.0]; 26.4 [4.0]	
Sn^{2+}	β_1 1.16 [0]; 0.74 [1.0]; 0.50 [2.0]; 0.75 [3.0]; 0.88 [4.0];	
	β_2 1.7 [0]; 0.9 [1.0]; 1.19 [2.0]; 1.15 [3.0]; 1.43 [4.0];	
	β_3 1.2 [3.0]; 1.5 [4.0]; β_4 0.4 [3.0]; 1.0 [4.0]	
Pb^{2+}	β_1 1.77 [0]; 1.06 [0.5]; 1.10 [1.0]; 1.20 [2.0]; 1.29 [3.0]; 1.48 [4.0]	
	β_2 2.6 [0]; 1.8 [0.5]; 1.8 [1.0]; 2.0 [2.0]; 2.2 [3.0]; 2.5 [4.0]	
	β_3 3.0 [0]; 2.0 [0.5]; 2.2 [1.0]; 2.5 [2.0]; 2.9 [3.0]; 3.5 [4.0];	
	β_4 2.3 [0]; 2.0 [1.0]; 2.6 [2.0]; 3.1 [3.0]; 3.5 [4.0];	
	β_5 1.6 [2.0]; 2.4 [3.0]; 2.7 [4.0]	
Bi^{3+}	β_1 3.06 [0]; 2.37 [0.5]; 2.22 [1.0]; 2.32 [2.0]; 2.63 [3.0]; 3.12 [4.0];	
	β_2 5.6 [0]; 4.2 [0.5]; 4.4 [1.0]; 4.4 [2.0]; 5.0 [3.0]; 5.7 [4.0];	
	β_3 7.4 [0]; 5.9 [0.5]; 6.2 [1.0]; 6.3 [2.0]; 6.7 [3.0]; 8.2 [4.0];	
	β_4 8.6 [0]; 7.3 [0.5]; 7.2 [1.0]; 7.8 [2.0]; 8.1 [3.0]; 10.0 [4.0];	
	β_5 9.2 [0]; 8.2 [0.5]; 8.7 [1.0]; 9.2 [2.0]; 9.0 [3.0]; 11.9 [4.0];	
	β_6 8.7 [0]; 8.3 [0.5]; 8.8 [1.0]; 9.5 [2.0]; 9.8 [3.0]; 11.8 [4.0]	

^{a} 20°C; ^{b} varied I; ^{c} temperature not given; ^{d} for log K_w = -13.78; ^{e} 18°C

REFERENCES

[1] W. M. Latimer, *Chem. Rev.*, 1936, **18**, 349.
[2] V. E. Mironov, F. Ya. Kul'ba, A. V. Fokina, V. S. Golubeva and V. A. Nazarov, *Zh. Neorgan. Khim.*, 1964, **9**, 2133.
[3] M. Shiloh and Y. Marcus, *Isr. J. Chem.*, 1965, **3**, 123.
[4] M. Shiloh and Y. Marcus, *J. Inorg. Nucl. Chem.*, 1966, **28**, 2725.
[5] J. R. Fryer and D. F. C. Morris, *Talanta*, 1968, **15**, 1309.
[6] J. E. Teggins, D. R. Gano, M. A. Tucker and D. S. Martin, Jr., *Inorg. Chem.*, 1967, **6**, 69.
[7] N. M. Nikolaeva and E. D. Pastukhova, *Zh. Neorgan. Khim.*, 1967, **12**, 1227.
[8] V. E. Mironov, Yu. I. Rutkovskii and E. I. Ignatenko, *Zh. Neorgan. Khim.*, 1965, **10**, 2339.

Table 6.20
Stability constants of iodo-complexes (L = I$^-$)

Metal ion	Equilibrium constants (Symbol for the constant or scheme of the equilibrium, value of log K or log β, $[I]$)	Ref.
K$^+$	β_1 -0.19 [0]	
Hf^{4+}	β_1 -0.5 [3.0]^a	
U^{4+}	β_1 0.18 [2.0]^b	[1]
Fe^{3+}	β_1 2.85, β_2 1.57 [0.1]	[2]
Pd^{2+}	β_4 24.5 [1.0]	
Pt^{2+}	M(OH)$_3$L/M(OH)$_4$.L.H 12, M(OH)$_2$L$_2$/M(OH)$_3$L.L.H 11.7, M(OH)L$_3$/M(OH)$_2$L$_2$.L.H 11, ML$_4$/M(OH)L$_3$.L.H 10 [^{c}]	[3]
Pt^{4+}	M(OH)$_5$L/M(OH)$_6$.L.H 12.2, M(OH)$_4$L$_2$/M(OH)$_5$L.L.H 12.0, M(OH)$_3$L$_3$/M(OH)$_4$L$_2$.L.H 11.9, M(OH)$_2$L$_4$/M(OH)$_3$L$_3$.L.H 11.8, M(OH)L$_5$/M(OH)$_2$L$_4$.L.H 11.4, ML$_6$/M(OH)L$_5$.L.H 10.4 [0.1]	[4]
Cu$^+$	β_2 8.9 [0]; β_3 9.4; β_4 9.7 [5.0]	
Ag$^+$	β_1 6.58 [0]^d; 8.1 [4.0]; β_2 11.7 [0]^d; 11.0 [4.0]; β_3 13.1 [0]^d; 13.6 [2.0]; 13.8 [4.0]; β_4 14.2 [2.0]; 14.3 [4.0]; β_{62} 29.7, β_{83} 46.4 [4.0]	
Zn^{2+}	β_1 -0.47, β_2 -2.00, β_3 -0.74, β_4 -1.25 [4.0]	[5]
Cd^{2+}	β_1 2.28 [0]; 1.86 [0.5]; 1.89 [1.0]; 1.99 [2.0]; 2.13 [3.0]; β_2 3.92 [0]; 3.2 [0.5]; 3.2 [1.0]; 3.4 [2.0]; 3.6 [3.0]; β_3 5.0 [0]; 4.4 [0.5]; 4.5 [1.0]; 4.8 [2.0]; 5.1 [3.0]; β_4 6.0 [0]; 5.5 [0.5]; 5.6 [1.0]; 6.1 [2.0]; 6.6 [3.0]	
Hg^{2+}	β_1 12.87, β_2 23.82, β_3 27.6, β_4 29.8, MOHL/ML.OH 9.74 [0.5]	
Sn^{2+}	β_1 0.70, β_2 11.13, β_3 2.1, β_4 2.3, β_6 2.6, β_8 2.1 [1.0]	
Pb^{2+}	β_1 1.26 [0]; 1.30 [2.0]; β_2 2.8 [0]; 2.4 [2.0]; β_3 3.4 [0]; 3.1 [2.0]; β_4 3.9 [0]; 4.4 [2.0]	
Ga^{3+}	β_1 -0.2 [0.7]^a	
In^{3+}	β_1 1.64 [0.7]^a; 0.99 [2.0]; β_2 2.56 [0.7]^a; 2.26 [2.0]	
Tl$^+$	β_1 0.74, β_2 0.90, β_3 1.06 [4.0]	
Tl^{3+}	β_4 35.7 [4.0]	
Bi^{3+}	β_1 3.63 [0.5]; β_4 15.0, β_5 16.8, β_6 18.8 [2.0]^a	

^{a} 20°C; ^{b} temperature not given; ^{c} I varied; ^{d} 18°C

REFERENCES

[1] L. G. Sillén and A. E. Martell, *Stability Constants of Metal-Ion Complexes*, p. 337. The Chemical Society, London 1964.
[2] R. I. Novoselov and B. V. Ptitsyn, *Zh. Neorgan. Khim.* 1965, **10**, 2282.
[3] N. M. Nikolaeva and E. D. Pastukhova, *Zh. Neorgan. Khim.*, 1967, **12**, 1510.
[4] N. M. Nikolaeva and E. D. Pastukhova, *Zh. Neorgan. Khim.*, 1967, **12**, 1227.
[5] V. E. Mironov, A. V. Fokina and Yu. I. Rutkovskii, *Zh. Neorgan. Khim.*, 1967, **12**, 2056.

CHAPTER 7

Stability constants of complexes with organic ligands

Only some of the more important complex-forming organic compounds are listed in the tables in this chapter. The selection from the wealth of published data on complexation systems of organic ligands and metal ions was based on consideration of possible applications of these constants. This choice of constants may be found of help in elucidating or optimizing various complexation reactions in analytical chemistry, namely those of masking or spectrophotometric reagents, titrants, metallochromic indicators, etc. In addition, these data may also be useful in biochemistry, medicine and various other fields of science and technology.

The arrangement of Tables 7.1–7.5 is similar to that of Chapter 5: the ligands are arranged according to their empirical formulae, the metals in sequence of the subgroups of the periodic system. To save space the protonation constants are also given for the ligands listed. The logarithmic values of both the protonation constants ($\log K_{Hn}$) and the stability constants ($\log \beta_{nm}$) refer to a temperature of 25°C (unless indicated otherwise by a footnote) and the ionic strength given in square brackets. If no reference to the literature is added, the data are quoted from the comprehensive tabulation of Critical Stability Constants [1–3].

REFERENCES

[1] A. E. Martell and R. M. Smith, *Critical Stability Constants*, Vol. 1, *Amino Acids*, Plenum Press, New York, 1974.
[2] R. M. Smith and A. E. Martell, *Critical Stability Constants*, Vol. 2, *Amines*, Plenum Press, New York, 1975.
[3] A. E. Martell and R. M. Smith, *Critical Stability Constants*, Vol. 3, *Other Organic Ligands*, Plenum Press, New York, 1976.

Table 7.1
Complexes of amines*

Formula Metal ion	Stability constants (Symbol for the constant or scheme of the equilibrium, value of log K or log β, [I])	Ref.
1. C_2H_7NO	2-aminoethanol $\quad H_2N.CH_2CH_2OH$ $\quad$ L	
H^+	K_{H1} 9.498 [0]; 9.52 [0.1]; 9.62 [0.5]	
Ni^{2+}	β_1 2.98 [0.1]; 3.06 [0.4]; β_2 5.33 [0.1]; 5.52 [0.4]; β_3 7.33 [0.1]; 6.95 [0.4]	
Cu^{2+}	β_1 5.7, β_2 9.8, β_3 13.0, β_4 15.2 [0.1]	
Ag^+	β_1 3.2 [0]; 3.13 [0.5]; β_2 6.76 [0]; 6.68 [0.5]	
Zn^{2+}	β_1 3.7, β_2 6.1, β_3 9.4 [0.1]	
Cd^{2+}	β_1 2.77, β_2 4.09, β_3 5.6 [0.1]	
Hg^{2+}	β_1 8.56 [0]; 8.51 [0.5]; β_2 17.33 [0]; 17.32 [0.5]	
Pb^{2+}	β_1 7.56 [0.1]	
2. $C_2H_8N_2$	ethylenediamine $\quad H_2N.CH_2CH_2.NH_2$ $\quad$ L en	
H^+	K_{H1} 9.928 [0]; 9.89 [0.1]; 10.18 [1.0]; K_{H2} 6.848 [0]; 7.08 [0.1]; 7.45 [1.0]	
Mg^{2+}	β_1 0.37 [1.4][a]	
Cr^{3+}	$ML_2/MOHL_2.H$ 4.86, $MOHL_2/M(OH)_2L_2.H$ 7.34 [0.1][b]	
Mn^{2+}	β_1 2.77, β_2 4.87, β_3 5.81 [1.4]	
Fe^{2+}	β_1 4.34, β_2 7.66, β_3 9.72 [1.4]	
Co^{2+}	β_1 (5.6) [0.1]; 5.96 [1.4]; β_2 10.5 [0.1]; 10.8 [1.4]; β_3 13.8 [0.1]; 14.1 [1.4]	
Co^{3+}	K_3 13.99 [1.0][a]; β_3 48.68 [1.4][a]; $ML_2/MOHL_2.H$ 5.80 [1.0]; $MOHL_2/M(OH)_2L_2.H$ 8.10 [1.0]	
Ni^{2+}	β_1 7.32 [1.0]; 7.35 [0.1]; 7.58 [1.0]; β_2 13.50 [0]; 13.54 [0.1]; 14.02 [1]; β_3 17.61 [0]; 17.71 [0.1]; 18.44 [1.0]	
Pd^{2+}	K_2 18.4 [1.0]	
Pt^{2+}	β_2 36.5 [1.0][d]	
Cu^+	β_2 11.2 [0.3]	
Cu^{2+}	β_1 10.48 [0]; 10.54 [0.1]; 10.71 [0.5]; 10.82 [1.0]; β_2 19.55 [0]; 19.6 [0.1]; 20.04 [0.5]; 20.2 [1.0]; $MOHL/ML.OH$ 0.73 [0.5]	
Ag^+	β_1 4.70, β_2 7.70, $MHL/ML.H$ 7.68, $M_2L/ML.M$ 1.8, $M_2L_2/(ML)^2$ 3.8 [0.1][c]	
Zn^{2+}	β_1 5.66 [0]; 5.7 [0.1]; 5.92 [1.4]; β_2 10.64 [0]; 10.62 [0.1]; 11.07 [1.4]; β_3 13.89 [0]; 13.23 [0.15]; 12.93 [1.4]	
Cd^{2+}	β_1 5.41 [0]; 5.45 [0.1]; 5.62 [1]; β_2 9.91 [0]; 9.98 [0.15]; 10.21 [1.0]; β_3 11.74 [0.1]; 12.30 [1.0]	

* All listed values of stability constants refer to 25°C unless stated otherwise in a note. The notes follow each section of the table separately.

Table 7.1 (Continued)

Formula Metal ion	Stability constants (Symbol for the constant or scheme of the equilibrium, value of log K or log β, $[I]$)	Ref.
Hg^{2+}	β_1 14.3; β_2 23.24 [0.1]; 23.42 [1.4]; MOHL/ML.OH 9.5; $MHL_2/ML_2.H$ 5.2, $MH_2L_2/MHL_2.H$ 4.2, $MH_2L_3/MHL_2.HL$ 3.6 [0.1]; MClL/MCl$_2$.L 5.54, $ML_2/MCl_2.L^2$ 9.73 [0]	
Pb^{2+}	β_1 7.0, β_2 8.45 [0.1]	

[a] 30°C; [b] 4°C; [c] 20°C; [d] 18°C

3. $C_3H_{10}N_2$	DL-1-methylethylenediamine $H_2N.CH(CH_3)CH_2.NH_2$ L pn	
H^+	K_{H1} 9.72 [0]; 9.78 [0.1]; 9.90 [0.5]; K_{H2} 6.61 [0]; 6.85 [0.1]; 7.06 [0.5]	
Ni^{2+}	β_1 7.29 [0]; 7.34 [0.1]; 7.50 [0.5]; β_2 13.43 [0]; 13.51 [0.1]; 13.86 [0.5]; β_3 (17.6) [0]; 17.82 [0.1]; 18.33 [0.5]	
Cu^{2+}	β_1 10.44 [0]; 10.58 [0.1]; 10.76 [0.5]; β_2 19.4 [0]; 19.7 [0.1]; 20.1 [0.5]	
Zn^{2+}	β_1 5.64 [0]; 5.72 [0.1]; 5.87 [0.5]; β_2 (10.58) [0]; (10.77) [0.1]; (11.20) [0.5]; β_3 (12.6) [0.65][a]	
Cd^{2+}	β_1 5.42; β_2 (9.97); β_3 (12.12) [0.65][a]	
Hg^{2+}	β_2 (23.51) [0.1]	
Pb^{2+}	β_2 8.62 [0.2]	

[a] 30°C

4. $C_3H_{10}N_2$	trimethylenediamine $H_2N.CH_2CH_2CH_2.NH_2$ L	
H^+	K_{H1} 10.49 [0]; 10.52 [0.1]; 10.77 [1.0]; K_{H2} 8.48 [0]; 8.74 [0.1]; 9.12 [1.0]	
Ni^{2+}	β_1 6.29 [0]; 6.31 [0.1]; 6.47 [1.0]; β_2 10.54 [0]; 10.6 [0.1]; 10.9 [1.0]; β_3 12.3 [0.15]; 12.0 [1.0]	
Cu^{2+}	β_1 9.61 [0]; 9.75 [0.1]; 10.00 [1.0]; β_2 16.65 [0]; 16.9 [0.1]; 17.3 [1.0]; ML/MOHL.H 7.42 [0]; 7.66 [0.1]; $ML/M(OH)_2L.H^2$ 19.10 [0]; 19.36 [0.1]; $(MOHL)_2/(MOHL)^2$ 2.17 [0]; 2.41 [0.1]	
Ag^+	β_1 5.71 [0]; 5.85 [0.1][a]; MHL/ML.H 7.34, $M_2L/ML.M$ 0.6 [0.1][a]	
Cd^{2+}	β_1 4.50 [0.1]; 4.72 [1.0][a]; β_2 7.20; β_3 8.0 [0.1]; MHL/ML.H 8.03 [1.0][a]	
Pb^{2+}	β_2 8.16 [0.2]	

[a] 20°C

Ch. 7] **Stability Constants of Complexes with Organic Ligands** 145

Table 7.1 (Continued)

Formula Metal ion	Stability constants (Symbol for the constant or scheme of the equilibrium, value of log K or log β, [I])	Ref.
5. $C_3H_{10}N_2O$	1,3-diamino-2-propanol HO.CH(CH$_2$NH$_2$)$_2$ L	
H$^+$	K_{H1} 9.55 [0]; 9.58 [0.1]; 9.81 [1.0]; K_{H2} 7.75 [0]; 7.98 [0.1]; 8.35 [1.0]	
Co^{2+}	β_1 3.90, β_2 7.14 [1.0]^a	
Ni^{2+}	β_1 5.49 [0]; 5.47 [0.16]^a; 5.64 [1.0]^a β_2 9.72 [0]; 9.61 [0.16]^a; 10.02 [1.0]^a	
Cu^{2+}	β_1 (9.7) [1.0]^a; MOHL.H/M.L 3.61 [0]; (4.34) [0.16]^a; (MOHL)$_2$.H^2/M^2.L^2 10.30 [0]; 10.48 [0.1]; 10.36 [1.0]	
Ag$^+$	β_1 5.47 [0]; 5.80 [1.0]^a	
Zn^{2+}	β_1 4.60; β_2 9.02 [1.0]^a	

^{a} 30°C

6. $C_3H_{11}N_3$	1,2,3-triaminopropane H$_2$N.CH$_2$CH(NH$_2$)CH$_2$NH$_2$ L	
H$^+$	K_{H1} 9.59 [0.1]^a; 9.63 [0.5]^a; K_{H2} 7.95 [0.1]^a; 8.08 [0.5]^a; K_{H3} 3.72 [0.1]^a; 3.99 [0.5]^a	
Co^{2+}	β_1 6.80, MHL/ML.H 6.89 [0.1]^a	
Ni^{2+}	β_1 9.30, MHL/ML.H 6.34 [0.1]^a	
Cu^{2+}	β_1 11.1, β_2 20.1, MHL/ML.H 7.30, MHL$_2$/ML$_2$.H 7.90, MH$_2$L$_2$/MHL$_2$.H 7.30 [0.1]^a	
Ag$^+$	β_1 5.7, MHL/ML.H 7.56, M$_2$L/ML.M 1.2 [0.1]^a	
Zn^{2+}	β_1 6.75, MHL/ML.H 7.09 [0.1]^a	
Cd^{2+}	β_1 6.45, MHL/ML.H 7.89 [0.1]^a	
Hg^{2+}	β_1 19.6, MHL/ML.H 7.93 [0.5]^a	

^{a} 20°C

7. $C_4H_{11}NO_3$	2-amino-2-(hydroxymethyl)-1,3-propanediol L (HOCH$_2$)$_2$.C(NH$_2$).CH$_2$OH THAM	
H$^+$	K_{H1} 8.075 [0]; 8.09 [0.1]; 8.15 [0.5]	
Ni^{2+}	β_1 2.63, β_2 4.5, M$_3$(OH)$_5$L$_3$.H^5/M^{3}L^3 -13.5, M$_2$(OH)$_3$L$_3$.H^3/M^2.L^3 -27.0 [0.1]	
Cu^{2+}	β_1 3.95, β_2 7.63, β_3 11.10, β_4 14.1, ML/MOHL.H 6.0, (MOHL)$_2$/(MOHL)^2 2.2, ML$_2$/MOHL$_2$.H 6.32, MOHL$_2$/M(OH)$_2$L$_2$.H 7.90 [0.1]	
Ag$^+$	β_1 3.14 [0.05]; 3.05 [0.5]; β_2 6.57 [0.05]; 6.53 [0.5]	

Table 7.1 (Continued)

Formula Metal ion	Stability constants (Symbol for the constant or scheme of the equilibrium, value of log K or log β, [I])	Ref.
8. $C_4H_{13}N_3$	iminobis(2-ethylamine) $HN(CH_2CH_2NH_2)_2$ dien	L
H^+	K_{H1} 9.80 [0]; 9.84 [0.1]; 9.92 [1.0]; K_{H2} 8.74 [0]; 9.02 [0.1]; 9.24 [1.0]; K_{H3} 3.64 [0]; 4.23 [0.1]; 4.73 [1.0]	
Mn^{2+}	β_1 3.99[a], β_2 6.91 [1.0]	
Fe^{2+}	β_1 6.23[a], β_2 10.53 [1.0]	
Co^{2+}	β_1 8.0 [0.1]; 8.57 [1.0]; β_2 13.9 [0.1]; 14.77 [1.0]	
Ni^{2+}	β_1 10.5 [0.1]; 10.96 [1.0]; β_2 18.6 [0.1]; 19.27 [1.0]	
Pd^{2+}	$MHL_2/ML_2 \cdot H$ 6.1, $MH_2L_2/MHL_2 \cdot H$ 2.5 [1.0]	
Cu^{2+}	β_1 15.6 [0]; 15.9 [0.1]; 16.34 [1.0]; β_2 20.9, $MHL/ML \cdot H$ 3.2, $MHL_2/ML_2 \cdot H$ 8.2, $MOHL/ML \cdot OH$ 4.6 [0.1]	
Ag^+	β_1 6.1, $MHL/ML \cdot H$ 7.0, $M_2L/ML \cdot M$ 1.4 [0.1][b]	
Zn^{2+}	β_1 8.7 [0]; 8.8 [0.1]; 9.22 [1.0]; β_2 14.3 [0.1]	
Cd^{2+}	β_1 8.4, β_2 13.8 [0.1][b]	
Hg^{2+}	β_1 21.8, β_2 29, $MHL/ML \cdot H$ 3.9, $MOHL/ML \cdot OH$ 6.3 [0.5][b]	
Pb^{2+}	β_1 8.50, β_2 10.37 [0.1]	

[a] 30°C; [b] 20°C

9. C_5H_5N	pyridine	L
H^+	K_{H1} 5.229 [0]; 5.24 [0.1]; 5.31 [0.5]; 5.39 [1.0]	
Mn^{2+}	β_1 0.14 [0.5]	
Fe^{2+}	β_1 0.6, β_2 0.9 [0.5]	
Co^{2+}	β_1 1.2 [0.1]; 1.19 [0.5]; β_2 1.8 [0.1]; 1.70 [0.5]	
Ni^{2+}	β_1 1.85 [0.1]; 1.87, β_2 3.10, β_3 3.71 [0.5]	
Cu^+	β_1 4.84, β_2 7.59, β_3 8.18, β_4 8.52 [1.0][a]	
Cu^{2+}	β_1 2.5 [0]; 2.54 [0.1]; 2.56 [0.5]; β_2 4.30 [0]; 4.45 [0.5]; β_3 5.16 [0]; 5.7 [0.5]; β_4 6.04 [0]; 6.5 [0.5]	
Ag^+	β_1 2.05 [0]; (1.93) [0.1]; 2.06 [0.5]; β_2 4.10 [0]; (4.22) [0.1]; 4.18 [0.5]	
Zn^{2+}	β_1 1.0 [0.1]; 0.99 [0.5]; β_2 1.6 [0.1][b]; 1.36 [0.5]; β_3 1.9 [0.1][b]; 1.55 [0.5]	
Cd^{2+}	β_1 1.28 [0.1]; 1.34 [0.5]; β_2 2.02 [0.1]; 2.13 [0.5]; β_3 2.3 [0.1][a]; 2.41 [0.5]	

Stability Constants of Complexes with Organic Ligands

Table 7.1 (Continued)

Formula Metal ion	Stability constants (Symbol for the constant or scheme of the equilibrium, value of log K or log β, $[I]$)	Ref.
	pyridine (continued)	
Hg^{2+}	β_1 5.1, β_2 10.0, β_3 10.3, β_4 10.6 [0.5]	

[a] 20°C; [b] 30°C

10. $C_6H_{15}NO_3$	triethanolamine $N(CH_2CH_2OH)_3$ L tris(2-hydroxyethyl)amine TEA	
H^+	K_{H1} 7.762 [0]; 7.8 [0.1]; 7.90 [0.5]; 7.99 [1.0]	
Co^{2+}	β_1 1.73 [0.5]	
Ni^{2+}	β_1 2.27, β_2 3.09 [0.5]	
Cu^{2+}	β_1 3.9 [0.1]; 4.23 [0.5]; β_2 6.0 [0.1]	
Ag^+	β_1 2.30 [0.5]; β_2 4.23 [0]^a; 3.64 [0.5]	
Cd^{2+}	β_1 2.70, β_2 4.60, β_3 5.21 [0.1]	
Hg^{2+}	β_1 6.90, β_2 13.08 [0.5]	

[a] 20°C

11. $C_6H_{18}N_4$	1,4,7,10-tetrazadecane $H_2NCH_2CH_2NHCH_2CH_2NHCH_2CH_2NH_2$ triethylenetetramine trien L	
H^+	K_{H1} 9.74 [0.1]; 9.87 [0.5]; 10.02 [1.0]; K_{H2} 9.08 [0.1]; 9.21 [0.5]; 9.39 [1.0]; K_{H3} 6.56 [0.1]; 6.87 [0.57]; 7.00 [1.0]; K_{H4} 3.25 [0.1]; 3.71 [0.5]; 4.00 [1.0]	
Mn^{2+}	β_1 4.90 [0.1]; 5.46 [1.0]	
Fe^{2+}	β_1 7.76 [0.1]; 8.39 [1.0]	
Co^{2+}	β_1 10.95 [0.1]; 11.35 [1.0]; MHL/ML.H 5.7 [0.1]^a	
Ni^{2+}	β_1 13.8 [0.1]; 14.4 [0.5]; (14.5) [1.0]; β_2 18.6, $M_2L_3/M^2.L^3$ 36.9 [0.5]; MHL/ML.H 4.8 [0.1]^a	
Cu^{2+}	β_1 20.1 [0.1]; (20.9) [1.0]; MHL/ML.H 3.5^a, ML/MOHL.H 10.8 [0.1]	
Ag^+	β_1 7.65, MHL/ML.H 8.0, $MH_2L/MHL.H$ 6.2, $M_2L/ML.M$ 2.4 [0.1]^a	
Zn^{2+}	β_1 12.03 [0.1]; (12.05) [1.0]; MHL/ML.H 5.1 [0.1]^a	
Cd^{2+}	β_1 10.63 [0.1]; 11.04 [1.0]; MHL/ML.H 6.2 [0.1]^a	
Hg^{2+}	β_1 25.0 [0.1]; 25.3 [0.5]^a; MHL/ML.H 5.5 [0.1]^a	
Pb^{2+}	β_1 10.4 [0.1]	

[a] 20°C

148 Stability Constants of Complexes with Organic Ligands [Ch. 7

Table 7.1 (Continued)

Formula Metal ion	Stability constants (Symbol for the constant or scheme of the equilibrium, value of log K or log β, [I])	Ref.
12. $C_6H_{18}N_4$	nitrilotris(2-ethylamine) $N(CH_2CH_2NH_2)_3$ L tren	
H^+	K_{H1} 10.03 [0]; 10.14 [0.1]; 10.29 [0.1]^a; K_{H2} 9.13 [0]; 9.43 [0.1]; 9.59 [0.1]^a; K_{H3} 7.85 [0]; 8.41 [0.1]; 8.56 [0.1]^a	
Mn^{2+}	β_1 5.8, 5.8^a [0.1]	
Fe^{2+}	β_1 8.7, 8.78^a [0.1]	
Co^{2+}	β_1 12.7, 12.8^a [0.1]	
Cu^{2+}	β_1 18.4 [0]; 18.5, 18.8^a [0.1]; MOHL/ML.OH 4.65, 4.70^a [0.1]	
Ag^+	β_1 7.8, MHL/ML.H 8.1, MH_2L/MHL.H 7.3, M_2L/ML.M 2.4 [0.1]^a	
Zn^{2+}	β_1 14.5, 14.65^a [0.1]	
Cd^{2+}	β_1 12.3 [0.1]^a	
Hg^{2+}	β_1 25.8 [0.5]^a; MHL/M.HL 4.5 [0.1]^a	

^{a} 20°C

13. $C_8H_{23}N_5$	1,4,7,10,13-pentazatridecane $HN(CH_2CH_2NHCH_2CH_2NH_2)_2$ tetraethylenepentamine tetren L	
H^+	K_{H1} 9.70, K_{H2} 9.14, K_{H3} 8.05, K_{H4} 4.70, K_{H5} 2.97 [0.1]	
Mn^{2+}	β_1 6.55 [0.1]	
Fe^{2+}	β_1 9.85, MHL/ML.H 7.1 [0.1]	
Co^{2+}	β_1 13.3, MH_2L/ML.H^2 10.4 [0.1]	
Ni^{2+}	β_1 17.4, MHL/ML.H 4.1, MH_2L/MHL.H 4.0 [0.1]	
Cu^{2+}	β_1 22.8, MHL/ML.H 5.2, MH_2L/MHL.H 3.8 [0.1]	
Ag^+	β_1 7.4, MHL/ML.H 8.3, MH_2L/MHL.H 7.5, MH_3L/MH_2L.H 5.5 [0.1]	
Zn^{2+}	β_1 15.1, MH_2L/ML.H^2 9.4 [0.1]	
Cd^{2+}	β_1 14.0 [0.1]	
Hg^{2+}	β_1 27.7 [0.1]	
Pb^{2+}	β_1 10.5 [0.1]	

14. $C_9H_6INO_4S$	7-iodo-8-hydroxyquinoline-5-sulphonic acid H_2L ferron	
H^+	K_{H1} 7.42 [0]; 7.08 [0.1]; 6.90 [1.0]; K_{H2} 2.51 [0]; 2.43 [0.1]; 2.43 [1.0]	
Ca^{2+}	β_1 3.07 [0]; 2.36 [0.1]; 2.01 [1.0]; β_2 4 [0]	
Fe^{3+}	β_1 8.9, β_2 17.3, β_3 25.2 [0.1]	

Table 7.1 (Continued)

Formula Metal ion	Stability constants (Symbol for the constant or scheme of the equilibrium, value of log K or log β, $[I]$)	Ref.
	ferron (continued)	
Co^{2+}	β_1 7.3, β_2 13.6, β_3 18.6 [0.1]	
Ni^{2+}	β_1 8.2, β_2 15.2, β_3 20.8 [0.1]	
Zn^{2+}	β_1 7.1, β_2 13.2 [0.1]	
Hg^{2+}	β_2 21.2 [0.1]	
Al^{3+}	β_1 7.6, β_2 14.7, β_3 20.3, $ML_2/MOHL_2$.H 5.0 [0.1]	
Ga^{3+}	β_1 14.7, β_2 23.9, β_3 29.6, MHL/ML.H 3.7, $M(OH)_2L/ML.(OH)^2$ 17.6, $MOHL_2/ML_2$.OH 7.1 [0.1]	
In^{3+}	β_1 15.1, β_2 24.2, β_3 30.7, $M(OH)_2L/ML.(OH)^2$ 19.2, $MOHL_2/ML_2$.OH 8.5 [0.1]	
Tl^{3+}	β_1 18.9, β_2 27.7, β_3 35.4, $M(OH)_2L/ML.(OH)^2$ 14.1 [0.1]	
$Ge(OH)_4$	$M(OH)_2L_2/M(OH)_4.(HL)^2$ 6.78 [0.5]	
15. C_9H_7NO	8-hydroxyquinoline oxine	HL
H^+	K_{H1} 9.81 [0]; 9.66 [0.1]; 9.62 [0.5]; 9.58 [1.0]; K_{H2} 4.91 [0]; 4.99 [0.1]; 5.09 [0.5]; 5.17 [1.0]	
Mg^{2+}	β_1 4.74 [0]; 4.31 [0.1]^a; 3.91 [1.0]^a	
Ca^{2+}	β_1 3.27 [0]; 2.82 [0.1]^a; 2.44 [1.0]^a	
Sr^{2+}	β_1 2.56 [0]; 2.11 [0.1]^a; 1.55 [1.0]^a	
Ba^{2+}	β_1 2.07 [0]; 1.62 [0.1]^a; 1.26 [1.0]^a	
La^{3+}	β_1 5.9, β_2 11.5, β_3 17.0 [0.1]	
Th^{4+}	β_1 10.5, β_2 20.4, β_3 29.9, β_4 38.8 [0.1]	
Mn^{2+}	β_1 6.24 [0.1]	
Fe^{3+}	β_1 14.52 [0]; 13.69 [0.1]; 13.0 [0.5]; β_2 26.3 [0.1]; 25.3 [0.5]; β_3 36.9 [0.5]	
Co^{2+}	β_1 8.65 [0]	
Ni^{2+}	β_1 9.27 [0]	
Cu^{2+}	β_1 12.56 [0]^a; 12.1, 23.0 [0.1]	
Ag^+	β_1 5.20, β_2 9.56 [0.1]^a	
Zn^{2+}	β_1 8.56 [0]^a, 8.52, β_2 15.8 [0.1]	
Cd^{2+}	β_1 7.78 [0]^a	
Ga^{3+}	β_1 14.5, β_2 28.0, β_3 40.5 [0.1]^a	
In^{3+}	β_1 12.0, β_2 23.9, β_3 35.4 [0.1]	
Pb^{2+}	β_1 9.02 [0]	
$Ge(OH)_4$	$M(OH)_2L_2/M(OH)_4.(HL)^2$ 6.61 [0.5]	

^{a} 20°C

Table 7.1 (Continued)

Formula Metal ion	Stability constants (Symbol for the constant or scheme of the equilibrium, value of log K or log β, [I])	Ref.
16. $C_9H_7NO_4S$	8-hydroxyquinoline-5-sulphonic acid sulfoxine	H_2L
H^+	K_{H1} 8.757 [0]; 8.42 [0.1]; 8.23 [0.5]; K_{H2} 4.112 [0]; 3.93 [0.1]; 3.86 [0.5]	
Mg^{2+}	β_1 4.79 [0]; 4.02 [0.1]; β_2 8.2 [0]; 7.63 [0.1]	
Ca^{2+}	β_1 3.52 [0]; 2.66 [0.1]	
Sr^{2+}	β_1 2.75 [0]; 1.98 [0.1]	
Ba^{2+}	β_1 2.31 [0]; 1.56 [0.1]	
La^{3+}	β_1 5.63 [0]; 5.42 [0.14]; 5.25 [0.41]; β_2 10.1 [0]; 9.89 [0.14]; 9.74 [0.41]; β_3 13.8 [0]; 13.41 [0.14]; 13.46 [0.41]	
$M^{3+}(4f)$	β_1 6.5 [0]; 6.3 [0.14]; 6.0 [0.41]; β_2 12.0 [0]; 11.6 [0.14]; 11.2 [0.41]; β_3 16.5 [0]; 16.2 [0.14]; 15.9 [0.41]	
Th^{4+}	β_1 9.56, β_2 18.29, β_3 25.91, β_4 32.02, $ML_3/MOHL_3 \cdot H$ 6.2, $(ML_3)^2/(MOHL_2)_2 \cdot (HL)^2$ 8.9 [0.1]	
Mn^{2+}	β_1 6.94 [0]; 5.67 [0.1]; β_2 10.72 [0.1]	
Co^{2+}	β_1 8.82 [0]; 8.11 [0.1]; β_2 15.9 [0]; 15.06 [0.1]	
Ni^{2+}	β_1 9.75 [0]; 9.02 [0.1]; β_2 18.5 [0]; 16.77 [0.1]	
Cu^{2+}	β_1 (11.53) [0]; 11.92 [0.1]; 11.57 [0.5]; β_2 (21.6) [0]; 21.87 [0.1]; 21.63 [0.5]	
UO_2^{2+}	β_1 8.52, β_2 15.68, $ML_2/MOHL_2 \cdot H$ 6.68, $(ML_2)^2/(MOHL_2)_2 \cdot H^2$ 11.7 [0.1]	
17. $C_{10}H_8N_2$	2,2'-bipyridyl bipy	L
H^+	K_{H1} 4.35 [0]; 4.42 [0.1]; 4.67 [1.0]; K_{H2} 1.5 [0.1]	
Mn^{2+}	β_1 2.62 [0.1]; 2.61 [1.0]; β_2 4.62 [0.1]; 4.47 [1.0]; β_3 5.6 [0.1]; 6.0 [1.0]	
Fe^{2+}	β_1 4.36 [0]; 4.20 [0.1]; 4.65 [1.0][a]; β_2 7.90 [0.1]; β_3 17.2 [0.1]; 17.49 [1.0]	
Fe^{3+}	$M_2(OH)_2L_4 \cdot H^2/M^2 \cdot L^4$ 16.29 [0.1][b]	
Co^{2+}	β_1 5.8 [0.1]; 5.81 [1.0]; β_2 11.24 [0.1]; 11.31 [1.0]; β_3 15.9 [0.1]; 16.18 [1.0]	
Ni^{2+}	β_1 7.04 [0.1]; 7.06 [1.0]; β_2 13.85 [0.1]; 14.01 [1.0]; β_3 20.16 [0.1]; 20.47 [1.0]	
Cu^+	β_2 12.95 [0.1][b]; 13.18 [0.3]	
Cu^{2+}	β_1 6.33, $ML/MOHL \cdot H$ (7.9), $ML/M(OH)_2L \cdot H^2$ 17.67, $(MOHL)_2 \cdot H^2/(ML)^2$ 10.81 [0.1]	

Ch. 7]   **Stability Constants of Complexes with Organic Ligands**   151

Table 7.1 (Continued)

Formula Metal ion	Stability constants (Symbol for the constant or scheme of the equilibrium, value of log K or log β, [I])	Ref.
	2,2'-bipyridyl (continued)	
Ag^+	β_1 3.03 [0.1]^c; 3.0 [1.0]; β_2 6.67 [0.1]^c; 7.11 [1.0]	
Zn^{2+}	β_1 5.13 [0.1]; 5.34 [1.0]; β_2 9.5 [0.1]; 9.96 [1.0]; β_3 13.2 [0.1]; 13.97 [1.0]	
Cd^{2+}	β_1 4.18, β_2 7.7, β_3 10.3 [0.1]	
Hg^{2+}	β_1 9.64, β_2 16.7, β_3 19.5 [0.1]^b	
Ga^{3+}	β_1 4.52, β_2 7.70 [1.0]	
In^{3+}	β_1 4.75, β_2 8.00 [1.0]	
Tl^{3+}	β_1 9.40, β_2 16.10, β_3 20.05 [1.0]	
Pb^{2+}	β_1 2.9 [0.1]	

^{a} 30°C; ^{b} 20°C; ^{c} 35°C

18. $C_{12}H_7N_2Cl$	2-chloro-1,10-phenanthroline	L
H^+	K_{H1} 4.18, K_{H2} 0.22 [0.1]	
Fe^{2+}	β_3 11.6 [0.1]	
Ni^{2+}	β_1 4.58, β_2 9.26, β_3 13.6 [0.1]	
Cu^+	β_2 14.6 [0.3]	
Cu^{2+}	β_1 5.07, β_2 10.07, β_3 13.9 [0.1]	
Zn^{2+}	β_1 3.3, β_2 6.6 [0.1]	

19. $C_{12}H_7N_2Cl$	5-chloro-1,10-phenanthroline	L
H^+	K_{H1} 4.07 [0.1]	
Fe^{2+}	β_1 5.71, β_2 10.72, β_3 19.7 [0.1]	
Ag^+	β_1 4.70, β_2 11.04 [0.1]	
Zn^{2+}	β_1 5.85 [0.1]	

20. $C_{12}H_7N_3O_2$	5-nitro-1,10-phenanthroline	L
H^+	K_{H1} (3.23) [0]; 3.22 [0.1]; 3.25 [0.3]	
Fe^{2+}	β_1 4.57 [0]^a; 5.06 [0.1]; β_3 15.6 [0]	
Co^{2+}	β_1 6.3, β_2 11.8, β_3 16.5 [0.1]	
Ni^{2+}	β_1 7.0, β_2 13.4, β_3 20.4 [0.1]	
Cu^{2+}	β_1 8.0, β_2 13.5, β_3 17.7 [0.1]	
Zn^{2+}	β_1 5.4 [0.1]	

^{a} 35°C

21. $C_{12}H_8N_2$	1,10-phenanthroline phen	L
H^+	K_{H1} 4.86 [0]; 4.93 [0.1]; 5.12 [1.0];	

152   Stability Constants of Complexes with Organic Ligands   [Ch. 7

Table 7.1 (Continued)

Formula Metal ion	Stability constants (Symbol for the constant or scheme of the equilibrium, value of log K or log β, [I])	Ref.
	1,10-phenanthroline (continued) K_{H2} 1.9 [0.1]	
Mg^{2+}	β_1 1.2 [0.1][a]	
Ca^{2+}	β_1 0.7 [0.1][a]	
Mn^{2+}	β_1 4.0, β_2 7.3, β_3 10.3 [0.1]	
Fe^{2+}	β_1 5.85, β_2 11.15, β_3 21.0 [0.1]	
Fe^{3+}	β_1 6.5, β_2 11.4 [0.1][a]; β_3 13.8 [0]; 14.1 [0.15]; $M_2L_4/M_2OHL_4 \cdot H$ 6.5, $M_2OHL_4/M_2(OH)_2L_4 \cdot H$ 4.4 [0.1][a]	
Co^{2+}	β_1 7.08, β_2 13.72, β_3 19.8 [0.1]	
Ni^{2+}	β_1 8.6 [0.1]; 8.65 [0.5]; β_2 16.7 [0.1]; 17.08 [0.5]; β_3 24.3 [0.1]; 24.91 [0.5]	
Cu^+	β_2 15.82 [0.3]	
Cu^{2+}	β_1 7.4, $ML/M(OH)_2L \cdot H^2$ 17.3, $(MOHL)_2 \cdot H^2/(ML)^2$ 10.69 [0.1]	
Ag^+	β_1 5.02 [0.1]; 5.0 [1.0]; β_2 12.06 [0.1]; 12.11 [1.0]	
Zn^{2+}	β_1 6.2 [0]; 6.4 [0.1]; β_2 (12.1) [0]; 12.2 [0.1]; β_3 (17.3) [0]; 17.1 [0.1]	
Cd^{2+}	β_1 5.8, β_2 10.6, β_3 14.6 [0.1]	
Hg^{2+}	β_2 19.65, β_3 23.25 [0.1][a]	
Pb^{2+}	β_1 4.65 [0.1][a]	
Ga^{3+}	β_1 5.58, β_2 9.21 [1.0]	
In^{3+}	β_1 5.70, β_2 10.04, β_3 14.0 [1.0]	
Tl^{3+}	β_1 11.57, β_2 18.30, β_3 24.3 [1.0]	

[a] 20°C

22. $C_{13}H_{10}N_2$	5-methyl-1,10-phenanthroline	L
H^+	K_{H1} 5.17 [0]; 5.27 [0.1]	
Mn^{2+}	β_1 4.28, β_2 7.6, β_3 15.3 [0.1]	
Fe^{2+}	β_1 6.00, β_2 11.45, β_3 22.1 [0.1]	
Co^{2+}	β_1 7.14, β_2 14.0, β_3 20.6 [0.1]	
Ni^{2+}	β_1 8.30, β_2 17.0, β_3 24.7 [0.1]	
Cu^{2+}	β_1 8.55, β_2 15.0, β_3 20.1 [0.1]	
Ag^+	β_1 7.30, β_2 13.39 [0.1]	
Zn^{2+}	β_1 6.62, β_2 12.6, β_3 18.3 [0.1]	
Cd^{2+}	β_1 6.13, β_2 11.0 [0.1]	

23. $C_{14}H_{12}N_2$	2,9-dimethyl-1,10-phenanthroline neocuproine	L
H^+	K_{H1} 5.85 [0.1]; 5.88 [0.3]	

Table 7.1 (Continued)

Formula Metal ion	Stability constants (Symbol for the constant or scheme of the equilibrium, value of log K or log β, [I])	Ref.
	2,9-dimethyl-1,10-phenanthroline (continued)	
Co^{2+}	β_1 4.2, β_2 7.0 [0.1]	
Ni^{2+}	β_1 5.0, β_2 8.5 [0.1]	
Cu^+	β_2 19.1 [0.3]	
Cu^{2+}	β_1 5.2 [0.1]; 6.1 [0.3]; β_2 11.0 [0.1]; 11.7 [0.3]	
Zn^{2+}	β_1 4.1, β_2 7.7 [0.1]	
Cd^{2+}	β_1 4.1, β_2 7.4, β_3 10.4 [0.1]	
24. $C_{14}H_{12}N_2$	4,7-dimethyl-1,10-phenanthroline	L
H^+	K_{H1} 5.95 [0.1]	
Fe^{2+}	β_1 5.60 [0.1]	
Co^{2+}	β_1 8.08, β_2 16.1, β_3 24.5 [0.1]	
Ni^{2+}	β_1 8.44, β_2 16.6, β_3 25.0 [0.1]	
Cu^{2+}	β_1 8.76, β_2 16.0, β_3 22.0 [0.1]	
Zn^{2+}	β_1 6.90, β_2 13.1, β_3 19.1 [0.1]	
25. $C_{14}H_{12}N_2$	5,6-dimethyl-1,10-phenanthroline	L
H^+	K_{H1} 5.60 [0.1]	
Fe^{2+}	β_1 6.37 [0.1]	
Co^{2+}	β_1 7.47, β_2 15.5, β_3 16.6 [0.1]	
Ni^{2+}	β_1 8.25, β_2 16.6, β_3 24.8 [0.1]	
Cu^{2+}	β_1 8.71, β_2 15.7, β_3 21.1 [0.1]	
Zn^{2+}	β_1 6.87, β_2 12.9, β_3 18.6 [0.1]	
26. $C_{18}H_{12}N_2$	5-phenyl-1,10-phenanthroline	L
H^+	K_{H1} 4.72 [0]; 4.03 [0.3; dioxan 50%]	[1, 3]
Fe^{2+}	β_3 21.1 [?]	[2]
Cu^{2+}	β_1 4.05 [0.3; dioxan 50%]	[3]
27. $C_{23}H_{16}N_2$	4,7-diphenyl-1,10-phenanthroline bathophenanthroline	L
H^+	K_{H1} 4.80 [C_2H_5OH 10%][a]; 4.30 [0.3; dioxan 50%]	[4, 3]
Fe^{2+}	β_2 21.8 [C_2H_5OH 10%][a]	[4]
Cu^{2+}	β_1 5.7; K_3 3.75 [0.3; dioxan 50%]	[3]

[a] 18°C

REFERENCES

[1] A. A. Schilt and G. F. Smith, *J. Phys. Chem.*, 1956, **60**, 1546.
[2] W. W. Brandt and D. K. Gullstrom, *J. Am. Chem. Soc.*, 1952, **74**, 3532.
[3] B. R. James and R. J. P. Williams, *J. Chem. Soc.*, **1961**, 2007.
[4] F. Nakashima and K. Sakai, *Bunseki Kagaku*, 1961, **10**, 89.

Table 7.2
Complexes of carboxylic acids*

Formula of the acid Metal ion	Stability constants (Symbol for the constant or scheme of the equilibrium, value of log K or log β, [I])	
1. CH_2O_2	formic acid H.COOH	HL
H^+	K_{H1} 3.745 [0]; 3.55 [0.1]; 3.49 [0.5]; 3.53 [1.0]; 3.90 [3.0]	
M^{2+}(IIA)	β_1 1.4 [0]	
Y^{3+}	β_1 1.12, β_2 1.95 [0.1]	
La^{3+}	β_1 1.10, β_2 2.09 [0.1]	
M^{3+}(4f)	β_1 (1.17), β_2 (1.72) [0.1]	
Th^{4+}	β_1 3.09, β_2 5.15, β_3 6.73 [1.0]^a	
UO_2^{2+}	β_1 1.86, β_2 3.05, β_3 3.52 [1.0]^a	
Fe^{3+}	β_1 3.1, $M_3(OH)_2L_6/M^3.(OH)^2.L^6$ 19.9 [1.0]^a	
Co^{2+}	β_1 0.73, β_2 1.18 [2.0]	
Ni^{2+}	β_1 (0.46) [2.0]	
Cu^{2+}	β_1 1.40, β_2 2.30, β_3 2.2, β_4 1.9 [2.0]	
Zn^{2+}	β_1 0.70, β_2 1.08, β_3 1.20 [2.0]	
Cd^{2+}	β_1 1.04, β_2 1.23, β_3 1.75 [2.0]	
Hg^{2+}	β_1 (5.43) [0.1]	
Al^{3+}	β_1 1.36 [1.0]	
In^{3+}	β_1 2.74, β_2 4.72, β_3 5.70, β_4 6.70 [2.0]^a	
Pb^{2+}	β_1 1.23, β_2 2.01, β_3 1.8 [2.0]	

^{a} 20°C

2. $C_2H_2O_4$	oxalic acid HOOC.COOH	H_2L
H^+	K_{H1} 4.266 [0]; 3.82 [0.1]; 3.55 [1.0]; K_{H2} 1.252 [0]; 1.04 [0.1; 1.0]	
Be^{2+}	β_1 4.08, β_2 5.38 [0.1]^a	
Mg^{2+}	β_1 2.76^a, β_2 4.24 [0.1]	
Ca^{2+}	β_1 1.66, β_2 2.69 [1.0]; MHL/M.HL 1.38; M(HL)$_2$/M.(HL)^2 1.8 [0.1]	
Sr^{2+}	β_1 1.25, β_2 1.90 [1.0]; MHL/M.HL 1.11; M(HL)$_2$/M.(HL)^2 1.7 [0.1]	
Ba^{2+}	β_1 2.31 [0]^b	

* The data listed refer to 25°C unless stated otherwise in a note. The notes follow each section of the table separately.

Table 7.2 (Continued)

Formula of the acid / Metal ion	Stability constants (Symbol for the constant or scheme of the equilibrium, value of log K or log β, [I])
	oxalic acid (continued)
Sc^{3+}	β_1 6.86, β_2 11.31, β_3 14.32, β_4 16.70 [1.0][a]
Y^{3+}	β_1 5.46, β_2 9.29 [0.1]
La^{3+}	β_1 4.3, β_2 7.9, β_3 10.3 [1.0]
Ce^{3+}	β_1 4.49, β_2 7.91, β_3 10.30, β_4 11.75 [1.0][a]
$M^{3+}(4f)$	β_1 (5.1), β_2 (8.7), β_3 (11.6), β_4 (13.2) [1.0][a]
Th^{4+}	β_1 8.23, β_2 16.8, β_3 22.8 [1.0]; MHL/M.HL 7.4 [0.05]
VO^{2+}	β_1 6.45, β_2 11.78 [1.0][a]
VO_2^+	β_1 5.0, β_2 8.49 [1.0][a]
Cr^{2+}	β_1 3.85, β_2 6.81 [0.1]
UO_2^{2+}	β_1 5.99, β_2 10.64, β_3 11.0 [1.0][a]
PuO_2^{2+}	β_2 9.4 [1.0][a]
Mn^{2+}	β_1 3.2, β_2 4.4 [0.1]
Mn^{3+}	β_1 9.98, β_2 16.57, β_3 18.42 [2.0]
Fe^{2+}	β_1 3.05, β_2 5.15 [1.0]
Fe^{3+}	β_1 7.53, β_2 13.64, β_3 18.49; MHL/M.HL 4.35 [0.5]
Co^{2+}	β_1 3.25, β_2 5.60 [1.0], MHL/M.HL 1.61, M(HL)$_2$/M.(HL)^2 2.89 [0.1]
Ni^{2+}	β_1 5.16 [0]
Cu^{2+}	β_1 4.84, β_2 9.21, MHL/M.HL 2.49 [0.1]
Ag^+	β_1 2.41 [1.0]
Zn^{2+}	β_1 3.88, β_2 6.40 [0.16], MHL/M.HL 1.72, M(HL)$_2$/M.(HL)^2 3.12 [0.1]
Cd^{2+}	β_1 2.75 [1.0]
Hg_2^{2+}	β_2 6.98 [2.5][c]
Hg^{2+}	β_1 9.66 [0.1]
Al^{3+}	β_1 6.1, β_2 11.09, β_3 15.12 [1.0]
Ga^{3+}	β_1 6.45, β_2 12.38, β_3 17.86 [1.0][a]
In^{3+}	β_1 5.30, β_2 10.52 [1.0], MHL/M.HL 3.08 [0.3]
Pb^{2+}	β_1 3.32, β_2 5.5 [1.0]

[a] 20°C; [b] 18°C; [c] 27°C

3. $C_2H_3ClO_2$	monochloroacetic acid	$ClCH_2.COOH$	HL
H^+	K_{H1} 2.865 [0]; 2.68 [0.1]; 2.64 [1.0]; 3.02 [3.0]		
Y^{3+}	β_1 1.06, β_2 1.93 [0.1]		
La^{3+}	β_1 1.12, β_2 1.86 [0.1]		
$M^{3+}(4f)$	β_1 (1.2), β_2 (2.1) [0.1]		
Th^{4+}	β_1 2.77, β_2 4.64, β_3 5.8, β_4 6.8 [1.0][a]		
UO_2^{2+}	β_1 1.43, β_2 2.28, β_3 2.7 [1.0]		
$MO_2^{2+}(5f)$	β_1 (1.2), β_2 (1.8), β_3 (2.4) [1.0]		
Fe^{3+}	β_1 2.1 [1.0][a]		

156 **Stability Constants of Complexes with Organic Ligands** [Ch. 7

Table 7.2 (Continued)

Formula of the acid Metal ion	Stability constants (Symbol for the constant or scheme of the equilibrium, value of log K or log β, $[I]$)
	monochloroacetic acid (continued)
Co^{2+}	β_1 0.2 $[2.0]^b$
Ni^{2+}	β_1 0.20 $[2.0]^b$
Cu^{2+}	β_1 0.91, β_2 1.09, β_3 1.45 $[1.0]^a$
Ag^+	β_1 0.64, β_2 0.7 $[0]$
Zn^{2+}	β_1 0.40 $[2.0]^b$
Cd^{2+}	β_1 0.84 $[2.0]^b$
Hg^{2+}	β_1 4.64 $[0.1]$
Pb^{2+}	β_1 1.50, β_2 1.8 $[2.0]^b$

^{a} 20°C; ^{b} 18°C

4. $C_2H_4O_2$	acetic acid	$CH_3.COOH$	HL
H^+	K_{H1} 4.757 $[0]$; 4.56 $[0.1]$; 4.50 $[0.5]$; 4.57 $[1.0]$; 4.80 $[2.0]$; 5.015 $[3.0]$		
Li^+	β_1 0.26 $[0]$		
Na^+	β_1 -0.18 $[0]$		
Be^{2+}	β_1 1.62, β_2 2.36 $[0.1]$		
Mg^{2+}	β_1 1.27 $[0]$; 0.51 $[0.2]^a$		
Ca^{2+}	β_1 1.18 $[0]$; 0.53 $[0.2]^a$		
Sr^{2+}	β_1 1.14 $[0]$; 0.43 $[0.2]^a$		
Ba^{2+}	β_1 1.07 $[0]$; 0.39 $[0.2]^a$		
Y^{3+}	β_1 1.59, β_2 2.73, β_3 3.5, β_4 3.3^b $[2.0]$		
La^{3+}	β_1 1.59, β_2 2.53, β_3 3.04, β_4 2.9^b $[2.0]$		
$M^{3+}(4f)$	β_1 (1.8), β_2 (2.9), β_3 (3.6), β_4 (3.6) $[2.0]$		
Th^{4+}	β_1 3.89, β_2 6.94, β_3 9.01, β_4 10.3, β_5 11.0 $[1.0]$		
Cr^{3+}	β_1 4.63, β_2 7.08, β_3 9.6 $[0.3]$		
UO_2^{2+}	β_1 2.44, β_2 4.42, β_3 6.43 $[1.0]$		
NpO_2^{2+}	β_1 2.31, β_2 4.23, β_3 6.00 $[1.0]^b$		
PuO_2^{2+}	β_1 2.13, β_2 3.49, β_3 5.01 $[1.0]$		
$M^{3+}(5f)$	β_1 (2.1) $[2.0]$		
Mn^{2+}	β_1 1.40 $[0]$; 0.69 $[1.0]$		
Fe^{2+}	β_1 1.40 $[0]$		
Fe^{3+}	β_1 3.38, β_2 6.5, β_3 8.3 $[0.1]^b$		
Co^{2+}	β_1 1.46 $[0]$; 0.81 $[1.0]$		
Ni^{2+}	β_1 1.43 $[0]$; 0.83 $[1.0]$		
Cu^{2+}	β_1 2.22 $[0]$; 1.83 $[0.1]$; 1.71 $[1.0]$; β_2 3.63 $[0]$; 3.09 $[0.1]$; 2.71 $[1.0]$; β_3 3.1, β_4 2.9 $[1.0]^b$		
Ag^+	β_1 0.73 $[0]$; 0.37 $[3.0]$; β_2 0.64 $[0]$; 0.14 $[3.0]$; β_3 -0.3 $[3.0]$		
Zn^{2+}	β_1 1.57 $[0]$; 1.1 $[0.1]$; 0.91 $[3.0]$; β_2 1.9 $[0.1]$; 1.36 $[3.0]$; β_3 1.57 $[3.0]$		
Cd^{2+}	β_1 1.93 $[0]$; 1.19 $[0.5]$; 1.32 $[3.0]$; β_2 3.15 $[0]$; 1.90 $[0.5]$; 2.26 $[3.0]$; β_3 2.17 $[0.5]$; 2.42 $[3.0]$; β_4 2.0 $[3.0]$		

Ch. 7]   Stability Constants of Complexes with Organic Ligands   157

Table 7.2 (Continued)

Formula of the acid Metal ion	Stability constants (Symbol for the constant or scheme of the equilibrium, value of log K or log β, [I])
	acetic acid (continued)
Hg^{2+}	β_1 5.55, β_2 9.30, β_3 13.28, β_4 17.06 [1.0][c]
Al^{3+}	β_1 1.51 [1.0]
In^{3+}	β_1 3.50, β_2 5.95, β_3 7.90, β_4 9.08 [1.0][b]
Tl^+	β_1 −0.11 [0]
Tl^{3+}	β_1 6.17, β_2 11.28, β_3 15.10, β_4 18.3, $MHL_2/M.H.L^2$ 12.99, $MOHL/M.OH.L$ 18.41, $M(OH)_2L/M.(OH)^2.L$ 30.1, $MOHL_2/M.OH.L^2$ 22.9 [3.0]
Sn^{2+}	β_1 3.3, β_2 6.0, β_3 7.3 [3.0]
Pb^{2+}	β_1 2.68 [0]; 2.15 [0.1]; 2.33 [3.0]; β_2 4.08 [0]; 3.5 [0.1]; 3.60 [3.0]; β_3 3.6, β_4 2.9 [3.0]

[a] Temperature not given; [b] 20°C; [c] 30°C

5. $C_2H_4O_3$	hydroxyacetic acid (glycollic acid) $CH_2(OH).COOH$ HL
H^+	K_{H1} 3.831 [0]; 3.63 [0.1]; 3.62 [1.0]; 3.74 [2.0]; 3.91 [3.0]
M^{2+}(IIA)	β_1 (0.9) [0.2][b]
Y^{3+}	β_1 2.47, β_2 4.40, β_3 5.7, β_4 6.3, β_5 6.3[a] [2.0]
La^{3+}	β_1 2.18, β_2 3.75, β_3 4.79, β_4 5.1, β_5 4.8[a] [2.0]
$M^{3+}(4f)$	β_1 (2.5), β_2 (4.5), β_3 (5.8), β_4 (6.3), β_5 (6.1)[a] [2.0]
Th^{4+}	β_1 3.98, β_2 7.36, β_3 9.95, β_4 11.95 [1.0][a]
$M^{3+}(5f)$	β_1 (2.6), β_2 (4.5) [2.0]
$MO_2^{2+}(5f)$	β_1 (2.3), β_2 (3.8), β_3 (4.8) [1.0][a]
Mn^{2+}	β_1 1.58 [0]
Fe^{2+}	β_1 1.33 [1.0]
Fe^{3+}	β_1 2.90, $MH_{-1}L/M.L.OH$ 15.39, $MH_{-1}L_2/MH_{-1}L.L$ 2.41, $MH_{-1}L_3/MH_{-1}L_2.L$ 1.5 [1.0]
Co^{2+}	β_1 1.48, β_2 2.29, β_3 2.52 [2.0]
Ni^{2+}	β_1 1.69, β_2 2.70, β_3 3.05 [2.0]
Ag^+	β_1 0.30, β_2 0.36 [3.0]
Cu^{2+}	β_1 2.31 [1.0]; 2.50 [3.0]; β_2 3.72 [1.0]; 4.20 [3.0]; β_3 4.27 [3.0]
Zn^{2+}	β_1 1.82, β_2 2.92, β_3 3.2 [2.0]
Cd^{2+}	β_1 1.47, β_2 2.0 [2.0]
In^{3+}	β_1 2.93, β_2 5.52, β_3 7.30, β_4 7.95 [2.0][a]
Pb^{2+}	β_1 2.01, β_2 2.94 [1.0]; β_1 2.23, β_2 3.24, β_3 3.2 [3.0]

[a] 20°C; [b] temperature not given

6. $C_3H_4O_4$	malonic acid $HOOC.CH_2.COOH$ H_2L
H^+	K_{H1} 5.696 [0]; 5.28 [0.1]; 5.07 [1.0]; 5.14 [2.0]; K_{H2} 2.847 [0]; 2.65 [0.1]; 2.60 [1.0]; 2.68 [2.0]

158 Stability Constants of Complexes with Organic Ligands [Ch. 7

Table 7.2 (Continued)

Formula of the acid Metal ion	Stability constants (Symbol for the constant or scheme of the equilibrium, value of $\log K$ or $\log \beta$, $[I]$)
	malonic acid (continued)
Be^{2+}	β_1 5.30, β_2 8.56 [0.1][a]
Mg^{2+}	β_1 2.11, MHL/M.HL 0.96 [0.1]
M^{2+}(IIA)	β_1 (1.3) [0.1]; β_2 (0.4) [0.2][d]
Sc^{3+}	β_1 5.87, β_2 10.12, β_3 13.07 [1.0]
Y^{3+}	β_1 4.40, β_2 7.04 [0.1]
La^{3+}	β_1 3.07, β_2 5.1, MHL/M.HL 1.24, $MHL_2/ML_2.H$ 4.1 [1.0]
$M^{3+}(4f)$	β_1 (3.7), β_2 (5.7), β_3 (7.7), MHL/M.HL (1.3), $MHL_2/ML_2.H$ (3.6) [1.0]
Th^{4+}	β_1 7.42, β_2 12.68 [1.0][a]
Cr^{2+}	β_1 3.92, β_2 7.13 [0.1]
Cr^{3+}	β_1 8.26 [2.0][b]
Mn^{2+}	β_1 3.28 [0]
Fe^{3+}	β_1 7.52 [0.5]
Co^{2+}	β_1 2.97, β_2 4.4, MHL/M.HL 0.82 [0.1]
Ni^{2+}	β_1 3.24, β_2 4.9, MHL/M.HL 1.04 [0.1]
Cu^{2+}	β_1 5.05, β_2 7.8, MHL/M.HL 2.15 [0.1]
Zn^{2+}	β_1 2.96, β_2 4.4, MHL/M.HL 0.99 [0.1]
Cd^{2+}	β_1 1.92, β_2 2.88, MHL/M.HL 0.69 [1.0][a]
Pb^{2+}	β_1 2.60, β_2 3.62, β_3 4.32 [2.0][c]

[a] 20°C; [b] 40°C; [c] 30°C; [d] temperature not given

7. $C_3H_6O_3$	D-2-hydroxypropionic acid $CH_3CH(OH).COOH$ HL lactic acid
H^+	K_{H1} 3.860 [0]; 3.66 [0.1]; 3.64 [1.0]; 3.81 [2.0]
M^{2+}(IIA)	β_1 (0.6), β_2 (0.9) [1.0]
Sc^{3+}	β_1 5.2 [0]
Y^{3+}	β_1 2.53, β_2 (4.70), β_3 (6.12) [2.0]; β_1 3.02, β_2 (5.33), β_3 (6.95) [0.1][a]
La^{3+}	β_1 2.27, β_2 (3.95), β_3 (5.1) [2.0]; β_1 2.60, β_2 (4.34), β_3 (5.74) [0.1][a]
$M^{3+}(4f)$	β_1 (3.0), β_2 (5.3), β_3 (6.7) [0.1][a]
Am^{3+}	β_1 2.52, β_2 (4.76), β_3 (6.0) [2.0]
Th^{4+}	β_1 4.21, β_2 (7.78), β_3 (10.54), β_4 (12.90) [1.0][a]
VO^{2+}	β_1 2.68, β_2 4.83 [1.0][a]
UO_2^{2+}	β_1 2.76, β_2 (4.43), β_3 (5.77) [1.0]
Mn^{2+}	β_1 0.92, β_2 (1.46), β_3 (1.6) [1.0]
Co^{2+}	β_1 1.38, β_2 2.37, β_3 (2.7) [1.0]
Ni^{2+}	β_1 1.64, β_2 (2.76), β_3 (3.1) [1.0]
Cu^{2+}	β_1 2.45, β_2 (4.08), β_3 (4.7) [1.0]
Zn^{2+}	β_1 1.7, β_2 (2.8), β_3 (3.4) [1.0]
Cd^{2+}	β_1 1.30, β_2 (2.1), β_3 (2.5) [1.0]

Stability Constants of Complexes with Organic Ligands

Table 7.2 (Continued)

Formula of the acid Metal ion	Stability constants (Symbol for the constant or scheme of the equilibrium, value of log K or log β, $[I]$)
	lactic acid (continued)
Pb^{2+}	β_1 1.99, β_2 (2.88), [1.0], β_3 (4.3) [2.0]

[a] 20°C

8. $C_4H_4O_4$	maleic acid HOOC.CH=CH.COOH H_2L
H^+	K_{H1} 6.332 [0]; 5.83 [0.1]; 5.62 [1.0];
	K_{H2} 1.910 [0]; 1.75 [0.1]; 1.63 [1.0]; 1.71 [2.0]
Na^+	β_1 0.7 [0]
M^{2+}(IIA)	β_1 (2.35) [0]
Y^{3+}	β_1 3.61, β_2 5.54 [0.1]
La^{3+}	β_1 3.45, β_2 5.43 [0.1]
M^{3+}(4f)	β_1 (3.7), β_2 (5.8) [0.1]
Th^{4+}	β_1 6.34, β_2 10.55 [1.0][a]
MO_2^{2+}(5f)	β_1 (4.35) [1.0]
Mn^{2+}	β_1 1.68 [0.16]
Ni^{2+}	β_1 2.0 [0.1]
Cu^+	β_1 3.05 [0.1]
Cu^{2+}	β_1 3.4 [0.1]; β_2 4.9, β_3 6.2 [0.2]
Zn^{2+}	β_1 2.0 [0.1]
Cd^{2+}	β_1 2.2, β_2 3.6, β_3 3.8 [0.2]
In^{3+}	β_1 5.0, β_2 7.1, β_3 6.2 [0.2]
Pb^{2+}	β_1 2.75, β_2 4.03, β_3 4.36, MHL/M.HL 0.58, M(HL)$_2$/M.(HL)^2 0.7 [1.0]

[a] 20°C

9. $C_4H_6O_4$	succinic acid HOOC.CH$_2$CH$_2$.COOH H_2L
H^+	K_{H1} 5.636 [0]; 5.24 [0.1]; 5.12 [1.0]; 5.21 [2.0];
	K_{H2} 4.207 [0]; 4.00 [0.1]; 3.95 [1.0]; 4.07 [2.0]
Be^{2+}	β_1 3.13, MHL/M.HL 1.44, MOHL/ML.OH 8.2 [1.0]
M^{2+}(IIA)	β_1 (1.1), MHL/M.HL 0.5 [0.1 – 0.2][a]
La^{3+}	β_1 3.96 [0]; MHL/M.HL 1.48, M(HL)$_2$/M.(HL)^2 2.7 [0.16]
M^{3+}(4f)	MHL/M.HL (1.8), M(HL)$_2$/M.(HL)^2 (3.1) [0.16]
Th^{4+}	β_1 6.23 [1.0][b]
UO_2^{2+}	β_1 3.68 [1.0]; 3.87, MHL/M.HL 2.13 [0.5]
Mn^{2+}	β_1 2.26, MHL/M.HL 1.2 [0]; β_1 1.48, MHL/M.HL 0.7 [0.2]
Fe^{2+}	β_1 1.4 [0.15][c]
Fe^{3+}	β_1 6.88 [0.5]
Co^{2+}	β_1 2.32 [0]; 1.70, MHL/M.HL 0.99 [0.1][b]
Ni^{2+}	β_1 2.34 [0]; 1.6 [0.1]; MHL/M.HL 1.3 [0]

Table 7.2 (Continued)

Formula of the acid / Metal ion	Stability constants (Symbol for the constant or scheme of the equilibrium, value of $\log K$ or $\log \beta$, $[I]$)
	succinic acid (continued)
Cu^{2+}	β_1 3.28 [0]; 2.6 [0.1]
Zn^{2+}	β_1 1.48, β_2 2.00 [1.0]; MHL/M.HL 0.96 [0.1][b]
Cd^{2+}	β_1 1.67, β_2 2.79, MHL/M.HL 0.99 [1.0][b]
Hg_2^{2+}	β_2 7.28, MOHL/M.OH.L 13.45 [2.5][d]
Pb^{2+}	β_1 2.40, β_2 3.73, β_3 4.11 [2.0][e]

[a] Temperature about 20–25°C; [b] 20°C; [c] 37°C; [d] 27°C; [e] 30°C

			OH		
			|		
10. $C_4H_6O_5$	L-malic acid	HOOC.CH$_2$.CH.COOH			H$_2$L

H^+	K_{H1} 5.097 [0]; 4.71 [0.1]; 4.45 [1.0]
	K_{H2} 3.459 [0]; 3.24 [0.1]; 3.11 [1.0]
M^+(IA)	β_1 (0.3) [0.1]
Be^{2+}	β_1 2.70, MHL/M.HL 1.21, $M_2L_2/M^2.L^2$ 8.48 [1.0]
M^{2+}(IIA)	β_1 (1.5), MHL/M.HL (0.8) [0.2][a]
Y^{3+}	β_1 4.91, β_2 (8.18) [0.1][b]
La^{3+}	β_1 4.37, β_2 (7.16) [0.1][b]
M^{3+}(4f)	β_1 (4.9), β_2 (8.1) [0.1][b]
UO_2^{2+}	$MH_{-1}L/M.L.OH$ 15.8, $M_2(H_{-1}L)_2/M^2.L^2(OH)^2$ 35.0[c] [1.0]
Mn^{2+}	β_1 2.24 [0.16]
Fe^{2+}	β_1 2.6 [0.1][b]
Fe^{3+}	β_1 7.1, $M_2(H_{-1}L)_2/M^2.L^2.(OH)^2$ 40.45, $M_2H_{-2}L_3/M_2(H_{-1}L)_2.L$ 5.0, $M_3H_{-4}L_5/M^3.L^5.(OH)^4$ 81.2 [0.1][b]
Co^{2+}	β_1 2.86, MHL/M.HL 1.64 [0.1][b]
Ni^{2+}	β_1 3.17, MHL/M.HL 1.83 [0.1][b]
Cu^{2+}	β_1 3.42, MHL/M.HL 2.00, $MH_{-1}L/ML.OH$ 9.26 [0.1][b]
Zn^{2+}	β_1 2.93, MHL/M.HL 1.66 [0.1][b]
Cd^{2+}	β_1 2.36, MHL/M.HL 1.34 [0.1][b]
Pb^{2+}	MHL/M.HL 2.45, $M(HL)_2/M.(HL)^2$ 3.70 [1.0]

[a] Temperature not given; [b] 20°C; [c] the ligand is a mixture of the D and L acids

11. $C_4H_6O_6$	D-tartaric acid HOOC.CH(OH).CH(OH).COOH H$_2$L
H^+	K_{H1} 4.366 [0]; 3.95 [0.1]; 3.73 [1.0]; 3.81 [2.0];
	K_{H2} 3.036 [0]; 2.82 [0.1]; 2.69 [1.0]; 2.83 [2.0]
Na^+	β_1 0.83, MHL/M.HL 0.22 [0]
Be^{2+}	β_1 1.74, MOHL/ML.OH 9.40 [1.0]
M^{2+}(IIA)	β_1 (1.6), MHL/M.HL (1.0) [0.2][a]
Sc^{3+}	β_2 (12.5) [0.1][b]

Stability Constants of Complexes with Organic Ligands

Table 7.2 (Continued)

Formula of the acid / Metal ion	Stability constants (Symbol for the constant or scheme of the equilibrium, value of $\log K$ or $\log \beta$, $[I]$)
	D-tartaric acid (continued)
Y^{3+}	β_1 4.03 [0.05]
La^{3+}	β_1 3.68, β_2 6.13, MHL/M.HL 2.44 [0.1]^c
$M^{3+}(4f)$	β_1 (4.2) [0.05]
UO_2^{2+}	$MH_{-1}L/M.L.OH$ 14.6, $M_2(H_{-1}L)_2/M^2.L^2.(OH)^2$ 32.44 [1.0]
VO^{2+}	β_1 4.9, β_2 9.0 [0.25]
Mn^{2+}	β_1 2.49 [0.1]
Fe^{2+}	β_1 2.2 [0.1]^b; 1.43, β_2 (2.5)^d [1.0]
Fe^{3+}	β_1 6.49, $M_2(H_{-1}L)_2/M^2.L^2.(OH)^2$ 39.5^d, $M_2H_{-3}L_2/M_2(H_{-1}L)_2OH$ 11.0^d, $M_3(H_{-2}L)_3/M^3.L^3.(OH)^6$ 92.3^d [0.1]^b, $M_2(H_{-2}L)_2/M_2(H_{-1}L)_2.(OH)^2$ (22.7)^d [1.0]
Zn^{2+}	β_1 3.82, β_2 (5.0)^d [0]
Hg^{2+}	β_1 7.0 [0.1]
Al^{3+}	β_1 5.32, β_2 (9.77)^d [1.0]
In^{3+}	β_1 4.44, β_2 (8.46)^d [1.0]
Sn^{2+}	β_1 5.2, β_2 (9.91)^d [0.1]^b
Pb^{2+}	β_1 2.60, β_2 (4.0)^d, MHL/M.HL 1.76, $MHL_2/ML_2.H$ (3.5)^d [1.0]
Sb(III)	$M_2L_2/[M(OH)_3]^2.(H_2L)^2.H^2$ 21.5, $(MOHL)^2/M_2L_2.(OH)^2$ 10.16 [0.1]^b

^{a} Temperature not given; ^{b} 20°C; ^{c} $(CH_3)_4N^+$ medium; ^{d} the ligand is a mixture of the D and L acids

12. $C_6H_8O_6$	L-ascorbic acid	HO.C=C.OH / HOCH$_2$.C.CH(O)C=O / OH H_2L

H^+	K_{H1} 11.56 [0]; 11.34 [0.1]; 11.35 [3.0]
	K_{H2} 4.17 [0]; 4.03 [0.1]; 4.37 [3.0]
Ca^{2+}	β_1 1.4, β_{12} 1.85, β_{33} 8.74, β_{43} 10.50, MHL/M.HL 0.03, $M_4OHL_3/M^4.L^3.OH$ 12.1, $M_4OHL_4/M^4.L^4.OH$ 14.7 [3.0]
Ti(IV)	$MO(HL)_2/MO.(HL)^2$ 24.8, $MO(H_2L)/MO.H_2L$ 3.1, $MO(H_2L)_2/MO.(H_2L)^2$ 6.25, $M(HL)_3/MO.H^2.(HL)^3$ 9.3 [0.1]^a
VO_2^+	$MH_2L/M.H_2L$ 2.69 [1.0]
UO_2^{2+}	MHL/M.HL 2.35, $M(HL)_2/M.(HL)^2$ 3.32, $M(OH)_2HL/M.HL.(OH)^2$ 19.4, $M(OH)(HL)_2/M.(HL)^2.OH$ 12.2 [0.1]
Mn^{2+}	MHL/M.HL 1.1 [0]
Fe^{2+}	MHL/M.HL 0.21, β_1 1.99 [3.0]
Co^{2+}	MHL/M.HL 1.4 [0]
Ni^{2+}	MHL/M.HL 1.1 [0]
Cu^{2+}	MHL/M.HL 1.57 [0.1]^b
Ag^+	β_1 3.66 [0.1]

Table 7.2 (Continued)

Formula of the acid Metal ion	Stability constants (Symbol for the constant or scheme of the equilibrium, value of $\log K$ or $\log \beta$, $[I]$)
	L-ascorbic acid (continued)
Zn^{2+}	MHL/M.HL 1.0 [0]
Cd^{2+}	MHL/M.HL 1.3 [0]; 0.42 [3.0]; $M_2L/M^2.L$ 5.83 [3.0]
Al^{3+}	MHL/M.HL 1.89, $MH_2L_2/M.(HL)^2$ 3.55 [0.1]
Pb^{2+}	MHL/M.HL 1.8 [0]; 1.77 [0.1]
As(III)	$M(OH)_2HL/M(OH)_2.HL$ 7.2 [0.1]

[a] 20°C; [b] 0°C

13. $C_6H_8O_7$ citric acid   HOOC.CH$_2$.C(OH).CH$_2$.COOH H_3L
$\qquad\qquad\qquad\qquad\qquad\qquad\quad$|
$\qquad\qquad\qquad\qquad\qquad\qquad$COOH

H^+	K_{H1} 6.396 [0]; 5.69 [0.1]; 5.83 [0.1][a]; 5.33 [1.0]; 5.18 [2.0],
	K_{H2} 4.761 [0]; 4.35 [0.1]; 4.35 [0.1][a]; 4.08 [1.0]; 4.16 [2.0],
	K_{H3} 3.128 [0]; 2.87 [0.1]; 2.87 [0.1][a]; 2.80 [1.0]; 2.90 [2.0]
M^+(IA)	β_1 (0.6) [0.1]
Be^{2+}	β_1 4.5, MHL/M.HL 2.2, $MH_2L/M.H_2L$ 1.4 [0.15][b]
M^{2+}(IIA)	β_1 (3.2), MHL/M.HL (1.9), $MH_2L/M.H_2L$ (0.9) [0.1][c]
Y^{3+}	β_1 7.87 [0.05]
La^{3+}	β_1 7.17, β_2 10.2, $MHL_2/ML.HL$ 2.2 [0.1]
Ce^{3+}	β_1 7.39, β_2 10.4, $MHL_2/ML.HL$ 2.4 [0.1]
M^{3+}(4f)	β_1 (7.9) [0.05]
UO_2^{2+}	β_1 7.4, β_{22} 18.87 [0.1]
M^{3+}(5f)	β_1 (7.8), β_2 (11.1), $MHL_2/ML.HL$ (2.6) [0.1]
Mn^{2+}	β_1 4.15, MHL/M.HL 2.16 [0.1][a]
Fe^{2+}	β_1 4.4, MHL/M.HL 2.65 [0.1][c]
Fe^{3+}	β_1 11.50, $M_2(H_{-1}L)_2/(ML)^2.(OH)^2$ 26.3 [0.1][c]
Co^{2+}	β_1 5.00, MHL/M.HL 3.02, $MH_2L/M.H_2L$ 1.25 [0.1][c]
Ni^{2+}	β_1 5.40, MHL/M.HL 3.30, $MH_2L/M.H_2L$ 1.75 [0.1][c]
Cu^{2+}	β_1 5.90, MHL/M.HL 3.42, $MH_2L/M.H_2L$ 2.26, $MH_{-1}L/ML.OH$ 9.61, β_{12} 8.10 [0.1][c], β_{22} 13.2, $M_2(H_{-1}L)_2/M_2L_2.(OH)^2$ 19.57 [1.0]
Zn^{2+}	β_1 4.27, β_2 5.90 [0.5]; β_1 4.98, MHL/M.HL 2.98, $MH_2L/M.H_2L$ 1.25 [0.1][c]
Cd^{2+}	β_1 3.15, β_2 4.54 [0.5]; β_1 3.75, MHL/M.HL 2.20, $MH_2L/M.H_2L$ 0.97 [0.1][c]
Hg^{2+}	β_1 10.9 [0.1]
Ga^{3+}	β_1 10.02, MOHL/ML.OH 10.9 [0.1]
Tl^+	β_1 1.04 [0.1]
Pb^{2+}	β_1 4.34, β_2 6.08, β_3 6.97 [3.0], β_1 4.08, β_2 6.1; MHL/M.HL 2.97, $MH_2L/M.H_2L$ 1.51, $MH_2L_2/ML_2.H^2$ 8.9, $MH_4L_2/MH_2L_2.H^2$ 6.7 [2.0]
As(III)	$M(OH)_2L/M(OH)_2.L$ 9.3, $M(OH)_2HL/M(OH)_2.HL$ 8.5 [0.1]

[a] $(CH_3)_4N^+$ medium; [b] 34°C; [c] 20°C

Stability Constants of Complexes with Organic Ligands

Table 7.2 (Continued)

Formula of the acid Metal ion	Stability constants (Symbol for the constant or scheme of the equilibrium, value of log K or log β, $[I]$)

14. $C_7H_6O_2$ benzoic acid $C_6H_5.COOH$ HL

- H^+ K_{H1} 4.202 [0]; 4.00 [0.1]; 3.97 [1.0]
- M^{2+}(IIA) β_1 (0.2) [0.4]^a
- Co^{2+} β_1 0.55 [0.4]^a
- Ni^{2+} β_1 0.9 [0.1]; 0.55 [0.4]^a
- Cu^{2+} β_1 1.6 [0.1]; 1.51 [0.4]^a
- Ag^+ β_1 0.52, β_2 0.54 [1.0]
- Zn^{2+} β_1 0.9 [0.1]; 0.74 [0.4]^a
- Cd^{2+} β_1 1.4 [0.1]; 1.15 [0.4]^a; 0.99 [1.0]^a; β_2 1.76 [1.0]^a
- Al^{3+} MOHL/M.OH.L 12.09 [0.5]
- Pb^{2+} β_1 2.0 [0.1]; 1.99 [0.4]^a; β_2 3.30 [1.0]^a

^{a} 30°C

15. $C_7H_6O_3$ salicylic acid $HO.C_6H_4.COOH$ H_2L
                 2-hydroxybenzoic acid

- H^+ K_{H1} 13.74 [0]; 13.4 [0.1]; 13.15 [1.0]; 13.12 [3.0];
 K_{H2} 2.97 [0]; 2.81 [0.1]; 2.78 [1.0]; 3.16 [3.0]
- Be^{2+} β_1 12.37, β_2 22.02 [0.1]^a
- M^{2+}(IIA)   MHL/M.HL (0.3) [0]
- La^{3+} MHL/M.HL 2.08 [0]
- Th^{4+} MHL/M.HL 4.25, $M(HL)_2/M(HL)^2$ 7.60, $M(HL)_3/M.(HL)^3$ 10.05, $M(HL)_4/M.(HL)^4$ 11.60 [0.1]
- VO^{2+} β_1 13.38 [0.1]; (12.7), β_2 (22.4) [0.1]^a
- Cr^{2+} β_1 8.41, β_2 15.36 [0.1]
- UO_2^{2+} β_1 12.08^a, β_2 20.83^a, MHL/M.HL 2.2 [0.1]
- Mn^{2+} β_1 5.90, β_2 9.8 [0.15]^a
- Fe^{2+} β_1 6.55, β_2 11.2 [0.15]^a
- Fe^{3+} β_1 17.44 [0]; 16.3 [0.1]; 16.35 [0.15]^a, 15.81, β_2 27.49, β_3 35.31 [3.0]; MHL/M.HL 4.4 [0.1]
- Co^{2+} β_1 6.72, β_2 11.4 [0.15]^a
- Ni^{2+} β_1 6.95, β_2 11.7 [0.15]^a
- Cu^{2+} β_1 10.62 [0.1]; 10.60, β_2 18.45 [0.15]^a
- Zn^{2+} β_1 6.85 [0.15]^a
- Cd^{2+} β_1 5.55 [0.15]^a
- B(III) $M(OH)_2L/M(OH)_3.HL$ 1.30, $ML_2/M(OH)_3.H_2L.HL$ 3.6 [0.1]
- Al^{3+} β_1 12.9, β_2 23.2, β_3 29.8 [0.1]^a

^{a} 20°C

Table 7.2 (Continued)

Formula of the acid / Metal ion	Stability constants (Symbol for the constant or scheme of the equilibrium, value of log K or log β, [I])
16. $C_7H_6O_4$	3,4-dihydroxybenzoic acid $(HO)_2.C_6H_3.COOH$ H_3L
H^+	K_{H1} 12.2 [0.1]^a; (12.80) [1.0]; K_{H2} 8.84, 8.70^a [0.1]; 8.68 [1.0]; K_{H3} 4.35, 4.34^a [0.1]; 4.34 [1.0]
Mg^{2+}	β_1 5.67, β_2 9.84 [0.1]^a
Ca^{2+}	β_1 3.71, β_2 6.36 [0.1]^a
Ti(IV)	$ML_3/MO.H^2.L^3$ 58.6, $MO(H_2L)_2/MO.(H_2L)^2$ 7.3 [0.1]^b
Nb(V)	$M(HL)_3/MO_2.(HL)^3.H^4$ 63.1, $MO_2(H_2L)/MO_2.H_2L$ 6.65, $MO(HL)_2/MO_2(H_2L).(H_2L)$ 5.65 [0.1]^b
Mn^{2+}	β_1 7.22, β_2 12.28 [0.1]^a
Co^{2+}	β_1 7.96, β_2 13.36, β_3 17.42 [0.1]^a
Ni^{2+}	β_1 8.27, β_2 12.98, β_3 16.87 [0.1]^a
Cu^{2+}	β_1 12.79, β_2 22.60 [0.1]^a
Zn^{2+}	β_1 8.91, β_2 15.62 [0.1]^a
B(III)	$M(OH)_2L/M(OH)_3.HL$ 3.83 [0.1]

^{a} 30°C; ^{b} 20°C

17. $C_7H_6O_4$	3,5-dihydroxybenzoic acid $(HO)_2.C_6H_3.COOH$ H_3L
H^+	K_{H1} 10.54 [0.2]; K_{H2} 9.08 [0]; 9.00 [0.2]; K_{H3} 4.12 [0]; 3.84 [0.2]
UO_2^{2+}	$MH_2L/M.H_2L$ 2.13, $M(OH)_2H_2L/M.HL$ 6.98 [0.2]

18. $C_7H_6O_5$	gallic acid $(HO)_3.C_6H_2.COOH$ H_4L 3,4,5-trihydroxybenzoic acid
H^+	K_{H2} 11.45 [0.1]^a; K_{H3} 9.11 [0]; 8.70^a, 8.68 [0.1] K_{H4} 4.43 [0]; 4.26^a, 4.27 [0.1]
Mo(VI)	$MO_2(HL)_2/MO_4.(H_2L)^2$ 5.4 [0.1]^a
UO_2^{2+}	$MH_3L/M.H_3L$ 2.3 [0.1]^a
B(III)	$M(OH)_2HL/M(OH)_3.H_2L$ 3.82 [0.1]

^{a} 20°C

19. $C_7H_6O_6S$	5-sulphosalicylic acid $HO_3S.C_6H_3(OH).COOH$ H_3L
H^+	K_{H1} 12.53 [0]; 11.72, 11.80^a [0.1]; 11.40 [1.0]; 11.74 [3.0] K_{H2} 2.84 [0]; 2.49, 2.49^a [0.1]; 2.32 [1.0]; 2.67 [3.0] K_{H3} −0.75 [1.0]
Be^{2+}	β_1 11.54, β_2 20.43 [0.1]^a
La^{3+}	β_1 5.92, β_2 10.73 [1.0]^a
$M^{3+}(4f)$	β_1 (6.9), β_2 (12.6) [1.0]^a

Ch. 7] **Stability Constants of Complexes with Organic Ligands** 165

Table 7.2 (Continued)

Formula of the acid Metal ion	Stability constants (Symbol for the constant or scheme of the equilibrium, value of log K or log β, $[I]$)
	5-sulphosalicylic acid (continued)
Ti(IV)	$ML_3/MO.H^2.L^3$ 42.2, $MO(HL)/MO.H.L$ 14.9, $MO(HL)_2/MO.H^2.L^2$ 29.0 [0.1][a]
Th^{4+}	β_1 12.30 [1.0]
VO^{2+}	β_1 11.71 [0.1]; 12.0, β_2 20.6 [0.1][a]; MOHL/M.OH.L 18.3, $(MOHL)_2/M^2.(OH)^2.L^2$ 41.9 [0.1]
Cr^{2+}	β_1 7.14, β_2 12.88 [0.1]
Cr^{3+}	β_1 9.56 [0.1]
UO_2^{2+}	β_1 11.14, 11.25[a] [0.1]; 10.44 [1.0]; β_2 19.20, 18.75[a] [0.1]
Mn^{2+}	β_1 5.24 [0.1]; 4.77 [1.0]; β_2 8.24 [0.1]; 8.19 [1.0]
Fe^{2+}	β_1 5.90, β_2 9.9 [0.15][a]
Fe^{3+}	β_1 14.60 [0.15][a]; 14.40 [0.5]; 14.42 [3.0]; β_2 25.15 [0.15][a]; (22.2) [0.5]; 25.2 [3.0]; β_3 (30.6) [0.5]; 32.2 [3.0]
Co^{2+}	β_1 6.13, β_2 9.82 [0.1]
Ni^{2+}	β_1 6.42, β_2 10.24 [0.1]
Cu^{2+}	β_1 10.74 [0]; 9.43 [0.1]; 8.91 [1.0]; β_2 17.17 [0]; 16.3 [0.1]; 15.86 [1.0]
Zn^{2+}	β_1 6.05, β_2 10.7 [0.15][a]
Cd^{2+}	β_1 4.64 [0.15][a]
Al^{3+}	β_1 12.3, β_2 20.0, β_3 25.8 [0.1][a]

[a] 20°C

20. $C_8H_6O_4$	phthalic acid $C_6H_4(COOH)_2$ H_2L benzene-1,2-dicarboxylic acid
H^+	K_{H1} 5.408 [0]; 4.93 [0.1]; 4.73 [0.5]; 4.71 [1.0]; 4.87 [3.0]; K_{H2} 2.950 [0]; 2.75 [0.1]; 2.66 [0.5]; 2.66 [1.0]; 2.99 [3.0]
Na^+	β_1 0.7 [0]
Be^{2+}	β_1 3.97, β_2 5.69 [0.1]
Ca^{2+}	β_1 2.42 [0]; 1.07 [0.1]
Ba^{2+}	β_1 2.33 [0]; 0.92 [0.1]
Y^{3+}	β_1 3.46 [0.1]
La^{3+}	β_1 4.74 [0]; 3.44 [0.1]
$M^{3+}(4f)$	β_1 (3.6) [0.1]
Th^{4+}	β_1 5.92, β_2 10.05 [1.0][a]
UO_2^{2+}	β_1 4.81 [0.1][b]; 4.38 [1.0]
NpO_2^+	β_1 2.22 [1.0][a]
PuO_2^{2+}	β_1 4.11 [1.0][a]
Mn^{2+}	β_1 2.74 [0]
Co^{2+}	β_1 2.83 [0]; 2.03 [0.1]; 1.53 [0.5]; 1.42 [1.0]; MHL/M.H.L 6.0 [0.5]
Ni^{2+}	β_1 2.95 [0]; 2.17 [0.1]; 1.70 [0.5]; 1.57 [1.0]; MHL/M.H.L 5.4 [0.5]

Formula of the acid Metal ion	Stability constants (Symbol for the constant or scheme of the equilibrium, value of log K or log β, [I])
	phthalic acid (continued)
Cu^{2+}	β_1 4.04 [0]; 3.15 [0.1]; 2.81 [0.5]; 2.69 [1.0]; MHL/M.H.L 5.9 [0.5] β_2 5.3 [0]; 3.73 [1.0]; 4.14 [2.0]
Zn^{2+}	β_1 2.91 [0]; 2.2 [0.1]; β_2 4.2 [0]
Cd^{2+}	β_1 2.5 [0.1]
Hg_2^{2+}	β_1 4.90 [0.1][c]
Al^{3+}	β_1 3.18, β_2 6.32 [0.5]
Ga^{3+}	β_1 5.15 [0.1]

[a] 20°C; [b] 31°C; [c] 18°C

21. $C_8H_8O_2S$	(phenylthio)acetic acid	$C_6H_5.S.CH_2.COOH$	HL
H^+	K_{H1} 3.33 [0.1]		
Mn^{2+}	β_1 0.72 [0.1]		
Co^{2+}	β_1 0.76 [0.1]		
Ni^{2+}	β_1 0.7 [0.1]		
Cu^{2+}	β_1 1.43 [0.1]		
Ag^+	β_1 2.83, 2.77[a], β_2 4.9[a] [0.1]; MHL/M.H.L 5.5 [0.1][ab]		
Zn^{2+}	β_1 0.8 [0.1]		
Cd^{2+}	β_1 1.2 [0.1]		
Pb^{2+}	β_1 1.8 [0.1]		

[a] Acetate buffer medium; [b] 20°C

22. $C_8H_8O_3$	mandelic acid L-phenylhydroxyacetic acid	$C_6H_5.CH(OH).COOH$	HL
H^+	K_{H1} 3.40 [0]; 3.19 [0.1]; 3.17 [1.0]; 3.31 [2.0]; 3.49 [3.0]		
Ca^{2+}	β_1 1.45 [0]		
Ba^{2+}	β_1 0.77 [0]		
Y^{3+}	β_1 2.56, β_2 (5.01) [0.1]		
La^{3+}	β_1 2.28 [0.1]; 1.93 [2.0]; β_2 (3.81) [0.1]		
$M^{3+}(4f)$	β_1 (2.6) [0.1]; (2.5) [2.0]; β_2 (4.9) [0.1]		
Th^{4+}	β_1 3.88, β_2 (6.89), β_3 (9.69), β_4 (11.98) [1.0][a]		
UO_2^{2+}	β_1 2.57, β_2 (4.10), β_3 (5.32) [1.0][a]		
Co^{2+}	β_1 1.22, β_2 1.74, β_3 2.67 [2.0][a]		
Ni^{2+}	β_1 1.41, β_2 2.26, β_3 2.9 [2.0][a]		
Zn^{2+}	β_1 1.51, β_2 2.58, β_3 3.36 [2.0][a]		

[a] 20°C

Ch. 7] Stability Constants of Complexes with Organic Ligands

Table 7.3
Complexes of amino acids*

Formula of the acid Metal ion	Stability constants (Symbol for the constant or scheme of the equilibrium, value of log K or log β, [I])	Ref.
1. $C_2H_5NO_2$	aminoacetic acid $H_2N.CH_2.COOH$ glycine	HL
H^+	K_{H1} 9.778 [0]; 9.57 [0.1]; 9.54 [0.5]; 9.75 [1.0]^a K_{H2} 2.350 [0]; 2.36 [0.1]; 2.36 [0.5]; 2.35 [1.0]^a	
Mg^{2+}	β_1 2.22 [0.1]	
M^{2+}(IIA)	β_1 1.0 [0]	
M^{3+}(4f)	β_1 (Ce, Pm, Eu) 0.6 [2.0]^a	
M^{3+}(5f)	β_1 (Am, Cm) 0.7–0.8 [2.0]^a	
Mn^{2+}	β_1 3.19 [0]; 2.80 [0.1]; 2.65 [0.5]; β_2 4.72 [0.15]^b; 4.7 [0.5]	
Fe^{2+}	β_1 4.31 [0]; (4.13) [0.1]; 3.83 [1.0]^a; β_2 (7.65) [0.1]	
Fe^{3+}	β_1 10.0 [1.0]^a	
Co^{2+}	β_1 5.07 [0]; 4.64 [0.1]; 4.57 [0.5]; β_2 9.04 [0]; 8.46 [0.1]; 8.28 [0.5]; β_3 11.63 [0]; 10.81 [0.1]; 10.80 [0.5]	
Ni^{2+}	β_1 6.18 [0]; 5.78 [0.1]; 5.64 [0.5]; β_2 11.13 [0]; 10.58 [0.1]; 10.50 [0.5]; β_3 (14.23) [0]; 14.00 [0.1]; 14.0 [0.5]	
Pd^{2+}	β_1 15.25; β_2 27.50 [1.0]^a	
Cu^+	β_1 10.1 [0.2]	
Cu^{2+}	β_1 8.56 [0]; 8.15 [0.1]; 8.14 [0.5]; β_2 15.64 [0]; 15.03 [0.1]; 14.97 [0.5]	
Ag^+	β_1 3.51 [0]; 3.20 [0.1]; 3.16 [0.5]; β_2 6.89 [0]; 6.63 [0.1]; 6.60 [0.5]	
Zn^{2+}	β_1 5.38 [0]; 4.96 [0.1]; 4.88 [0.5]; β_2 9.81 [0]; 9.19 [0.1]; 9.06 [0.5]; β_3 12.33 [0]; 11.6 [0.1]; 11.56 [0.5]	
Cd^{2+}	β_1 4.69 [0]; 4.22 [0.1]; 4.14 [1.0]; β_2 8.40 [0]; 7.69 [0.1]; 7.60 [1.0]; β_3 10.68 [0]; 9.74 [1.0]	
Hg^{2+}	β_1 10.3, β_2 19.2 [0.5]^a; MClL.Cl/MCl$_2$.L 3.42, ML$_2$.Cl^2/MCl$_2$.L^2 6.03 [0]	
Ga^{3+}	β_1 9.33 [0.1]^c	
Pb^{2+}	β_1 (5.47); β_2 (8.86) [0]; 7.7 [0.1]	

^{a} 20°C; ^{b} 37°C; ^{c} 22°C

2. $C_3H_7NO_2$	L-2-aminopropionic acid $\quad CH_3.CH.COOH$ alanine $\qquad\qquad\qquad\qquad\;\; |$ $\qquad\qquad\qquad\qquad\qquad\quad NH_2$	HL
H^+	K_{H1} 9.867 [0]; 9.69 [0.1]; 9.66 [0.5]; 9.84 [1.0]^a K_{H2} 2.348 [0]; 2.30 [0.1]; 2.29 [0.5]; 2.31 [1.0]^a	

* All listed values of the constants refer to 25°C unless stated otherwise in a note. The notes follow each section of the table separately.

168   Stability Constants of Complexes with Organic Ligands   [Ch. 7

Table 7.3 (Continued)

Formula of the acid Metal ion	Stability constants (Symbol for the constant or scheme of the equilibrium, value of log K or log β, [I])	Ref.
	alanine (continued)	
Mg^{2+}	β_1 1.96 [0]	
M^{2+}(IIA)	β_1 (0.9) [0]	
Mn^{2+}	β_1 3.02 [0]; 2.5 [0.05]; 2.47 [0.15][b]; β_2 4.26, β_3 5.7 [0.15][b]	
Fe^{2+}	β_1 3.54 [1.0][a]	
Fe^{3+}	β_1 (10.4) [1.0][a]	
Co^{2+}	β_1 4.72 [0]; 4.31 [0.1]; β_2 (8.40) [0]; (7.8) [0.1]; β_3 (10.15) [0]; (9.5) [0.16]	
Ni^{2+}	β_1 5.83 [0]; 5.40 [0.1]; 5.31 [0.5]; β_2 (10.48) [0]; 9.87 [0.1]; 9.73 [0.5]; β_3 (12.73) [0.5]	
Cu^+	β_2 (9.6) [0.3]	
Cu^{2+}	β_1 8.55 [0]; 8.13 [0.1]; 8.14 [0.5]; β_2 (15.50) [0]; 14.92 [0.1] 14.90 [0.5]	
Ag^+	β_1 3.64; β_1 (7.18) [0]	
Zn^{2+}	β_1 4.95 [0]; 4.58 [0.1]; 4.56 [0.5]; β_2 (9.23) [0]; (8.6) [0.1]; 8.54 [0.5]; β_3 10.7 [0.15][b]; 10.57 [0.5]	
Cd^{2+}	β_1 3.80; β_2 7.10; β_3 (9.09) [1.0]	
Pb^{2+}	β_1 (5.00) [0]; 4.15 [0.37][a]; β_2 (8.24) [0]	

[a] 20°C; [b] 37°C

3. $C_3H_7NO_2$	3-aminopropionic acid $H_2N.CH_2.CH_2.COOH$ HL β-alanine	
H^+	K_{H1} 10.295 [0]; 10.10 [0.1]; 10.06 [0.5]; K_{H2} 3.551 [0]; 3.53 [0.1]; 3.47 [0.5]	
Co^{2+}	β_1 4.21 [0]; 3.58, β_2 6.14 [0.2]	
Ni^{2+}	β_1 4.99 [0]; 4.58 [0.1]; 4.46 [0.5]; β_2 7.95 [0.1]; 7.84 [0.5]; β_3 9.55 [0.5]	
Cu^{2+}	β_1 7.04, β_2 12.54 [0.1]	
Ag^+	β_1 3.33, β_2 7.12 [0.5]	
Zn^{2+}	β_1 4.10 [0.2][a]; 3.90, β_2 7.20, β_3 10.40 [0.5]	

[a] 15°C

4. $C_4H_7NO_4$	iminodiacetic acid HN$\begin{array}{c}\diagup CH_2.COOH\\ \diagdown CH_2.COOH\end{array}$ IDA	
H^+	K_{H1} 9.79 [0]; 9.44[a], 9.34 [0.1]; 9.17 [0.5]; 9.35 [1.0]; K_{H2} 2.84 [0]; 2.62[a], 2.61 [0.1]; 2.56 [0.5]; 2.56 [1.0]; K_{H3} 1.82 [0.1]; 1.76 [0.5]; 1.88 [1.0]	

Table 7.3 (Continued)

Formula of the acid / Metal ion	Stability constants (Symbol for the constant or scheme of the equilibrium, value of $\log K$ or $\log \beta$, $[I]$)	Ref.
	iminodiacetic acid (continued)	
Li^+	β_1 0.96 $[0.1]^a$	
Na^+	β_1 0.36 $[0.1]^a$	
Mg^{2+}	β_1 3.66 $[0]^a$; 2.94^a, 2.98 $[0.1]$	
Ca^{2+}	β_1 3.41 $[0]^a$; 2.59^a, 2.59 $[0.1]$	
Sr^{2+}	β_1 2.23^a, 2.23 $[0.1]$	
Ba^{2+}	β_1 1.67^a, 1.67 $[0.1]$	
Sc^{3+}	β_1 9.8, β_2 15.9, MHL/M.H.L 10.6, $MH_2L/M.H^2.L$ 13.5 $[1.0]$	
Y^{3+}	β_1 6.78, β_2 12.03 $[0.1]$	
La^{3+}	β_1 5.88^a, 5.88, β_2 9.97^a, 9.97 $[0.1]$	
$M^{3+}(4f)$	β_1 (6.4) $[0.1]$; (6.6) $[1.0]$; β_2 (12.2) $[0.1]$; (11.8) $[1.0]$; β_3 (15.5), MHL/M.H.L (10.7), $MH_2L/M.H^2.L$ (12.8) $[1.0]$	
Cr^{3+}	β_1 10.9, β_2 21.4 $[0.1]$	
UO_2^{2+}	β_1 8.96 $[0.1]$; 8.71 $[1.0]$	
NpO_2^+	β_1 6.3, MHL/M.L.H 10.8 $[0.1]$	
PuO_2^+	β_1 6.2	
Fe^{2+}	β_1 5.8 $[0.1]^a$; 5.54 $[0.5]$; β_2 10.1 $[0.1]^a$; 9.81 $[0.5]$	
Fe^{3+}	β_1 10.72 $[0.5]$	
Co^{2+}	β_1 6.94, 6.97^a $[0.1]$; β_2 12.23, 12.31^a $[0.1]$	
Co^{3+}	β_1 29.6 $[0.1]$	
Ni^{2+}	β_1 9.24 $[0]$; 8.13, 8.19^a $[0.1]$; β_2 15.71 $[0]$; 14.1, 14.3^a $[0.1]$	
Pd^{2+}	β_1 (17.5), MHL/ML.H 0.75 $[1.0]^a$	
Cu^{2+}	β_1 10.63^a, 10.57 $[0.1]$; β_2 16.68^a, 16.54 $[0.1]$; MOHL/ML.OH 6.26 $[0.05]$	
Zn^{2+}	β_1 7.27^a, 7.24 $[0.1]$; β_2 12.60^a, 12.52 $[0.1]$	
Cd^{2+}	β_1 5.73^a, 5.71; β_2 10.19^a, 10.12 $[0.1]$	
Hg_2^{2+}	β_1 10.81 $[0.1]$	
Hg^{2+}	β_1 (11.76) $[0.1]$	
Al^{3+}	β_1 8.10, β_2 15.07 $[0.5]$	
Pb^{2+}	β_1 7.45^a, 7.41 $[0.1]$; 7.31, MHL/ML.H 3.05, $MH_2L/MHL.H$ 2.3 $[0.5]$	

^{a} 20°C

5. $C_4H_7NO_4$	L-aminosuccinic acid aspartic acid	$HOOC.CH_2.CH.COOH$ $\vert$ NH_2	H_2L
H^+	K_{H1} 10.002 $[0]$; 9.63 $[0.1]$; 9.62 $[1.0]^a$; 10.01 $[3.0]$; K_{H2} 3.900 $[0]$; 3.70 $[0.1]$; 3.67 $[1.0]^a$; 4.07 $[3.0]$; K_{H3} 1.990 $[0]$; 1.93 $[0.1]$; 2.00 $[1.0]^a$; 2.35 $[3.0]$		
Mg^{2+}	β_1 2.43 $[0.1]$		

Table 7.3 (Continued)

Formula of the acid / Metal ion	Stability constants (Symbol for the constant or scheme of the equilibrium, value of $\log K$ or $\log \beta$, $[I]$)	Ref.
	L-aminosuccinic acid (continued)	
M^{2+}(IIA)	β_1 (1.4) [0.1]	
La^{3+}	β_1 4.84, β_2 (8.26)^b [0.1]	
M^{3+}(4f)	β_1 5.8, (5.25)^c, β_2 10.4, (9.1)^c, β_3 (11.9)^c [0.1]	
UO_2^{2+}	MHL/M.HL 2.61 [0.2]	
Mn^{2+}	β_1 (3.7) [0.1]	
Fe^{2+}	β_1 4.34 [1.0]^a	
Fe^{3+}	β_1 11.4 [1.0]^a	
Co^{2+}	β_1 5.95, β_2 10.23 [0.1]	
Ni^{2+}	β_1 7.16, β_2 12.40 [0.1]	
Cu^{2+}	β_1 8.57, β_2 (15.35) [0.1]^b	
Zn^{2+}	β_1 5.84, β_2 (10.15) [0.1]^b	
Cd^{2+}	β_1 4.39, β_2 7.55 [0.1]	

^{a} 20°C; ^{b} 30°C; ^{c} 30°C ($Ce^{3+} - Nd^{3+}$)

6. $C_4H_8N_2O_3$	glycylglycine $H_2N.CH_2.CO.NH.CH_2.COOH$ HL	
H^+	K_{H1} 8.252 [0]; 8.07 [0.1]; 8.10 [1.0]; K_{H2} 3.144 [0]; 3.13 [0.1]; 3.16 [1.0]	
Ca^{2+}	β_1 1.24 [0]	
Mn^{2+}	β_1 2.15 [0]	
Fe^{2+}	β_1 2.62 [1.0]^a	
Fe^{3+}	β_1 9.10 [1.0]^a	
Co^{2+}	β_1 3.49 [0]; 3.01 [0.1]; (2.73) [1.0]; β_2 5.88 [0]; 5.35 [0.1]; (5.02) [1.0]	
Ni^{2+}	β_1 4.49 [0]; 4.05 [0.1]; 4.03 [1.0]; β_2 7.91 [0]; 7.22 [0.1]; 7.24 [1.0]; β_3 9.4 [0.1]; 9.41 [1.0]; MHL/ML.H 6.29 [0.1]; M(H$_{-1}$L)L.H/ML$_2$ -9.35 [0.1]; -9.31 [1.0]; M(H$_{-1}$L)$_2$.H/M(H$_{-1}$L)L -9.95 [0.1]; -10.08 [1.0]	
Cu^{2+}	β_1 6.04 [0]; 5.50 [0.1]; 5.51 [1.0]; M(H$_{-1}$L).H/ML -4.07 [0.1]; -4.27 [1.0]; M(H$_{-1}$L)L/M(H$_{-1}$L).L 3.14 [0.1]; 3.05 [1.0]; MOH(H$_{-1}$L).H/M(H$_{-1}$L) -9.28 [0.1]; -9.45 [1.0]; M(OH)$_2$(H$_{-1}$L).H/MOH(H$_{-1}$L) -12.8 [1.0]; M$_2$OH(H$_{-1}$L)$_2$/M(H$_{-1}$L).MOH(H$_{-1}$L) 2.16 [0.1]; 2.14 [1.0]	
Ag^+	β_1 2.72, β_2 4.98 [0]	
Zn^{2+}	β_1 3.80 [0]; 3.44 [0.1]; (3.1) [0.8]; β_2 6.57 [0]; 6.31 [0.1]; MHL/ML.H 5.6 [0.8]	
Cd^{2+}	β_1 3.33 [0]; 2.90 [0.1]; 2.8 [0.8]; β_2 6.03 [0]; 5.36 [0.1]; MHL/ML.H 6.4 [0.8]	
Ga^{3+}	β_1 7.57 [0.1]^b	

Stability Constants of Complexes with Organic Ligands

Table 7.3 (Continued)

Formula of the acid Metal ion	Stability constants (Symbol for the constant or scheme of the equilibrium, value of log K or log β, [I])	Ref.
Pb^{2+}	glycylglycine (continued) β_1 3.23 [0]; 3.0 [0.8]; β_2 5.93 [0]; MHL/ML.H 6.4 [0.8]	

[a] 20°C; [b] 22°C

7. $C_5H_9NO_4$	N-methyliminodiacetic acid $CH_3N\begin{smallmatrix}\diagup CH_2.COOH\\ \diagdown CH_2.COOH\end{smallmatrix}$ H_2L
	MIDA
H^+	K_{H1} 10.088 [0][a]; 9.65[a], 9.56 [0.1]; K_{H2} 2.146 [0][a]; 2.12[a], 2.12 [0.1]
Li^+	β_1 1.20 [0.1][a]
Na^+	β_1 0.61 [0.1][a]
Mg^{2+}	β_1 4.18 [0][a]; 3.44[a], 3.48 [0.1]; β_2 5.80[a], 5.83 [0.1]
Ca^{2+}	β_1 4.51 [0][a]; 3.81[a], 3.79 [0.1]; β_2 6.60[a], 6.57 [0.1]
Sr^{2+}	β_1 3.67 [0][a]; 2.92[a], 2.90 [0.1]; β_2 (4.76) [0.1]
Ba^{2+}	β_1 3.45 [0][a]; 2.59[a], 2.60 [0.1]; β_2 (4.94) [0.1]
Y^{3+}	β_1 7.02, β_2 12.56 [0.1]
La^{3+}	β_1 6.23, β_2 10.93 [0.1]
$M^{3+}(4f)$	β_1 (7.1), β_2 (12.7) [0.1]
UO_2^{2+}	β_1 9.70[a], 9.71 [0.1]; MOHL/ML.OH 7.88, $(MOHL)_2/(ML)^2.(OH)^2$ 19.17 [1.0]
NpO_2^+	β_1 7.37, MHL/ML.H 3.54 [0.1]
Mn^{2+}	β_1 5.87 [0]; 5.40[a], 5.39; β_2 9.56[a], 9.55 [0.1]
Fe^{2+}	β_1 6.65, β_2 12.02 [0.1][a]
Co^{2+}	β_1 7.62[a], 7.60; β_2 13.91[a], 13.84 [0.1]
Ni^{2+}	β_1 8.73[a], 8.67; β_2 15.95[a], 15.85 [0.1]
Cu^{2+}	β_1 11.09[a], 11.04, β_2 17.92[a], 17.76, MOHL/ML.OH 5.1[a] [0.1]
Zn^{2+}	β_1 7.66[a], 7.63, β_2 14.09[a], 14.01 [0.1]
Cd^{2+}	β_1 6.77[a], 6.75, β_2 12.52[a], 12.43 [0.1]
Hg^{2+}	β_1 5.47, β_2 9.15, MOHL/ML.OH 4.77 [0.1][a]
Pb^{2+}	β_1 8.02[a], 7.97, β_2 12.12[a], MOHL/ML.OH 4.92[a] [0.1]

[a] 20°C

8. $C_6H_5NO_2$	pyridine-2-carboxylic acid $C_5H_4N.COOH$ HL
	picolinic acid
H^+	K_{H1} 5.39 [0]; 5.23[a]; 5.21 [0.1]; 5.17 [0.5]; K_{H2} 1.01 [0]; 1.04[a]; 1.03 [0.1]; 0.86 [0.5]
Mg^{2+}	β_1 2.58 [0]; 2.20 [0.1][a]; β_2 3.95 [0]
M^{2+}(IIA)	β_1 (1.9) [0]; (1.7) [0.1][a]; β_2 3−4 [0]
Y^{3+}	β_1 4.03, β_2 7.36, β_3 10.0 [0.1]

172 Stability Constants of Complexes with Organic Ligands [Ch. 7

Table 7.3 (Continued)

Formula of the acid / Metal ion	Stability constants (Symbol for the constant or scheme of the equilibrium, value of $\log K$ or $\log \beta$, $[I]$)	Ref.
	pyridine-2-carboxylic acid (continued)	
La^{3+}	β_1 3.51, β_2 6.46, β_3 8.74, β_4 10.0 [0.1]	
$M^{3+}(4f)$	β_1 (4.3), β_2 (7.6), β_3 (10.4), β_4 (12.5) [0.1]	
UO_2^{2+}	β_1 4.51, MHL/ML.H (2) [0.1]	
PuO_2^{2+}	β_1 4.58, MHL/ML.H 5.22 [0.1]	
Mn^{2+}	β_1 3.88 [0]; 3.57 [0.1]^a; β_2 7.08 [0]; 6.32, β_3 8.1 [0.1]^a	
Fe^{2+}	β_1 4.90 [0]; β_2 9.00, β_3 12.30 [0.1]^a	
Fe^{3+}	β_2 12.80, MOHL$_2$/ML$_2$.OH 11.0, (MOHL)$_2$/(MOHL)^2 3.06 [0.1]^a	
Co^{2+}	β_1 5.74, β_2 10.44, β_3 14.09 [0.1]	
Ni^{2+}	β_1 7.63 [0]; 6.80 [0.1]^a; β_2 (12.45) [0]; 12.58, β_3 17.22 [0.1]^a	
Cu^{2+}	β_1 7.95 [0.1]^a; 7.76 [0.5]; β_2 14.95 [0.1]^a; 14.70 [0.5]	
Ag^+	β_1 3.40, β_2 5.9 [0.1]^a	
Zn^{2+}	β_1 5.75 [0]; 5.30 [0.1]^a; β_2 10.01 [0]; 9.62, β_3 12.92 [0.1]^a	
Cd^{2+}	β_1 4.79 [0]; 4.55 [0.1]^a; β_2 8.25 [0]; 8.16, β_3 10.76 [0.1]^a	
Hg^{2+}	β_1 7.70, β_2 15.55 [0.1]	
Pb^{2+}	β_1 5.07 [0]; 4.58 [0.1]^a; β_2 8.57 [0]; 7.92 [0.1]^a	

^{a} 20°C

9. $C_6H_9NO_6$	nitrilotriacetic acid HOOC.CH$_2$.N(CH$_2$.COOH)(CH$_2$.COOH) NTA	H_3L
H^+	K_{H1} 10.334 [0]^a; 9.71^a, 9.65 [0.1]; 9.33 [0.5]; 8.96 [1.0]^a; K_{H2} 2.940 [0]^a; 2.48^a, 2.48 [0.1]; 2.43 [0.5]; 2.27 [1.0]^a; K_{H3} 1.650 [0]^a; 1.8^a, 1.8 [0.1]; 1.97 [0.5]; 1.98 [1.0]^a; K_{H4} 0.8 [0.1]^a; 1.1 [1.0]^a	
Li^+	β_1 2.51 [0.1]^a	
Na^+	β_1 1.22 [0.1]^a	
Be^{2+}	β_1 7.11 [0.1]^a	
Mg^{2+}	β_1 6.500 [0]^a; 5.41^a, 5.47 [0.1]	
Ca^{2+}	β_1 7.608 [0]^a; 6.41^a, 6.39 [0.1]; β_2 8.86^a, 8.76 [0.1]^a	
Sr^{2+}	β_1 4.98^a, 4.97 [0.1]	
Ba^{2+}	β_1 5.875 [0]^a; 4.82^a, 4.80 [0.1]	
Sc^{3+}	β_1 12.7, β_2 24.1^a [0.1]; MOHL/ML.OH (7.44) [0.2]^a	
Y^{3+}	β_1 11.41^a, 11.42; β_2 20.43^a, (20.41); MOHL/ML.OH (6.39) [0.1]	
La^{3+}	β_1 10.47^a, 10.47; β_2 17.84^a, (17.83); MOHL/ML.OH (5.9) [0.1]	
$M^{3+}(4f)$	β_1 (11.5)^a, (11.5); β_2 (20.6)^a, (20.5), MOHL/ML.OH (6.4)^a [0.1]	
Zr^{4+}	β_1 20.8 [0.1]; 20.8 [0.2]^a; 19.5 [1.0]^a	
Hf^{4+}	β_1 20.3 [0.2]^a	
Th^{4+}	β_1 (13.3), M(OH)$_2$L/ML.OH^2 19.0 [0.1]	

Ch. 7] **Stability Constants of Complexes with Organic Ligands** 173

Table 7.3 (Continued)

Formula of the acid / Metal ion	Stability constants (Symbol for the constant or scheme of the equilibrium, value of log K or log β, $[I]$)	Ref.
	nitrilotriacetic acid (continued)	
V^{3+}	β_1 13.41, β_2 23.09, MOHL/ML.OH 7.79 $[0.1]^a$	
VO^{2+}	MOHL/ML.OH 6.40 $[0.1]$	
Cr^{3+}	MOHL/ML.OH 7.72, M(OH)$_2$L/MOHL.OH 5.50 $[0.1]^a$	
$M^{3+}(5f)$	β_1 (11.7), β_2 (20.7) $[0.1]$	
Np^{4+}	β_1 17.28, β_2 32.06 $[1.0]$	
NpO_2^+	β_1 6.80, MHL/ML.H 4.5, MOHL/ML.OH 2.3 $[0.1]$	
PuO_2^+	β_1 6.91 $[0.1]$	
UO_2^{2+}	β_1 9.56^a, 9.50 $[0.1]$	
Mn^{2+}	β_1 8.573 $[0]^a$; 7.44^a, 7.46 $[0.1]$; β_2 10.99^a, 10.94 $[0.1]$	
Mn^{3+}	β_1 20.25 $[1.0]^a$	
Fe^{2+}	β_1 8.33 $[0.1]^a$; β_2 12.8 $[0.2]^a$; MHL/ML.H 1.9 $[0.2]^a$; MOHL/ML.OH 3.4 $[0.1]^a$	
Fe^{3+}	β_1 15.9^a, 15,9, β_2 24.3, MOHL/ML.OH (9.9)^a $[0.1]$; (8.8) $[1.0]^a$ M(OH)$_2$L/MOHL.OH (6.1) $[0.1]^a$; (MOHL)$_2$/(MOHL)^2 (4.0); (MOHL)$_2$/(ML)^2.OH^2 (21.6) $[1.0]$	
Co^{2+}	β_1 10.38^a, 10.38, β_2 14.39^a, 14.33, MOHL/ML.OH 3.0 $[0.1]$	
Co^{3+}	MOHL/ML.OH 7.11, M(OH)$_2$L/MOHL.OH 4.29 $[0.1]^a$	
Ni^{2+}	β_1 11.53^a, 11.50, β_2 16.42^a, 16.32, MOHL/ML.OH 2.92 $[0.1]$	
Pd^{2+}	β_1 19.30, MHL/ML.H 0.50 $[1.0]^a$	
Cu^{2+}	β_1 12.96^a, 12.94, β_2 17.43^a, 17.42, MOHL/ML.OH 4.64 $[0.1]$	
Ag^+	β_1 (5.16) $[0.1]^a$	
Zn^{2+}	β_1 10.67^a, 10.66, β_2 14.29^a, 14.24, MOHL/ML.OH 3.72 $[0.1]$	
Cd^{2+}	β_1 9.83^a, 9.78, β_2 14.61^a, 14.39, MOHL/ML.OH 2.53 $[0.1]$	
Hg^{2+}	β_1 14.6 $[0.1]$	
Al^{3+}	β_1 11.4, MHL/ML.H 1.90, MOHL/ML.OH 8.7, M(OH)$_2$L/MOHL.OH 5.5 $[0.2]$	
Ga^{3+}	β_1 13.6 $[0.1]^a$	
In^{3+}	β_1 16.9 $[0.1]^a$	
Tl^+	β_1 4.75 $[0.1]^a$	
Tl^{3+}	β_1 20.9, β_2 32.5 $[1.0]^a$	
Pb^{2+}	β_1 11.39^a, 11.34 $[0.1]$	
As(III)	M(OH)$_2$HL/M(OH)$_2$.HL 15.3 $[0.1]$	
Bi^{3+}	β_1 17.5, β_2 26.0 $[1.0]^a$	

^{a} 20°C

10. $C_7H_5NO_4$	pyridine-2,6-dicarboxylic acid dipicolinic acid	H_2L
H^+	K_{H1} 5.07 $[0]^a$; 4.68^a, 4.68 $[0.1]$; 4.51 $[0.5]$; K_{H2} 2.24 $[0]^a$; 2.10^a, 2.09 $[0.1]$; 2.13 $[0.5]$	

Table 7.3 (Continued)

Formula of the acid Metal ion	Stability constants (Symbol for the constant or scheme of the equilibrium, value of $\log K$ or $\log \beta$, $[I]$)	Ref.
	pyridine-2,6-dicarboxylic acid (continued)	
Mg^{2+}	β_1 2.30[a], 2.32, β_2 3.0 [0.1]	
Ca^{2+}	β_1 (4.40)[a], (4.60), β_2 7.58 [0.1]	
Sr^{2+}	β_1 3.89[a], 3.80, β_2 5.76 [0.1]	
Ba^{2+}	β_1 3.46[a], 3.43, β_2 4.9 [0.1]	
Sc^{3+}	β_1 11.2, β_2 18.9 [0.5]	
Y^{3+}	β_1 8.44, β_2 15.66, β_3 21.18 [0.5]	
La^{3+}	β_1 7.94, β_2 13.71, β_3 17.95 [0.5]	
$M^{3+}(4f)$	β_1 (8.7), β_2 (15.9), β_3 (21.2) [0.5]	
Mn^{2+}	β_1 5.01, β_2 8.49 [0.1][a]	
Fe^{2+}	β_1 5.71, β_2 10.36 [0.1][a]	
Fe^{3+}	β_1 10.91, β_2 17.13 [0.1][a]	
Co^{2+}	β_1 6.65, β_2 12.70 [0.1][a]	
Ni^{2+}	β_1 6.95, β_2 13.50 [0.1][a]	
Cu^{2+}	β_1 9.14 [0.1][a]; 8.88 [0.5]; β_2 16.52 [0.1][a]; 16.17 [0.5]	
Zn^{2+}	β_1 6.35; β_2 11.88 [0.1][a]	
Cd^{2+}	β_1 6.75 [0.1][a]; 6.51 [0.5]; β_2 11.15 [0.1][a]; 10.77 [0.5]	
Hg^{2+}	β_2 20.28 [0.1][a]	
Al^{3+}	β_1 4.87, β_2 8.32 [0.5]	
Pb^{2+}	β_1 8.70 [0.1][a]; 8.66 [0.5]; β_2 11.60 [0.1][a]; 11.55 [0.5]	

[a] 20°C

11. $C_7H_7NO_2$	2-aminobenzoic acid anthranilic acid	HL
H^+	K_{H1} 4.96 [0]; 4.79 [0.1]; K_{H2} 2.08 [0]; 2.00 [0.1]	
M^{2+}(IIA)	β_1 (0.6) [0]	
$M^{3+}(4f)$	β_1 3.2 [0.1; La, Ce, Nd, Pr][a]	[1]
Mn^{2+}	β_1 0.99, β_2 (2.87) [0]	
Co^{2+}	β_1 1.56 [0]	
Ni^{2+}	β_1 2.12, β_2 3.59 [0]	
Cu^{2+}	β_1 4.25 [0]	
Ag^+	β_1 1.86 [0]	
Zn^{2+}	β_1 2.57 [0]	
Cd^{2+}	β_1 1.83 [0]	
Pb^{2+}	β_1 2.82 [0]	

[a] 30°C

Ch. 7]   **Stability Constants of Complexes with Organic Ligands**   175

Table 7.3 (Continued)

Formula of the acid Metal ion	Stability constants (Symbol for the constant or scheme of the equilibrium, value of log K or log β, [I])	Ref.
12. $C_{10}H_7NO_2$	quinoline-2-carboxylic acid HL quinaldinic acid	
H^+	K_{H1} 4.97 [0]; 4.75 [0.1]; 4.68 [1.0]; K_{H2} 1.9 [0.02]	[2]
M^{2+}(IIA)	β_1 (1.3), β_2 (2.8) [0]	
Mn^{2+}	β_1 2.96, β_1 5.92 [0]	
Fe^{2+}	β_1 3.92, β_2 7.67 [0]	
Co^{2+}	β_1 4.49, β_2 8.23 [0]	
Ni^{2+}	β_1 4.95, β_2 8.65 [0]	
Cu^{2+}	β_1 5.91 [0]	
Fe^{3+}	β_2 7.58 [0.1]^a	[3]
Zn^{2+}	β_1 4.17 [0]	
Cd^{2+}	β_1 4.12, β_2 6.83 [0]	
Pb^{2+}	β_1 3.95, β_2 7.02 [0]	

^{a} 20°C

13. $C_{10}H_7NO_2$	quinoline-8-carboxylic acid HL	
H^+	K_{H1} 6.87 [0]; 6.76 [0.1]; K_{H2} 1.80 [0]; 2 [0.1]	[4, 5, 6]
M^{2+}(IIA)	β_1 (1.2); β_2 (5.0) [0]	[4]
Fe^{2+}	β_1 3.68, β_2 10.25 [0]	[4]
Mn^{2+}	β_1 2.11, β_2 6.97 [0]	[4]
Ni^{2+}	β_1 4.46 [0]; 4.0 [0.1]; β_2 12.59 [0]	[4, 6]
Co^{2+}	β_1 3.61, β_2 10.39 [0]	[4]
Cu^{2+}	β_1 6.08 [0]	[4]
Ag^+	β_1 2.13 [0]	[5]
Zn^{2+}	β_1 3.05 [0]; 2.7 [0.1]; β_2 8.94 [0]	[4, 6]
Cd^{2+}	β_1 2.27 [0]; 2 [0.1]; β_2 6.96 [0]	[4, 6]
Pb^{2+}	β_1 2.45, β_2 8.38 [0]	[4]
14. $C_{10}H_{16}N_2O_8$	ethylenediaminetetra-acetic acid H_4L EDTA	
	K_{H1} 11.014 [0]^a; 10.24^a, 10.17 [0.1]; 9.95 [1.0]^a; K_{H2} 6.320 [0]^a; 6.16^a, 6.11 [0.1]; 6.27 [1.0]^a; K_{H3} 2.66^a, 2.68 [0.1]; 2.3 [1.0]^a; K_{H4} 2.0^a, 2.0 [0.1]; 2.2 [1.0]^a; K_{H5} 1.5^a, 1.5 [0.1]; 1.4 [1.0]^a; K_{H6} −0.1^a, 0.0 [1.0]	
Li^+	β_1 2.79^a, 2.79 [0.1]	
Na^+	β_1 1.66^a, 1.64 [0.1]	
K^+	β_1 0.8^a [0.1]	
Be^{2+}	β_1 9.2^a [0.1]	

Formula of the acid Metal ion	Stability constants (Symbol for the constant or scheme of the equilibrium, value of log K or log β, [I])	Ref.
	EDTA (continued)	
Mg^{2+}	β_1 9.12 [0][a]; 8.79[a], 8.83 [0.1]; MHL/ML.H 3.85 [0.1][a]	
Ca^{2+}	β_1 11.00 [0][a]; 10.69[a], 10.61 [0.1]; MHL/ML.H 3.18 [0.1][a]	
Sr^{2+}	β_1 (8.80) [0][a]; 8.73[a], 8.68 [0.1]; MHL/ML.H 3.93 [0.1][a]	
Ba^{2+}	β_1 (7.78) [0][a]; 7.86[a], 7.80 [0.1]; MHL/ML.H 4.57 [0.1][a]	
Ra^{2+}	β_1 7.1 [0.1][a]	
Sc^{3+}	β_1 23.1, MHL/ML.H 2.0, MOHL/ML.OH 3.29 [0.1][a]	
Y^{3+}	β_1 18.09[a], 18.08 [0.1]	
La^{3+}	β_1 15.50[a], 15.46, MHL/ML.H 2.0[a], 2.24 [0.1]	
$M^{3+}(4f)$	β_1(17.9)[a], (17.9) [0.1]	
Ti^{3+}	β_1 (21.3) [0.1]	
TiO^{2+}	β_1 17.5 [0.1]	[7]
Zr^{4+}	β_1 29.5[a]; 29.4, MOHL/ML.OH 7.6, $(MOHL)_2/(MOHL)^2$ 3.5 [0.1]	
Hf^{4+}	β_1 29.5 [0.2][a]	
Th^{4+}	β_1 23.2[a], MHL/ML.H 1.98[a], MOHL/ML.OH 6.74, $(MOHL)_2/(ML)^2.(OH)^2$ 17.87[a], 17.78, $(MOHL)_2/(MOHL)^2$ 4.3 [0.1]	
V^{2+}	β_1 12.7, MHL/ML.H 3.5 [0.1][a]	
V^{3+}	β_1 (26.0), MOHL/ML.OH 4.41 [0.1][a]	
VO^{2+}	β_1 18.8, MOHL/ML.OH 10.95 [0.1][a]	
VO_2^+	β_1 15.55, MHL/ML.H 4.31, MH_2L/MHL.H 3.49, MH_3L/MH_2L.H 1.4 [0.1][a]	
Cr^{2+}	β_1 (13.6), MHL/ML.H 3.00 [0.1][a]	
Cr^{3+}	β_1 (23.4), MHL/ML.H 1.95, MOHL/ML.OH 6.56 [0.1][a]	
Mo^{5+}	β_1 (6.4)[b]	[8]
Mo(VI)	$MO_3L/MO_4.HL$.H 8.8, MO_3HL/MO_3L.H 7.5, $(MO_3)_2L/(MO_4)^2.L.H^4$ 35.1, $MO_3L/MO_3.L$ 10.7, $(MO_3)_2L/(MO_3)^2.L$ 19.5, $(MO_3HL)^2/(MO_3)_2L.H_2L$ 0.26[b,c]	[9]
U^{4+}	β_1 25.8[a], 25.7 [0.1]; 23.2 [1.0]; MOHL/ML.OH 9.06, $(MOHL)_2/(ML)^2.(OH)^2$ 21.03, $(MOHL)_2/(MOHL)^2$ 2.9 [0.1]	
UO_2^{2+}	β_1 (19.7) [1.0]; MHL/M.HL 7.36[a], 7.40, $M_2L/M^2.L$ 17.87 [0.1]; MOHL/ML.OH 8.17, $(MHL)^2/(ML)_2.H^2$ −7.97, $(ML)_2/(ML)^2$ 3.27 [1.0]; $(MOH)_2L/M_2L.(OH)^2$ 16.4 [0.15]	[10]
$M^{3+}(5f)$	β_1 (18.3) [0.1]	
Np^{4+}	β_1 24.6 [1.0]	
NpO_2^+	β_1 7.33, MHL/M.HL 5.30, MOHL/ML.OH 2.27 [0.1]	
PuO_2^+	MHL/M.HL 4.80 [0.1]	
Mn^{2+}	β_1 13.87[a], 13.81, MHL/ML.H 3.1[a], 3.1 [0.1]	
Mn^{3+}	β_1 (25.3) [0.1]; (27.0) [1.0][a]	

Stability Constants of Complexes with Organic Ligands

Table 7.3 (Continued)

Formula of the acid Metal ion	Stability constants (Symbol for the constant or scheme of the equilibrium, value of log K or log β, $[I]$)	Ref.
	EDTA (continued)	
Fe^{2+}	β_1 14.32[a], 14.27, MHL/ML.H 2.75[a], 2.7, MOHL/ML.OH 4.88[a], $M(OH)_2L/MOHL.OH$ 4.11[a] [0.1]	
Fe^{3+}	β_1 25.1[a], 25.0 [0.1]; (25.15) [1.0][a]; MHL/ML.H 1.3, MOHL/ML.OH 6.46 [0.1][a]; 6.21 [1.0]; $M(OH)_2L/MOHL.OH$ 4.54 [0.1][a]; $(MOHL)_2/(MOHL)^2$ 2.95 [1.0]	
Co^{2+}	β_1 16.31[a], 16.26, MHL/ML.H 3.0[a], 3.0 [0.1]; 2.93 [1.0]	
Co^{3+}	β_1 41.4 [0.1]; 40.7 [1.0]; MHL/ML.H 2.98 [0.1][a]	
Ni^{2+}	β_1 18.62[a], 18.52, MHL/ML.H 3.2[a], 3.2 [0.1]; 3.12 [1.0]; MOHL/ML.OH 1.8 [0.1][a]	
Pd^{2+}	β_1 (18.5) [0.2]; MHL/ML.H 3.1, MH_2L/MHL.H 0.90 [1.0][a]	
Pt^{2+}	MHL/ML.H 2.88, MH_2L/MHL.H 2.18, MH_3L/MH_2L.H 0.5, MOHL/ML.OH 4.87 [0.1][a]	
Cu^{2+}	β_1 18.80[a], 18.70, MHL/ML.H 3.0[a], 3.0 [0.1]; 2.87 [1.0]; MOHL/ML.OH 2.5 [0.1][a]; 2.4 [1.0]	
Ag^+	β_1 7.32, MHL/ML.H 6.01 [0.1][a]	
Zn^{2+}	β_1 16.50[a], 16.44, MHL/ML.H 3.0[a], 3.0; MOHL/ML.OH 2.1[a] [0.1]	
Cd^{2+}	β_1 16.46[a], 16.36, MHL/ML.H 2.9[a], 2.9 [0.1]	
Hg^{2+}	β_1 21.7[a], 21.5, MHL/ML.H 3.1[a], 3.1, MOHL/ML.OH 4.84[a] [0.1]	
Al^{3+}	β_1 16.3[a], 16.5, MHL/ML.H 2.5[a], 2.5, MOHL/ML.OH 8.06[a], 7.95, $M(OH)_2L/MOHL.OH$ 3.98[a] [0.1]; 3.47 [0.2]	
Ga^{3+}	β_1 20.3, MHL/ML.H 1.83, MOHL/ML.OH 8.31 [0.1][a]	
In^{3+}	β_1 25.0[a], 24.9, MHL/ML.H 1.5[a], 1.5, MOHL/ML.OH 5.32[a] [0.1]	
Tl^+	β_1 6.54, MHL/ML.H 5.77 [0.1][a]	
Tl^{3+}	β_1 (35.3) [0.1]; 37.8 [1.0][a]; MOHL/ML.OH 7.96 [1.0][a]	
Ge(IV)	$ML/M(OH)_4.H_4L$ 4.80, $MOHL/M(OH)_4.H_3L$ 4.52, MOHL/ML.OH 11.38 [0.1]	
Sn^{2+}	β_1 18.3, MHL/ML.H 2.5, MH_2L/MHL.H 1.5 [1.0][a]	
Pb^{2+}	β_1 18.04[a], 17.88, MHL/ML.H 2.8[a] [0.1]; 2.49 [1.0]	
As(III)	$M(OH)_2HL/M(OH)_2L.H$ 7.3, $M(OH)_2HL/M(OH)_2.HL$ 9.2, $M(OH)_2H_2L/M(OH)_2HL.H$ 3.4 [0.1][a]	
Sb(III)	$ML/M(OH)_3.H_3L$ 31.54[a], MHL/ML.H 1.02, MOHL/ML.OH 8.24 [0.1]; 8.69 [1.0]; $M(OH)_2L/MOHL.OH$ 7.46 [0.1]	
Bi^{3+}	β_1 27.8 [0.1][a]; 26.7 [1.0][a]; MHL/ML.H 1.43 [0.1][a]; 1.7, MOHL/ML.OH 2.9 [1.0][a]	

[a] 20°C; [b] ionic strength is not given; [c] about 35°C

Table 7.3 (Continued)

Formula of the acid Metal ion	Stability constants (Symbol for the constant or scheme of the equilibrium, value of log K or log β, $[I]$)	Ref.
15. $C_{10}H_{18}N_2O_7$	N-(2-hydroxyethyl)ethylenedinitrilotriacetic acid H_3L HEDTA	
H^+	K_{H1} 9.89[a], 9.81 [0.1]; 8.65 [1.0]; K_{H2} 5.41[a], 5.37 [0.1]; 5.11 [1.0]; K_{H3} 2.6[a], 2.6 [0.1]; 2.3 [1.0]	
Mg^{2+}	β_1 7.0[a], 7.0 [0.1]	
Ca^{2+}	β_1 8.3[a], 8.2 [0.1]	
Sr^{2+}	β_1 6.9[a], 6.8 [0.1]	
Ba^{2+}	β_1 6.3[a], 6.2 [0.1]	
Sc^{3+}	β_1 17.3 [0.1]	
Y^{3+}	β_1 14.78[a], 14.75, MOHL/ML.OH 4.76[a] [0.1]	
La^{3+}	β_1 13.61[a], 13.56, MOHL/ML.OH 3.46[a] [0.1]	
$M^{3+}(4f)$	β_1 (15.4)[a], (15.3), MOHL/ML.OH (4.5)[a] [0.1]	
Th^{4+}	β_1 18.5, MOHL/ML.OH 8.4, $(MOHL)_2/(MOHL)^2$ 5.2 [0.1]	
Np^{4+}	β_1 20.82, β_2 33.59 [1.0]	
NpO_2^+	β_1 6.9, MHL/M.HL 4.06, MOHL/ML.OH 2.41 [0.1]	
PuO_2^+	MHL/M.HL 4.46 [0.1]	
$M^{3+}(5f)$	β_1 (16.0), β_2 (27.7) [0.1]	
Mn^{2+}	β_1 10.9[a]; 10.8 [0.1]	
Mn^{3+}	β_1 22.7 [0.2]	
Fe^{2+}	β_1 12.3[a]; 12.2 [0.1]; MHL/ML.H 2.7 [0.2]; MOHL/ML.OH 5.03, $M(OH)_2L/MOHL.OH$ 3.97 [0.1]	
Fe^{3+}	β_1 19.8, MOHL/ML.OH 9.90 [0.1]; 9.68 [1.0]; $M(OH)_2L/MOHL.OH$ 4.98 [0.1]; 5.10 [1.0]; $M(OH)_3L/M(OH)_2L.OH$ 3.78 [0.1]; $(MOHL)_2/(MOHL)^2$ −2.38 [1.0]	
Co^{2+}	β_1 14.6[a], 14.5 [0.1]	
Co^{3+}	β_1 37.4 [0.1]	
Ni^{2+}	β_1 17.3[a], 17.1 [0.1]; MHL/ML.H 2.54 [1.25]	
Cu^{2+}	β_1 17.6[a], 17.5, MHL/ML.H 2.42 [0.1]; 2.32 [1.25]	
Ag^+	β_1 6.71 [0.1]	
Zn^{2+}	β_1 14.7[a], 14.6 [0.1]	
Cd^{2+}	β_1 13.3[a], 13.1 [0.1]	
Hg^{2+}	β_1 20.30[a], 20.05. MOHL/ML.OH 5.4 [0.1]	
Al^{3+}	β_1 14.3[a], 14.4 [0.1], MHL/ML.H 2.14, MOHL/ML.OH 8.88, $M(OH)_2L/M(OH)L.OH$ 4.58 [0.2]	
Ga^{3+}	β_1 16.9; MOHL/ML.OH 9.78 [0.1][a]	
In^{3+}	β_1 20.2 [0.1][a]	
Ge(IV)	$ML/M(OH)_4.H_4L$ 4.44 [0.1]	
Pb^{2+}	β_1 15.7[a], 15.5 [0.1]	

Formula of the acid Metal ion	Stability constants (Symbol for the constant or scheme of the equilibrium, value of log K or log β, $[I]$)	Ref.
	HEDTA (continued)	
Sb(III)	MOHL/M(OH)$_3$.H$_2$L 13.12^a, ML/MOHL.H 3.1^a, 3.05 [0.1]; 3.2, M(OH)$_2$L/MOHL.OH 8.13 [1.0]	
Bi^{3+}	β_1 22.3, MOHL/ML.OH 8.00 [1.0]	

^{a} 20°C

16. C$_{13}$H$_{20}$N$_2$O$_8$	trans-1,2-cyclopentylenedinitrilotetra-acetic acid CPDTA	H$_4$L
H$^+$	K_{H1} 10.09^a, 10.01 [0.1]; 10.13 [1.0]^a; K_{H2} 7.48, K_{H3} 2.44, K_{H4} 1.87 [0.1]^a	
Mg^{2+}	β_1 9.24^a, 9.30 [0.1]; 9.29 [1.0]^a; MHL/ML.H 5.36 [0.1]^a	
Ca^{2+}	β_1 11.26^a, 11.20 [0.1]; 11.30 [1.0]^a; MHL/ML.H 3.93 [0.1]^a	
Sr^{2+}	β_1 9.42^a, 9.42 [0.1]; 9.47 [1.0]^a; MHL/ML.H 5.51 [0.1]^a	
Ba^{2+}	β_1 8.57^a, 8.54 [0.1]; 8.62 [1.0]^a; MHL/ML.H 6.42 [0.1]^a	
La^{3+}	β_1 17.01^a, 16.99 [0.1]	
M^{3+}(4f)	β_1 (18.7)^a, (18.7) [0.1]	
Co^{2+}	β_1 (12.07), MHL/ML.H (3.37) [0.1]^a	
Cd^{2+}	β_1 18.25 [0.1]^a	
Hg^{2+}	β_1 23.97, MHL/ML.H 3.68 [0.1]^a	
Pb^{2+}	β_1 18.88 [0.1]^a	

^{a} 20°C

17. C$_{14}$H$_{22}$N$_2$O$_8$	cis-1,2-cyclohexylenedinitrilotetra-acetic acid cis-DCTA	H$_4$L
H$^+$	K_{H1} 10.70, K_{H2} 5.21, K_{H3} 3.50, K_{H4} 2.44 [0.1]^a	
Mg^{2+}	β_1 8.38, MHL/ML.H 4.44 [0.1]^a	
Ca^{2+}	β_1 9.45, MHL/ML.H 4.11 [0.1]^a	
Sr^{2+}	β_1 7.33, MHL/ML.H 5.32 [0.1]^a	

^{a} 20°C

18. C$_{14}$H$_{22}$N$_2$O$_8$	trans-1,2-cyclohexylenedinitrilotetra-acetic acid DCTA	H$_4$L
H$^+$	K_{H1} 12.4^a, 12.3 [0.1]; 9.30 [1.0]^a; K_{H2} 6.15^a, 6.12 [0.1]; 5.87 [1.0]^a; K_{H3} 3.53 [0.1]^a; 3.52 [1.0]^a; K_{H4} 2.42 [0.1]^a; 2.41 [1.0]^a; K_{H5} 1.72 [1.0]^a	
Be^{2+}	β_1 11.51 [0.1]^a	
Mg^{2+}	β_1 11.02^a, 11.07 [0.1]	

Table 7.3 (Continued)

Formula of the acid / Metal ion	Stability constants (Symbol for the constant or scheme of the equilibrium, value of log K or log β, [I])	Ref.
	DCTA (continued)	
Ca^{2+}	β_1 13.20^a, 13.15 [0.1]	
Sr^{2+}	β_1 10.59^a, 10.58 [0.1]	
Ba^{2+}	β_1 8.69^a, 8.6, MHL/ML.H 6.86^a [0.1]	
Sc^{3+}	β_1 26.1, MOHL/ML.OH 2.6 [0.1]^a	
Y^{3+}	β_1 19.85, MHL/ML.H 2.18 [0.1]^a	
La^{3+}	β_1 16.96^a, 16.98, MHL/ML.H 2.24^a [0.1]	
$M^{3+}(4f)$	β_1 (19.9), MHL/ML.H (2.3) [0.1]^a	
Zr^{4+}	β_1 (29.9) [2.0, $HClO_4$]^a	
Th^{4+}	β_1 25.6^a, MHL/ML.H 2.50^a, MOHL/ML.OH 5.93, $(MOHL)_2/(MOHL)^2$ 4.3 [0.1]	
VO^{2+}	β_1 20.10 [0.1]^a	
U^{4+}	β_1 27.6^a, MOHL/ML.OH 8.93, $(MOHL)_2/(MOHL)^2$ 3.50 [0.1]	
UO_2^{2+}	MHL/M.HL 5.27 [0.1]^a	
$M^{3+}(5f)$	β_1 (19.9) [0.1]	
Mn^{2+}	β_1 17.48^a, 17.43, MHL/ML.H 2.8^a [0.1]	
Fe^{2+}	β_1 19.0^a, 18.90 [0.1]; MHL/ML.H 2.7 [0.2]	
Fe^{3+}	β_1 30.0 [0.1]; 30.1 [0.2]^a; MOHL/ML.OH 4.25 [0.1]^a; 4.5, $(MOHL)_2/(MOHL)^2$ 1.01 [1.0]^a	
Mn^{3+}	β_1 28.9 [0.2]	
Co^{2+}	β_1 19.62^a, 19.58, MHL/ML.H 2.9 [0.1]	
Ni^{2+}	β_1 20.3^a, 20.2 [0.1]; MHL/ML.H 2.74 [1.25]	
Pd^{2+}	MHL/ML.H 3.60 [1.0]^a	
Cu^{2+}	β_1 22.00^a, 21.92, MHL/ML.H 3.1 [0.1]; 2.68 [1.25]	
Ag^+	β_1 9.03 [0.1]	
Zn^{2+}	β_1 19.37^a, 19.35, MHL/ML.H 2.9^a [0.1]	
Cd^{2+}	β_1 19.93^a, 19.84, MHL/ML.H 3.0^a [0.1]	
Hg^{2+}	β_1 25.00^a, 24.79, MHL/ML.H 3.1^a, MOHL/ML.OH 3.49^a [0.1]	
Al^{3+}	β_1 19.5^a, 19.6, MHL/ML.H 2.0^a, MOHL/ML.OH 6.37^a, 5.96 [0.1]	
Ga^{3+}	β_1 23.2, MHL/ML.H (2.42), MOHL/ML.OH 6.46 [0.1]^a	
In^{3+}	β_1 28.8, MOHL/ML.OH 5.00 [0.1]^a	
Tl^+	β_1 6.7 [0.1]^a	
Tl^{3+}	β_1 38.3 [1.0]^a	
Sn^{2+}	β_1 17.8, MHL/ML.H 3.1, MH_2L/MHL.H 2.5 [1.0]^a	
Pb^{2+}	β_1 20.38^a, 20.24, MHL/ML.H 2.8^a [0.1]	
Sb(III)	ML/M(OH)$_3$.H_3L 29.5 [0.1]^a	
Bi^{3+}	β_1 32.3 [0.1]^a; 31.9 [0.5]; MHL/ML.H 1.25; MOHL/ML.OH 3.0 [1.0]^a	

^{a} 20°C

Table 7.3 (Continued)

Formula of the acid Metal ion	Stability constants (Symbol for the constant or scheme of the equilibrium, value of $\log K$ or $\log \beta$, $[I]$)	Ref.
19. $C_{14}H_{23}N_3O_{10}$	diethylenetrinitrilopenta-acetic acid $\qquad$ H_5L DTPA	
H^+	K_{H1} 10.55^a, 10.45 [0.1]; 9.48 [1.0]^a; K_{H2} 8.59^a, 8.53 [0.1]; 8.26 [1.0]^a; K_{H3} 4.30^a, 4.28 [0.1]; 4.17 [1.0]^a; K_{H4} 2.66^a, 2.65 [0.1]; 2.6 [1.0]^a; K_{H5} 1.82^a, 1.82 [0.1]; 2.3 [1.0]^a	
Li^+	β_1 3.1 [0.1]	
Mg^{2+}	β_1 9.30^a, 9.34, MHL/ML.H 7.09^a, 6.85 [0.1]	
Ca^{2+}	β_1 10.83^a, 10.75, MHL/ML.H 6.11^a, 6.11, $M_2L/ML.M$ 1.98^a, 1.6 [0.1]	
Sr^{2+}	β_1 9.77^a, 9.68, MHL/ML.H 5.69^a, 5.4 [0.1]	
Ba^{2+}	β_1 8.87^a, 8.78, MHL/ML.H 5.55^a, 5.34 [0.1]	
Sc^{3+}	β_1 (24.5)^a, (24.4) [0.1]	
Y^{3+}	β_1 22.13^a, 22.05, MHL/ML.H 1.91^a [0.1]	
La^{3+}	β_1 19.54^a, 19.48, MHL/ML.H 2.60^a [0.1]	
$M^{3+}(4f)$	β_1 (22.3)^a, (22.2), MHL/ML.H (2.2)^a, $M_2L/ML.M$ (Nd, Sm, Eu) 3.5^a [0.1]	
Zr^{4+}	β_1 35.8 [0.23]^a, 36.9, MOHL/ML.OH 8.1 [1.0]^a	
Hf^{4+}	β_1 35.4 [0.23]^a	
Th^{4+}	β_1 (28.78), MHL/ML.H 2.16, MOHL/ML.OH 4.9 [0.1]^a	
U^{4+}	MOHL/ML.OH 6.26 [0.1]^a	
Np^{4+}	β_1 30.3 [1.0]	
$M^{3+}(5f)$	β_1 (22.8) [0.1]	
Mn^{2+}	β_1 15.60^a, 15,51, MHL/ML.H 4.64^a, 4.40, $M_2L/ML.M$ 2.09^a [0.1]	
Mn^{3+}	β_1 31.1 [1.0]^a	
Fe^{2+}	β_1 16.5^a, 16.4, MHL/ML.H 5.35^a, 5.30, MOHL/ML.OH 5.01, $M(OH)_2L/MOHL.OH$ 4.37 [0.1]	
Fe^{3+}	β_1 28.0^a, MHL/ML.H 3.58^a, 3.56, MOHL/ML.OH 3.9^a, 4.12 [0.1]	
Co^{2+}	β_1 19.27^a, 19.15, MHL/ML.H 4.74^a, 4.94, $MH_2L/MHL.H$ 3.22, $M_2L/ML.M$ 3.51^a, 3.74 [0.1]	
Ni^{2+}	β_1 20.32^a, 20.17, MHL/ML.H 5.62^a, 5.67, $MH_2L/MHL.H$ 3.02, $M_2L/ML.M$ 5.41^a, 5.59 [0.1]	
Pd^{2+}	MHL/ML.H 3.67 [1.0]^a	
Cu^{2+}	β_1 21.55^a, 21.38, MHL/ML.H 4.74^a, 4.81, $MH_2L/MHL.H$ 3.04, $M_2L/ML.M$ 5.54^a, 6.79 [0.1]	
Ag^+	β_1 8.61 [0.1]	
Zn^{2+}	β_1 18.40^a, 18.29, MHL/ML.H 5.43^a, 5.60, $MH_2L/MHL.H$ 3.06, $M_2L/ML.M$ 4.36^a, 4.48 [0.1]	

Table 7.3 (Continued)

Formula of the acid Metal ion	Stability constants (Symbol for the constant or scheme of the equilibrium, value of log K or log β, $[I]$)	Ref.
	DTPA (continued)	
Cd^{2+}	β_1 19.2[a], 19.0, MHL/ML.H 4.06[a], 4.17, MH$_2$L/MHL.H 3.32, M$_2$L/ML.M 2.96[a], 2.3 [0.1]	
Hg^{2+}	β_1 26.70[a], 26.40, MHL/ML.H 4.24[a] [0.1]	
Al^{3+}	β_1 18.6[a], 18.7, MHL/ML.H 4.63[a], 4.3, MOHL/ML.OH 6.6 [0.1]	
Ga^{3+}	β_1 25.54, MHL/ML.H 4.35, MOHL/ML.OH 6.52 [0.1][a]	
In^{3+}	β_1 29.0, MOHL/ML.OH 2.06 [0.1][a]	
Tl^+	β_1 5.97, MHL/ML.H 8.8 [0.1][a]	
Tl^{3+}	β_1 46.0 [1.0][a]	
Sn^{2+}	β_1 20.7, MHL/ML.H 4.1, MH$_2$L/MHL.H 2.5 [1.0][a]	
Pb^{2+}	β_1 18.80[a], 18.66, MHL/ML.H 4.52[a], M$_2$L/ML.M 3.41[a] [0.1]	
Sb(III)	ML/M(OH)$_3$.H$_3$L 24.83[a], MHL/ML.H 3.57[a], 3.31 [0.1]	
Bi^{3+}	β_1 35.6 [1.0][a]; MHL/ML.H 2.4 [0.1]; 2.6 [1.0][a]; MH$_2$L/MHL.H 1.8 [0.1]; MOHL/ML.OH 2.7 [1.0][a]	

[a] 20°C

20. $C_{14}H_{24}N_2O_{10}$ ethylene-*bis*(oxyethylenenitrilo)tetra-acetic acid H$_4$L EGTA

H^+	K_{H1} 9.47[a], 9.40 [0.1]; 9.22 [1.0][a]; K_{H2} 8.85[a], 8.78 [0.1]; 8.67 [1.0][a]; K_{H3} 2.66 [0.1][a]; 2.5 [1.0][a]; K_{H4} 2.0 [0.1][a]; 2.4 [1.0][a]	
Na^+	β_1 1.38 [0.1]	[11]
Mg^{2+}	β_1 5.21[a], 5.28, MHL/ML.H 7.62[a] [0.1]	
Ca^{2+}	β_1 10.97[a], 10.86, MHL/ML.H 3.79[a] [0.1]	
Sr^{2+}	β_1 8.50[a], 8.43, MHL/ML.H 5.33[a] [0.1]	
Ba^{2+}	β_1 8.41[a], 8.30, MHL/ML.H 5.31[a] [0.1]	
Sc^{3+}	β_1 (18.2) [0.1][a]	
Y^{3+}	β_1 17.16 [0.1][a]	
La^{3+}	β_1 15.84[a], 15.77 [0.1]	
M^{3+}(4f)	β_1 (17.5) [0.1][a]	
Th^{4+}	MOHL/ML.OH 6.5 [0.1]	
UO_2^{2+}	MHL/M.HL 9.38[a], 9.41, M$_2$L/M^2.L 17.66 [0.1]; MOHL/ML.OH 7.8, (MHL)^2/(ML)$_2$.H^2 −8.48, (ML)$_2$/(ML)^2 3.48 [1.0]	
Mn^{2+}	β_1 12.28[a], 12.18, MHL/ML.H 4.1[a] [0.1]	
Mn^{3+}	β_1 22.7 [0.2]	
Fe^{2+}	β_1 11.87[a], 11.80, MHL/ML.H 4.3[a] [0.1]	
Fe^{3+}	β_1 20.5 [0.1]	
Co^{2+}	β_1 12.39[a], 12.35, MHL/ML.H 5.1[a] [0.1]; 4.9 [0.2]; M$_2$L/ML.M 3.3 [0.1][a]	

Stability Constants of Complexes with Organic Ligands   183

Table 7.3 (Continued)

Formula of the acid / Metal ion	Stability constants (Symbol for the constant or scheme of the equilibrium, value of log K or log β, [I])	Ref.
	EGTA (continued)	
Ni^{2+}	β_1 13.55[a], 13.50, MHL/ML.H 5.1[a], M_2L/ML.M 4.9[a] [0.1]	
Cu^{2+}	β_1 17.71[a], 17.57, MHL/ML.H 4.36[a], 4.28, M_2L/ML.M 4.31, M_2OHL/M_2L.OH 6.9, $M_2(OH)_2$L/M_2OHL.OH 5.8 [0.1]	
Ag^+	β_1 6.88[a], (7.06), MHL/ML.H 7.51[a] [0.1]	
Zn^{2+}	β_1 12.7[a], 12.6, MHL/ML.H 4.96[a], M_2L/ML.M 3.3[a] [0.1]	
Cd^{2+}	β_1 16.7[a], 16.5, MHL/ML.H 3.47[a] [0.1]	
Hg^{2+}	β_1 23.2[a], 22.9, MHL/ML.H 3.02[a], 3.06 [0.1]	
Al^{3+}	β_1 13.90, MHL/ML.H 3.97, MOHL/ML.OH 8.6, $M(OH)_2$L/MOHL.OH 5.33 [0.2]	
Tl^+	β_1 4.38 [0.1][a]; 4.0 [0.3]; MHL/ML.H 8.93 [0.1][a]; 9.09 [0.3]	
Sn^{2+}	β_1 18.7, MHL/ML.H 2.7, MH_2L/MHL.H 1.8 [0.1][a]	
Pb^{2+}	β_1 14.71[a], 14.54, MHL/ML.H 5.16[a], M_2L/ML.M 4.6[a] [0.1]	

[a] 20°C

21. $C_{18}H_{30}N_4O_{12}$	triethylenetetranitrilohexa-acetic acid TTHA	H_6L
H^+	K_{H1} 10.65[a], 10.5, 10.33[b], K_{H2} 9.54[a], 9.44; 9.35[b], K_{H3} 6.10[a], 6.16; 5.98[b], K_{H4} 4.03[a], 4.08, 4.00[b], K_{H5} 2.7[a], 2.7; (2.5[b]), K_{H6} 2.3[a], 2.4, 2.5[b] [0.1]	
Mg^{2+}	β_1 8.43, 8.47[b], MHL/ML.H 9.31, 9.25[b]; MH_2L/MHL.H 4.65, (4.0)[b], M_2L/ML.M 5.5, 5.9[b], M_3L/M_2L.M 5.3 [0.1]	
Ca^{2+}	β_1 10.52[a], 9.89, 10.06[b], MHL/ML.H 8.53, 8.56[a], 8.34[b], MH_2L/MHL.H 4.75[a], 4.87, (3.7)[b], M_2L/ML.M 4.3[a], 4.3, 4.1[b], M_3L/M_2L.M 4.0 [0.1]	
Sr^{2+}	β_1 9.26, MHL/ML.H 7.78, MH_2L/MHL.H (4.2), M_2L/ML.M 3.4 [0.1][b]	
Ba^{2+}	β_1 8.22, MHL/ML.H 7.66, MH_2L/MHL.H (5.5), M_2L/ML.M 3.4 [0.1][b]	
La^{3+}	β_1 22.3, MHL/ML.H 3.51, MH_2L/MHL.H 3.11, M_2L/ML.M 3.4 [0.1]	
$M^{3+}(4f)$	β_1 (23), MHL/ML.H (4), MH_2L/MHL.H (3), M_2L/ML.M (4) [0.1] (Nd, Sm, Ho, Er)	
Zr^{4+}	MH_2L/M.H_2L 19.7 [0.5][a]	
Hf^{4+}	MH_2L/M.H_2L 19.1 [0.5][a]	
Th^{4+}	β_1 31.9, MHL/ML.H 3.05 [0.1]	
U^{4+}	MHL/ML.H 2.28 [0.1]	
Am^{3+}	β_1 (26.6) [0.1]	
Mn^{2+}	β_1 16.0, MHL/ML.H 8.74, MH_2L/MHL.H 3.45, M_2L/ML.M 6.54 [0.1]	

Table 7.3 (Continued)

Formula of the acid / Metal ion	Stability constants (Symbol for the constant or scheme of the equilibrium, value of log K or log β, $[I]$)	Ref.
	TTHA (continued)	
Fe^{2+}	β_1 17.0, MHL/ML.H 8.56, MH$_2$L/MHL.H 3.8, MOHL/ML.OH 4.98, M(OH)$_2$L/MOHL.OH 4.19, M$_2$OHL/M$_2$L.OH 5.27, M$_2$L/ML.M 9.36, (MOH)$_2$L/M$_2$OHL.OH 5.18 [0.1]	
Fe^{3+}	β_1 26.8, MHL/ML.H 7.55, MH$_2$L/MHL.H 2.68, MOHL/ML.OH 4.20, M(OH)$_2$L/MOHL.OH 3.50, M$_2$L/ML.M 13.7, (MOH)$_2$L/M$_2$L.(OH)^2 21.2 [0.1]	
Co^{2+}	β_1 18.4, MHL/ML.H 8.05, MH$_2$L/MHL.H 4.03, MH$_3$L/MH$_2$L.H 2.65, MH$_4$L/MH$_3$L.H 1.57, M$_2$L/ML.M 9.7, M$_2$HL/M$_2$L.H 3.0, M$_2$H$_2$L/M$_2$HL.H 2.6 [0.1]	
Co^{3+}	β_1 39.9 [0.1]	
Ni^{2+}	β_1 19.4, MHL/ML.H 7.98, MH$_2$L/MHL.H (4.86), MH$_3$L/MH$_2$L.H 2.74, MH$_4$L/MH$_3$L.H 1.15, M$_2$L/ML.M 13.0, M$_2$HL/M$_2$L.H 2.6, M$_2$H$_2$L/M$_2$HL.H 2.3 [0.1]	
Cu^{2+}	β_1 20.5, 21.8[a], MHL/ML.H 8.03[a], 7.98, MH$_2$L/MHL.H 4.05, MH$_3$L/MH$_2$L.H 2.86, MH$_4$L/MH$_3$L.H 2.04, M$_2$L/ML.M 13.64[a], 13.4, M$_2$HL/M$_2$L.H 3.0, M$_2$H$_2$L/M$_2$HL.H 2.7 [0.1]	
Ag^+	β_1 9.0, MHL/ML.H 9.11, M$_2$L/ML.M 5.22 [0.1]	
Zn^{2+}	β_1 18.1[a], 18.0, MHL/ML.H 8.03[a], 8.13, MH$_2$L/MHL.H 4.6, MH$_3$L/MH$_2$L.H 3.2, M$_2$L/ML.M 11.9[a], 11.9, M$_2$HL/M$_2$L.H 3.0, M$_2$H$_2$L/M$_2$HL.H 2.6 [0.1]	
Cd^{2+}	β_1 18.6, MHL/ML.H 8.5, MH$_2$L/MHL.H 3.1, MH$_3$L/MH$_2$L.H 2.7, M$_2$L/ML.M 8.3 [0.1]	
Hg^{2+}	β_1 26.1, MHL/ML.H 6.3, MH$_2$L/MHL.H 3.5, MH$_3$L/MH$_2$L.H 3.0, M$_2$L/ML.M 12.3, M$_2$HL/M$_2$L.H 3.6, M$_2$H$_2$L/M$_2$HL.H 2.7, (MOH)$_2$L/M$_2$L.(OH)^2 12.8 [0.1]	
Al^{3+}	β_1 21.0, MHL/ML.H 5.85, M$_2$L/ML.M 9.2, M(OH)$_2$L/M$_2$L.(OH)^2 15.9 [0.1]	
Ga^{3+}	MHL/ML.H 4.52, MH$_2$L/MHL.H 3.54, MH$_3$L/MH$_2$L.H 2.29, M$_2$L/ML.M 10.0 [0.1]	
Pb^{2+}	β_1 18.5, MHL/ML.H 8.15, MH$_2$L/MHL.H 3.8, MH$_3$L/MH$_2$L.H 2.8, M$_2$L/ML.M 10.8, M$_2$HL/M$_2$L.H 3.0, M$_2$H$_2$L/M$_2$HL.H 2.6 [0.1]	
Bi^{3+}	MHL/ML.H 4.16, MH$_2$L/MHL.H 2.84, MH$_3$L/MH$_2$L.H 2.11 [0.1]	

[a] 20°C; [b] 30°C

REFERENCES

[1] M. Cefola, A. S. Tompa, A. V. Celiano and P. S. Gentile, *Inorg. Chem.*, 1962, **1**, 290.
[2] F. Holmes and W. R. C. Crimmin, *J. Chem. Soc.*, **1955**, 1175.
[3] P. E. Wenger, D. Monnier and L. Epars, *Helv. Chim. Acta*, 1952, **35**, 396, 569.
[4] P. Lumme, *Ann. Acad. Sci. Fennicae* AII, 1955, **68**, 7.
[5] P. O. Lumme, *Suom. Kemistilehti*, 1958, **B31**, 250.
[6] M. Yasuda, private communication, cf. *Stability Constants of Metal-Ion Complexes*, L. G. Sillén and A. E. Martell (eds.), p. 627, The Chemical Society, London, 1964.
[7] J. Starý, *Anal. Chim. Acta*, 1963, **28**, 132.
[8] I. Sajó, *Acta Chim. Acad. Sci. Hung.*, 1968, **16**, 115.
[9] R. J. Kula, *Anal. Chem.*, 1965, **37**, 989.
[10] T. R. Bhat and M. Krishnamurthy, *J. Inorg. Nucl. Chem.*, 1964, **26**, 587.
[11] M. Tanaka, S. Funabashi and K. Shirai, *Inorg. Chem.*, 1968, **7**, 573.

Table 7.4
Complexes of other organic ligands*

Formula Metal ion	Stability constants (Symbol for the constant or scheme of the equilibrium, value of log K or log β, [I])	Ref.
1. CH_4N_2S	thiourea $H_2N.\overset{\overset{\displaystyle S}{\|}}{C}.NH_2$	L
Cu^{2+}	β_4 15.4 [0.5]	
Ag^+	β_1 7.11, β_2 10.61, β_3 12.73, β_4 13.57 [0.5]	
Pd^{2+}	β_4 30.1 [0.5]	
Zn^{2+}	β_1 0.5, β_2 0.8, β_3 0.9 [2.0]	
Cd^{2+}	β_1 1.5 [0]; 1.33 [0.1]; 1.82 [1.0]; 2.12 [3.0]; β_2 2.2 [0]; 2.18 [0.1]; 2.64 [1.0]; 3.65 [3.0]; β_3 2.6 [0]; 2.7 [0.1]; 3.0 [1.0]; 4.0 [3.0]; β_4 3.1 [0]; 3.3 [0.1]; 3.8 [1.0]; 5.5 [3.0]	
Hg^{2+}	β_2 21.3 [0.1][a]; 22.1 [1.0]; β_3 24.2 [0.1][a]; 24.7 [1.0]; β_4 25.8 [0.1][a]; 26.5 [1.0]	
Pb^{2+}	β_1 0.09 [0]; 0.17 [0.1]; 0.63 [1.0]; 1.68 [3.0]; β_2 0.83 [0]; 0.86 [0.1]; 1.37 [1.0]; 2.54 [3.0]; β_3 1.3 [0]; 1.4 [0.1]; 1.8 [1.0]; 2.9 [3.0]; β_4 1.5 [0]; 1.5 [0.1]; 2.0 [1.0]; 3.1 [3.0]; β_5 1.5 [0]; 1.4 [0.1]; 2.0 [1.0]; 2.9 [3.0]; β_6 1.7 [0]; 1.8 [0.1]; 2.0 [1.0]; 3.3 [3.0]	

[a] 20°C

* All listed values of constants refer to 25°C unless stated otherwise in a note. The notes follow each section of the table separately.

186 Stability Constants of Complexes with Organic Ligands [Ch. 7

Table 7.4 (Continued)

Formula Metal ion	Stability constants (Symbol for the constant or scheme of the equilibrium, value of log K or log β, $[I]$)	Ref.

2. CH_5N_3O semicarbazide $H_2N.NH.\overset{\overset{O}{\|}}{C}.NH_2$ L

H^+ K_{H1} 3.53 $[0.1]^a$
Cu^{2+} β_1 4.00, β_2 6.94 $[0.1]^a$
Ag^+ β_1 1.95, β_2 2.70 $[0.1]^a$
Zn^{2+} β_1 2.3, β_2 3.7 $[0.1]^a$
Cd^{2+} β_1 1.26, β_2 2.79 $[0.1]^a$
Hg^{2+} β_2 11.6, β_3 15.2 $[0.1]^a$
Pb^{2+} β_1 2.11, β_2 2.86 $[0.1]^a$

^{a} 30°C

3. CH_5N_3S thiosemicarbazide $H_2N.\overset{\overset{S}{\|}}{C}.NHNH_2$ L

H^+ K_{H1} 12.6 $[0.1]$; 12.5 $[1.0]$; 12.6 $[2.0]$; $K_{H2}(L)$ 11.9 $[0.1]$; 12.0 $[1.0]$;
    12.1 $[2.0]$; $K_{H3}(L + H^+ = LH^+)$ 1.6 $[0.1]^a$; 1.8 $[1.0]$; 2.0 $[1.0]$
Cu^{2+} β_1 6.11, β_2 11.59 $[0.1]^a$
Ag^+ β_3 12.9 $[0.8]$
Zn^{2+} β_2 2.8 $[0.1]^a$
Cd^{2+} β_1 2.28 $[0.1]^a$; 2.57 $[1.0]$; β_2 4.40 $[0.1]^a$; 4.70, β_3 5.86 $[1.0]$
Hg^{2+} β_2 22.4, β_3 24.8 $[0.1]^a$; β_4 25.8 $[0.8]$
Pb^{2+} β_1 2.89 $[0.1]^a$

^{a} 30°C

4. $C_2H_5NO_2$ acetohydroxamic acid $CH_3.\overset{\overset{O}{\|}}{C}.NHOH$ HL

H^+ β_1 9.36 $[0.1]^a$
Ca^{2+} β_1 2.4 $[0.1]^a$
La^{3+} β_1 5.16, β_2 9.33, β_3 11.88 $[0.1]^a$
$M^{3+}(4f)$ β_1 (6.1), β_2 (11.1), β_3 12.8–16.5 $[0.1]^a$ (Ce, Sm, Gd, Dy, Yb)
Mn^{2+} β_1 4.0, β_2 6.9 $[0.1]^a$
Fe^{2+} β_1 4.8, β_2 8.5 $[0.1]^a$
Fe^{3+} β_1 11.42, β_2 21.10, β_3 28.33 $[0.1]^a$
Co^{2+} β_1 5.1, β_2 8.9 $[0.1]^a$
Ni^{2+} β_1 5.3, β_2 9.3 $[0.1]^a$
Cu^{2+} β_1 7.9 $[0.1]^a$

Table 7.4 (Continued)

Formula Metal ion	Stability constants (Symbol for the constant or scheme of the equilibrium, value of $\log K$ or $\log \beta$, $[I]$)	Ref.
	acetohydroxamic acid (continued)	
Zn^{2+}	β_1 5.4, β_2 9.6 $[0.1]^a$	
Cd^{2+}	β_1 4.5, β_2 7.8 $[0.1]^a$	
Al^{3+}	β_1 7.95, β_2 15.29, β_3 21.47 $[0.1]^a$	
Pb^{2+}	β_1 6.7, β_2 10.7 $[0.1]^a$	

^{a} 20°C

5. C_2H_6OS	2-mercaptoethanol HS.CH$_2$CH$_2$OH HL	
H^+	K_{H1} 9.72 [0]; 9.40 [0.1]; 9.34 [0.5]	
Ni^{2+}	β_{22} 10.73, β_{43} 23.76, β_{64} 36.79 $[0.5]^a$	
Ag^+	β_1 13.0, β_2 17.9, β_{21} 19 $[0.1]^b$	
Zn^{2+}	β_{23} 18.32, β_{63} 38.52, β_{94} 57.80, $\beta_{12,5}$ 77.20, $\beta_{15,6}$ 95.92 [0.5]	
In^{3+}	β_1 9.1, β_2 17.2, β_3 24.1, β_4 29.9 $[0.1]^b$	
Ge(IV)	MOH(H$_{-1}$L)$_2$.H/H$_2$MO$_3$.(HL)^2 -4.21 [0.1]	
Pb^{2+}	β_{12} 8.94, β_{22} 15.77, β_{32} 22.03, β_{43} 32.65, β_{53} 38.50 [0.5]	
Sb^{3+}	M(H$_{-1}$L)$_2$.H/M(OH)$_3$.(HL)^2 -1.73; MHL$_2$/ML$_2$.H 7.98 $[0.1]^b$	

^{a} The values of $\log K$ were calculated with the aid of the expression $\log K = 6.515n - 2.3$, where n is the number of ligands in a molecule of the complex (cf. ref. [1]); ^{b} 20°C

6. $C_3H_8OS_2$	2,3-dimercaptopropanol HS.CH$_2$CH(SH).CH$_2$OH H$_2$L BAL	
H^+	K_{H1} 10.68, 10.61^a, K_{H2} 8.58, 8.58^a [0.1]	
Mn^{2+}	β_1 5.23, β_2 10.43 $[0.1]^a$	
Fe^{2+}	β_2 15.8, β_{32} 28 $[0.1]^a$	
Fe^{3+}	MOHL/M.OH.L 30.7 [0.1]	
Ni^{2+}	β_2 22.8, M$_2$OHL$_3$/M^2.OH.L^3 45.6 $[0.1]^a$	
Zn^{2+}	β_1 13.5, β_2 23.3, β_{32} 40.4 $[0.1]^a$	

^{a} 30°C

7. $C_3H_8O_3$	glycerol HO.CH$_2$CH(OH)CH$_2$OH L	
Y^{3+}	MH$_{-1}$L.H/M.L -6.80 [0.1]	
La^{3+}	MH$_{-1}$L.H/M.L -8.30 [0.1]	
$M^{3+}(4f)$	MH$_{-1}$L.H/M.L (-7.1) [0.1]	
B(III)	M(OH)$_2$(H$_{-2}$L).H/M(OH)$_3$.L -7.54, M(H$_{-2}$L)$_2$.H/M(OH)$_3$.L^2 -7.17 [0.1]	

Table 7.4 (Continued)

Formula Metal ion	Stability constants (Symbol for the constant or scheme of the equilibrium, value of log K or log β, [I])	Ref.
	glycerol (continued)	
Ge(IV)	$HMO_2(H_{-2}L).H/M(OH)_4.L$ -7.32, $HMO(H_{-2}L)_2.H/M(OH)_4.L^2$ -6.64 [0.1]	
Pb^{2+}	$MOH(H_{-2}L)/M(OH)_3.L$ 1.2 [1.0]	
Te(VI)	$H_3MO_4(H_{-2}L).H/M(OH)_6.L$ -5.63 [0.1]	

8. $C_3H_{12}NO_9P_3$ nitrilotris(methylenephosphonic) acid $N(CH_2PO_3H_2)_3$ H_6L

H^+	K_{H1} 12.3, K_{H2} 6.66, K_{H3} 5.46, K_{H4} 4.30 [1.0]	
Mg^{2+}	β_1 6.49, MHL/ML.H 9.1, MH_2L/MHL.H 6.2, MH_3L/MH_2L.H 4.6 [1.0]	
Ca^{2+}	β_1 6.68, MHL/ML.H 8.5, MH_2L/MHL.H 6.1, MH_3L/MH_2L.H 4.9 [1.0]	

9. $C_4H_8N_2O_2$ butane-2,3-dione dioxime $CH_3.C-C.CH_3$ H_2L
   biacetyl dioxime   ‖ ‖
   dimethylglyoxime   HO-N N-OH

H^+	K_{H1} 12.0 [0]; 11.9 [0.1]; K_{H2} 10.66 [0]; 10.45 [0.1]	
Co^{2+}	MHL/M.HL 8.35, $M(HL)_2/M.(HL)^2$ 16.98 [0.1]	[2]
Ni^{2+}	$M(HL)_2/M.(HL)^2$ 17.84 [0]; 17.24 [0.1]; $M(HL)_2/M(HL)_{2(s)}$ -6.03 [0.1]	
Pd^{2+}	$M(HL)_2/M.(HL)^2$ 34.1 [0.1]; 34.3 [1.0]; $MOH(HL)_2/M(HL)_2.OH$ 5.50 [0.1]; 5.6 [1.0]	
Cu^{2+}	$M(HL)_2/M.(HL)^2$ 19.24, $MOH(HL)_2/M.(HL)^2.OH$ 22.42 [0.1]	

10. $C_5H_8O_2$ pentane-2,4-dione $CH_3.CO.CH_2.CO.CH_3$ HL
    acetylacetone

H^+	K_{H1} 8.99 [0]; 8.80 [0.1]; 8.80 [1.0]	
Be^{2+}	β_1 7.90 [0]; 7.55 [1.0]; β_2 14.59 [0]; 14.35 [0.1]; MOHL/M.OH.L 15.50 [0]; $M(OH)_2L/M.(OH)^2.L$ 19.70 [0]	
Mg^{2+}	β_1 3.65, β_2 6.25 [0]	
Sc^{3+}	β_1 8.0, β_2 15.2 [0][a]	
Y^{3+}	β_1 6.4 [0][a]; 5.89 [0.1]; β_2 11.1 [0][a]; 10.85 [0.1][a]; β_3 (13.9) [0][a]; 14.1 [0.1][a]	
$M^{3+}(4f)$	β_1 (5.9), β_2 (10.3)[a], β_3 (13.6)[a] [0.1]	
Ti^{3+}	β_1 10.43, β_2 18.82, β_3 24.9 [1.0]	
Zr^{4+}	β_1 11.25 [1.0][b]	
Th^{4+}	β_1 8.8 [0][a]; 7.7 [0.1]; β_2 16.2 [0][a]; 14.9 [0.1]; β_3 22.5 [0][a]; 20.8 [0.1]; β_4 26.7 [0][a]; 25.8 [0.1]	
V^{2+}	β_1 5.38, β_2 10.19, β_3 14.70 [1.0]	
VO^{2+}	β_1 8.68 [0]; 8.40 [0.1]; β_2 15.84 [0]; 15.40 [0.1]	

Table 7.4 (Continued)

Formula Metal ion	Stability constants (Symbol for the constant or scheme of the equilibrium, value of log K or log β, [I])	Ref.
	acetylacetone (continued)	
Cr^{2+}	β_1 5.96, β_2 11.70 [1.0]	
U^{4+}	β_1 8.6, β_2 17.0, β_3 23.4, β_4 29.5 [0.1]	
UO_2^{2+}	β_1 7.7 [0]; 6.8 [0.1]; β_2 14.1 [0]; 13.1 [0.1]; $MHL_2/ML_2 \cdot H$ 4.4, $MHL_3/ML_2 \cdot HL$ 1.7 [0.1]	
Np^{4+}	β_1 8.58, β_2 17.23, β_3 23.94, β_4 30.22 [1.0]	
Pu^{4+}	β_1 10.5, β_2 19.7, β_3 28.1, β_4 34.1 [0.1]	
Mn^{2+}	β_1 4.21 [0]; 4.07 [0.1]; β_2 7.30 [0]	
Mn^{3+}	$ML_3/ML_2 \cdot L$ 3.86 [0.2]	
Fe^{2+}	β_1 5.07, β_2 8.67 [0]^a	
Fe^{3+}	β_1 9.8, β_2 18.8, β_3 26.2 [0]^a	
Co^{2+}	β_1 5.40, β_2 9.54 [0]	
Ni^{2+}	β_1 6.00 [0]; 5.72 [0.1]; β_2 10.60 [0]; 9.66 [0.1]	
Pd^{2+}	β_1 16.4, β_2 27.5 [0]	
Cu^{2+}	β_1 8.25 [0]; 8.16 [0.1]; 8.22 [1.0]; β_2 15.05 [0]; 14.76 [0.1]; 14.81 [1.0]	
Zn^{2+}	β_1 5.06 [0]; 4.68 [0.1]; β_2 9.00 [0]; 7.92 [0.1]	
Cd^{2+}	β_1 3.83, β_2 6.65 [0]; 6.12 [0.7]^a	
Hg^{2+}	β_2 21.5 [0.5]^a	
Al^{3+}	β_1 8.6, β_2 16.5, β_3 22.3 [0]^a	
Ga^{3+}	β_1 9.4, β_2 17.8, β_3 23.7 [0]^a	
In^{3+}	β_1 8.0, β_2 15.1 [0]^a	
Pb^{2+}	β_2 6.32 [0.7]^a	

^{a} 30°C; ^{b} 19°C

11. $C_5H_{11}NS_2$	N,N-diethyldithiocarbamic acid $(C_2H_5)_2N \cdot CS \cdot SH$ HL	
H^+	K_{H1} 3.42 [0], 3.31 [0.1]; 3.22 [0.5]; 3.16 [1.0]	
UO_2^{2+}	β_4 (17.5) [1.0]	[3]
Cu^{2+}	β_1 14.9, β_2 28.8 [75% EtOH]^a	[4]
Ag^+	β_1 8.3 [0.1; 75% dioxane]	[5]
Cd^{2+}	β_1 14.9, β_2 28.8 [0.01; 75% EtOH]^b	[4]
Hg^{2+}	β_1 22.3, β_2 38.1, β_3 39.1	[6]
Tl^+	β_1 4.3, β_2 5.3	[7]
M^{n+}	$K[M(HL)_n \cdot (H_2A)^n/M(HA)_n \cdot (HL)^n]$; [$CCl_4$; $n = 1-4$; H_2A = dithizone; M = Ag, As, Bi, Cd, Cu, Mg, Pd, Pb, Sb, Se, Te, Tl, Zn]^c	[8]

^{a} 20°C; ^{b} 25 ± 2°C; ^{c} temperature not given

Table 7.4 (Continued)

Formula Metal ion	Stability constants (Symbol for the constant or scheme of the equilibrium, value of $\log K$ or $\log \beta$, $[I]$)	Ref.
12. $C_6H_2Cl_2O_4$	2,5-dichloro-3,6-dihydroxy-1,4-benzoquinone chloranilic acid	H_2L
H^+	K_{H1} 2.60 [0.15]; 2.37 [0.5]; K_{H2} 0.69 [0.15]; 0.66 [0.5]	
Zr^{4+}	β_1 (9.27), β_2 (16.56) [2.0]	
Hf^{4+}	β_1 (6.7), β_3 (20.7) [3.0]	
Fe^{3+}	β_1 5.8, β_2 9.6 [0.15]	
Ni^{2+}	β_1 4.3 [0.15]	
Ge(IV)	$M(OH)_2L_2/M(OH)_4.(HL)^2$ 6.57, $HM(OH)_2L_2/M(OH)_2L_2.H$ 0.8 [0.5]	
13. $C_6H_6O_2$	1,2-dihydroxybenzene	H_2L
H^+	K_{H1} (12.8) [0]; 13.0 [0.1]; 13.0 [1.0]; K_{H2} 9.40 [0]; 9.23 [0.1], [1.0]	
Be^{2+}	β_1 13.52, β_2 23.35, $MHL/M.HL$ 5.0, $MHL_2/ML.HL$ 2.8 [0.1]^a	
Mg^{2+}	β_1 5.7 [0.1]^b	
Sc^{3+}	β_1 17.04 [0.1]	
La^{3+}	β_1 9.46 [0.1]	
$M^{3+}(4f)$	β_1 (11.2) [0.1]	
Ti(IV)	$MO(HL)_2/MO.(HL)^2$ 24.3, $ML_3/MO(HL)_2.L$ 21.5 [0.1]^a	
V^{3+}	β_1 18.3, $M(OH)_2L/ML.(OH)^2$ 19.4, $M(OH)_2L_2/M(OH)_2L.L$ 10.1 [0.5]^c	
VO^{2+}	β_1 17.7, β_2 33.5 [0.1]^a	
Mo(VI)	$HMO_3L/MO_4.H_2L.H$ 8.4^d, $MO_2L_2/MO_4.(H_2L)^2$ 5.1^a, $HMO_2L_2/MO_2L_2.H$ 3.65 [0.1]	
W(VI)	$HMO_3L/MO_4.H_2L.H$ 4.0^d, $MO_2L_2/MO_4.(H_2L)^2$ 6.7^a [0.1]	
UO_2^{2+}	β_1 15.9, $MHL/M.HL$ 6.3, $MHL_2/ML.HL$ 4.9, $MHL_3/MHL_2.HL$ 3.7 [0.1]^a	
Mn^{2+}	β_1 7.47 [0]; 7.72 [0.1]; β_2 12.8 [0], 13.6 [0.1]	
Fe^{2+}	β_1 7.95, β_2 13.5 [0]	
Co^{2+}	β_1 8.32 [0]; 8.60 [0.1]; β_2 14.7 [0]; 15.0 [0.1]	
Ni^{2+}	β_1 8.77 [0]; 8.92; β_2 14.4 [0.1]	
Cu^{2+}	β_1 (14.1) [0]; 13.90 [0.1]; 13.62 [1.0]; β_2 (24.6) [0]; 24.9 [0.1]; 24.94 [1.0]	
Zn^{2+}	β_1 9.90 [0.1]; 9.54 [1.0]; β_2 17.4 [0.1]; 17.5 [1.0]	
Cd^{2+}	β_1 8.2 [0.1]^b	
B(III)	$M(OH)_2L/M(OH)_3.HL$ 4.10, $ML_2/M(OH)_3.H_2L.HL$ 4.39 [0.1]	
Al^{3+}	β_1 16.3, β_2 29.3, β_3 37.6 [0.1]^a	
Ca^{3+}	β_1 18.9 [0.1]^e	
Ge(IV)	$ML_3/M(OH)_4.H_2L.(HL)^2$ 17.26 [0.1]	

Table 7.4 (Continued)

Formula Metal ion	Stability constants (Symbol for the constant or scheme of the equilibrium, value of log K or log β, $[I]$)	Ref.
	1,2-dihydroxybenzene (continued)	
As(III)	$M(OH)_2L/M(OH)_3 \cdot HL$ 2.34, $ML_2/M(OH)_3 \cdot H_2L \cdot HL$ 2.83 [0.1]	
Sb(III)	$ML_2/M(OH)_3 \cdot HL \cdot H_2L$ 10.2, $MOHL/M(OH)_3 \cdot H_2L$ 3.3 [0.1]^a	

^{a} 20°C; ^{b} 30°C; ^{c} 21°C; ^{d} 20–25°C; ^{e} 22°C

14. $C_6H_6O_2$	1,3-dihydroxybenzene	H_2L
H^+	K_{H1} 11.06; K_{H2} 9.30 [0.1]^a	
UO_2^{2+}	β_1 16.9, $MHL/M \cdot H \cdot L$ 17.0 [0.1]^a	

^{a} 20°C

15. $C_6H_6O_3$	1,2,3-trihydroxybenzene	H_3L
H^+	K_{H1} 14^a, K_{H2} 11.08^a, K_{H3} 8.94^a, 8.98 [0.1]	
Be^{2+}	$MHL/M \cdot HL$ 13.5, $MH_2L/M \cdot H_2L$ 4.6 [0.1]^a	
V^{3+}	β_1 18.1, $M(OH)L/M \cdot L \cdot (OH)^2$ 38.0, $M(OH)_2L_2/M \cdot L^2(OH)^2$ 49.0 [0.5]^c	
VO^{2+}	$MHL/M \cdot HL$ 15.0 [0.1]	
Mo(VI)	$HMO_3L/MO_4 \cdot H \cdot H_3L$ 8.7, $MO_2(HL)_2/MO_4 \cdot (H_3L)^2$ 5.5, $HMO_2(HL)_2/MO_2(HL)_2 \cdot H$ 3.70 [0.1]	
W(VI)	$MO_2(HL)_2/MO_4 \cdot (H_3L)^2$ 7.0^a, $HMO_3(HL)/MO_4 \cdot H_3L \cdot H$ 9.4^b [0.1]	
UO_2^{2+}	β_{12} 27.2^b, $MH_2L_2/ML \cdot H_2L$ 4.29 [0.1]	
B(III)	$M(OH)_2HL/M(OH)_3 \cdot H_2L$ 3.93, $M(OH)(HL)_2/M(OH)_3 \cdot (H_2L)^2$ 13.56 [0.1]	
Ge(IV)	$M(HL)_3/H_2MO_3 \cdot H_3L \cdot (H_2L)^2$ 17.36 [0.1]	
As(III)	$M(OH)_2HL/M(OH)_3 \cdot H_2L$ 2.66, $M(HL)_2 \cdot H/M(OH)_3 \cdot (H_3L)^2$ −6.05 [0.1]	

^{a} 20°C; ^{b} 20–25°C; ^{c} 21°C

16. $C_6H_6O_4$	5-hydroxy-2-hydroxymethyl-4-pyrone kojic acid	HL
H^+	K_{H1} 7.88 [0]; 7.66 [0.1]; 7.67 [1.0]; 7.88 [2.0]	
Mg^{2+}	β_1 2.92, β_2 5.11 [0.1]	
Ca^{2+}	β_1 2.5 [0.1]^a	
Y^{3+}	β_1 6.18, β_2 11.37, β_3 15.52 [0.1]^b	
La^{3+}	β_1 5.38, β_2 9.56^b, β_3 12.87^b [0.1]	
$M^{3+}(4f)$	β_1 (6.16), β_2 (11.3)^b, β_3 (15.5)^b [0.1]	
Hf^{4+}	β_1 12.04, β_2 22.59 [1.0]^a	

Formula Metal ion	Stability constants (Symbol for the constant or scheme of the equilibrium, value of log K or log β, $[I]$)	Ref.
	kojic acid (continued)	
UO_2^{2+}	β_1 7.1, β_2 12.6, β_3 16.1 [0.1][a]	
Mn^{2+}	β_1 3.95 [0.1]; 3.66 [2.0]; β_2 6.78 [0.1]; 6.65 [2.0]; β_3 8.5 [2.0]	
Fe^{3+}	β_1 9.2[a], β_2 17.2[a], β_3 24.4[a], MOHL/ML.OH 10.62, $M(OH)_2L/ML.(OH)^2$ 20.2, $M_2(OH)_2L_2/(ML)^2.(OH)^2$ 23.6, $M_2(OH)_4L_2/(ML)^2.(OH)^4$ 45.2 [0.1]	
Co^{2+}	β_1 4.55, β_2 8.26, β_3 10.70 [2.0]	
Ni^{2+}	β_1 4.9 [0.1][a]; 4.86 [2.0]; β_2 8.7 [0.1][a]; 8.81 [2.0]; β_3 11.62 [2.0]	
Cu^{2+}	β_1 6.6 [0.1][a]; 6.6 [2.0]; β_2 11.8 [0.1][a]; 11.7 [2.0]	
Zn^{2+}	β_1 4.9 [0.1][a]; 5.03 [2.0]; β_2 9.1 [0.1][a]; 9.34 [2.0]; β_3 12.4 [2.0]	
Cd^{2+}	β_1 4.6 [0.1][a]	
Al^{3+}	β_1 7.7, β_2 14.2, β_3 19.5 [0.1][a]	
Ge(IV)	$M(OH)_2L_2/M(OH)_4.(HL)^2$ 2.81 [0.5]	

[a] 20°C; [b] 30°C

17. $C_6H_6O_5S$	1,2-dihydroxybenzene-4-sulphonic acid	H_3L
H^+	K_{H1} 12.8[a], (12.16)[b], K_{H2} 8.50[a], 8.26[b] [0.1]	
Mg^{2+}	β_1 6.27, β_2 10.41 [0.1][b]	
Ca^{2+}	β_1 4.40, β_2 7.99 [0.1][b]	
Sr^{2+}	β_1 3.61 [0.1][b]	
VO^{2+}	β_1 16.7, β_2 31.2 [0.1][a]	
Mo(VI)	$MO_2L_2/MO_4.(H_2L)^2$ 5.28 [0.1][a]	
UO_2^{2+}	$MHL/M.HL$ 6.4, $M_2(OH)L_2/M^2.OH\ L^2$ 43.3 [0.1][a]	
Mn^{2+}	β_1 7.87, β_2 12.53 [0.1][b]	
Fe^{3+}	$K_1 > 17$, K_2 14.0, K_3 9.1 [0.1][b]	
Co^{2+}	β_1 8.54, β_2 14.40, β_3 17.48 [0.1][b]	
Ni^{2+}	β_1 8.85, β_2 14.41, β_3 19.14 [0.1][b]	
Cu^{2+}	β_1 13.29, β_2 23.52 [0.1][b]	
Zn^{2+}	β_1 9.40, β_2 16.60 [0.1][b]	
B(III)	$M(OH)_2L/M(OH)_3.HL$ 3.78 [0.1][a]	
Al^{3+}	β_1 16.6, β_2 29.9, β_3 39.2 [0.1][a]	
Ce(IV)	$ML_3/M(OH)_4.H_2L.(HL)^2$ 17.9 [0.1][a]	

[a] 20°C; [b] 30°C

18. $C_6H_6O_8S_2$	1,2-dihydroxybenzene-3,5-disulphonic acid tiron	H_4L
H^+	K_{H1} 13.3 [0]; 12.5 [0.1]; 11.8 [1.0]; K_{H2} 8.31 [0]; 7.61 [0.1]; 7.20 [1.0]	

Table 7.4 (Continued)

Formula Metal ion	Stability constants (Symbol for the constant or scheme of the equilibrium, value of log K or log β, [I])	Ref.
	tiron (continued)	
Be^{2+}	β_1 12.88, β_2 22.25, MHL/M.HL 4.2, MHL$_2$/M.L.HL 15.18 [0.1][a]	
Mg^{2+}	β_1 6.86, MHL/M.HL 1.98 [0.1][a]	
M^{2+}(IIA)	β_1 (4.8), MHL/M.HL (2.0) [0.1][a]	
Sc^{3+}	β_1 18.96, MHL/M.HL 8.94 [0.1]	
Y^{3+}	β_1 13.72, MHL/M.HL 5.13 [0.1]	
M^{3+}(4f)	β_1 (14.1), MHL/M.HL (5.6) [0.1]	
Ti(IV)	ML$_3$/MO.H^2.L^3 57.6; MO(HL)$_2$/MO(HL)^2 14.9 [0.1][a]	
Th^{4+}	M$_2$(OH)$_2$L$_3$/(ML)^2.(OH)^2.L 35.8, M$_2$L$_3$/M$_2$(OH)$_2$L$_3$.H^2 12.8 [0.1]	
VO^{2+}	β_1 16.74, β_2 32.79, MOHL/ML.OH 7.5, (MOHL)$_2$/(MOHL)^2 4.3 [0.1]	
Mo(VI)	MO$_2$L$_2$/MO$_4$.(H$_2$L)^2 6.44 [0.1][a]	
UO_2^{2+}	β_1 (15.9), MHL/M.HL (6.4) [0.1][a], M$_2$(OH)L$_2$/(MHL)^2(OH)^3 33[a], M$_3$(OH)$_2$L$_3$/(ML)^3(OH)^2 (30.4), M$_3$(OH)$_2$L$_4$/M$_3$(OH)$_2$L$_3$.L (8.96) [0.1]	
Mn^{2+}	β_1 7.20, β_2 12.75, β_3 16.28, MHL/M.HL 2.05 [1.0]	
Fe^{3+}	β_1 20.4 [0.1]; 19.0 [1.0]; β_2 35.4 [0.1]; 33.9 [1.0]; β_3 45.8 [0.1]; 45.3 [1.0]; MHL/M.HL 9.7 [0.1]; 9.1 [1.0]	
Co^{2+}	β_1 10.78 [0]; 9.49 [0.1][a]; 8.19 [1.0]; β_2 14.41 [1.0]; MHL/M.HL 3.08 [0.1][a]	
Ni^{2+}	β_1 11.24 [0]; 9.96 [0.1][a]; 8.56 [1.0]; β_2 14.90 [1.0]; MHL/M.HL 3.00 [0.1][a]	
Cu^{2+}	β_1 15.62 [0]; 14.27 [0.1]; 12.76 [1.0]; β_2 25.46 [0.1]; 23.73 [1.0]; MHL/M.HL 5.48 [0.1][a]	
Zn^{2+}	β_1 11.68 [0]; 10.41 [0.1][a]; 9.00 [1.0]; β_2 18.52 [0.1][b]; 16.91 [1.0]; MHL/M.HL 3.30 [0.1][a]	
Cd^{2+}	β_1 10.29 [0]; 7.69, β_2 13.29 [1.0]	
Hg^{2+}	β_1 19.86 [0.1]	
B(III)	M(OH)$_2$L/M(OH)$_3$.HL 3.97 [0]; 3.9 [0.1]; 4.0 [1.0]	
Al^{3+}	β_1 19.06 [0]; 16.6 [0.1]; 14.9 [1.0]; β_2 31.10 [0]; 30.0 [0.1]; 28.4 [1.0]; β_3 33.5 [0]; 40.0 [0.1][a]	
Ga^{3+}	β_1 19.24 [0.1]	
In^{3+}	β_1 16.34 [0.1]	
Ge(IV)	ML$_3$/M(OH)$_4$.(HL)^3.H 25.1 [0.1]; 23.6 [1.0]	
As(III)	ML$_2$/M(OH)$_3$.(HL)^2.H 8.43 [0]	
Sb(III)	ML$_2$/M(OH)$_3$(HL)^2.H 17.9[a], K_2 14.4 [0.1]; 13.0 [1.0]; MHL$_2$/ML$_2$.H 2.00 [0.1]; 2.04 [1.0]; MOHL.H$_2$L/ML$_2$.H 1.23 [0.1][a]	

[a] 20°C; [b] 30°C

Table 7.4 (Continued)

Formula Metal ion	Stability constants (Symbol for the constant or scheme of the equilibrium, value of log K or log β, $[I]$)	Ref.
19. $C_6H_{10}N_2O_2$	cyclohexane-1,2-dione dioxime nioxime	H_2L
H^+	K_{H1} 12.1, K_{H2} 10.57 [0.1]	
Ni^{2+}	MHL/M.HL 10.8, $M(HL)_2/M.(HL)^2$ 21.5 [0.1]	
20. $C_6H_{12}O_6$	D($-$)-fructose	L
H^+	$L/H_{-1}L.H$ 12.03 [0]	
B(III)	$M(OH)_2H_{-2}L/M(OH)_3.OH.L$ 8.1, $M(H_{-2}L)_2/M(OH)_3.OH.L^2$ 9.7 [0.1]	
As(III)	$M(OH)_2(H_{-2}L)/M(OH)_3.OH.L$ 5.4 [0.1]	
Ge(IV)	$HMO(H_{-2}L)_2/M(OH)_4.OH.L^2$ 10.7 [0.1]	
Te(VI)	$H_3MO_4(H_{-2}L)/M(OH)_6.OH.L$ 8.3 [0.1]	
21. $C_6H_{12}O_6$	α-D($+$)-glucose	L
H^+	$L/H_{-1}L.H$ 12.28 [0]	
B(III)	$M(OH)_2(H_{-2}L)/M(OH)_3.OH.L$ 6.7, $M(H_{-2}L)_2/M(OH)_3.OH.L^2$ 7.3 [0.1]	
22. $C_6H_{14}O_6$	D($-$)-mannitol $\quad HOCH_2\underset{\underset{HO}{\mid}}{CH}.\underset{\underset{OH}{\mid}}{CH}.\overset{\overset{HO}{\mid}}{CH}.\overset{\overset{OH}{\mid}}{CH}.CH_2OH$	L
Mo(VI)	$H_2(MO_4)_2L/H^2.(MO_4)^2.L$ 17.48, $H_3(MO_4)_2L/H_2(MO_4)_2L.H$ 4.07 [3.0]	
B(III)	$M(OH)_2(H_{-2}L)/M(OH)_3.OH.L$ 7.8 [0.1]; 8.19 [3.0] $M(H_{-2}L)_2/M(OH)_3.OH.L^2$ 9.9 [0.1]; 10.1 [3.0] $HM(OH)_2(H_{-2}L)/M(OH)_2(H_{-2}L).H$ 5.87 [3.0] $HM(OH)_2(H_{-2}L)_2/M(OH)_2(H_{-2}L)_2.H$ 2.70 [0.1] $M_2(OH)_4(H_{-4}L)/[M(OH)_3]^2.(OH)^2.L$ 14.0 [0.1]; 14.79 [3.0] $M_2(OH)_2(H_{-2}L)(H_{-4}L)/[M(OH_3]^2.(OH)^2.L^2$ 17.6 [3.0]	
Ce(IV)	$HMO_2(H_{-2}L)/M(OH)_4.OH.L$ 7.3 [0.5]; $HMO(H_{-2}L)_2/M(OH)_4.OH.L^2$ 9.7 [0.1]; 9.8 [0.5]; $H_2M_2O_3(H_{-2}L)(H_{-4}L)/[M(OH)_4]^2.(OH)^2.L^2$ 16.9 [0.5]	
Pb^{2+}	$MOH(H_{-2}L)/M(OH)_3.L$ 2.78 [1.0]	
As(III)	$M(OH)_2(H_{-2}L)/M(OH)_3.OH.L$ 5.7 [0.1]; $HM(OH)_2(H_{-2}L)/M(OH)_3.L$ 0.38 [0.1]	
Te(VI)	$H_3MO_4(H_{-2}L)/M(OH)_6.OH.L$ 9.6 [0.1]	

Stability Constants of Complexes with Organic Ligands

Table 7.4 (Continued)

Formula Metal ion	Stability constants (Symbol for the constant or scheme of the equilibrium, value of log K or log β, $[I]$)	Ref.
23. $C_7H_5NS_2$	2-mercaptobenzo-1,3-thiazole	HL
H^+	K_{H1} 6.93 $[0.1]^a$	
Zn^{2+}	β_1 3.25, β_2 5.74 $[0.1]^a$	
^{a} 20°C		
24. $C_7H_6O_2$	2-hydroxybenzaldehyde salicylaldehyde	HL
H^+	K_{H1} 8.37 $[0]$; 8.13 $[0.1]$; 8.07 $[0.5]$	
Ca^{2+}	β_1 1.1 $[0.1]$	
Y^{3+}	β_1 4.34 $[0.1]$	
La^{3+}	β_1 3.40 $[0.1]$	
Mn^{2+}	β_1 2.15, β_2 4.0 $[0.5]$	
Fe^{3+}	β_1 8.75, β_2 15.55 $[3.0]$	
Ni^{2+}	β_1 3.58, β_2 6.5 $[0.5]$	
Cu^{2+}	β_1 5.56 $[0.1]$; 5.36, β_2 10.11 $[0.5]$	
Zn^{2+}	β_1 2.87, β_2 5.00 $[0.5]$	
25. $C_7H_6O_2$	1-hydroxycyclohepta-3,5,7-trien-2-one tropolone	HL
H^+	K_{H1} 6.88 $[0]$; 6.70 $[0.1]$; 6.64 $[0.5]$; 6.75 $[1.0]^a$	
Be^{2+}	β_1 7.40 $[0.1]$	
M^{2+}(IIA)	β_1 (3.1) $[0.1]$	
Y^{3+}	β_1 7.18, β_2 13.26, β_3 18.07, β_4 21.49 $[0.1]$	
La^{3+}	β_1 6.19, β_2 11.12, β_3 15.31 $[0.1]$	
M^{3+}(4f)	β_1 (7.2), β_2 (13.2), β_3 (18.1), β_4 (Gd–Lu) (22.6) $[0.1]$	
Th^{4+}	β_1 9.61, β_2 18.24, β_3 25.89, β_4 32.56, β_5 34.85, β_6 36.72 $[0.1]$	
UO_2^{2+}	β_1 8.18, β_2 15.07 $[1.0]^a$	
NpO_2^+	β_1 5.45, β_2 9.81 $[1.0]^a$	
Pu^{3+}	β_1 7.20 $[1.0]^a$	
Mn^{2+}	β_1 4.60 $[0.1]$	
Fe^{3+}	β_1 10.50 $[2.0]$	
Co^{2+}	β_1 5.59 $[0.1]$	
Ni^{2+}	β_1 5.97 $[0.1]$	
Cu^{2+}	β_1 8.35 $[0.1]$	
Zn^{2+}	β_1 5.84 $[0.1]$	
Cd^{2+}	β_1 4.60 $[0.1]$	
Ge(IV)	$M(OH)_2L_2/M(OH)_4 \cdot (HL)^2$ 8.03, $ML_3/M(OH)_4 \cdot (HL)^3 \cdot H$ 13.3 $[0.5]$	

196   Stability Constants of Complexes with Organic Ligands   [Ch. 7

Table 7.4 (Continued)

Formula Metal ion	Stability constants (Symbol for the constant or scheme of the equilibrium, value of log K or log β, $[I]$)	Ref.
Pb^{2+}	tropolone (continued) β_1 6.64 [0.16][b]	

[a] 20°C; [b] 30°C

26. $C_7H_6O_5S$	5-sulphosalicylaldehyde	H_2L
H^+	K_{H1} 7.32 [0]; 6.93 [0.1]; 6.82 [0.5]; 6.89 [1.0] K_{H2} 0.24 [0.5]	
Be^{2+}	β_1 3.40, $M_2(OH)L/M_2(OH).L$ 7.98, $M_3(OH)_3L/M_3(OH)_3.L$ 3.26, $M_3(OH)_3L_2/M_3(OH)_3.L^2$ 6.56, $M_3(OH)_3L_3/M_3(OH)_3.L^3$ 8.15 [0.5]	
Cu^{2+}	β_1 5.75 [0]; 5.01 [0.1]; 4.76 [0.5]; 4.75 [1.0] β_2 9.61 [0]; 8.78 [0.1]; 8.46 [0.5]; 8.50 [1.0]	

27. $C_8H_5F_3O_2S$	1-(2-thienyl)-4,4,4-trifluorobutane-1,3-dione thenoyltrifluoroacetone	HL
H^+	K_{H1} 6.53 [1.0]	
Fe^{3+}	β_1 7.18 [1.0]	
Co^{2+}	β_1 3.68 [1.0]	
Ni^{2+}	β_1 4.45 [1.0]	
Cu^{2+}	β_1 5.68 [1.0]	
Zr^{4+}	β_1 10.98, β_2 21.88, β_3 32.24, β_4 42.17 [1.0]	
Hf^{4+}	β_1 10.60, β_2 21.44, β_3 31.50, β_4 41.52 [1.0]	

28. $C_{10}H_7NO_2$	2-nitroso-1-naphthol	HL
H^+	K_{H1} 7.47 [0]; 7.25 [0.1]; 7.21 [0.5]; 7.22 [1.0]	
Th^{4+}	β_1 8.30, β_2 15.54, β_3 23.04, β_4 29.26 [0.1]	
Zn^{2+}	β_1 3.91 [0.1]	
Cd^{2+}	β_1 3.33 [0.1]	

29. $C_{10}H_7NO_2$	1-nitroso-2-naphthol	HL
H^+	K_{H1} 7.87 [0]; 7.65 [0.1]; 7.59 [0.5]; 7.62 [1.0]	
Gd^{3+}	β_1 4.70 [0.1]	
Th^{4+}	β_1 8.50, β_2 16.13, β_3 24.02, β_4 30.28 [0.1]	
Zn^{2+}	β_1 4.63 [0.1]	

30. $C_{10}H_7NO_5S$	2-nitroso-1-naphthol-4-sulphonic acid	H_2L
H^+	K_{H1} 6.62 [0]; 6.20 [0.1]; K_{H2} 1.94 [0]; 2.63 [0.1]	
Y^{3+}	β_1 3.97 [0]; 2.87 [0.1]	
La^{3+}	β_1 4.56 [0]; 3.46 [0.1]; β_3 9.9 [0.1]	

Table 7.4 (Continued)

Formula Metal ion	Stability constants (Symbol for the constant or scheme of the equilibrium, value of log K or log β, $[I]$)	Ref.
	2-nitroso-1-naphthol-4-sulphonic acid (continued)	
$M^{3+}(4f)$	β_1 (4.4) [0]; (3.3) [0.1]	
Mn^{2+}	β_1 2.07 [0]	
Co^{2+}	β_1 4.34 [0.1]	
Ni^{2+}	β_1 6.28 [0]; 5.50 [0.1]; β_2 11.26 [0]; 10.42 [0.1]; β_3 14.55 [0]; 17.25 [0.1]	
Cu^{2+}	β_1 7.8, β_2 13.0 [0]	
Zn^{2+}	β_1 3.86 [0]; 3.06 [0.1]	
Cd^{2+}	β_1 3.12 [0]; 2.36 [0.1]	
Pb^{2+}	β_1 4.74 [0]; 3.84 [0.1]	
31. $C_{10}H_7NO_8S_2$	1-nitroso-2-naphthol-3,6-disulphonic acid nitroso-R acid	H_3L
H^+	K_{H1} 7.52 [0]; 6.88 [0.1]; 6.60 [0.5]; 6.55 [1.0]	
Y^{3+}	β_1 6.24 [0]; 4.48 [0.1]; 3.59 [0.5]; β_2 7.83 [0.1]; 6.83 [0.5]; β_3 11.29 [0.1]; 10.43 [0.5]	
La^{3+}	β_1 6.19 [0]; 4.37 [0.1]; 3.49 [0.5]; β_2 7.83 [0.1]; 6.67 [0.5]; β_3 11.24 [0.1]; 10.19 [0.5]	
$M^{3+}(4f)$	β_1 (4.8); β_2 (Eu) 7.68 [0.1]	
VO^{2+}	β_1 7.96 [0]; 6.71 [0.1]	
Mn^{2+}	β_1 3.73 [0]; 2.69 [0.1]	
Co^{2+}	β_1 5.40 [0.1]	
Ni^{2+}	β_1 8.3 [0]; 6.9, β_2 12.5, β_3 17.3 [0.1]	
Cu^{2+}	β_1 9.9 [0]; 7.7 [0.1]; β_2 15.6 [0]; 15.0 [0.1]	
Zn^{2+}	β_1 5.73 [0]; 4.46 [0.1]; β_2 7.63 [0]; 7.10 [0.1]	
Cd^{2+}	β_1 4.65 [0]; 3.42 [0.1]; β_2 6.59 [0]; 6.00 [0.1]	
Pb^{2+}	β_1 6.07 [0]; 4.64 [0.1]; β_2 8.34 [0]; 7.37 [0.1]	
32. $C_{10}H_8N_2O_2$	1,2-di(2'-furyl)ethane-1,2-dione dioxime 2,2'-furil dioxime	H_2L
H^+	K_{H1} 11.1 [0.1]; 11.4 [1.0]; K_{H2} 9.73 [0.1]; 9.72 [1.0]	[9]
Ni^{2+}	$MHL/M.HL$ 8.2, $M(HL)_2/M.(HL)^2$ 14.90 [0.1]	
Pd^{2+}	$M(HL)_2/M.(HL)^2$ (43.7), $MOH(HL)_2/M(HL)_2.OH$ 3.4 [1.0]	
33. $C_{10}H_8O_8S_2$	1,8-dihydroxynaphthalene-3,6-disulphonic acid chromotropic acid	H_4L
H^+	K_{H1} 15.6[a]; K_{H2} 5.35 [0.1]; 5.13 [0.5]; K_{H3} (0.73), K_{H4} (0.61) [0.1][a]	

Formula Metal ion	Stability constants (Symbol for the constant or scheme of the equilibrium, value of log K or log β, [I])	Ref.
	chromotropic acid (continued)	
Be^{2+}	β_1 16.34, β_2 28.19, MHL/M.HL 2.9 [0.1][a]	[9, 10]
Ti(IV)	$MOL_2/MO.L^2$ 40.5, $MOL_3/MO.L^3$ 56.4, $ML_3/MO.H^2.L^3$ 60.5, $MO(HL)_2/MOL_2.H^2$ 4.4 [0.1][a]	
VO^{2+}	β_1 17.17[b], MOHL/M.OH.L 25.0 [0.1]	[11]
Nb(V)	$ML_3/MO_2.H.(HL)^3$ 17.9 [0.1][a]; $MOL_2/MO_2.(HL)^2$ 11.3 [3.0][a]	
Mo(VI)	$MO_2L_2/MO_4.(H_2L)^2$ 4.1, $HMO_3L/H_2MO_4.HL$ 4.6, $MO_2L_2/H_2MO_4.(HL)^2$ 7.1 [0.1][a]	
UO_2^{2+}	β_1 16.6, β_2 28.1, MHL/M.HL 4.0, $MHL_2/ML.HL$ 1.5 [0.1][a]	
Cu^{2+}	β_1 13.5, β_2 23.2, MOHL/M.OH.L 19.0 [0.1]	
B(III)	$M(OH)_2L/M(OH)_3.HL$ 3.8, $ML_2/M(OH)_3.H_2L.HL$ 3.0 [0.1][a]	[12]
Al^{3+}	β_1 17.1, β_2 29.9 [0.1][a]	

[a] 20°C; [b] 30°C

34. $C_{10}H_{10}O_2$	1-phenylbutane-1,3-dione benzoylacetone	HL	
H^+	K_{H1} 8.89 [0.1]; 8.24 [1.0]		
Ce^{3+}	β_1 6.19, β_2 12.10, β_3 16.99 [0.1][a]		
Zr^{4+}	β_1 12.71, β_2 24.57, β_3 35.91, β_4 47.00 [1.0]		
Hf^{4+}	$ML_4/M(OH)_2.H^2.L^4$ 41.8 [1.0]		
VO^{2+}	β_1 10.49, β_2 20.54 [0.1]		
$VOOH^{2+}$	β_2 21.3 [0.1]		

[a] 20°C

35. $C_{12}H_{11}N_3O_4S$	2-mercapto-N-(2'-naphthyl)acetamide thionalide	HL	[13]
H^+	K_{H1} 10.20 [0.1; dioxane 75%]		
Mn^{2+}	β_1 4.4, β_2 8.8 [0.1; dioxane 75%]		
Fe^{2+}	β_1 7.0, β_2 13.7, β_3 19.2 [0.1; dioxane 75%]		
Co^{2+}	β_1 7.3, β_2 14.1, β_3 20.1 [0.1; dioxane 75%]		
Ni^{2+}	β_1 6.9, β_2 13.4 [0.1; dioxane 75%]		
Cu^{2+}	β_1 10.3, β_2 20.0, β_3 28.4 [0.1; dioxane 75%]		
Zn^{2+}	β_1 7.8, β_2 14.9 [0.1; dioxane 75%]		

36. $C_{12}H_{12}N_2OS_2$	5-(4'-dimethylaminobenzylidene)rhodanine	HL	[14]
H^+	K_{H1} 8.20 [0.1][a]		
Ag^+	β_1 9.15 [0.1][a]		

Formula Metal ion	Stability constants (Symbol for the constant or scheme of the equilibrium, value of log K or log β, $[I]$)	Ref.
	5-(4′-dimethylaminobenzylidene)rhodanine (continued)	
Cu^{2+}	β_1 6.08 $[0.1]^a$	

^{a} 20°C

37. $C_{14}H_{12}N_2O_2$	1,2-diphenylethane-1,2-dione dioxime α-benzil dioxime	H_2L
H^+	K_{H1} 11.2 $[0.1]$, 11.8 $[1.0]^a$; K_{H2} 10.3 $[0.1]$, 10.15 $[1.0]^a$	
Ni^{2+}	$M(HL)_2/M.(HL)^2$ 26.2 $[0.1]$	
Pd^{2+}	$M(HL)_2/M.(HL)^2$ 34.6, $MOH(HL)_2/M(HL)_2.OH$ 4.9 $[1.0]^a$	

^{a} 20°C

REFERENCES

[1] H. F. DeBrabander, L. C. VanPoucke and Z. Eeckhaut, *Inorg. Chim. Acta*, 1972, **6**, 459.
[2] C. V. Banks and S. Anderson, *Inorg. Chem.*, 1963, **2**, 112.
[3] R. A. Zingaro, *J. Am. Chem. Soc.*, 1956, **78**, 3568.
[4] J. Jansen, *Rec. Trav. Chim. Pays-Bas*, 1956, **75**, 1411.
[5] I. M. Bhatt and P. K. Soni, *Indian J. Chem.*, 1968, **6**, 115.
[6] W. Kemula, A. Hulanicki and W. Nawrot, *Rocz. Chem.*, 1962, **36**, 1717; 1964, **38**, 1065.
[7] A. Hulanicki, *Thesis, Warsaw*, 1967.
[8] J. Starý and J. Růžička, *Talanta*, 1968, **15**, 505.
[9] L. Sommer, *Collection Czech. Chem. Commun.*, 1963, **28**, 2393.
[10] M. Bartušek and L. Sommer, *J. Inorg. Nucl. Chem.*, 1965, **27**, 2397.
[11] K. Lal and R. P. Agarwal, *J. Less-Common Metals*, 1967, **12**, 269.
[12] M. Bartušek and L. Havelková, *Collection Czech. Chem. Commun.*, 1967, **32**, 3853.
[13] K. Burger, L. Korecz and A. Tóth, *Acta Chim. Acad. Sci. Hung.*, 1968, **55**, 1.
[14] O. Navrátil and J. Kotas, *Collection Czech. Chem. Commun.*, 1965, **30**, 2736.

Table 7.5
Stability of complexes with compounds useful as complexometric indicators*

Metal ion	Equilibrium constants (Symbol for the constant or scheme of the equilibrium, value of log K or log β, [I])	Ref.
1. CH_4N_4S	thiourea (see Table 7.4, part 1)	
2. $C_6H_6O_8S_2$	tiron (see Table 7.4, part 18)	
3. $C_7H_6O_3$	salicylic acid (see Table 7.2, part 15)	
4. $C_7H_6O_6S$	5-sulphosalicylic acid (see Table 7.2, part 19)	
5. $C_8H_5O_6N_5$	purpuric acid (murexide = ammonium salt) H_3L	
H^+	K_{H1} 10.9, K_{H2} 9.2, K_{H3} (0) [0.1]^a	
Mg^{2+}	$MH_2L/M \cdot H_2L$ 2.2 [0.1]^a	
Ca^{2+}	β_1 5.0, MHL/M·HL 3.6, $MH_2L/M \cdot H_2L$ 2.6 [0.1]^b; 2.68 [0.2]	
Sc^{3+}	$MH_2L/M \cdot H_2L$ 4.50 [0.1]^d	
La^{3+}	$MH_2L/M \cdot H_2L$ 3.43 [0.1]^d	
Ce^{3+}	$MH_2L/M \cdot H_2L$ 3.65 [0.1]^d	
$M^{3+}(4f)$	$MH_2L/M \cdot H_2L$ (4) [0.1]^d	
Co^{2+}	$MH_2L/M \cdot H_2L$ 2.46 [0.1]^d	
Ni^{2+}	$MH_2L/M \cdot H_2L$ 3.36^d, 4.6^a [0.1]	
Cu^{2+}	$MH_2L/M \cdot H_2L$ 3.5 [0]; 5 [0.1]^a	
Zn^{2+}	$MH_2L/M \cdot H_2L$ 3.1 [0.1]^a; $M(HL)_2/M \cdot (HL)^2$ 9.3 [^{e}]	
Cd^{2+}	$MH_2L/M \cdot H_2L$ 4.2 [0.1]^a	
In^{3+}	$MH_2L/M \cdot H_2L$ 4.6 [0.1]^d	[1]

^{a} 20°C; ^{b} about 20°C; ^{c} temperature not given; ^{d} 12°C; ^{e} I not given

6. $C_9H_6INO_4S$	ferron (see Table 7.1, part 14)	
7. C_9H_7NO	8-hydroxyquinoline (see Table 7.1, part 15)	
8. $C_9H_7NO_4S$	8-hydroxyquinoline-5-sulphonic acid (see Table 7.1, part 16)	
9. $C_{10}H_7NO_8S_2$	nitroso-R acid (see Table 7.4, part 31)	
10. $C_{10}H_8O_8S_2$	chromotropic acid (see Table 7.4, part 33)	
11. $C_{11}H_9N_3O_2$	1,3-dihydroxy-4-(2-pyridylazo)benzene H_2L PAR	
H^+	K_{H1} 12.31; 12.31^a [0.1]; 12.31 in 50% dioxane [0.005]; K_{H2} 5.50; 5.71^a [0.1]; 6.9 in 50% dioxane [0.005]; K_{H3} 2.69; 2.75^a [0.1]; 2.3 in 50% dioxane [0.005]	[2, 3]
Sc^{3+}	MHL/M·HL 6.25 [0.1]^a	
La^{3+}	MHL/M·HL 2.69 [0.1]^a	
$M^{3+}(4f)$	MHL/M·HL (3.5) [0.1]^a	
UO_2^{2+}	MHL/M·HL 12.5, $M(HL)_2/M \cdot (HL)^2$ 20.9 [0.1]	

* All the constants given are for 25°C, unless otherwise stated in notes given at the end of each part of the table

Stability Constants of Complexes with Organic Ligands 201

Table 7.5 (Continued)

Metal ion	Equilibrium constants (Symbol for the constant or scheme of the equilibrium, value of log K or log β, $[I]$)	Ref.
	PAR (continued)	
Co^{2+}	MHL/M.HL 10.0 [0.1]; (12) in 50% dioxane; $M(HL)_2/M.(HL)^2$ 17.1 [0.1]	
Ni^{2+}	MHL/M.HL 13.2, $M(HL)_2/M.(HL)^2$ 26.0 in 50% dioxane [0.1]	
Cu^{2+}	MHL/M.HL 14.8, $M(HL)_2/M.(HL)^2$ 23.9 [0.1]	
Zn^{2+}	MHL/M.HL 10.5, $M(HL)_2/M.(HL)^2$ 17.1 [0.1]	
Pb^{2+}	MHL/M.HL 8.6, $M(HL)_2/M.(HL)^2$ 15.7 [0.1]	

[a] 20°C

12. $C_{12}H_8N_2$	1,10-phenanthroline (see Table 7.1, part 21)	

13. $C_{13}H_9N_3OS$	1-(1',3'-thiazolyl-2'-azo)-2-naphthol TAN	HL
H^+	K_{H1} 9.10; K_{H2} 0.88 $[0.05]^a$	
Co^{2+}	β_1 9.50, β_2 19.00 $[0.05]^b$	
Cu^{2+}	β_1 10.92, β_2 22.52 $[0.05]^a$	
Ag^+	β_1 8.67 $[0.05]^b$	
Zn^{2+}	β_1 9.87, β_2 19.74 $[0.05]^a$	
Cd^{2+}	β_1 9.18, β_2 17.88 $[0.05]^b$	[4–6]

[a] 20°C; [b] about 20°C

14. $C_{13}H_{12}N_4O$	1,5-diphenylcarbazone	L
H^+	K_{H1} 9.26 in 50% dioxane [0.1]	
Ni^{2+}	β_1 6.02 in 50% dioxane [0.1]	
Cu^{2+}	β_1 9.8, β_2 19.5, β_3 29 $[0]^a$	
Zn^{2+}	β_1 5.76 in 50% dioxane [0.1]	[7, 8]

[a] 28°C

15. $C_{13}H_{12}N_4S$	1,5-diphenylthiocarbazone dithizone	HL
H^+	K_{H1} 4.45; 5.80 in 50% dioxane [0.1]	
Ni^{2+}	β_1 5.83 [0.1]	
Cu^{2+}	β_2 22.3 [0.1]	
Zn^{2+}	β_1 7.75 [1.0]; 6.18 in 50% dioxane [0.1]; β_2 15.05 [1.0]	
Pb^{2+}	β_2 15.2 [0.1]	[8]

Metal ion	Equilibrium constants (Symbol for the constant or scheme of the equilibrium, value of log K or log β, $[I]$)	Ref.
16. $C_{14}H_8O_7S$	1,2-dihydroxyanthraquinone-3-sulphonic acid Alizarin Red S	H_3L
H^+	K_{H1} 11.36 $[0]^a$; 10.9 $[0.1]^a$; 10.85 $[0.5]$; K_{H2} 6.17 $[0]^a$; 5.77 $[0.1]^a$; 5.49 $[0.5]$; K_{H3} 0.97 $[0]^a$	
Be^{2+}	β_1 10.96 $[0.1]^a$	
Zr^{4+}	$M(OH)_2L/M.(OH)^2.L$ 49.0 $[1.6]$	[9]
Mo(VI)	$MO_2L_2/MO_4.H_2L$ 9.2 $[0.1]$	[10]
V(IV)	$MO_2L/MO_3.H_2L$ 8.4 $[^b]$	[11]
V(V)	$MO_2L/MO_3.H_2L$ 8.5 $[0.1]$	[12]
B(III)	$M(OH)_2HL.H/M(OH)_3.H_3L$ -3.4 $[0.1]^a$	

^{a} 20°C; ^{b} I not given

Metal ion	Equilibrium constants	Ref.
17. $C_{15}H_{11}N_3O$	1-(2'-pyridylazo)-2-naphthol PAN	HL
H^+	K_{H1} 11.2^a; 12.2 in 20% dioxane $[0.1]$; 12.3^b in 50% dioxane K_{H2} 2.9^a; 2.3 in 20% dioxane $[0.1]$; 2^b in 50% dioxane	[13, 14]
$M^{3+}(4f)$	β_1 (12.5); β_2 (24); β_3 (34.5); β_4 (44) (Eu, Ho) $[0.05]^d$	[5, 6]
Mn^{2+}	β_1 8.5; β_2 16.4 in 50% dioxane; 15.3^a $[0.1]$	
Ni^{2+}	β_1 12.7; β_2 25.3 in 50% dioxane $[^b]$	
Co^{2+}	β_1 (12^b) in 50% dioxane; 12.15, β_2 24.16 $[0.05]$	
Cu^{2+}	β_1 16^b in 20% dioxane; 12.6, MOHL/ML.OH 6.9 $[0.1]^a$	
Zn^{2+}	β_1 12.72^c $[0.05]$; 11.2^b in 50% dioxane; β_2 24.54^c $[0.05]$; 21.7^b in 50% dioxane; 21.8 $[0.1]^a$	[13, 14]

^{a} Temperature about 30°C; ^{b} I not given; ^{c} 20°C; ^{d} temperature about 20°C

Metal ion	Equilibrium constants	Ref.
18. $C_{15}H_{11}N_3O$	1-(2'-pyridylazo)-4-naphthol p-PAN	HL
H^+	K_{H1} 9.1 $[0.1]^a$; 10.74^a in 50% dioxane; K_{H2} 3.0 $[0.1]^a$; 2.54^a in 50% dioxane	
Ni^{2+}	β_2 23 in 50% dioxane $[0.1]$	
Cu^{2+}	β_2 20 in 50% dioxane $[0.1]$	
Zn^{2+}	β_2 19 in 50% dioxane $[0.1]$	[15]

^{a} Temperature about 33°C

Metal ion	Equilibrium constants	Ref.
19. $C_{16}H_{13}AsN_2O_{11}S_2$	2-arsonophenylazochromotropic acid Arsenazo I, Neothorin	H_4L

Stability Constants of Complexes with Organic Ligands

Table 7.5 (Continued)

Metal ion	Equilibrium constants (Symbol for the constant or scheme of the equilibrium, value of $\log K$ or $\log \beta$, $[I]$)	Ref.
	Arsenazo I (continued)	
H^+	K_{H2} 10.07, K_{H3} 7.61, K_{H4} 2.95 [0.1], K_{H5} 2.16, K_{H6} 0.07 [a]	
Mg^{2+}	MHL/M.HL 5.57 [0.1]	
Ca^{2+}	MHL/M.HL 5.51 [0.1]	
Sr^{2+}	MHL/M.HL 4.40 [0.1]	
Ba^{2+}	MHL/M.HL 4.19 [0.1]	
La^{3+}	$MH_2L/M.L.H^2$ 28.8 [a]^b	
Pu^{4+}	$MH_3L/M.H_3L$ 7.7, $MOH(H_4L)/MOH.H_4L$ 6.6 [a]^b	
Th^{4+}	$MH_4L/M.H_4L$ 6.8 [a]^b	[16, 17]

^{a} I not given; ^{b} 20°C

20. $C_{17}H_{14}N_2O_5S$ 3′-methyl-6′-hydroxybenzeneazo-2-naphthol-4-sulphonic acid
Calmagite H_3L

H^+	K_{H1} 12.4, K_{H2} 8.1 [0.1]	
Mg^{2+}	β_1 8.1 [0.1]	
Ca^{2+}	β_1 6.1 [0.1]	[18]

21. $C_{19}H_{14}O_7S$ 3,3′,4′-trihydroxyfuchsone-2″-sulphonic acid
Pyrocatechol Violet H_4L

H^+	K_{H1} 12.8, K_{H2} 9.76, K_{H3} 7.80, K_{H4} 0.8 [0.1]	
Th^{4+}	$MH_2L/M.H_2L$ 9.83, $MH_3L_2/MH_2L.HL$ 15.38, $MH_4L_3/MH_3L_2.HL$ 13.62, $M_2HL/M^2.HL$ 20.04 [0.1]	
Cr^{3+}	$MH_2L/M.H_2L$ 8.70, $MH_3L_2/MH_2L.HL$ 11.15, $MH_4L_3/MH_3L_2.HL$ 8.96 [0.1]	
UO_2^{2+}	$MH_2L/M.H_2L$ 7.05, $M(H_2L)_2/M.(H_2L)^2$ 12.65, $M_2HL/M^2.HL$ 14.7 [0.1]	
Ga^{3+}	β_1 22.2, β_{21} 26.8 [~0.1]	[19]
Bi^{3+}	β_1 27.1, β_{21} 32.3 [~0.1]	[20, 21]

22. $C_{20}H_{13}O_7N_3S$ 3-hydroxy-4-(1′-hydroxy-2′-naphthylazo)-8-nitro-naphthalene-1-sulphonic acid
Eriochrome Black T H_3L

H^+	K_{H1} 11.95 [0]; 11.39 [0.1]; K_{H2} (5.81) [0]; 6.80 [0.1]	
Mg^{2+}	β_1 7.0 [0.08]^a	
Ca^{2+}	β_1 5.4 [0.02]^a	[22]
Ba^{2+}	β_1 3.0 [0.1]	
Mn^{2+}	β_1 9.6, β_2 17.6 [0.1]	[23]
Co^{2+}	β_1 20.0 [0.3]^a	
Cu^{2+}	β_1 21.38 [0.3]^a	

Table 7.5 (Continued)

Metal ion	Equilibrium constants (Symbol for the constant or scheme of the equilibrium, value of log K or log β, $[I]$)	Ref.
	Eriochrome Black T (continued)	
Zn^{2+}	β_1 12.9 [0.1]; 12.31 [0.3][a]; β_2 20.0 [0.1]	[24]
Cd^{2+}	β_1 12.74 [0.3][a]	
Pb^{2+}	β_1 13.19 [0.3][a]	

[a] 20°C

23. $C_{20}H_{14}N_2O_5S$	3-hydroxy-4-(2'-hydroxy-1'-naphthylazo)naphthalene-1-sulphonic acid Calcon, Eriochrome Black R	H_3L
H^+	K_{H1} 13.80 [0]; 13.5 [0.1]; K_{H2} 7.31 [0]; 7.36 [0.1]	
Mg^{2+}	β_1 7.6 [0.1]	
Ca^{2+}	β_1 5.6 [0.1]	
Zn^{2+}	β_1 12.5 [0.1]; $Zn(NH_3)L/Zn \cdot NH_3 \cdot L$ 16.4 [[a]]	[25]

[a] Temperature and I not given

24. $C_{20}H_{16}N_4O_6S$	2-carboxy-2'-hydroxy-5'-sulphoformazylbenzene Zincon	H_4L
H^+	K_{H1} 14, K_{H2} 8.3, K_{H3} 4.5 [~0.1]	
Zn^{2+}	$ML \cdot H/M \cdot HL$ 0.1, $Zn(OH)L/Zn \cdot OH \cdot L$ 13 [~0.1]	[26]

25. $C_{22}H_{16}N_4O_{14}S_2$	2,7-bis(2'-sulphophenylazo)chromotropic acid Sulphonazo III	H_6L
H^+	K_{H1} 14.5, K_{H2} 11.7, K_{H3} 2.9, K_{H4} 2.3, K_{H5} 1.9, K_{H6} 0.9, K_{H7} 0.3, K_{H8} -2.0 [0.2][a]	
Ba^{2+}	$MH_2L/M \cdot H_2L$ 25.9 [0.2][a]	[27]

[a] Temperature about 20°C

26. $C_{22}H_{16}Cl_2N_4O_{14}P_2S_2$	2,7-bis(4'-chloro-2'-phosphonophenylazo)chromotropic acid Chlorophosphonazo III	H_8L
H^+	K_{H1} 14.6, K_{H2} 11.1, K_{H3} 9.4, K_{H4} 7.0, K_{H5} 4.2, K_{H6} 1.5, K_{H7} 0.6, K_{H8} 0.3, K_{H9} -0.5, K_{H10} -2.1 [0.2]	
Mg^{2+}	$MH_4L/M \cdot H_4L$ 5.3 [0.2]	
Ca^{2+}	$M(H_4L)_2/M \cdot (H_4L)^2$ 9.8 [0.2]	
Sr^{2+}	$M(H_4L)_2/M \cdot (H_4L)^2$ 11.4 [0.2]	
Ba^{2+}	$M(H_4L)_2/M \cdot (H_4L)^2$ 12.3 [0.2]	
UO_2^{2+}	$M(H_6L)_2/M \cdot H^{12} \cdot L^2$ 47.7 [0.2]	[28]

Table 7.5 (Continued)

Metal ion	Equilibrium constants (Symbol for the constant or scheme of the equilibrium, value of log K or log β, $[I]$)	Ref.

27. $C_{22}H_{18}As_2N_4O_{14}S_2$ 2,7-bis(2'-arsonophenylazo)chromotropic acid H_8L
    Arsenazo III

H^+	K_{H1} 12.33, K_{H2} 7.48, K_{H3} 5.35, K_{H4} 2.41, K_{H5} 2.41 $[0.2]^a$	
La^{3+}	β_{22} 81.2, $M_2H_4L/M^2 \cdot H_4L$ 14.9, $M_2(H_4L)_2/M^2(H_4L)^2$ 28.4 $[0.2]^a$	
$M^{3+}(4f)$	β_{22} (82) (Dy, Gd, Sm, Yb) $[0.2]^a$	
Zr^{4+}	$M_2(H_9L)_2/M^2 \cdot H^{10}(H_4L)^2$ 32.0 $[0.2]^a$	[29, 30]

^{a} Temperature about 20°C

28. $C_{23}H_{16}Cl_2O_9S$ 2″,6″-dichloro-4′-hydroxy-3,3′-dimethyl-3″-sulpho-
    fuchsone-5,5′-dicarboxylic acid H_4L
    Chrome Azurol S

H^+	K_{H1} 11.80^a, 11.75, K_{H2} 4.80^a, 4.88, K_{H3} 2.25^a, 2.25 $[0.1]$
Be^{2+}	$MHL/M \cdot HL$ 4.66, β_{21} 15.8, β_{22} 26.8^a $[0.1]$
UO_2^{2+}	$MHL/M \cdot HL$ 5.35, β_{21} 18.3 $[0.1]^a$
Cu^{2+}	$MHL/M \cdot HL$ 4.02, β_{21} 13.7 $[0.1]$
Fe^{3+}	β_1 15.6, β_{21} 20.2, β_{22} 36.02 $[0.1]^a$

^{a} 20°C

29. $C_{23}H_{18}O_9S$ 4′-hydroxy-3,3′-dimethyl-2″-sulphofuchsone-5,5′-dicarboxylic
    acid H_4L
    Eriochrome Cyanine R

H^+	K_{H1} 11.85, K_{H2} 5.47, K_{H3} 2.30, K_{H4} −4.9 $[0.1]^a$
Be^{2+}	$MHL/M \cdot HL$ 5.49, β_{22} 28.3 $[0.1]^a$
Fe^{3+}	β_1 17.9, β_{21} 22.5, β_{22} 37.9 $[0.1]^a$

^{a} Temperature about 20°C

30. $C_{30}H_{18}N_6O_{21}S_6$ cyclo-tris-7-(1-azo-8-hydroxynaphthalene-3,6-disulphonic
    acid H_9L
    Calcichrome

H^+	K_{H1} 11.5, K_{H2} 7.10 $[0.2]^a$	
Ca^{2+}	$MH_2L/M \cdot H_2L$ 7.85 $[0.2]^a$	[31]

^{a} Temperature about 20°C

Table 7.5 (Continued)

Metal ion	Equilibrium constants (Symbol for the constant or scheme of the equilibrium, value of log K or log β, $[I]$)	Ref.
31. $C_{31}H_{32}N_2O_{13}S$	5,5'-bis[N,N-bis(carboxymethyl)amino]methyl-4'-hydroxy-3,3'-dimethylfuchsone-2''-sulphonic acid H_6L Xylenol Orange	
H^+	K_{H1} 12.28, K_{H2} 10.46, K_{H3} 6.4, K_{H4} 3.23, K_{H5} 2.58, K_{H6} −1.09, K_{H7} −1.74 $[0.2]^a$	
Sc^{3+}	MHL/M.HL 18.82, β_{22} 61.2 $[0.2]^a$	
Y^{3+}	MHL/M.HL 12.81 $[0.2]^b$	
La^{3+}	MHL/M.HL 11.67 $[0.2]^b$	
$M^{3+}(4f)$	β_{22} (45) $[0.2]^a$	
Ti(IV)	β_{22} 57.8, $MO(H_5L).H/MO.H_6L$ 3.46 $[0.5]$	
Th^{4+}	β_{22} 52.5 $[0.2]^a$, $MH_2L/M.H_2L$ 12.35, $M(H_2L)_2/M.(H_2L)^2$ 19.31, $M_2H_2L/MH_2L.M$ 9.7 $[0.1]$	
Zr^{4+}	β_{22} 31.0 $[0.2]^a$	
VO_2^+	β_{22} 63.1 $[0.2]^a$	
UO_2^{2+}	β_{22} 38.57 $[0.2]^a$	
Fe^{3+}	β_{21} 39.80 $[0.2]^a$	
Cd^{2+}	β_1 16.06 $[0.3]$	
Ga^{3+}	$MH_2L/M.H_2L$ 13.36 $[0.2]^b$, $MH_5L/M.H_5L$ 7.0, $M(H_5L)_2/M.(H_5L)^2$ 11.5 $[0.01]^c$	
Bi^{3+}	β_{22} 75.6 $[0.2]^a$	[32]

[a] Temperature about 20°C; [b] 20°C; [c] temperature not stated

32. $C_{32}H_{32}N_2O_{12}$	3',3''-bis{[bis(carboxymethyl)amino]-methyl}-5',5''-dimethylphenolphthalein H_6L Phthaleincomplexone, Metalphthalein	
H^+	K_{H1} 12.0, K_{H2} 11.4, K_{H3} 7.8, K_{H4} 7.0, K_{H5} 2.9, K_{H6} 2.2 $[0.1]^a$	[33]
Mg^{2+}	β_1 8.9, MHL/ML.H 10.6, $MH_2L/MHL.H$ 7.4, $MH_3L/MH_2L.H$ 6, β_{22} 14.1, $M_2HL/M_2L.H$ 6, $[0.1]^a$	
Ca^{2+}	β_1 7.8, MHL/ML.H 11.1, $MH_2L/MHL.H$ 7.6, $MH_3L/MH_2L.H$ 8.9, β_{21} 12.8, $M_2HL/M_2L.H$ 7 $[0.1]^a$	
Zn^{2+}	β_1 15.1, MHL/ML.H 10.7, $MH_2L/MHL.H$ 7.7, $MH_3L/MH_2L.H$ 3.2, β_{21} 24.9, $M_2HL/M_2L.H$ 5.9 $[0.1]^a$	
Ba^{2+}	β_1 6.2, MHL/ML.H 10.6, $MH_2L/MHL.H$ 6.7, $MH_3L/MH_2L.H$ 6.8, β_{21} 9.2, $M_2HL/M_2L.H$ 8.6 $[0.1]^a$	

[a] 20°C

Stability Constants of Complexes with Organic Ligands

Table 7.5 (Continued)

Metal ion	Equilibrium constants (Symbol for the constant or scheme of the equilibrium, value of log K or log β, [I])	Ref.
33. $C_{37}H_{44}N_2O_{13}S$	3′,3″-bis{[bis(carboxymethyl)amino]methyl}-5′,5″-diisopropyl-2′,2″-dimethylphenolsulphophthalein H_6L Methylthymol Blue	
H^+	K_{H1} 13.4, K_{H2} 11.15, K_{H3} 7.4, K_{H4} 3.8, K_{H5} 3.3, K_{H6} 3.0 [0.2]^a	[34]
Y^{3+}	β_{22} 32.8, $M_2(HL)_2/M^2.(HL)^2$ 23.6, $M_2HL_2/M^2.H.L^2$ 42.4 [0.2]^a	
La^{3+}	β_{22} 35.8, $M_2(OH)_2L_2/M^2.(OH)^2.L^2$ 23.2 [0.2]^a; $MH_2L/M.H_2L$ 5.35, $M_2H_2L/M^2.H_2L$ 5.0 [0.4]^a	
Mn^{2+}	$MHL/M.HL$ 9.2 [0.1]	
Fe^{3+}	MH_2L/M H_2L 20.56, $MH_6L_2/M.(H_3L)^2$ 21.7 [0.1]^a	[34, 35]

^{a} Temperature not stated

34. $C_{38}H_{44}N_2O_{12}$	3,3′-bis[N,N-bis(carboxymethyl)aminomethyl]thymolphthalein H_6L Thymolphthalexone, Thymolphthaleinkomplexon	
H^+	K_{H1} 12.25, K_{H2} 7.35 [0.2]^a	
Ca^{2+}	β_{22} 42.74 [0.2]^a	[36]

^{a} Temperature about 20°C

REFERENCES

[1] G. Geier, *Z. Elektrochem.*, 1965, **69**, 617.
[2] A. Corsini, I. M. L. Yih, Q. Fernando and H. Freiser, *Anal. Chem.*, 1962, **34**, 1090.
[3] A. Corsini, Q. Fernando and H. Freiser, *Inorg. Chem.*, 1963, **2**, 224.
[4] O. Navrátil, *Collection Czech. Chem. Commun.*, 1964, **29**, 2490.
[5] O. Navrátil, *Collection Czech. Chem. Commun.*, 1966, **31**, 2492.
[6] O. Navrátil, *Collection Czech. Chem. Commun.*, 1967, **32**, 2004.
[7] H. R. Geering and J. F. Hodgson, *Anal. Chim. Acta*, 1966, **36**, 537.
[8] K. S. Math, Q. Fernando and H. Freiser, *Anal. Chem.*, 1964, **36**, 1762.
[9] H. E. Zittel and T. M. Florence, *Anal. Chem.*, 1967, **39**, 320.
[10] R. L. Seth and A. K. Dey, *Z. Anorg. Chem.*, 1963, **321**, 278.
[11] P. Sanyal and S. P. Mushran, *Anal. Chim. Acta*, 1966, **35**, 400.
[12] S. K. Banerji and A. K. Dey, *Z. Anorg. Chem.*, 1961, **309**, 226.
[13] L. G. Sillén and A. E. Martell, *Stability Constants of Metal–Ion Complexes, Supplement No. 1*, p. 739. The Chemical Society, London, 1971.
[14] D. Betteridge, Q. Fernando and H. Freiser, *Anal. Chem.*, 1963, **35**, 294.

[15] D. Betteridge, P. K. Todd, Q. Fernando and H. Freiser, *Anal. Chem.*, 1963, **35**, 729.
[16] A. E. Klygin and V. K. Pavlova, *Zh. Neorgan. Khim.*, 1960, **5**, 1516.
[17] A. E. Klygin and V. K. Pavlova, *Zh. Neorgan. Khim.*, 1961, **6**, 1050.
[18] F. Lindstrom and H. Diehl, *Anal. Chem.*, 1960, **32**, 1123.
[19] O. Ryba, J. Cífka, D. Ježková, M. Malát and V. Suk, *Collection Czech. Chem. Commun.*, 1958, **23**, 71.
[20] O. Ryba, J. Cífka, V. Suk and M. Malát, *Collection Czech. Chem. Commun.*, 1956, **21**, 349.
[21] P. Waldvogel, *Thesis*, ETH Zürich, 1959 (in G. Schwarzenbach and H. Flaschka, *Complexometric Titrations*, 2nd Ed., p. 82, Methuen, London, 1969.
[22] G. Schwarzenbach and W. Biedermann, *Helv. Chim. Acta*, 1948, **31**, 678.
[23] A. Ringbom, *Complexation in Analytical Chemistry*, p. 362, Interscience, New York, 1963.
[24] M. Kodama and H. Ebine, *Bull. Chem. Soc. Japan*, 1967, **40**, 1857.
[25] L. G. Sillén and A. E. Martell, *Stability Constants of Metal–Ion Complexes*, p. 724, The Chemical Society, London, 1964.
[26] A. Ringbom, G. Pensar and E. Wänninen, *Anal. Chim. Acta*, 1958, **19**, 525.
[27] B. Buděšínský and D. Vrzalová, *Z. Anal. Chem.*, 1965, **210**, 161.
[28] B. Buděšínský, K. Maas and A. Bezděková, *Collection Czech. Chem. Commun.*, 1967, **32**, 1528.
[29] B. Buděšínský, *Collection Czech. Chem. Commun.*, 1963, **28**, 2902.
[30] B. Buděšínský, *Z. Anal. Chem.*, 1964, **202**, 96, 206, 401.
[31] A. Bezděková and B. Buděšínský, *Collection Czech. Chem. Commun.*, 1965, **30**, 811.
[32] L. G. Sillén and A. E. Martell, see ref. 13, p. 788.
[33] G. Anderegg, H. Flaschka, R. Sallmann and G. Schwarzenbach, *Helv. Chim. Acta*, 1954, **37**, 113.
[34] L. G. Sillén and A. E. Martell, see ref. 13, p. 791.
[35] J. Körbl and B. Kakáč, *Collection Czech. Chem. Commun.*, 1958, **23**, 889.
[36] A. Bezděková and B. Buděšínský, *Collection Czech. Chem. Commun.*, 1965, **30**, 818.

CHAPTER 8

Solubility products of sparingly soluble salts

The heterogeneous equilibrium of a saturated solution in presence of a sparingly soluble metal salt (e.g. of formula $M_m A_n$) is characterized by the solubility product, which is defined either as a thermodynamic constant K_s^0 in terms of activities of the ions [cf. expression in Eq. (2.77), p. 40] or as a concentration solubility product K_s [Eq. (2.70)] written in terms of ion concentrations; the constant K_s holds for a certain ionic strength of the solution.

In various brief tables in the literature the thermodynamic values of solubility products are commonly listed; only in more comprehensive tabulations are the relevant ionic strength and temperature stated (e.g. ref. [1, 2]). Thus in a practical case of calculation of solution composition the effect of ionic strength should be taken into consideration; the necessary correction is introduced by estimating the activity coefficients from one of the Debye–Hückel equations, e.g. the Davies expression in Eq. (2.19), p. 25. However, the solubility of a metal salt is affected to a far greater extent by various side-reactions, some of which may change the solubility by several orders of magnitude. First, an increase in solubility can be caused by protonation of the anion A, as expressed by the protonation coefficient $\alpha_{A(H)}$, or by hydrolysis of the metal ion M, as expressed by $\alpha_{M(OH)}$. The solubility may also be influenced by formation of complexes (of the type $M_x A_y$) with the precipitate ions present in the mother liquor. The presence of an interfering metal ion can result in formation of complexes (NA_q etc.) with the anion of the precipitate, again affecting the solubility. In addition, the effect of complex-forming components of buffers, masking agents, and even ions of an electrolyte added to adjust the ionic strength of the solution should be taken into account, as should the effect of ion-pair formation (MA). All these effects can be involved in calculation of the side-reaction coefficients α_M and α_A (see Chapters 10 and 11). The solubility is also significantly influenced by the modification, purity and particle size of the precipitate, but these effects cannot be assessed and evaluated by a simple calculation (for further information see ref. [3]); they should be taken into consideration in the experimental estimation of the constant. A published value of the solubility product should thus refer to the equilibrium of a well-defined solid with its saturated solution: this means a certain pure crystalline modification in a defined form of dispersion. Solubility calculations are then valid for a precipitate of identical properties.

The methods for determination of solubility products need not be reviewed here. The basic information is available in textbooks on physical chemistry or electrochemistry (e.g. [4]). If the solid phase is separated before the measurement, it is necessary to take care that no particles of the precipitate, for example even in the form of a colloidal dispersion, are carried over into the solution being separated. Any reliable method applicable for the concentration range of the ions to be determined can then be used for determination of the solubility. If it is preferred to make the measurement in the presence of the solid phase, the method employed should not effect the equilibrium; otherwise a correction for the effect must be included in the calculation. Conductometry and potentiometry are the major methods which do not disturb the equilibrium of a saturated solution. A potentiometric measurement is obviously more selective but sometimes not adequately sensitive.

The table of solubility products given in this chapter presents the thermodynamic constants pK_s^0 (for ionic strength approaching zero) or the concentration constants pK_s (for the ionic strength given in square brackets); all data are for a temperature of 25°C unless stated otherwise.

The solubility products are arranged in alphabetical order of the chemical symbols (or formulae) of the cations, with salts of inorganic anions (listed in sequence of subgroups of the periodic system) given before compounds with organic precipitants (arranged in order of increasing complexity of empirical formula).

REFERENCES

[1] L. G. Sillén and A. E. Martell, *Stability Constants of Metal-Ion Complexes*, The Chemical Society, London, 1964.

[2] A. Ringbom, *Complexation in Analytical Chemistry*, p. 339, Interscience, New York, 1963.

[3] H. A. Laitinen, *Chemical Analysis*, 1st Ed., p. 117, McGraw-Hill, New York, 1960.

[4] G. Kortüm, *Treatise on Electrochemistry*, 2nd Ed., pp. 264, 314, Elsevier, Amsterdam, 1965.

Table 8.1
Solubility products of sparingly soluble compounds*

Metal ion	Formula of the compound†, value of pK_s, [*I*]	Ref.§
Ag^+	Ag_2CO_3 11.19 [a]	[3]
	AgCN 15.66 [0]; 15.4 [1.0]	
	AgCNO 6.64 [0][b]	
	AgCNS 11.97 [0]; 11.80 [0.1]	
	$Ag_4Fe(CN)_6$ 44.07 [0]	
	AgN_3 8.56 [0]; 8.80 [4.0]	
	$AgNO_2$ 4.13 [0]; 3.57 [1.0][c]; 3.36 [2.0][c]	
	$Ag_2N_2O_2$ ($2Ag^+ + N_2O_2^{2-}$) 18.89 [0]	
	Ag_3PO_4 17.55 [0]	
	Ag_3AsO_4 22.0 [a]	[4]
	$\frac{1}{2}Ag_2O$ ($Ag^+ + OH^-$) 7.71 [0]; 7.18 [1.0]; 7.42 [3.0]	
	Ag_2S 50.1 [0]; 49.7 [1.0][c]	
	Ag_2SO_3 13.82 [0]	
	Ag_2SO_4 4.83 [0]	
	Ag_2Se 63.7 [0]; 53.8 [1.0]	
	Ag_2SeO_3 15.55 [0]	
	Ag_2SeO_4 8.91 [0]	
	$Ag_2H_2TeO_4$ ($2Ag^+ + H_2TeO_4^{2-}$) 2.43 [0]	
	Ag_2HVO_4 ($2Ag^+ + HVO_4^{2-}$) 13.7 [a]	[5]
	AgH_2VO_4 ($Ag^+ + H_2VO_4^-$) 6.3 [a]	[5]
	Ag_2CrO_4 11.92 [0]	
	Ag_2MoO_4 11.55 [0]	
	Ag_2WO_4 11.26 [0]	
	AgCl 9.74 [0]; 9.62 [0.5]; 9.74 [1.0]; 10.05 [3.0]; 10.40 [4.0][c]	
	AgBr 12.30 [0]; 12.10 [0.1]; 11.92 [1.0]	
	$AgBrO_3$ 4.26 [0]	
	AgI 16.08 [0]; 16.35 [4.0]	
	$AgIO_3$ 7.51 [0]; 7.08 [1.0]	
	$Ag_2C_2O_4$ 11.0 [0]	[6]

[a] Temperature and ionic strength not stated; [b] 19°C; [c] 20°C

* Values for 25°C are given unless otherwise stated in footnotes.
† Products of the solubility equilibrium are given in parentheses after the formula of the compound, if there may be doubt of their composition.
§ If no reference is given, the values are taken from Smith and Martell [1] or (for values printed in italics) are calculated from data given by Baes and Mesmer [2].

Table 8.1 (Continued)

Metal ion	Formula of the compound, value of pK_s, $[I]$	Ref.
Al^{3+}	$AlPO_4$ 18.24 [a]	
	$Al(OH)_3$ (amorphous) 32.34 [0]; 31.6 [0.1]	[7]
	$Al(OH)_3$ (hydrargillite) 36.30 [0]	[7]
	$\alpha\text{-}Al(OH)_3$ (gibbsite) 33.5 [0]; 31.79 [1.0]	
	$\gamma\text{-}AlO(OH)$ (boehmite) 34.02 [0]	[7]
As(III)	As_4O_6 (cubic) ($\frac{1}{4}As_4O_{6(s)} \to As(OH)_{3(l)}$) 0.68 [0] – [1.0]	
Au^{3+}	$Au(OH)_3$ *47.5* [0]; *46.9* [1.0]	
Ba^{2+}	$BaCO_3$ 8.3 [0]	
	$Ba_3(PO_4)_2$ 29.3 [0]; $BaHPO_4$ 7.40 [0]^c	
	$Ba(OH)_2 \cdot 8 H_2O$ 3.6 [0]	
	$BaSO_4$ 9.96 [0]	
	BaS_2O_3 4.79 [0]	
	$BaSeO_3$ 6.57 [0]	
	$BaSeO_4$ 7.46 [0]	
	$BaCrO_4$ 9.67 [0]; 8.96 [0.1]; 8.39 [1.0]	
	BaF_2 5.76 [0]	
	$Ba(BrO_3)_2 \cdot H_2O$ 5.11 [0.5]	
	$Ba(IO_3)_2$ 7.60 [0]; 7.76 [0.5]; 7.60 [1.0]; 7.43 [2.0]	
	BaC_2O_4 6.0 [0.1]	[8]
	$Ba(C_9H_6NO_2)_2$ (oxinate) 7.7 [0.1]	[6]
Be^{2+}	$Be(OH)_2$ (amorphous) 21.0 [0]; $\alpha\text{-}Be(OH)_2$ 21.31 [0]; 20.6 [1.0]; $\beta\text{-}Be(OH)_2$ 21.7 [0]	
Bi^{3+}	$BiONO_3$ ($Bi^{3+} + 2 OH^- + NO_3^-$) 31.20 [0]	
	$BiPO_4$ 22.9 [a]	[9]
	$\frac{1}{2}\alpha\text{-}Bi_2O_3$ ($Bi^{3+} + 3 OH^-$) *38.53* [0]; *36.91* [1.0]	
	Bi_2S_3 100 [0]	
	$BiOCl$ ($Bi^{3+} + 2 OH^- + Cl^-$) 35.8 [0]	
	$BiOBr$ ($Bi^{3+} + 2 OH^- + Br^-$) 34.16 [2.0]	
	BiI_3 18.09 [2.0]^c	
Ca^{2+}	$CaCO_3$ (calcite) 8.35 [0]; 8.01 [0.15]^d	
	$CaCO_3$ (aragonite) 8.22 [0]	
	$CaSiO_3$ 7.2 [0]	
	$CaHPO_4 \cdot 2 H_2O$ 6.58 [0]	
	$Ca_2P_2O_7$ 7.9 [0]	

^{a} Temperature and ionic strength not stated; ^{c} 20°C; ^{d} 22°C

Solubility Products of Sparingly Soluble Salts

Table 8.1 (Continued)

Metal ion	Formula of the compound, value of pK_s, $[I]$	Ref.
	Ca^{2+} (continued)	
	$Ca(OH)_2$ 5.19 [0]; *4.29* [1.0]	
	$CaSO_3$ 6.5 [0]; 4.80 [1.0]	
	$CaSO_4$ 4.62 [0]; 2.92 [1.0]; 3.16 [3.0]	
	$CaSeO_4$ 4.40 [0]	
	$CaMoO_4$ 8.0 [0]	
	CaF_2 10.41 [0]	
	$Ca(IO_3)_2$ 6.15 [0]; 5.07 [0.5]; 4.89 [1.0]; 4.70 [2.0]	
	CaC_2O_4 7.9 [0.1][c]	[8]
	$Ca(C_9H_6NO)_2$ (oxinate) 10.4 [0.1]	[6]
Cd^{2+}	$CdCO_3$ 13.74 [0]	
	β-$Cd(OH)_2$ 14.35 [0]; 14.29 [3.0]; γ-$Cd(OH)_2$ 14.10 [3.0]	
	CdS 27.0 [0]; 25.8 [1.0]	
	$CdSe$ 35.2 [0]	
	CdC_2O_4 7.82 [a]	[6]
	$Cd(C_5H_{10}NS_2)_2$ (diethyldithiocarbamate) 22.0 [0.1]	[10]
	$Cd(C_7H_6NO_2)_2$ (anthranilate) 8.39 [0]	[11]
Ce^{3+}	CeP_2O_7 23.46 [0.1]	
	$Ce(OH)_3$ 23.2 [0]; 19.9 [1.0]	
	CeF_3 15.1 [a]	
	$Ce(IO_3)_3$ 10.86 [0]	
	$Ce_2(C_2O_4)_3$ 25.5 [2.0][c]	[6]
Ce^{4+}	CeO_2 ($Ce^{4+} + 4\,OH^-$) 65.0 [3.0]	
	$Ce(IO_3)_4$ 9.50 [0]	
Co^{2+}	$CoCO_3$ 9.98 [0]	
	$Co(OH)_2$ (blue) 14.2 [e]; (active pink) 14.8; (inactive pink) 15.7 [e]	[13]
	$Co(OH)_2$ 14.9 [0]; 14.6 [1.0]	
	α-CoS 21.3 [0]; β-CoS 25.6 [0]	
	$CoSe$ 31.2 [0]	
	$Co(C_7H_6NO_2)_2$ (anthranilate) 10.97 [0]	[11]
	$Co(C_9H_6NO)_2$ (oxinate) 24.2 [0.1]	[6]
Co^{3+}	$Co(OH)_3$ 44.5 [0][b]	
Cs^+	$CsIO_4$ 2.65 [0]	

[a] Temperature and ionic strength not stated; [b] 19°C; [c] 20°C; [e] dil. soln.

Table 8.1 (Continued)

Metal ion	Formula of the compound, value of pK_s, [I]	Ref.
Cu^+	CuCNS 13.40 [5.0] $\frac{1}{2}Cu_2O$ ($Cu^+ + OH^-$) 14.7 [0] Cu_2S 48.5 [0] Cu_2Se 60.8 [0] CuCl 7.38 [5.0] CuBr 8.3 [0] CuI 12.0 [0]	
Cu^{2+}	$CuCO_3$ 9.63 [0] $Cu_2(OH)_2CO_3$ ($Cu^{2+} + 2 OH^- + CO_3^{2-}$) 31.90 [0] $Cu_3(OH)_2(CO_3)_2$ (azurite) 45.96 [0] $Cu_2(OH)_2CO_3$ (malachite) 33.78 [0] $K_2Cu(HCO_3)_4$ ($2 K^+ + Cu^{2+} + 4 HCO_3^-$) 11.5 [1.0] $Cu(OH)_2$ 19.32 [0]; 18.9 [1.0]; CuO 20.35 [0]; *19.66* [1.0] CuS 36.1 [0] CuSe 48.1 [0] CuC_2O_4 7.54 [0]; 6.7 [0.1] $Cu(C_5H_{10}NS_2)_2$ (diethyldithiocarbamate) 29.6 [0.1] $Cu(C_6H_5N_2O_2)_2$ (cupferronate) 16.03 [^{a}] $Cu(C_7H_6NO_2)_2$ (anthranilate) 14.18 [0] $Cu(C_9H_6NO)_2$ (oxinate) 29.1 [0.1]	 [14] [6] [10] [15] [11] [6]
Fe^{2+}	$FeCO_3$ 10.68 [0] $Fe(OH)_2$ (active) 15.1 [0]; *14.28* [1.0] FeS 18.1 [0] FeSe 26.0 [0]	
Fe^{3+}	$FePO_4 \cdot 2 H_2O$ 26.4 [0] $Fe_4(P_2O_7)_3$ (22.6) [^{a}] $Fe(OH)_3$ 38.8 [0]; 38.6 [3.0]; $\frac{1}{2}Fe_2O_3$ ($Fe^{3+} + 3 OH^-$) 42.7 [0]; α-FeO(OH) 41.5 [0]; *40.2* [1.0]; 41.1 [3.0] FeO(OH) (amorphous) *39.5* [0]; *38.9* [1.0] $Fe(C_6H_5N_2O_2)_3$ (cupferronate) 25 [^{a}] $Fe(C_9H_6NO)_3$ (oxinate) 43.5 [0.1]	 [6] [15] [16]
Ge^{2+}	GeO ($Ge^{2+} + 2 OH^-$) (brown modification) 3.7 [0]	
Ga^{3+}	$GaPO_4$ 21.0 [1.0] $Ga(OH)_3$ 37 [0] (amorphous) GaO(OH) 39.1 [0]; 37.2 [1.0]	

^{a} Temperature and ionic strength not stated

Solubility Products of Sparingly Soluble Salts

Table 8.1 (Continued)

Metal ion	Formula of the compound, value of pK_s, $[I]$	Ref.
Hf^{4+}	HfO_2 ($Hf^{4+} + 4\,OH^-$) 54.8 [0]; *52.7* [1.0]	
Hg_2^{2+}	Hg_2CO_3 16.05 [0]	
	$Hg_2(CN)_2$ 39.3 [0]	
	$Hg_2(SCN)_2$ 19.52 [0]; 19.00 [1.0]	
	Hg_2SO_4 6.13 [0]; 4.46 [3.0]	
	Hg_2CrO_4 8.70 [0]	
	Hg_2WO_4 16.96 [0.1]	
	Hg_2Cl_2 17.91 [0]; 16.88 [0.5]	
	Hg_2Br_2 22.25 [0]; 21.29 [0.5]	
	Hg_2I_2 28.33 [0]; 27.47 [0.5]	
Hg^{2+}	$Hg(SCN)_2$ 19.56 [1.0]	
	HgO ($Hg^{2+} + 2\,OH^-$) (red modification) 25.44 [0]; *24.90* [1.0]; 26.0 [3.0]	
	HgS (black modification) 52.7 [0]; 51.0 [1.0]^c	
	HgS (red modification) 53.3 [0]	
	$HgSe$ 64.5 [0]; 56.6 [1.0]	
	$HgBr_2$ 18.9 [0.5]	
	HgI_2 27.95 [0.5]	
	$Hg(C_5H_{10}NS_2)$ (diethyldithiocarbamate) 43.5 [0.1]	[17]
In^{3+}	$InPO_4$ 21.63 [1.0]	
	$In(OH)_3$ 36.9 [0]; 35.54 [1.0]; $\frac{1}{2}In_2O_3$ ($In^{3+} + 3\,OH^-$); 35.9 [0]	
	In_2S_3 73.2 [0]	
	$In(C_5H_{10}NS_2)_3$ (diethyldithiocarbamate) 25.0 [0.1]	[18]
	$In(C_9H_6NO)_3$ (oxinate) 31.34 [^{a}]	[19]
K^+	KIO_4 3.43 [0]	
La^{3+}	$La_2(CO_3)_3$ 33.4 [0]	
	$LaPO_4$ 22.43 [0.5]	
	$La(OH)_3$ 20.7 [0]; *19.6* [1.0]	
	$La(IO_3)_3$ 10.99 [0]	
	$La_2(C_2O_4)_3$ 25.0 [0.1]^c	[8]
Li^+	LiF 2.77 [0]	
Ln^{3+} (M^{3+}, $4f$)	$Ln_2(CO_3)_3$ 32 [0] (Nd, Sm, Gd, Dy, Yb)	
	$Ln(IO_3)_3$ 11 [0] (Pr–Lu)	
	$Ln(OH)_3$ 25 [0]; *23.4* [1.0] (Pr–Lu)	

^{a} Temperature and ionic strength not stated; ^{c} 20°C

Table 8.1 (Continued)

Metal ion	Formula of the compound, value of pK_s, [I]	Ref.
Mg^{2+}	$MgCO_3$ 7.46 [0]; $MgCO_3 \cdot 3H_2O$ 4.67 [0]	
	$Mg_3(PO_4)_2 \cdot 8H_2O$ 25.20 [0]; $MgHPO_4 \cdot 3H_2O$ 5.82 [0];	
	$MgNH_4PO_4$ ($Mg^{2+} + NH_4^+ + PO_4^{3-}$) 12.6 [0]	
	$Mg(OH)_2$ 11.15 [0]; *10.25* [1.0]	
	$MgSeO_3 \cdot 6H_2O$ 5.36 [0][c]	
	MgF_2 8.18 [0]	
	MgC_2O_4 4.07 [a]	[6]
	$Mg(C_9H_6NO)_2$ (oxinate) 15.4 [0]; 14.8 [0.1]	[6]
Mn^{2+}	$MnCO_3$ 9.30 [0]; 9.68 [3.0]	
	$Mn(OH)_2$ (active) 12.8 [0]; *11.94* [1.0]	
	MnS (pink) 10.5 [0]; MnS (green) 13.5 [0]	
	$Mn(C_9H_6NO)_2$ (oxinate) 21.7 [0]	
Ni^{2+}	$NiCO_3$ 6.87 [0]	
	$Ni(OH)_2$ 15.2 [0]; *16.3* [1.0]	
	α-NiS 19.4 [0]; β-NiS 24.9 [0]; γ-NiS 26.6 [0]	
	NiSe (32.7) [0]	
	$Ni(C_4H_7N_2O_2)_2$ (dimethylglyoximate) 23.66 [0.1]	[20]
	$Ni(C_7H_6NO_2)_2$ (anthranilate) 11.72 [0]	[11]
	$Ni(C_9H_6NO)_2$ (oxinate) 25.5 [0.1]	[6]
Np^{4+}	NpO_2 60 [0]; *57.9* [1.0]	
	$Np(C_2O_4)_2$ 22.07 [f][c]	[30]
NpO_2^+	NpO_2OH (amorphous) ($NpO_2^+ + OH^-$) 9.3 [0]; 9.1 [1.0]	
NpO_2^{2+}	$NpO_2(OH)_2$ (amorphous) ($NpO_2^{2+} + 2OH^-$) 21.4 [0]; 20.7 [1.0]	
Pa^{4+}	PaO_2 *53.4* [0]; *53.3* [1.0]	
Pa(V)	$\frac{1}{2}Pa_2O_5$ ($PaO_2^+ + OH^-$) *16* [0]; *15.8* [1.0]	
Pb^{2+}	$PbCO_3$ 13.13 [0]; 11.01 [1.0]	
	$Pb_2Fe(CN)_6$ 18.02 [0]	
	$Pb_3(PO_4)_2$ 43.53 [0][g]; $PbHPO_4$ 11.43 [0]	
	PbO (red) 15.3 [0]; PbO (yellow) 15.1 [0];	
	$\frac{1}{2}Pb_2O(OH)_2$ ($Pb^{2+} + 2OH^-$) 14.9 [0]	
	PbS 27.5 [0]	
	$PbSO_4$ 7.79 [0]; 6.20 [1.0]	

[a] Temperature and ionic strength not stated; [c] 20° C; [f] ionic strength not stated; [g] 38° C

Table 8.1 (Continued)

Metal ion	Formula of the compound, value of pK_s, [I]	Ref.
	Pb^{2+} (continued)	
	PbSe 42.1 [0]	
	$PbSeO_4$ 6.84 [0]	
	$PbCrO_4$ 13.75 [0]; 12.9 [0.1]	[6]
	$PbMoO_4$ (13.0) [0]	
	PbF_2 7.44 [0]; 6.26 [1.0]; 6.60 [2.0]	
	$PbCl_2$ 4.78 [0]; 5.0 [3.0]	
	$PbBr_2$ 5.68 [4]	
	$Pb(BrO_3)_2$ 5.10 [0]	
	PbI_2 8.10 [0]; 7.61 [2.0]	
	PbC_2O_4 10.5 [0]; 9.7 [0.1]	[6]
	$Pb(C_5H_{10}NS_2)_2$ (diethyldithiocarbamate) 21.7 [0.1]	[22]
	$Pb(C_7H_6NO_2)_2$ (anthranilate) 9.81 [0]	[11]
	$Pb(C_{13}H_{11}N_4S)_2$ (dithizonate) 23.7 [0.1]^h	[23]
Pb^{4+}	PbO_2 *64* [0]	
Pu^{3+}	$Pu(OH)_3$ 19.7 [0]	
Pu^{4+}	$Pu(OH)_4$ 47.3 [0]	
	$Pu(C_2O_4)_2$ 21.3 [0.5−1.0]^c	[21]
PuO_2^+	$PuO_2(OH)$ (amorphous) 8.6 [0]; 8.4 [1.0]	
PuO_2^{2+}	$PuO_2C_2O_4$ ~9.23 [^{f}]^c	[24]
Pd^{2+}	α-$Pd(N_3)_2$ 8.57 [0]	
	$Pd(OH)_2$ 28.5 [0]	
Sb^{3+}	$\frac{1}{2}Sb_2O_3$ ($SbO^+ + OH^-$) (rhombic) 17.66 [5.0]; (cubic) 17.78 [5.0]	
	Sb_2S_3 93 [0]	[25]
Sc^{3+}	$ScO(OH)$ ($Sc^{3+} + 3 OH^-$) 32.7 [0]; 30.9 [1.0]	
Sn^{2+}	SnO 26.2 [0]; *25.43* [1.0]	
	SnS 25.9 [0]	
	SnSe 38.4 [0]	
	SnI_2 5.08 [0]	
Sn^{4+}	SnO_2 64.4 [0]	

^{c} 20°C; ^{f} ionic strength not stated: ^{h} temperature not stated

Table 8.1 (Continued)

Metal ion	Formula of the compound, value of pK_s, [I]	Ref.
Sr^{2+}	$SrCO_3$ 9.03 [0] $SrHPO_4$ 6.92 [0]^c $Sr_2P_2O_7$ 7.5 [0] $SrSO_4$ 6.50 [0] SrF_2 8.54 [0] SrC_2O_4 6.4 [0.1]^c $Sr(C_9H_6NO)_2$ (oxinate) 8.7 [0.1]	 [8] [6]
Th^{4+}	$Th(OH)_4$ 44.7 [0]^i; ThO_2 49.7 [0]; 47.6 [1.0] ThF_4 28.3 [3.0] $Th(IO_3)_4$ 14.62 [0.5] $Th(C_2O_4)_2$ 24.96 [0]; 21.38 [1.0]	 [26]
Ti^{4+}	$Ti(OH)_4$ 53.10 [0] $TiO(OH)_2$ (TiO^{2+} + 2 OH^-) 32.8 [0]; *31.9* [1.0]	[27]
Tl^+	$TlSCN$ 3.79 [0]; 3.16 [4.0] TlN_3 3.66 [0] Tl_2S 21.2 [0] $Tl_2S_2O_3$ 4.54 [4.0] Tl_2CrO_4 12.01 [0] $TlCl$ 3.74 [0] $TlBr$ 5.44 [0] $TlBrO_3$ 3.78 [0]^k TlI 7.23 [0]; 6.73 [1.0]	
Tl^{3+}	$\frac{1}{2}Tl_2O_3$ (Tl^{3+} + 3 OH^-) 45.2 [0]; *44.4* [1.0]; 45.0 [3.0] $Tl(C_9H_6NO)_3$ (oxinate) 32.4 [^{a}]	 [19]
U^{4+}	UO_2 56.2 [0] UF_4 21.24 [^{a}]	
UO_2^{2+}	$(UO_2)_3(PO_4)_2$ 49.7 [0.3]^c; UO_2HPO_4 12.17 [0.3]^c $UO_2(OH)_2$ 22.4 [0] $UO_2(IO_3)_2$ 7.01 [0.2] $UO_2C_2O_4$ 8.66 [1.0]^c	 [28]
V^{3+}	$V(OH)_3$ 34.3 [0]	

^{a} Temperature and ionic strength not stated; ^{c} 20° C; ^{i} 22° C; ^{k} 30° C

Table 8.1 (Continued)

Metal ion	Formula of the compound, value of pK_s, [I]	Ref.
VO^{2+}	$(VO)_3(PO_4)_2$ 25.1 [0] $VO(OH)_2$ 23.5 [0]; $\frac{1}{2}\alpha\text{-}V_2O_4$ (VO^{2+} + 2 OH^-) 23.71 [0]; 22.95 [1.0]	
V(V)	$NH_4H_2VO_4$ (NH_4^+ + $H_2VO_4^-$) 3.5 [0] $\frac{1}{2}V_2O_5$ (VO_2^+ + OH^-) 13.34 [0]; 13.14 [1.0]	
W(VI)	H_2WO_4 (WO_4^{2-} + 2 H^+) 14.05 [0]	
Y^{3+}	$Y_2(CO_3)_3$ 30.6 [0] $Y(OH)_3$ 23.2 [0]; 22.4 [1.0] $Y(IO_3)_3$ 10.15 [0]	
Zn^{2+}	$Zn(CN)_2$ 15.5 [3.0] $ZnCO_3$ 10.0 [0] $Zn_2Fe(CN)_6$ 15.68 [0] $Zn_3(PO_4)_2 \cdot 4\,H_2O$ 35.3 [0] $Zn(OH)_2$ (amorphous) 15.52 [0]; $\beta\text{-}Zn(OH)_2$ 16.2 [0]; 15.5 [0.2] ZnO 16.66 [0]; 16.14 [1.0] $\alpha\text{-}ZnS$ 24.7 [0]; 24.4 [1.0]; $\beta\text{-}ZnS$ 22.5 [0] $ZnSe$ 29.4 [0] ZnC_2O_4 8.89 [0]; 8.1 [0.1] $Zn(C_9H_6NO)_2$ (oxinate) 23.7 [0.1]	[6] [6]
Zr^{4+}	$ZrO(H_2PO_4)_2$ 17.64 [a] ZrO_2 54.1 [0]; 52.0 [1.0]	[29]

[a] Temperature and ionic strength not stated

REFERENCES

[1] R. M. Smith and A. E. Martell, *Critical Stability Constants*, Vol. 4. Plenum Press, New York, 1976.
[2] C. F. Baes, Jr. and R. E. Mesmer, *The Hydrolysis of Cations*, Wiley, New York, 1976.
[3] J. M. Spencer and M. Le Pla, *Z. Anorg. Chem.*, 1909, **65**, 10.
[4] V. G. Chukhlantsev, *Zh. Neorgan. Khim.*, 1956, **1**, 1975.
[5] H. T. S. Britton and R. A. Robinson, *J. Chem. Soc.*, **1930**, 2328.
[6] A. Ringbom, *Complexation in Analytical Chemistry*, pp. 339–347, Interscience, New York, 1963.
[7] L. G. Sillén and A. E. Martell, *Stability Constants of Metal-Ion Complexes*, p. 66. The Chemical Society, London, 1964.

[8] J. Starý, *Anal. Chim. Acta*, 1963, **28**, 132.
[9] F. G. Zharovskii, *Tr. Komis. po Anal. Khim. Akad. Nauk SSSR*, 1951, **3**, 101.
[10] A. Hulanicki, *Acta Chim. Acad. Sci. Hung.*, 1961, **27**, 411.
[11] P. O. Lumme, *Kemistil. Suomen*, 1957, **30B**, 176; 1958, **31B**, 232, 250; 1960, **33B**, 69, 85.
[12] J. L. Weaver and W. C. Purdy, *Anal. Chim. Acta*, 1959, **20**, 376.
[13] W. Feitknecht and L. Hartmann, *Chimia*, 1954, **8**, 95.
[14] L. Meites, *J. Am. Chem. Soc.*, 1950, **72**, 184.
[15] W. F. Linke and A. Seidell, *Solubilities*, Van Nostrand, New York, 1958.
[16] J. Tsau, S. Matsuo, P. Clerc and R. Benoit, *Bull. Soc. Chim. France*, **1967**, 1039.
[17] W. Kemula, A. Hulanicki and W. Nawrot, *Roczniki Chem.*, 1962, **36**, 1717.
[18] R. Staroszcik and H. Siaglo, *Chem. Analit. (Warsaw)*, 1965, **10**, 265.
[19] I. V. Pyastniskii and A. P. Kostyshina, *Ukr. Khim. Zh.*, 1957, **23**, 1957.
[20] C. V. Banks and S. Anderson, *Inorg. Chem.*, 1963, **2**, 112.
[21] A. I. Moskvin and A. D. Gel'man, *Zh. Neorgan. Khim.*, 1958, **3**, 956.
[22] H. Irving and F. de Silva, *J. Chem. Soc.*, **1963**, 448, 458.
[23] O. B. Mathre and E. B. Sandell, *Talanta*, 1964, **11**, 295.
[24] A. D. Gel'man, L. E. Drabkina and A. I. Moskvin, *Zh. Neorgan. Khim.*, 1958, **3**, 1546, 1934.
[25] See ref. 7, p. 221.
[26] E. G. Moorhead and N. Sutin, *Inorg. Chem.*, 1966, **5**, 1866.
[27] N. N. Mironov and A. I. Odnosevtsev, *Zh. Neorgan. Khim.*, 1957, **2**, 2202.
[28] A. I. Moskvin and F. A. Zakharova, *Zh. Neorgan. Khim.*, 1959, **4**, 2151.
[29] W. H. Burgus, *U.S. At. Energy Commission Rept.*, CN-2560, 1945.
[30] A. D. Gel'man and P. I. Kondratova, see ref. 21.

CHAPTER 9

Oxidation–reduction potentials

Practically all calculations and many applications of equilibrium systems involving oxidation–reduction (redox) equilibria are based on values of standard (or conditional) redox potentials for the redox pairs which take part in the reactions. The redox potentials for two redox pairs can be combined [cf. expressions in Eqs. (2.55) and (2.56)] to define the value of the equilibrium constant for the overall redox reaction, which makes possible further calculations in terms of concentrations (or activities) of the reacting species. Various effects of other chemical reactions, which may occur in the system under consideration, can be included in the values of the conditional redox potentials: the expressions in Eqs. (3.23)–(3.33) indicate the calculations needed to allow for protonation, complex-formation, and precipitation side-reactions.

Approaches to determination of redox potentials (see e.g. ref. [1]) are described in monographs and numerous original papers; electrometric methods (potentiometry, voltammetry) are mainly used, but some other techniques (kinetic reaction studies, chemical analysis of the system, thermodynamic calculations) are also applicable. Tables of standard redox potentials and conditional redox potentials are commonly found in monographs on electrochemistry, as well as in various text-books on chemistry. More extensive tabulations of redox potentials are found in compilations of electrochemical data and equilibrium constants [2–10].

The tables in this chapter represent a certain selection of standard and conditional redox potentials of inorganic and organic systems. In Table 9.1 the standard redox potentials are listed for simple redox pairs (i.e. of uncomplexed or non-protonated forms) in alphabetical order of the symbols of the elements. This table also presents data for redox pairs, one or both forms of which are oxides; potentials are also included for some redox reactions which involve formation of sparingly soluble compounds or complexes of the redox forms. The applicability of these data in various fields of chemistry was considered in compiling the table. All the redox pairs are formulated in terms of reduction, so the signs of the redox potentials agree with the 1953 Stockholm convention.

The values of equilibrium constants for the partial redox reactions [cf. Eq. (2.59), p. 37] are given in parentheses after the relevant redox potentials in Tables 9.1 and 9.2. The equilibrium constants for redox pairs can easily be combined with constants for other equilibria in order to define constants for overall reactions under consideration (see examples given in Chaper 14 or elsewhere). All listed data refer to a temperature of $25°C$ (unless stated otherwise in a note); the ionic strength I of the medium or the electrolyte composition and concentration is given in square brackets. The data thought to be reliable are underlined.

Table 9.2 has a similar arrangement of the redox data for organic compounds which are of some potential interest for analytical chemistry, electrochemistry or biochemistry. The compounds are arranged in sequence of increasing complexity of the composition of the oxidized form.

REFERENCES

[1] G. Kortüm, *Treatise on Electrochemistry*, 2nd Ed., p. 308, Elsevier, Amsterdam, 1965.
[2] A. J. De Bethune and N. A. Swendeman-Loud, *Standard Aqueous Electrode Potentials and Temperature Coefficients at 25° C*, Hampel-Skokie, Illinois, 1964.
[3] L. G. Sillén and A. E. Martell, *Stability Constants of Metal-Ion Complexes*, The Chemical Society, London, 1964.
[4] L. G. Sillén and A. E. Martell, *Stability Constants of Metal-Ion Complexes*, Supplement No. 1, The Chemical Society, London, 1971.
[5] M. Pourbaix, *Atlas d'équilibres électrochimiques à 25° C*, Gauthier-Villars, Paris, 1963.
[6] G. Charlot, A. Collumeau and M. J. C. Marchon, *Selected Constants; Oxidation-Reduction Potentials of Inorganic Substances in Aqueous Solution*, Butterworths, London, 1971.
[7] D. Dobos, *Electrochemical Data*, Akadémiai Kiadó, Budapest, 1975.
[8] G. Milazzo and S. Caroli, *Tables of Standard Electrode Potentials*, Wiley, New York, 1978.
[9] W. M. Latimer, *The Oxidation States of the Elements and their Potentials in Aqueous Solutions*, 2nd Ed., Prentice-Hall, Englewood Cliffs, 1956.
[10] W. M. Clark, *Oxidation-Reduction Potentials of Organic Systems*, Baillière, Tindall and Cox, London, 1960.

Table 9.1

Oxidation–reduction potentials (V) of inorganic systems*

Element Redox couple	Oxidation–reduction system	Values of potentials†	(log K†)	[I]	Ref.§
Ag					
Ag(III)/Ag(II)	$\frac{1}{2}Ag_2O_{3(s)} + H^+ + e = AgO_{(s)} + \frac{1}{2}H_2O$	+1.71	(28.9)	[0]	
Ag(III)/Ag(II)	$\frac{1}{2}Ag_2O_{3(s)} + 3H^+ + 2e = Ag^+ + \frac{3}{2}H_2O$	+1.76	(59.5)	[0]	
Ag(II)/Ag(I)	$Ag^{2+} + e = Ag^+$	+2.00	(33.8)	[4M HClO$_4$]	
	$AgO_{(s)} + H^+ + e = \frac{1}{2}Ag_2O + \frac{1}{2}H_2O$	+1.43	(24.2)	[0]	[2]^a
	$AgO_{(s)} + \frac{1}{2}H_2O + e = \frac{1}{2}Ag_2O + OH^-$	+0.60	(10.1)	[0]	
	$Ag(phen)_2^{2+} + e = Ag(phen)_2^+$	+2.22	(37.5)	[^{b}]	[3]
Ag(I)/Ag	$Ag^+ + e = Ag_{(s)}$	+0.7994	(13.51)	[0]	
	$\frac{1}{2}Ag_2O_{(s)} + \frac{1}{2}H_2O + e = Ag_{(s)} + OH^-$	+0.34	(5.7)	[0]	
	$AgCl_{(s)} + e = Ag_{(s)} + Cl^-$	+0.2223	(3.758)	[0]	
		+0.232	(3.92)	[1.0]	[4]
		+0.220	(3.72)	[2.0]	[4]
		+0.204	(3.45)	[3.0]	[4]
	$AgBr_{(s)} + e = Ag_{(s)} + Br^-$	+0.071	(1.20)	[0]	
	$AgI_{(s)} + e = Ag_{(s)} + I^-$	−0.152	(−2.57)	[0]	
	$Ag_2S_{(s)} + 2e = 2Ag_{(s)} + S^{2-}$	−0.71	(−24)	[0]	
Al					
Al(III)/Al	$Al^{3+} + 3e = Al_{(s)}$	−1.66	(−84.2)	[0]	[5]
Al(I)/Al	$Al^+ + e = Al_{(s)}$	−0.550	(−9.3)	[0]	

^{a} log K calculated from the value given in ref. [2], log $K_w = -14.00$; ^{b} I not given

* Values for aqueous solutions at 25°C are given unless otherwise stated in a footnote.

† Values of the standard redox potentials ($I = 0$) and/or the formal redox potentials ($I \neq 0$, medium specified); values of log K for the redox half-reaction (always given in parentheses) are valid under the same conditions as the relevant potentials.

§ If no reference to the literature is given, the values of potentials and/or log K originate from the list published by Charlot et al. [1].

Table 9.1 (Continued)

Element Redox couple	Oxidation–reduction system	Values of potentials	(log K)	[I]	Ref.
Am					
Am(VI)/Am(V)	$AmO_2^{2+} + e = AmO_2^+$	$+1.6$	(27)	[0]	
Am(VI)/Am(III)	$AmO_2^{2+} + 4\,H^+ + 3e = Am^{3+} + 2\,H_2O$	$+1.75$	(88.7)	[0]	
Am(V)/Am(III)	$AmO_2^+ + 4\,H^+ + 2e = Am^{3+} + 2\,H_2O$	$+1.83$	(61.9)	[0]	
Am(IV)/Am(III)	$Am^{4+} + e = Am^{3+}$	$+2.4$	(41)	[0]	
Am(III)/Am	$Am^{3+} + 3e = Am_{(s)}$	-2.38	(-121)	[0]	
As					
As(V)/As(III)	$H_3AsO_4 + 2\,H^+ + 2e = HAsO_2 + H_2O$	$+0.56$	(18.9)	[0]	
	$AsO_4^{3-} + 2\,H_2O + 2e = AsO_2^- + 4\,OH^-$	-0.67	(-22.7)	[0]	
As(III)/As	$\tfrac{1}{2}\,As_2O_{3(s)} + 3\,H^+ + 3e = As_{(s)} + \tfrac{3}{2}\,H_2O$	$+0.235$	(11.9)	[0]	[5]
	$HAsO_2 + 3\,H^+ + 3e = As_{(s)} + 2\,H_2O$	$+0.23$	(11.7)	[0.2 – 1M HClO$_4$]	[5]
	$AsO_2^- + 2\,H_2O + 3e = As_{(s)} + 4\,OH^-$	$+0.2475$	(12.55)	[0]	[5]
As/As($-$III)	$As_{(s)} + 3\,H^+ + 3e = AsH_{3(g)}$	-0.66	(-33.5)	[1M NaOH]	
		-0.61	(-31)	[0]	
Au					
Au(III)/Au(I)	$AuCl_4^- + 2e = AuCl_2^- + 2\,Cl^-$	$+0.92$	(31.1)	[0]	
	$AuBr_4^- + 2e = AuBr_2^- + 2\,Br^-$	$+0.80$	(27.0)	[0]	[5]
Au(III)/Au	$Au^{III} + 3e = Au_{(s)}$	$+1.50$	(76.0)	[0]	
	$Au(OH)_{3(s)} + 3\,H^+ + 3e = Au_{(s)} + 3\,H_2O$	$+1.45$	(73.5)	[0]	
	$AuCl_4^- + 3e = Au_{(s)} + 4\,Cl^-$	$+1.00$	(50.7)	[0]	
	$AuBr_4^- + 3e = Au_{(s)} + 4\,Br^-$	$+0.85$	(43)	[0]	
Au(I)/Au	$AuCl_2^- + e = Au_{(s)} + 2\,Cl^-$	$+1.15$	(19.4)	[0]	
	$Au(CN)_2^- + e = Au_{(s)} + 2\,CN^-$	-0.6	(-10)	[0][c]	

[c] Lab. temp. (about 20° C)

Table 9.1 (Continued)

Element Redox couple	Oxidation–reduction system	Values of potentials	(log K)	[I]	Ref.
B					
B(III)/B	$H_3BO_3 + 3H^+ + 3e = B_{(s)} + 3H_2O$	-0.87	(-44)	[0]	
	$BF_4^- + 3e = B_{(s)} + 4F^-$	-1.04	(-52.7)	[0]	
Ba					
Ba(II)/Ba	$Ba^{2+} + 2e = Ba_{(s)}$	-2.91	(-98.4)	[0]	
Be					
Be(II)/Be	$Be^{2+} + 2e = Be_{(s)}$	-1.85	(-62.5)	[0]	
Bi					
Bi(IV)/Bi(III)	$BiO_{2(s)} + \frac{1}{2}H_2O + e = \frac{1}{2}Bi_2O_{3(s)} + OH^-$	$+0.55$	(9.4)	[1M NaOH]^c	
Bi(III)/Bi	$\frac{1}{2}Bi_2O_{3(s)} + 3e = Bi_{(s)} + 3OH^-$	-0.46	(-23.3)	[0]	
	$BiOCl_{(s)} + 2H^+ + 3e = Bi_{(s)} + 3Cl^-$	$+0.16$	(8.1)	[0]	
	$BiCl_4^- + 3e = Bi_{(s)} + 4Cl^-$	$+0.16$	(8.1)	[0]	
Bk					
Bk(IV)/Bk(III)	$Bk^{4+} + e = Bk^{3+}$	$+1.6$	(27)	[1M HClO$_4$]	
Bk(III)/Bk	$Bk^{3+} + 3e = Bk_{(s)}$	-2.4	(-122)	[0]	
Br					
Br(V)/Br	$BrO_3^- + 6H^+ + 5e = \frac{1}{2}Br_{2(l)} + 3H_2O$	$+1.5$	(127)	[0]	
Br(V)/Br(−I)	$BrO_3^- + 3H_2O + 6e = Br^- + 6OH^-$	$+0.61$	(62)	[0]	
Br(I)/Br	$HBrO + H^+ + e = \frac{1}{2}Br_{2(l)} + H_2O$	$+1.6$	(27)	[0]	
Br(I)/Br(−I)	$BrO^- + H_2O + 2e = Br^- + 2OH^-$	$+0.76$	(26)	[0]	
	$BrCl + 2e = Br^- + Cl^-$	$+1.35$	(45.6)	[0]	[5]

^{c} Lab. temp. (about 20°C)

Table 9.1 (Continued)

Element Redox couple	Oxidation–reduction system	Values of potentials	(log K)	[I]	Ref.
Br (continued)					
Br/Br(−I)	$\frac{1}{2}Br_{2(l)} + e = Br^-$	+1.06	(17.9)	[0]	[5]
	$\frac{1}{2}Br_{2(aq)} + e = Br^-$	+1.08	(18.3)	[0]	[4]
Combined systems	$BrO_3^- + 5\,Br^- + 6\,H^+ = 3\,Br_{2(aq)} + 3\,H_2O$		(33.14)	[0]	[5]
	$HBrO + Br^- + H^+ = Br_2 + H_2O$		(8.24)	[0]	[5]
C					
C(IV)/C(III)	$HCNO + H^+ + e = \frac{1}{2}(CN)_{2(g)} + \frac{1}{2}H_2O$	+0.33	(5.6)	[0]	[6]
C(IV)/C(II)	$CO_{2(g)} + 2\,H^+ + 2e = CO_{(g)} + H_2O$	−0.103	(−3.5)	[0]	[6]
C(IV)/C	$CO_{2(g)} + 4\,H^+ + 4e = C_{(s)} + H_2O$	+0.207	(14.0)	[0]	
C(III)/C(II)	$\frac{1}{2}CN_{2(g)} + H^+ + e = HCN$	+0.37	(6.3)	[0]	
C(II)/C	$CO_{(g)} + 2\,H^+ + 2e = C_{(s)} + H_2O$	+0.518	(17.5)	[0]	[6]
Ca					
Ca(II)/Ca	$Ca^{2+} + 2e = Ca_{(s)}$	−2.87	(−97.0)	[0]	[4]
	$Ca^{2+} + (Hg) + 2e = CaHg_x$	−1.996	(−67.48)	[0]	
Cd					
Cd(II)/Cd	$Cd^{2+} + 2e = Cd_{(s)}$	−0.403	(−13.63)	[0]	[4]
		−0.411	(−13.90)	[1M − 3M HClO_4]	
	$Cd^{2+} + (Hg) + 2e = CdHg_{x(sat.)}$	−0.352	(−11.90)	[0]	
Ce					
Ce(IV)/Ce(III)	$Ce^{4+} + e = Ce^{3+}$	+1.74	(29.4)	[1M HClO_4]	[5]
		+1.45	(24.5)	[0.5M H_2SO_4]	[5]
	$Ce(IV) + e = Ce(III)$	+1.443	(24.4)	[1M H_2SO_4]	
		+1.43	(24.2)	[2M H_2SO_4]	

Table 9.1 (Continued)

Element Redox couple	Oxidation–reduction system	Values of potentials	(log K)	[I]	Ref.
Ce (continued)					
	$Ce(IV) + e = Ce(III)$	+1.28	(21.6)	[1M HCl]	
		+1.61	(27.2)	[1M HNO_3]	
Ce(III)/Ce	$Ce^{3+} + 3e = Ce_{(s)}$	−2.33	(−118)	[0]	
Cf					
Cf(III)/Cf	$Cf^{3+} + 3e = Cf_{(s)}$	−2.1	(−106)	[0]	
Cl					
Cl(VII)/Cl(V)	$ClO_4^- + 2H^+ + 2e = ClO_3^- + H_2O$	+1.19	(40.2)	[0]	
Cl(V)/Cl(IV)	$ClO_3^- + 2H^+ + e = ClO_{2(g)} + H_2O$	+1.15	(19.5)	[0]	
Cl(V)/Cl(III)	$ClO_3^- + H_2O + 2e = ClO_2^- + 2OH^-$	+0.33	(11)	[0]	
Cl(V)/Cl	$ClO_3^- + 6H^+ + 5e = \frac{1}{2}Cl_{2(g)} + 3H_2O$	+1.47	(124)	[0]	
Cl(IV)/Cl(III)	$ClO_{2(g)} + H^+ + e = HClO_2$	+1.27	(21.5)	[0]	
	$ClO_2 + e = ClO_2^-$	+0.93	(15.7)	[0]	
Cl(III)/Cl(I)	$HClO_2 + 2H^+ + 2e = HClO + H_2O$	+1.64	(55.4)	[0]	
Cl(I)/Cl	$HClO + H^+ + e = \frac{1}{2}Cl_{2(g)} + H_2O$	+1.63	(27.6)	[0]	
Cl/Cl(−1)	$\frac{1}{2}Cl_{2(g)} + e = Cl^-$	+1.358	(22.96)	[0]	
		+1.3699	(23.158)	[1M HClO_4]	[4]
		+1.3411	(22.671)	[3M HClO_4]	[4]
	$\frac{1}{2}Cl_{2(aq)} + e = Cl^-$	+1.39	(23.5)	[0]	
Cm					
Cm(III)/Cm	$Cm^{3+} + 3e = Cm_{(s)}$	−2.70	(−137)	[0]	

Table 9.1 (Continued)

Element Redox couple	Oxidation-reduction system	Values of potentials	(log K)	[I]	Ref.
Co					
Co(III)/Co(II)	$Co^{3+} + e = Co^{2+}$	+1.95	(33)	[$4M$ HClO$_4$]	[4]
	$CoOOH_{(s)} + e = Co(OH)_{2(s)}^d$	+0.17	(2.9)	[0]	
	$Co(NH_3)_6^{3+} + e = Co(NH_3)_6^{2+}$	+0.1	(1.7)	[0]	
	$Co(NH_3)_5^{3+} + e = Co(NH_3)_5^{2+}$	+0.37	(6.25)	[$1M$ NH$_4$NO$_3$]	
	$Co(CN)_6^{3-} + e = Co(CN)_6^{4-}$	−0.83	(−15)	[$0.2M$ KOH]^e	
	$Co(bipy)_3^{3+} + e = Co(bipy)_3^{2+}$	+0.31	(5.2)	[dil., pH = 2]	
	$Co(EDTA)^- + e = Co(EDTA)^{2-}$	+0.37	(6.25)	[$0.2M$ KNO$_3$]	[6]
	$Co(phen)_3^{3+} + e = Co(phen)_3^{2+}$	+0.42	(7.1)	[^{b}]	[3]
Co(II)/Co	$Co^{2+} + 2e = Co_{(s)}$	−0.29	(−9.8)	[0]	[6]
	$Co(alanine)_2^{2+} + 2e = Co_{(s)} + 2$ alanine	−0.527	(−17.8)	[dil. NaOH]	
Cr					
Cr(VI)/Cr(III)	$\frac{1}{2}Cr_2O_7^- + 7H^+ + 3e = Cr^{3+} + \frac{7}{2}H_2O$	+1.33	(67.4)	[0]	[5]
	$HCrO_4^- + 7H^+ + 3e = Cr^{3+} + 4H_2O$	+1.20	(60.8)	[0]	
Cr(III)/Cr(II)	$Cr^{3+} + e = Cr^{2+}$	−0.41	(−6.9)	[0]	
	$Cr(III) + e = Cr(II)$	−0.38	(−6.4)	[$1M$ HCl]	
	$Cr(CN)_6^{3-} + e = Cr(CN)_6^{4-}$	1.14	(19.3)	[$1M$ KCN]	
Cr(III)/Cr	$Cr^{3+} + 3e = Cr_{(s)}$	−0.74	(−37.5)	[0]	
Cr(II)/Cr	$Cr^{2+} + 2e = Cr_{(s)}$	−0.86	(−29.0)	[0]^f	
Cs					
Cs(I)/Cs	$Cs^+ + e = Cs_{(s)}$	−2.92	(−49.3)	[0]	
	$Cs^+ + (Hg) + e = CsHg_x$	−1.78	(−30.1)	[0]	

^{b} I not given; ^{d} blue modification; ^{e} $T = 2°C$; ^{f} temperature not stated

Table 9.1 (Continued)

Element Redox couple	Oxidation–reduction system	Values of potentials	(log K)	[I]	Ref.
Cu					
Cu(III)/Cu(II)	$Cu^{3+} + e = Cu^{2+}$	+2.3	(42.4)	[0][g]	
	$\frac{1}{2}Cu_2O_{3(s)} + H^+ + e = CuO_{(s)} + \frac{1}{2}H_2O$	+1.6	(29.5)	[var][g]	
Cu(II)/Cu(I)	$Cu^{2+} + e = Cu^+$	+0.17	(2.9)	[0]	
	$CuO_{(s)} + H^+ + e = \frac{1}{2}Cu_2O_{(s)} + \frac{1}{2}H_2O$	+0.64	(11.8)	[0][g]	
	$Cu(phen)_2^{2+} + e = Cu(phen)_2^+$	+0.17	(2.9)	[b]	[3]
Cu(II)/Cu	$Cu^{2+} + 2e = Cu_{(s)}$	+0.34	(11.5)	[0]	
	$Cu^{2+} + (Hg) + 2e = CuHg_x$	+0.3435	(11.61)	[0]	[5]
	$CuEDTA^{2-} + 2e = Cu_{(s)} + EDTA^{4-}$	−0.216	(−7.30)	[0.25M EDTA]	[6]
Cu(I)/Cu	$Cu^+ + e = Cu_{(s)}$	+0.52	(8.8)	[0]	
	$\frac{1}{2}Cu_2O_{(s)} + \frac{1}{2}H_2O + e = Cu_{(s)} + OH^-$	−0.36	(−6.1)	[0]	[5, 6]
	$CuCl_{(s)} + e = Cu_{(s)} + Cl^-$	+0.14	(2.4)	[0]	
Combined systems	$Cu^{2+} + Cu_{(s)} = 2Cu^+$		(−6.0)	[0]	[4, 5]
F					
F/F(−1)	$\frac{1}{2}F_{2(g)} + e = F^-$	+2.87	(+48.5)	[0]	
Fe					
Fe(VI)/Fe(III)	$FeO_4^{2-} + 3H_2O + 3e = FeOOH_{(s)} + 5OH^-$	+0.7	(35.5)	[0]	
Fe(III)/Fe(II)	$Fe^{3+} + e = Fe^{2+}$	+0.771	(13.0)	[0]	
		+0.767	(13.0)	[1M HClO$_4$]	[5]
		+0.7375	(12.47)	[2M HClO$_4$]	
	$FeOOH_{(s,amorphous)} + 3H^+ + e = Fe^{2+} + 2H_2O$	+0.923	(15.6)	[0]	[4]

[b] I not given; [g] $T = 0°C$

Table 9.1 (Continued)

Element Redox couple	Oxidation–reduction system	Values of potentials	(log K)	[I]	Ref.
Fe (continued)	$Fe_3O_{4(s)} + 8 H^+ + 2e = 3 Fe^{2+} + 4 H_2O$	+1.23	(41.6)	[0]	
	$Fe(CN)_6^{3-} + e = Fe(CN)_6^{4-}$	+0.355	(6.00)	[0]	
	$Fe(bipy)_3^{3+} + e = Fe(bipy)_3^{2+}$	+1.120	(18.9)	[dil]	
	$FeEDTA^- + e = FeEDTA^{2-}$	+0.1172	(1.981)	[0.1M KCl]	
	$Fe(phen)_3^{3+} + e = Fe(phen)_3^{2+}$	+1.06	(18.2)	[0.1M KCl][h]	
Fe(II)/Fe	$Fe^{2+} + 2e = Fe_{(s)}$	−0.44	(−14.9)	[0]	[6]
Fm					
Fm(III)/Fm	$Fm^{3+} + 3e = Fm_{(s)}$	−2.1	(−106)	[0]	
Ga					
Ga(III)/Ga(I)	$Ga^{3+} + 2e = Ga^+$	⩽−0.40	(−13.5)	[0]	[5]
Ga(III)/Ga	$Ga^{3+} + 3e = Ga_{(s)}$	−0.56	(−28.4)	[0]	
Ga(I)/Ga	$Ga^+ + e = Ga_{(s)}$	⩾−0.79	(−13.4)	[0]	[5]
Ge					
Ge(IV)/Ge(II)	$Ge^{4+} + 2e = Ge^{2+}$	0	(0)	[0]	
	$GeO_{2(s,hexagonal)} + 2 H^+ + 2e = GeO_{(s,brown)} + H_2O$	−0.12	(−4.0)	[0]	[5]
	$= GeO_{(s,yellow)} + H_2O$	−0.27	(−9.1)	[0]	[5]
	$GeO_{2(s,hexagonal)} + 4 H^+ + 2e = Ge^{2+} + 2 H_2O$	−0.25	(−8.4)	[0]	
	$GeO_{2(s,tetragonal)} + 4 H^+ + 2e = Ge^{2+} + 2 H_2O$	−0.34	(−11.5)	[0]	
Ge(IV)/Ge	$GeO_{2(s,hexagonal)} + 4 H^+ + 4e = Ge_{(s)} + 2 H_2O$	−0.01	(−0.7)	[0]	
	$GeO_{2(s,tetragonal)} + 4 H^+ + 4e = Ge_{(s)} + 2 H_2O$	−0.05	(−3.4)	[0]	

[h] 20°C

Table 9.1 (Continued)

Element Redox couple	Oxidation–reduction system	Values of potentials	(log K)	[I]	Ref.
Ge (continued)					
Ge(II)/Ge	$H_2GeO_3 + 4H^+ = Ge_{(s)} + 3H_2O$	+0.01	(0.7)	[0]	
Ge/Ge(−IV)	$Ge^{2+} + 2e = Ge_{(s)}$	+0.23	(7.8)	[0]	
	$Ge_{(s)} + 4H^+ + 4e = GeH_{4(g)}$	−0.3	(−20)	[0][c]	
H					
H(I)/H	$H^+ + e = \frac{1}{2}H_{2(g)}$	0.0000	(0.00)	[0]	[5]
	$H^+ + e_{(aq)} = H_{(aq)}$	+0.57	(9.6)	[0]	
	$D^+ + e = \frac{1}{2}D_{2(g)}$	+0.029	(0.49)	[0]	
Hf					
Hf(IV)/Hf	$Hf^{4+} + 4e = Hf_{(s)}$	−1.58	(−107)[a]	[0]	[5]
	$HfO_{2(s)} + 4H^+ + 4e = Hf_{(s)}$	−1.57	(−106)	[0]	
Hg					
Hg(II)/Hg(I)	$2Hg^{2+} + 2e = Hg_2^{2+}$	+0.907	(30.7)	[0]	[5]
		+0.920	(31.1)	[2M HClO$_4$]	
Hg(II)/Hg	$HgO_{(s)} + 2H^+ + 2e = Hg_{(l)}$	+0.926	(31.3)	[0]	[5]
Hg(I)/Hg	$Hg_2^{2+} + 2e = 2Hg_{(l)}$	+0.792	(26.8)	[0]	
		+0.770	(26.0)	[2M HClO$_4$]	
	$Hg_2Cl_{2(s)} + 2e = 2Hg_{(l)} + 2Cl^-$	+0.268	(9.06)	[0]	[5]
		+0.334	(11.3)	[0.1M KCl]	
		+0.281	(9.50)	[1M KCl]	
		+0.241	(8.15)	[sat. KCl]	
	$Hg_2Br_{2(s)} + 2e = 2Hg_{(l)} + 2Br^-$	+0.1392	(4.706)	[0]	
	$Hg_2SO_{4(s)} + 2e = 2Hg_{(l)} + SO_4^{2-}$	+0.614	(20.8)	[0]	

[c] Lab. temp. (about 20°C)

Table 9.1 (Continued)

Element Redox couple	Oxidation-reduction system	Values of potentials	(log K)	[I]	Ref.
I					
I(VII)/I(V)	$H_5IO_6 + H^+ + 2e = IO_3^- + 3 H_2O$	+1.6	(54)	[0]	
I(V)/I	$IO_3^- + 6 H^+ + 5e = \frac{1}{2} I_{2(s)} + 3 H_2O$	+1.19	(101)	[0]	
I(V)/I(−I)	$IO_3^- + 6e = I^- + 6 OH^-$	+0.26	(26)	[0]	[5]
I(III)/I	$ICl_{3(s)} + 3e = \frac{1}{2} I_{2(s)} + 3 Cl^-$	+1.28	(64.9)	[0]	
I(I)/I	$HIO + H^+ + e = \frac{1}{2} I_{2(s)} + H_2O$	+1.45	(24.5)	[0]	[5]
	$ICN + H^+ + e = \frac{1}{2} I_{2(s)} + HCN$	+0.63	(10.7)	[0]	
	$ICl + e = \frac{1}{2} I_{2(s)} + Cl^-$	+1.19	(20.1)	[0]	
	$IBr + e = \frac{1}{2} I_{2(s)} + Br^-$	+1.02	(17.2)	[0]	
I(I)/I(−I)	$IO^- + 2e = I^- + 2 OH^-$	+0.49	(16.6)	[0]	[5]
I/I(−I)	$\frac{1}{2} I_{2(s)} + e = I^-$	+0.535	(9.04)	[0]	
	$\frac{1}{2} I_{2(aq)} + e = I^-$	+0.621	(10.5)	[0]	
		+0.6276	(10.61)	[0.5M H_2SO_4]	[5]
		+0.536	(18.1)	[0]	[5]
	$I_3^- + 2e = 3 I^-$	+0.545	(18.4)	[0.5M H_2SO_4]	
In					
In(III)/In(I)	$In^{3+} + 2e = In^+$	−0.4042	(−13.7)	[0]	[5]
		−0.43	(−14.5)	[3M $NaClO_4$]	
In(III)/In	$In^{3+} + 3e = In_{(s)}$	−0.34	(−17)	[0]	
		−0.343	(−17.4)	[3M $NaClO_4$]	[5]
In(I)/In	$In^+ + e = In_{(s)}$	−0.126[h]	(−2.17)	[0.7M $HClO_4$]	[4]
		−0.18	(−3.0)	[3M $NaClO_4$]	[5]
Combined system	$In^{3+} + 2 In_{(s)} = 3 In^+$		(−6.94)	[0]	

[h] 20°C

Table 9.1 (Continued)

Element Redox couple	Oxidation–reduction system	Values of potentials	(log K)	[I]	Ref.
Ir					
Ir(VI)/Ir(IV)	$IrO_4^{2-} + 4H^+ + 2e = IrO_{2(s)}$	< +1.3	(< 44)	[0]	[5]
Ir(IV)/Ir(III)	$IrCl_6^{2-} + e = IrCl_6^{3-}$	+0.87	(14.7)	[0]	
		+0.93	(15.7)	[1M HCl]	
Ir(IV)/Ir	$IrO_{2(s)} + 4H^+ + 4e = Ir_{(s)} + 2H_2O$	+0.93	(62.9)	[0]	[5]
Ir(IV)/Ir	$IrCl_6^{2-} + 4e = Ir_{(s)} + 6Cl^-$	+0.835	(56.5)	[0]	[5]
Ir(III)/Ir	$IrCl_6^{3-} + 3e = Ir_{(s)} + 6Cl^-$	+0.77	(39.0)	[0]	
K					
K(I)/K	$K^+ + e = K_{(s)}$	−2.925	(−49.45)	[0]	
	$K^+ + (Hg) + e = KHg_x$	−1.972	(−33.34)	[0]	
La					
La(III)/La	$La^{3+} + 3e = La_{(s)}$	−2.52	(−128)	[0]	
Li					
Li(I)/Li	$Li^+ + e = Li_{(s)}$	−3.03	(−51.2)	[0]	
	$Li^+ + (Hg) + e = LiHg_x$	−2.00	(−33.8)	[0]	
Ln (Pr – Lu)					
Ln(III)/Ln	$Ln^{3+} + 3e = Ln_{(s)}$	−2.37	(−120)	[0]	[5][a]
Mg					
Mg(II)/Mg	$Mg^{2+} + 2e = Mg_{(s)}$	−2.37	(−80.0)	[0]	
Mn					
Mn(VII)/Mn(VI)	$MnO_4^- + e = MnO_4^{2-}$	+0.57	(9.6)	[0]	

[a] log K calculated from the values given in ref. [5]

Table 9.1 (Continued)

Element Redox couple	Oxidation–reduction system	Values of potentials	(log K)	[I]	Ref.
Mn (continued)					
Mn(VII)/Mn(IV)	$MnO_4^- + 4H^+ + 3e = MnO_{2(s,\beta)} + 2H_2O$	+1.68	(85.1)	[0]	
Mn(VII)/Mn(II)	$MnO_4^- + 8H^+ + 5e = Mn^{2+} + 4H_2O$	+1.51	(127.5)	[0]	
Mn(VI)/Mn(V)	$MnO_4^{2-} + e = MnO_4^{3-}$	+0.27	(4.56)	[0]	
Mn(VI)/Mn(IV)	$MnO_4^{2-} + 2H_2O + 2e = MnO_{2(s)} + 4OH^-$	+0.511	(17.7)	[0]^i	[5]
Mn(IV)/Mn(III)	$Mn(IV) + e = Mn(III)$	+1.652	(27.93)	[7.5M H$_2$SO$_4$]	[5]
Mn(IV)/Mn(II)	$MnO_{2(s)} + 4H^+ + 2e = Mn^{2+} + 2H_2O$	+1.23	(41.6)	[0]	
	$Mn_3O_{4(s)} + 8H^+ + 2e = 3Mn^{2+} + 4H_2O$	+1.755	(59.34)	[0]	[4]
Mn(III)/Mn(II)	$Mn(III) + e = Mn(II)$	+1.488	(25.15)	[7.5M H$_2$SO$_4$]	[5]
	$Mn(OH)_{3(s)} + 3H^+ + e = Mn^{2+} + 3H_2O$	+1.841	(31.12)	[0]	[4]
	$Mn(OH)_{3(s)} + e = Mn(OH)_{2(s)} + OH^-$	+0.154	(2.6)	[0]	[5]
Mn(II)/Mn(I)	$Mn(CN)_6^{3-} + e = Mn(CN)_6^{4-}$	−0.24	(−4.05)	[1.5M NaCN]	[5]
	$Mn(CN)_6^{4-} + e = Mn(CN)_6^{5-}$	−1.05	(−17.7)	[2M NaCN]	[5]
Mn(II)/Mn	$Mn^{2+} + 2e = Mn_{(s)}$	−1.17	(−39.5)	[0]	[5]
	$Mn(OH)_{2(s)} + 2e = Mn_{(s)} + 2OH^-$	−1.55	(−52.4)	[0]	[5]
Mo					
Mo(VI)/Mo(V)	$Mo(VI) + e = Mo(V)$	+0.53	(9.1)	[5M HCl]^h	
	$MoO_2^{2+} + 2H^+ + e = MoO^{3+}$	+0.405	(7.0)	[0.5M H$_2$SO$_4$]^h	[5]
	$MoO_{12}O_{40}^{4-} + 4H^+ + 4e = H_4SiMo_{12}O_{40}^{4-}$	+0.48	(8.1)	[0]	
Mo(VI)/Mo	$H_2MoO_4 + 6H^+ + 6e = Mo_{(s)} + 4H_2O$	+0.59	(40)	[0.5M H$_2$SO$_4$]^c	
Mo(V)/Mo(IV)	$Mo(CN)_8^{3-} + e = Mo(CN)_8^{4-}$	+0.816	(13.8)	[0.5M H$_2$SO$_4$]	[4]

^{c} Lab. temp. (about 20°C), ^{h} 20°C, ^{i} 18°C

Table 9.1 (Continued)

Element Redox couple	Oxidation–reduction system	Values of potentials	(log K)	[I]	Ref.
Mo (continued)					
Mo(V)/Mo(III)	$Mo(V) + 2e = Mo(III)$	0.23	(7.9)	[8.75M H_2SO_4][h]	[5]
		−0.01	(−0.3)	[0.45M H_2SO_4][h]	[5]
	$MoO_2^+ + 4H^+ + 2e = Mo^{3+} + 2H_2O$	0	(0)	[0]	[5]
Mo(IV)/Mo(III)	$Mo(IV) + e = Mo(III)$	0.1	(1.7)	[4.5M H_2SO_4][f]	
Mo(III)/Mo	$Mo^{3+} + 3e = Mo_{(s)}$	−0.2	(−10)	[0]	
Md					
Md(III)/Md	$Md^{3+} + 3e = Md_{(s)}$	−2.2	(−112)	[0]	
N					
N(V)/N(IV)	$NO_3^- + 2H^+ + e = \frac{1}{2}N_2O_{4(g)} + H_2O$	+0.80	(13.5)	[0]	[5]
N(V)/N(III)	$NO_3^- + 3H^+ + 2e = HNO_2 + H_2O$	+0.94	(31.8)	[0]	[4]
N(V)/N(II)	$NO_3^- + 4H^+ + 3e = NO_{(g)} + 2H_2O$	+0.96	(48.7)	[0]	[4]
N(IV)/N(III)	$N_2O_{4(g)} + 2H^+ + 2e = 2HNO_2$	+1.07	(36.2)	[0]	[5]
N(III)/N(II)	$NO^+ + e = NO_{(g)}$	+1.46	(24.7)	[0]	[5]
	$HNO_2 + H^+ + e = NO_{(g)} + H_2O$	+0.983	(16.6)	[0]	
N(II)/N(I)	$2NO_{(g)} + 2H^+ + 2e = H_2N_2O_2$	+0.71	(24)	[0]	
	$2NO_{(g)} + 2H^+ + 2e = N_2O_{(g)} + H_2O$	+1.59	(53.8)	[0]	[5]
N(I)/N	$H_2N_2O_2 + 2H^+ + 2e = N_{2(g)} + 2H_2O$	+2.65	(89.6)	[0]	[5]
	$N_2O_{(g)} + 2H^+ + 2e = N_{2(g)} + H_2O$	+1.77	(59.8)	[0]	
3N/N(−I)$_3$	$3N_{2(g)} + 2H^+ + 2e = 2HN_3$	−3.1	(−105)	[0]	
N/N(−I)	$N_{2(g)} + 2H_2O + 4H^+ + 2e = 2NH_3OH^+$	−1.87	(−63.2)	[0]	
N/N(−II)	$N_{2(g)} + 5H^+ + 4e = N_2H_5^+$	−0.23	(−15.6)	[0]	[5]

[f] $T = ?$ [h] $20°C$

Table 9.1 (Continued)

Element Redox couple	Oxidation–reduction system	Values of potentials	(log K)	[I]	Ref.
N (continued)					
N(−I)/N(−II)	$2\,NH_3OH^+ + H^+ + 2e = N_2H_5^- + 2\,H_2O$	+1.42	(48.0)	[0]	[5]
N(−I)/N(−III)	$NH_3OH^+ + 2\,H^+ + 2e = NH_4^+ + H_2O$	+1.35	(45.6)	[0]	
N(−II)/N(−III)	$N_2H_5^+ + 3\,H^+ + 2e = 2\,NH_4^+$	+1.27	(42.9)	[0]	
Na					
Na(I)/Na	$Na^+ + e = Na_{(s)}$	−2.713	(−45.88)	[0]	
	$Na^+ + (Hg) + e = NaHg_x$	−1.84	(−31.10)	[0]	
Nb					
Nb(V)/Nb	$\frac{1}{2}Nb_2O_{5(s)} + 5\,H^+ + 5e = Nb_{(s)} + \frac{5}{2}H_2O$	−0.65	(−55)	[0]	[5]
	$NbO^{3+} + 2\,H^+ + 2e = Nb^{3+} + H_2O$	−0.34	(−11.5)	[HCl + H_2SO_4]^i	
Nb(III)/Nb	$Nb^{3+} + 3e = Nb_{(s)}$	−1.1	(−56)	[0]	
Ni					
Ni(IV)/Ni(III)	$Ni(OH)_{4(s)} + e = Ni(OH)_{3(s)} + OH^-$	+0.6	(10)	[0]	
Ni(IV)/Ni(II)	$NiO_{2(s)} + 4\,H^+ + 2e = Ni^{2+} + 2\,H_2O$	~ +2.0	(68)	[0]	
Ni(III)/Ni(II)	$Ni(OH)_{3(s)} + 3\,H^+ + e = Ni^{2+} + 3\,H_2O$	+2.08	(35.2)	[0]	
	$Ni(OH)_{3(s)} + e = Ni(OH)_{2(s)} + OH^-$	0.48	(8.1)	[0]	
Ni(II)/Ni	$Ni^{2+} + 2e = Ni_{(s)}$	−0.25	(−8.45)	[0]	
	$Ni(OH)_2 + 2e = Ni_{(s)} + 2\,OH^-$	−0.72	(−24.3)	[0]	
No					
No(III)/No	$No^{3+} + 3e = No_{(s)}$	−2.5	(−127)	[0]	

^{i} 18°C

Table 9.1 (Continued)

Element Redox couple	Oxidation–reduction system	Values of potentials	(log K)	[I]	Ref.
Np					
Np(VI)/Np(V)	$NpO_2^{2+} + e = NpO_2^+$	$+1.14$	(19.3)	[1M HClO$_4$]	
Np(V)/Np(IV)	$NpO_2^+ + 4H^+ + e = Np^{4+} + 2H_2O$	$+0.74$	(12.5)	[1M HClO$_4$]	
Np(IV)/Np(III)	$Np^{4+} + e = Np^{3+}$	$+0.15$	(2.5)	[1M HClO$_4$]	
		$+0.14$	(2.4)	[1M HCl]	
Np(III)/Np	$Np^{3+} + 3e = Np_{(s)}$	-1.85	(-93.8)	[0]	
O					
O/O($-$II)	$O_{3(g)} + 2H^+ + 2e = O_{2(g)} + H_2O$	$+2.07$	(70.0)	[0]	
	$O_{2(g)} + 4H^+ + 4e = 2H_2O$	$+1.229$	(83.1)	[0]	
	$O_{2(g)} + 2H_2O + 4e = 4OH^-$	$+0.401$	(27.1)	[0]	
O/O($-$I)	$O_{2(g)} + H_2O + 2e = HO_2^- + OH^-$	-0.076	(-2.6)	[0]	[5]
	$O_{2(g)} + 2H^+ + 2e = H_2O_2$	$+0.69$	(23.4)	[0]	[5]
O($-$I)/O($-$II)	$H_2O_2 + 2H^+ + 2e = 2H_2O$	$+1.77$	(59.8)	[0]	[5]
	$HO_2^- + H_2O + 2e = 3OH^-$	$+0.88$	(29.7)	[0]	
Os					
Os(VIII)/Os(VI)	$HOsO_5^- + 2e = OsO_4^{2-} + OH^-$	$+0.30$	(10)	[0]	[5]
Os(VIII)/Os(IV)	$OsO_{4(aq)} + 4H^+ + 4e = OsO_2 \cdot aq_{(s)} + 2H_2O$	$+0.964$	(65.4)	[0]	[5]
OsO(VIII)/Os	$OsO_{4(s,yellow)} + 8H^+ + 8e = Os_{(s)} + 4H_2O$	$+0.85$	(115)	[0]	
Os(VI)/Os(IV)	$OsO_4^{2-} + 2H_2O + 2e = OsO_{2(s)} + 4OH^-$	$+0.1$	(3)	[0]	[5]
Os(IV)/Os(III)	$Os(IV) + e = Os(III)$	$+0.446$	(7.67)	[0.1M HCl]	
		$+0.42$	(7.2)	[1M HCl]	
	$OsCl_6^{2-} + e = OsCl_6^{3-}$	$+0.85$	(14)	[1M HCl]	
Os(IV)/Os	$OsO_{2(s)} + 2H_2O + 4e = Os_{(s)} + 4OH^-$	-0.15	(-10)	[0]	[5]
Os(II)/Os	$Os^{2+} + 2e = Os_{(s)}$	$+0.85$	(29)	[1M HCl]	[5]

Table 9.1 (Continued)

Element Redox couple	Oxidation–reduction system	Values of potentials	(log K)	[I]	Ref.
P					
P(V)/P(III)	$H_3PO_4 + 2H^+ + 2e = H_3PO_3 + H_2O$	-0.28	(-9.4)	[0]	
P(III)/P(I)	$H_3PO_3 + 2H^+ + 2e = H_3PO_2 + H_2O$	-0.50	(-16.9)	[0]	
P(I)/P	$H_3PO_2 + H^+ + e = \frac{1}{4}P_4 + 2H_2O$	-0.51	(-8.6)	[0]	
P/P(−II)	$P_{4(s)} + 2H^+ + 2e = H_2P_{4(g)}$	-0.35	(-12)	[0]	
P/P(−III)	$\frac{1}{4}P_{4(s,white)} + 3H^+ + 3e = PH_{3(g)}$	$\underline{0.06}$	(3.0)	[0]	
Pa					
Pa(V)/Pa	$PaO_2^+ + 4H^+ + 5e = Pa_{(s)} + 2H_2O$	~ -1	(-85)	[0]	
Pa(IV)/Pa(III)	$PaO_{2(s)} + 4H^+ + e = Pa^{3+} + 2H_2O$	-0.5	(-8.5)	[0]	
	$Pa^{4+} + e = Pa^{3+}$	-1.0	(-17)	[0]	
Pa(IV)/Pa	$Pa^{4+} + 4e = Pa_{(s)}$	-1.7	(-115)	[0]	
Pa(III)/Pa	$Pa^{3+} + 3e = Pa_{(s)}$	-1.95	(-98.9)	[0]	
Pb					
Pb(IV)/Pb(II)	$3\,PbO_{2(s)} + 2H_2O + 4e = Pb_3O_{4(s)} + 4OH^-$	$+0.295$	(19.95)	[0]	[5]
	$PbO_{2(s)} + 4H^+ + 2e = Pb^{2+} + 2H_2O$	$+1.455$	(49.19)	[0]	[5]
	$Pb(IV) + 2e = Pb^{2+}$	$+1.655$	(56.0)	1.1M $HClO_4$	[5]
	$PbO_{2(s)} + 2H^+ + 2e = PbO_{(s)} + H_2O$	$+0.28$	(9.5)	[0]	
	$PbO_{2(s)} + 4H^+ + SO_4^{2-} + 2e = PbSO_{4(s)} + 2H_2O$	$+1.69$	(57.1)	[0]	[5]
	$Pb(OH)_6^{2-} + 2e = Pb(OH)_3^- + 3OH^-$	$+0.30$	(10)	1.7–3.6M NaOH	[5]
	$Pb_3O_{4(s)} + H_2O + 2e = 3\,PbO_{(s)} + 2OH^-$	$+0.2488$	(8.41)	[0]	[5]
Pb(II)/Pb	$Pb^{2+} + 2e = Pb_{(s)}$	-0.126	(-4.3)	[0]	
	$PbO_{(s,red)} + 2H^+ + 2e = Pb_{(s)} + H_2O$	$+0.242$	(8.42)	[0]	[5]

Table 9.1 (Continued)

Element Redox couple	Oxidation–reduction system	Values of potentials	(log K)	[I]	Ref.
Pd					
Pd(VI)/Pd(IV)	$PdO_{3(s)} + 2H^+ + 2e = PdO_{2(s)} + H_2O$	+1.22	(42.3)	[0][i]	
Pd(IV)/Pd(II)	$PdO_{2(s)} + 2H^+ + 2e = PdO_{(s)} + H_2O$	+0.95	(32.9)	[0][i]	
	$PdCl_6^{2-} + 2e = PdCl_4^{2-} + 2Cl^-$	+1.29	(43.5)	[1M HCl]	[5]
Pd(II)/Pd	$Pd^{2+} + 2e = Pd_{(s)}$	+0.92	(31)	[0]	
	$Pd(OH)_{2(s)} + 2e = Pd_{(s)} + 2OH^-$	−0.19	(−6.4)	[0.1M K$_2$SO$_4$]	
	$PdCl_4^{2-} + 2e = Pd_{(s)} + 4Cl^-$	+0.62	(21)	[0]	
	$Pd(CN)_4^{2-} + 2e = Pd_{(s)} + 4CN^-$	−1.53	(−51.7)	[1M KCN]	
Pt					
Pt(IV)/Pt(II)	$PtO_{2(s)} + 2H^+ + 2e = Pt(OH)_{2(s)}$	+1.1	(37)	[0]	[6]
	$PtO_{2(s)} + 2H^+ + 2e = PtO_{(s)} + H_2O$	+1.045	(35.3)	[0]	[6]
	$Pt(OH)_6^{2-} + 2e = Pt(OH)_{2(s)} + 4OH^-$	+0.2	(6.8)	[0]	[5]
	$PtCl_6^{2-} + 2e = PtCl_4^{2-} + 2Cl^-$	+0.73	(24.7)	[0]	
	$Pt(CN)_4Cl_2^{2-} + 2e = Pt(CN)_4^{2-} + 2Cl^-$	+0.79	(26.7)	[0]	
Pt(II)/Pt	$Pt(OH)_{2(s)} + 2H^+ + 2e = Pt_{(s)} + 2H_2O$	+0.89	(30)	[0.1M KCl]	
		+0.98	(33)	[1M KCl]	
	$PtCl_4^{2-} + 2e = Pt_{(s)} + 4Cl^-$	+0.73	(25)	[0]	
Pu					
Pu(VI)/Pu(V)	$PuO_2^{2+} + e = PuO_2^+$	+0.929	(15.7)	[0]	[5]
		+0.9164	(15.49)	[1M HClO$_4$]	[5]
	$Pu(VI) + e = Pu(V)$	+0.91	(15.4)	[1M HCl]	
		+0.92	(15.5)	[1M HClO$_4$]	

[i] 18°C

Table 9.1 (Continued)

Element Redox couple	Oxidation–reduction system	Values of potentials	(log K)	[I]	Ref.
Pu (continued)					
Pu(VI)/Pu(IV)	$Pu(VI) + 2e = Pu(IV)$	+0.925	(31.3)	[0.1M HNO_3]	[5]
		+1.05	(35.5)	[1M HNO_3]	
				[1M HCl]	
	$PuO_2^{2+} + 4H^+ + 2e = Pu^{4+} + 2H_2O$	+1.04	(35.2)	[1M $HClO_4$]	[5]
		+1.040	(35.2)	[0]	[5]
		+1.043	(35.3)	[1M $HClO_4$]	[5]
Pu(IV)/Pu(III)	$Pu^{4+} + e = Pu^{3+}$	+0.967	(16.4)	[0]	
		+0.972	(16.4)	[1M $HClO_4$]	
Pu(III)/Pu	$Pu^{3+} + 3e = Pu_{(s)}$	−2.03	(−103)	[0]	
Ra					
Ra(II)/Ra	$Ra^{2+} + 2e = Ra_{(s)}$	−2.92	(−98.7)	[0]	
Rb					
Rb(I)/Rb	$Rb^+ + e = Rb_{(s)}$	−2.93	(−49.5)	[0]	
	$Rb^+ + (Hg) + e = RbHg_x$	−1.81	(−30.6)	[0]	
Re					
Re(VII)/Re(VI)	$ReO_4^- + 2H^+ + e = ReO_{3(s)} + H_2O$	+0.77	(13)	[0]	
Re(VII)/Re(IV)	$ReO_4^- + 4H^+ + 3e = ReO_{2(s)} + 2H_2O$	+0.51	(26)	[0]^f	
Re(VII)/Re	$ReO_4^- + 8H^+ + 7e = Re_{(s)} + 4H_2O$	+0.37	(44)	[0]^c	
Re(VI)/Re(IV)	$ReO_{3(s)} + 2H^+ + 2e = ReO_{2(s)} + H_2O$	+0.4	(13)	[0]^c	
Re(IV)/Re	$ReO_{2(s)} + 4H^+ + 4e = Re_{(s)} + 2H_2O$	+0.26	(17.6)	[0]	

^{c} Lab. temp. (about 20° C); ^{f} temperature not stated

Table 9.1 (Continued)

Element Redox couple	Oxidation–reduction system	Values of potentials	(log K)	[I]	Ref.
Re(continued)					
Re(III)/Re	$Re^{3+} + 3e = Re_{(s)}$	+0.3	(15)	[0]^c	
Re(I)/Re(−I)	$Re^+ + 2e = Re^-$	−0.23	(−7.8)	[0]	
Re/Re(−I)	$Re_{(s)} + e = Re^-$	−0.14	(−2.4)	[0]^g	
Rh					
Rh(VI)/Rh(IV)	$Rh(VI) + 2e = Rh(IV)$	+1.5	(51)	[0.1M H_2SO_4]^j	
Rh(VI)/Rh(III)	$Rh(VI) + 3e = Rh(III)$	+1.5	(78)	[1M $HClO_4$]^i	
Rh(IV)/Rh(III)	$Rh(IV) + e = Rh(III)$	+1.43	(24.2)	[0.5M H_2SO_4]	
Rh(III)/Rh	$RhCl_6^{3-} + 3e = Rh_{(s)} + 6 Cl^-$	+0.44	(22)	[0]	
Ru					
Ru(VIII)/Ru(VII)	$RuO_4 + e = RuO_4^-$	+1.00	(16.9)	[0]	
Ru(VII)/Ru(VI)	$RuO_4^- + e = RuO_4^{2-}$	+0.59	(10)	[0]	
Ru(VI)/Ru(IV)	$RuO_4^{2-} + 2 H_2O + 2e = RuO_2 \cdot aq_{(s)} + 4 OH^-$	+0.349	(11.8)	[0]	[5]
Ru(IV)/Ru(III)	$Ru(IV) + e = Ru(III)$	+0.908	(15.3)	[0.5M HCl]	[5]
		+0.86	(14.5)	[2M HCl]	
Ru(IV)/Ru	$RuO_{2(s)} + 4 H^+ + 4e = Ru_{(s)} + 2 H_2O$	+0.788	(53.3)	[0]	[5]
Ru(III)/Ru(II)	$Ru(III) + e = Ru(II)$	~0	(0)	[H^+, ClO_4^- or Cl^- or SO_4^{2-}]	
	$Ru^{3+} + e = Ru^{2+}$	+0.2487	(4.20)	[0]	[4]
		+0.18	(3.0)	[1M H_2SO_4]	[4]
S					
S(VI)/S(VI)*	$S_2O_8^{2-} + 2e = 2 SO_4^{2-}$	+2.0	(68)	[0]	

^{i} 18°C; ^{j} 22°C;

* Sulphur remains in oxidation state +VI; the peroxo link is reduced.

Table 9.1 (Continued)

Element Redox couple	Oxidation–reduction system	Values of potentials	(log K)	[I]	Ref.
S (continued)					
S(VI)/S(IV)	$SO_4^{2-} + 4H^+ + 2e = H_2SO_3 + H_2O$	+0.17	(57)	[0]	
S(IV)/S(III)	$2SO_3^{2-} + 2H_2O + 2e = S_2O_4^{2-} + 4OH^-$	−1.12	(−37.9)	[0]	
S(IV)/S(II)	$2H_2SO_3 + 2H^+ + 4e = S_2O_3^{2-} + 3H_2O$	+0.40	(27.0)	[0]	
S(IV)/S	$H_2SO_3 + 4H^+ + 4e = S_{(s)} + 3H_2O$	+0.45	(30.4)	[0]	
S(x)/S(II)	$S_4O_6^{2-} + 2e = 2S_2O_3^{2-}$	+0.09	(3)	[0]	
S(II)/S	$S_2O_3^{2-} + 6H^+ + 4e = 2S_{(s)} + 3H_2O$	+0.5	(34)	[0]	
S(I)/S	$(SCN)_2 + 2e = 2SCN^-$	+0.77	(26)	[0]^f	
S/S(−II)	$S_{(s)} + 2H^+ + 2e = H_2S$	+0.141	(4.7)	[0]	
	$S_{(s)} + 2H^+ + 2e = H_2S_{(g)}$	+0.171	(5.8)	[0]	
	$S_{(s,rhomb)} + 2e = S^{2-}$	−0.48	(−16.5)	[0]^h	[4]
	$nS_{(s)} + 2e = S_n^{2-}$ n = 1	−0.46	(−15.7)	[0]	[4]
	n = 2	−0.49	(−16.6)	[0]	[4]
	n = 3	−0.45	(−15.4)	[0]	[4]
	n = 4	−0.36	(−12.1)	[0]	[4]
	n = 5	−0.34	(−11.5)	[0]	[4]
	n = 6	−0.36	(−12.1)	[0]	[5]
Combined systems	$S_{(s)} + SO_3^{2-} = S_2O_3^{2-}$		(5.8)	[0]	
Sb					
Sb(V)/Sb(III)	$Sb(V) + 2e = Sb(III)$	+0.75	(25)	[3.5M HCl]	
		+0.82	(28)	[6M HCl]	
		−0.43	(−15)	[3M KOH]	
	$\tfrac{1}{2}Sb_2O_{5(s)} + 2H^+ + 2e = \tfrac{1}{2}Sb_2O_{3(s)} + H_2O$	+0.69	(23)	[0]	
	$SbO_3^- + H_2O + 2e = SbO_2^- + 2OH^-$	−0.59	(−20)	[10M KOH]	

^{f} temperature not stated; ^{h} 20° C

Table 9.1 (Continued)

Element Redox couple	Oxidation-reduction system	Values of potentials	(log K)	[I]	Ref.
Sb (continued)					
Sb(III)/Sb	$\frac{1}{2}Sb_2O_{3(s)} + 3H^+ + 3e = Sb_{(s)} + \frac{3}{2}H_2O$	+0.15	(7.6)	[0]	
	$SbO^+ + 2H^+ + 3e = Sb_{(s)} + H_2O$	+0.21	(10.7)	[0]	
Sb/Sb(−III)	$Sb_{(s)} + 3H^+ + 3e = SbH_{3(g)}$	−0.51	(−26)	[0]	
Sc					
Sc(III)/Sc	$Sc^{3+} + 3e = Sc_{(s)}$	−2.1	(−106)	[0]	
Se					
Se(VI)/Se(IV)	$SeO_4^{2-} + 4H^+ + 2e = H_2SeO_3 + H_2O$	+1.15	(38.9)	[0]	
Se(IV)/Se	$H_2SeO_3 + 4H^+ + 4e = Se_{(s)} + 3H_2O$	+0.74	(50)	[0]	
Se/Se(−II)	$Se_{(s)} + 2H^+ + 2e = H_2Se_{(g)}$	−0.37	(−12.5)	[0]	[5]
	$Se_{(s)} + 2H^+ + 2e = H_2Se$	−0.40	(−13.5)	[0]	
Si					
Si(IV)/Si	$SiO_{2(s)} + 4H^+ + 4e = Si_{(s)} + 2H_2O$	−0.86	(−58)	[0]	
	$SiF_6^{2-} + 4e = Si_{(s)} + 6F^-$	−1.2	(−84)	[0]	
Si/Si(−IV)	$Si_{(s)} + 4H^+ + 4e = SiH_{4(g)}$	+0.10	(6.8)	[0]	[5]
Sn					
Sn(IV)/Sn(II)	$Sn(IV) + 2e = Sn(II)$	+0.144	(4.87)	[0.53M HCl]	[5]
		+0.14	(4.7)	[1M HCl]	
		+0.1325	(4.48)	[2M HCl]	[5]
Sn(II)/Sn	$Sn^{2+} + 2e = Sn_{(s)}$	−0.14	(−4.7)	[0]	
Sr					
Sr(II)/Sr	$Sr^{2+} + 2e = Sr_{(s)}$	−2.89	(−97.7)	[0]	

Table 9.1 (Continued)

Element / Redox couple	Oxidation-reduction system	Values of potentials	(log K)	[I]	Ref.
Ta					
Ta(V)/Ta	$\frac{1}{2}Ta_2O_{5(s)} + 5H^+ + 5e = Ta_{(s)} + \frac{5}{2}H_2O$	-0.81	(-68.6)	[0]	
Te					
Te(VI)/Te(IV)	$Te(OH)_{6(s)} + 2H^+ + 2e = TeO_{2(s)} + 4H_2O$	$+1.02$	(34.5)	[0]	
Te(IV)/Te	$TeO_{2(s)} + 4H^+ + 4e = Te_{(s)} + 2H_2O$	$+0.59$	(40)	[0]	
	$TeCl_6^{2-} + 4e = Te_{(s,monocryst)} + 6Cl^-$	$+0.63$	(43)	[dil. HCl]	
Te/Te(−II)	$\frac{1}{2}Te_{2(s)} + 2H^+ + 2e = H_2Te_{(g)}$	-0.50	(-17)	[0]^k	
	$Te_{2(s)} + 2H^+ + 2e = H_2Te_{2(g)}$	-0.36	(-12)	[0]^k	
Th					
Th(IV)/Th	$Th^{4+} + 4e = Th_{(s)}$	-1.90	(-128.5)	[0]	
	$ThO_2 + 4H^+ + 4e = Th + 2H_2O$	-1.789	(-121.0)	[0]	[6]
Ti					
Ti(IV)/Ti(III)	$Ti(IV) + e = Ti(III)$	$+0.130$	(2.2)	[4M H$_2$SO$_4$]	[5]
		$+0.125$	(2.1)	[4M HCl]	[5]
	$TiO^{2+} + 2H^+ + e = Ti^{3+} + H_2O$	$+0.100$	(1.7)	[0]	[6]
	$Ti(OH)^{3+} + H^+ + e = Ti^{3+} + H_2O$	-0.055	(-0.93)	[0]	[6]
	$TiO_2 \cdot aq_{(s)} + 4H^+ + 4e = Ti_{(s)} + 2H_2O$	-0.86	(-58)	[0]	
Ti(III)/Ti(II)	$Ti^{3+} + e = Ti^{2+}$	-0.368	(-6.22)	[0]	[6]
Ti(II)/Ti	$Ti^{2+} + 2e = Ti_{(s)}$	-1.630	(-55.11)	[0]	[6]
Tl					
Tl(III)/Tl(I)	$Tl^{3+} + 2e = Tl^+$	$+1.26$	(42.6)	[0]	[5]
		$+1.28$	(43.3)	[3M HClO$_4$]	

^{k} 30°C

Table 9.1 (Continued)

Element Redox couple	Oxidation–reduction system	Values of potentials	(log K)	[I]	Ref.
Tl (continued)					
Tl(I)/Tl	$\frac{1}{2}Tl_2O_{3(s)} + 2e = Tl^+ + 3\,OH^-$	+0.02	(0.7)	[0]	[5]
	$Tl^+ + e = Tl_{(s)}$	−0.336	(−5.68)	[0]	
	$Tl(OH) + e = Tl_{(s)} + OH^-$	−0.34	(−5.7)	[0]	
	$Tl^+ + (Hg) + e = TlHg_x$	−0.3335	(−5.64)	[0]	[5]
		−0.3572	(−6.04)	[1M HClO$_4$]	[4]
	$TlCl_{(s)} + e = Tl_{(s)} + Cl^-$	−0.557	(−9.42)	[0]	
		−0.546	(−9.23)	[1M HClO$_4$]	
		−0.558	(−9.43)	[2M HClO$_4$]	
		−0.575	(−9.72)	[3M HClO$_4$]	
	$TlCl_{(s)} + (Hg) + e = TlHg_x + Cl^-$	−0.5203	(−8.796)	[0]	[4]
U					
U(VI)/U(V)	$UO_2^{2+} + e = UO_2^+$	+0.06	(1)	[0.1M HCl]	[5]
U(VI)/U(IV)	$UO_2^{2+} + 2e = UO_{2(s)}$	+0.447	(15.1)	[0]	
	$UO_2^{2+} + 4H^+ + 2e = U^{4+} + 2H_2O$	+0.33	(11)	[0]	
U(V)/U(IV)	$UO_2^+ + 4H^+ + e = U^{4+} + 2H_2O$	+0.55	(9.3)	[0]	[5]
U(IV)/U(III)	$U^{4+} + e = U^{3+}$	−0.609	(−10.3)	[0]	[5]
U(III)/U	$U^{3+} + 3e = U_{(s)}$	−1.8	(−91)	[0]	[5]
V					
V(V)/V(IV)	$VO_2^+ + 2H^+ + e = VO^{2+} + H_2O$	+0.999	(16.9)	[0]	
V(V)/V	$VO_2^+ + 4H^+ + 5e = V_{(s)} + 4H_2O$	−0.25	(−21)	[0]	
V(IV)/V(III)	$VO^{2+} + 2H^+ + e = V^{3+} + H_2O$	+0.34	(5.7)	[0]	

Table 9.1 (Continued)

Element Redox couple	Oxidation-reduction system	Values of potentials	(log K)	[I]	Ref.
V (continued)					
V(III)/V(II)	$V(III) + e = V(II)$	−0.27	(−4.7)	$[1M\ HClO_4]^m$	
		−0.28	(−4.9)	$[1M\ HCl]^m$	
	$V^{3+} + e = V^{2+}$	−0.255	(−4.31)	[0]	
V(II)/V	$V^{2+} + 2e = V_{(s)}$	−1.2	(−40.6)	[0]	
W					
W(VI)/W(V)	$WO_{3(s)} + H^+ + e = \frac{1}{2}W_2O_{5(s)} + \frac{1}{2}H_2O$	−0.03	(−0.5)	[0]	
	$W(VI) + e = W(V)$	0.26	(4.4)	$[12M\ HCl]^i$	[5]
W(VI)/W	$WO_{3(s)} + 6H^+ + 6e = W_{(s)} + 3H_2O$	−0.09	(−9)	[0]	
	$WO_4^{2-} + 4H_2O + 6e = W_{(s)} + 8OH^-$	−1.01	(−102)	[0]	
	$H_2WO_4 + 6H^+ + 6e = W_{(s)} + 4H_2O$	−0.03	(−3)	[0]	
W(V)/W(IV)	$W(V) + e = W(IV)$	−0.3	(−5)	$[12M\ HCl]$	
	$\frac{1}{2}W_2O_{5(s)} + H^+ + e = WO_{2(s)} + \frac{1}{2}H_2O$	−0.04	(−0.7)	[0]	
W(V)/W(III)	$W(V) + 2e = W(III)$ (colourless)	−0.1	(−3.4)	$[12M\ HCl]$	$[5]^a$
	$W(V) + 2e = W(III)$ (red)	−0.31	(−10.5)	$[12M\ HCl]$	[5]
W(IV)/W	$WO_{2(s)} + 4H^+ + 4e = W_{(s)} + 2H_2O$	−0.12	(−8.1)	[0]	[5]
Xe					
Xe(VIII)/Xe(VI)	$H_4XeO_6 + 2H^+ + 2e = XeO_3 + 3H_2O$	∼ +3.0	(100)	[0]	
Xe(VI)/Xe	$XeO_3 + 6H^+ + 6e = Xe_{(g)} + 3H_2O$	∼ +1.8	(180)	[0]	
Y					
Y(III)/Y	$Y^{3+} + 3e = Y_{(s)}$	−2.37	(−120)	[0]	

^{a} Calculated from the value given in ref. [5]. $-\log L_w = 14.0$; ^{i} 18° C; ^{m} 17° C

Table 9.1 (Continued)

Element Redox couple	Oxidation–reduction system	Values of potentials	(log K)	[I]	Ref.
Zn					
Zn(II)/Zn	$Zn^{2+} + 2e = Zn_{(s)}$	-0.7628	(-25.78)	[0]	
Zr					
Zr(IV)/Zr	$Zr^{4+} + 4e = Zr_{(s)}$	-1.392	(-94.1)	[0]	
	$ZrO_{2(s)} + 4 H^+ = Zr_{(s)} + 2 H_2O$	-1.43	(-96.0)	[0]	[5][a]

[a] Calculated from the value given in ref. [5]; $-\log K_w = 14.00$

REFERENCES

[1] G. Charlot, A. Collumeau and M. J. C. Marchon, *Selected Constants. Oxidation-Reduction Potentials of Inorganic Substances in Aqueous Solution*, Butterworths, London, 1971.
[2] T. P. Dirkse, *J. Electrochem. Soc.*, 1962, **109**, 173.
[3] Dwyer F. P. and D. P. Mellor, *Chelating Agents and Metal Chelates*, p. 267, Academic Press, New York, 1964.
[4] L. G. Sillén and A. E. Martell, *Stability Constants of Metal-Ion Complexes*, Supplement No. 1, The Chemical Society, London, 1971.
[5] L. G. Sillén and A. E. Martell, *Stability Constants of Metal-Ion Complexes*, The Chemical Society, London, 1964.
[6] G. Milazzo and S. Caroli, *Tables of Standard Electrode Potentials*, Wiley, New York, 1978.

Table 9.2
Oxidation–reduction potentials (V) of organic systems*

Formula of the oxidized form	Oxidation–reduction system	Values† of potentials	(log K)‡ [medium]	Ref.§
C	$C + 4H^+ + 4e = CH_4$	+0.1316	(−8.8) [0]	
CO_2	$CO_{2(g)} + 2H^+ + 2e = HCOOH$	−0.20	(−6.8) [0]	[2]
	$2CO_{2(g)} + 2H^+ + 2e = H_2C_2O_4$	−0.49	(−16.6) [0]	[2]
CH_2O	$HCHO + 2H^+ + 2e = CH_3OH$	+0.232	(7.84) [0]	
CH_2O_2	$HCOOH + 2H^+ + 2e = HCHO + H_2O$	+0.056	(1.9) [0]	
	$HCOO^- + 3H^+ + 2e = HCHO + H_2O$	+0.167	(5.65) [0]	
CH_2O_3	$2H_2CO_3 + 2H^+ + 2e = H_2C_2O_4 + 2H_2O$	−0.386	(−13.0) [0]	
$C_2H_2O_4$	$H_2C_2O_4 + 2H^+ + 2e = 2HCOOH$	+0.074	(2.5) [0]	
C_2H_4O	$CH_3CHO + 2H^+ + 2e = C_2H_5OH$	+0.19	(6.4) [0]	[3]
$C_2H_4O_2$	$CH_3COOH + 2H^+ + 2e = CH_3CHO + H_2O$	−0.13	(−4.4) [0]	[4]

* Values for aqueous solutions at 25°C are given unless otherwise stated in footnotes.
† Values of the standard redox potentials ($I = 0$) and/or the conditional redox potentials ($I \neq 0$, medium specified); values of log K of the redox half-reaction (always given in parentheses) are valid under the same conditions as the potentials.
§ If no reference to the literature is given, the values of potentials and/or log K originate from the monograph by Milazzo and Caroli [1].

Table 9.2 (Continued)

Formula of the oxidized form	Oxidation-reduction system	Values of potentials	(log K) [medium]	Ref.

$C_3H_4O_3$	$CH_3COCOOH + H^+ + e = CH_3C.COOH$
	          $\|$
	         OH
	pyruvic acid → lactic acid
		−0.175	(−2.96)
			[pH = 7]
				[5]

$C_6H_4O_2$	$C_6H_4O_2 + 2H^+ + 2e = C_6H_4(OH)_2$ p-quinone → hydroquinone	+0.6997	(+23.66) [0]	[6]
		+0.6902	(+23.34) [1M NaClO$_4$]	[6]
		+0.6846	(+23.15) [2M NaClO$_4$]	[6]
		+0.6778	(+22.91) [3M NaClO$_4$]	[6]
$C_6H_6O_6$	$C_6H_6O_6 + 2e = C_6H_6O_6^{2-}$ dehydroascorbic acid → ascorbic acid	−0.07	(−2.4) [0]	[3]
$C_6H_{12}N_2O_4S_2$	$C_6H_{12}N_2O_4S_2 + 2H^+ + 2e = 2C_3H_7NO_2S$ cystine → cysteine	−0.33	(−11) [pH = 7]	[5]
$C_{12}H_7Cl_2NO_2$	$C_{12}H_7Cl_2NO_2 + 2H^+ + 2e = C_{12}H_9Cl_2NO_2$ 2,6-dichlorophenolindophenol → leuco-base	+0.33	(11) [0]	[3]
$C_{16}H_{18}N_3S^+$	$C_{16}H_{18}N_3S^+ + H^+ + 2e = C_{16}H_{19}N_3S$ Methylene Blue → leuco-form	+0.22	(7.4) [0]	[3]
—	ferricinium + e = ferrocene	+0.4	(6.8) [0]	
		+0.1	(1.7) [0.1]	
—	Fe(III)-protoporphyrin + e + H_2O = Fe(II)-protoporphyrin + OH^-	−0.15	(−2.5) [pH = 7]	[7]

REFERENCES

[1] G. Milazzo and S. Caroli, *Tables of Standard Electrode Potentials*, Wiley, New York, 1978.
[2] G. Charlot, A. Collumeau and M. J. C. Marchon, *Selected Constants. Oxidation-Reduction Potentials of Inorganic Substances in Aqueous Solution*, Butterworths, London, 1971.
[3] J. Inczédy, *Analytical Application of Complex Equilibria*, p. 378, Horwood, Chichester, 1976.
[4] D. Dobos, *Electrochemical Data*, Akadémiai Kiadó, Budapest, 1975.
[5] O. Tomíček, *Chemické indikátory*, p. 152, Jednota českých matematiků a fysiků, Prague, 1946; *Chemical Indicators*, Butterworths, London, 1951.
[6] L. G. Sillén and A. E. Martell, *Stability Constants of Metal-Ion Complexes*, Supplement No. 1, The Chemical Society, London, 1971.
[7] F. P. Dwyer and D. P. Mellor, *Chelating Agents and Metal Chelates*, p. 469, Academic Press, New York, 1964.

CHAPTER 10

Side-reaction coefficients for protonation equilibria

Protonation of ligands is a common side-reaction which affects complexation equilibria in aqueous solution. The extent of a protonation side-reaction depends on the pH of the solution and on the basicity of the ligand, which is characterized by the protonation constants [cf. Section 2.4.1, Eqs. (2.32–2.35)]. The protonation coefficient $\alpha_{L(H)}$ (more simply denoted as α_H) represents a convenient auxiliary function expressing the effect of protonation of a ligand. Rewriting Eq. (3.14), p. 48, for the protonation equilibrium $L \to HL \to H_2L \dots$ gives the expression

$$\alpha_{L(H)} = 1 + \beta_{H1}[H^+] + \beta_{H2}[H^+]^2 + \dots \qquad (10.1)$$

where the β_{Hk} values ($k = 1, 2, 3, \dots$) are the overall protonation constants [cf. Eq. (2.35), p. 32] for consecutive species in a polyprotic system of the ligand L. Some ligands are co-ordinated to the central ion even in a partially protonated form, to give protonated complexes. The protonation coefficient with respect to a protonated ligand species can be expressed, for example, as follows:

$$\alpha_{HL(H)} = \frac{[L']}{[HL]} = \alpha_{L(H)}/\beta_{H1}[H^+]$$
$$= 1/K_{H1}[H^+] + 1 + K_{H2}[H^+] + K_{H2}K_{H3}[H^+]^2 + \dots \qquad (10.2)$$

or

$$\alpha_{H_2L(H)} = \frac{[L']}{[H_2L]} = \alpha_{L(H)}/\beta_{H2}[H^+]^2$$
$$= 1/K_{H1}K_{H2}[H^+]^2 + 1/K_{H2}[H^+] + 1 + K_{H3}[H^+]$$
$$+ K_{H3}K_{H4}[H^+]^2 + \dots \qquad (10.3)$$

A known value of the protonation coefficient for a certain side-reaction of a complex-forming reagent at a given pH of the solution allows easy evaluation of the conditional equilibrium constant [cf. Eqs. (3.20) and

(3.21), p. 50] by which the complex is characterized for a particular application (masking, complexometric titration, photometric determination, extraction procedure, etc.). It may also be used for calculation of the effect of protonation of an anionic precipitant on solubility [see Eq. (3.22), p. 52].

Tables 10.1 and 10.2 should make such calculations easy, since they list the logarithms of the protonation coefficients of some 32 inorganic and 200 organic ligands which are of interest in applications of complexation equilibria in aqueous solutions in both chemistry and biochemistry. As the number of columns in a table is necessarily limited, the protonation coefficients are calculated for all integral pH values between 2 and 12, and at interval of 0.5 within the main pH region from 3 to 11. The values of $\log \alpha_H$ for pH < 2 or > 12 can sometimes be obtained by extrapolation. The protonation coefficient can also be read conveniently from a diagram of $\log \alpha_H$ vs. pH. A collection of such curves is available in the book by Kragten [1].

The values of $\log \alpha_H$ were calculated by computer on the basis of the protonation constants given in Tables 5.1 and 5.2 for ionic strength $I = 0.1$ and a temperature of $25°C$. Exceptions are noted in the list of organic ligands. The arrangement of inorganic ligands follows the sequence of the central atoms in subgroups of the periodic system. Organic ligands are arranged according to the empirical formula of the electrically neutral form. Incompletely protonated forms are marked with asterisks, the number of which gives the number of protons bound to the ligand in a complex. As a key to Table 10.2 there is a numerical list of the names (and abbreviations) of the compounds, and an alphabetical list. The formula H_nL denotes the composition corresponding to the empirical formula given. As mentioned above, a different definition of the protonation coefficient is taken to represent further protonation of the ligand species to give HL, H_2L, etc. The number of hydrogen ions in a resulting hydrogen complex is marked with asterisks. The symbol α_H generally refers to values calculated for the ligand species L.

REFERENCES

[1] J. Kragten, *Atlas of Metal-Ligand Equilibria in Aqueous Solution*, Horwood, Chichester, 1978.

Table 10.1

List of inorganic ligands given in the table (some synonyms are in parentheses)

1. cyanide, CN^-;
2. thiocyanate, SCN^-;
3. carbonate, CO_3^{2-};
4. hydrogen carbonate, HCO_3^-;
5. azide, N_3^-;
6. ammonia, NH_3;
7. nitrite, NO_2^-;
8. phosphate(3−), PO_4^{3-};
9. hydrogen phosphate(2−), HPO_4^{2-};
10. dihydrogen phosphate(1−), $H_2PO_4^-$;
11. diphosphate(4−), $P_2O_7^{4-}$;
12. hydrogen diphosphate(3−), $HP_2O_7^{3-}$;
13. dihydrogen diphosphate(2−), $H_2P_2O_7^{2-}$;
14. triphosphate(5−), $P_3O_{10}^{5-}$;
15. hydrogen triphosphate(4−), $HP_3O_{10}^{4-}$;
16. tetraphosphate(6−), $P_4O_{13}^{6-}$;
17. hydrogen tetraphosphate(5−), $HP_4O_{13}^{5-}$;
18. cyclo-triphosphate(3−) (trimetaphosphate), $P_3O_9^{3-}$;
19. cyclo-tetraphosphate(4−) (tetrametaphosphate), $P_4O_{12}^{4-}$;
20. arsenate(3−), AsO_4^{3-};
21. tetrahydroxoarsenite(1−), $As(OH)_4^-$;
22. hydrogen peroxide (anion), HO_2^-;
23. sulphide, S^{2-};
24. hydrosulphide, HS^-;
25. sulphite(2−), SO_3^{2-};
26. hydrogen sulphite, HSO_3^-;
27. sulphate(2−), SO_4^{2-};
28. thiosulphate(2−), $S_2O_3^{2-}$;
29. chromate(2−), CrO_4^{2-};
30. molybdate(2−), MoO_4^{2-};
31. tungstate(2−), WO_4^{2-};
32. fluoride, F^-.

Table 10.1
Values of side-reaction coefficients $\alpha_{L(H)}$ for inorganic ligands

Ligand	pH 2.0	3.0	3.5	4.0	4.5	5.0	5.5	6.0
1. CN^-	7.01	6.01	5.51	5.01	4.51	4.01	3.51	3.01
2. SCN^-	0.03	0.00	0.00	0.00	0.00	0.00	0.00	0.00
3. CO_3^{2-}	12.16	10.16	9.16	8.16	7.17	6.19	5.25	4.39
4. HCO_3^-	4.16	3.16	2.66	2.16	1.67	1.19	0.75	0.39
5. N_3^-	2.45	1.47	1.00	0.58	0.28	0.11	0.04	0.01
6. NH_3	7.29	6.29	5.79	5.29	4.79	4.29	3.79	3.29
7. NO_2^-	1.04	0.30	0.12	0.04	0.01	0.00	0.00	0.00
8. PO_4^{3-}	13.76	11.50	10.48	9.47	8.49	7.54	6.66	5.92
9. HPO_4^{2-}	4.02	2.76	2.24	1.73	1.25	0.80	0.42	0.18
10. $H_2PO_4^-$	0.30	0.04	0.02	0.01	0.03	0.08	0.20	0.46
11. $P_2O_7^{4-}$	10.72	8.44	7.42	6.42	5.42	4.45	3.52	2.69
12. $HP_2O_7^{3-}$	4.26	3.07	2.55	2.05	1.55	1.08	0.65	0.32
13. $H_2P_2O_7^{2-}$	0.22	0.03	0.01	0.01	0.01	0.04	0.11	0.28
14. $P_3O_{10}^{5-}$	10.20	7.65	6.56	5.53	4.55	3.62	2.80	2.12
15. $HP_3O_{10}^{4-}$	4.20	2.65	2.06	1.53	1.05	0.62	0.30	0.12
16. $P_4O_{13}^{6-}$	11.46	9.04	7.99	6.98	5.98	4.98	4.00	3.06
17. $HP_4O_{13}^{5-}$	5.12	3.70	3.15	2.64	2.14	1.64	1.16	0.72
18. $P_3O_9^{3-}$	0.09	0.01	0.00	0.00	0.00	0.00	0.00	0.00
19. $P_4O_{12}^{4-}$	0.13	0.01	0.00	0.00	0.00	0.00	0.00	0.00
20. AsO_4^{3-}	14.90	12.53	11.48	10.47	9.46	8.47	7.48	6.51
21. $As(OH)_4^-$	7.13	6.13	5.63	5.13	4.63	4.13	3.63	3.13
22. HO_2^-	9.65	8.65	8.15	7.65	7.15	6.65	6.15	5.65
23. S^{2-}	16.41	14.41	13.41	12.41	11.41	10.42	9.44	8.51
24. HS^-	4.61	3.61	3.11	2.61	2.11	1.62	1.14	0.71
25. SO_3^{2-}	4.43	3.35	2.84	2.34	1.85	1.36	0.90	0.50
26. HSO_3^-	0.09	0.01	0.00	0.00	0.01	0.02	0.06	0.16
27. SO_4^{2-}	0.13	0.02	0.00	0.00	0.00	0.00	0.00	0.00
28. $S_2O_3^{2-}$	0.15	0.02	0.01	0.00	0.00	0.00	0.00	0.00
29. CrO_4^{2-}	3.74	2.74	2.24	1.75	1.26	0.81	0.44	0.19
30. MoO_4^{2-}	3.21	1.31	0.57	0.18	0.05	0/02	0.00	0.00
31. WO_4^{2-}	4.10	2.11	1.16	0.41	0.09	0.02	0.00	0.00
32. F^-	0.97	0.26	0.10	0.03	0.01	0.00	0.00	0.00

Side-Reaction Coefficients for Protonation Equilibria

$\log \alpha_{L(H)}$

pH 6.5	7.0	7.5	8.0	8.5	9.0	9.5	10.0	10.5	11.0	12.0
2.51	2.01	1.52	1.05	0.63	0.31	0.12	0.04	0.01	0.00	0.00
0.00	0.00	0.00	0.00	0.00	0.00	0.00	0.00	0.00	0.00	0.00
3.66	3.06	2.52	2.01	1.52	1.04	0.62	0.30	0.12	0.04	0.00
0.16	0.06	0.02	0.01	0.02	0.04	0.12	0.30	0.62	1.04	2.00
0.00	0.00	0.00	0.00	0.00	0.00	0.00	0.00	0.00	0.00	0.00
2.79	2.29	1.80	1.31	0.86	0.47	0.21	0.08	0.03	0.01	0.00
0.00	0.00	0.00	0.00	0.00	0.00	0.00	0.00	0.00	0.00	0.00
5.31	4.76	4.25	3.74	3.24	2.74	2.24	1.75	1.26	0.81	0.19
0.07	0.02	0.01	0.00	0.00	0.00	0.00	0.01	0.02	0.07	0.45
0.85	1.30	1.79	2.28	2.78	3.28	3.78	4.29	4.80	5.35	6.73
2.00	1.43	0.94	0.53	0.24	0.09	0.03	0.01	0.00	0.00	0.00
0.13	0.06	0.07	0.16	0.37	0.72	1.16	1.64	2.13	2.63	3.63
0.59	1.02	1.53	2.12	2.83	3.68	4.62	5.60	6.59	7.59	9.59
1.55	1.05	0.62	0.30	0.12	0.04	0.01	0.00	0.00	0.00	0.00
0.05	0.05	0.12	0.30	0.62	1.04	1.51	2.00	2.50	3.00	4.00
2.21	1.51	0.95	0.52	0.23	0.09	0.03	0.01	0.00	0.00	0.00
0.37	0.17	0.11	0.18	0.39	0.75	1.19	1.67	2.16	2.66	3.66
0.00	0.00	0.00	0.00	0.00	0.00	0.00	0.00	0.00	0.00	0.00
0.00	0.00	0.00	0.00	0.00	0.00	0.00	0.00	0.00	0.00	0.00
5.59	4.78	4.11	3.54	3.01	2.51	2.01	1.51	1.04	0.62	0.12
2.63	2.13	1.64	1.16	0.72	0.37	0.15	0.05	0.02	0.01	0.00
5.15	4.65	4.15	3.65	3.15	2.65	2.15	1.66	1.18	0.74	0.16
7.66	6.95	6.35	5.82	5.31	4.80	4.30	3.80	3.30	2.80	1.81
0.36	0.15	0.05	0.02	0.01	0.00	0.00	0.00	0.00	0.00	0.01
0.23	0.09	0.03	0.01	0.00	0.00	0.00	0.00	0.00	0.00	0.00
0.39	0.75	1.19	1.67	2.16	2.66	3.16	3.66	4.16	4.66	5.66
0.00	0.00	0.00	0.00	0.00	0.00	0.00	0.00	0.00	0.00	0.00
0.00	0.00	0.00	0.00	0.00	0.00	0.00	0.00	0.00	0.00	0.00
0.07	0.02	0.01	0.00	0.00	0.00	0.00	0.00	0.00	0.00	0.00
0.00	0.00	0.00	0.00	0.00	0.00	0.00	0.00	0.00	0.00	0.00
0.00	0.00	0.00	0.00	0.00	0.00	0.00	0.00	0.00	0.00	0.00
0.00	0.00	0.00	0.00	0.00	0.00	0.00	0.00	0.00	0.00	0.00

Table 10.2
Values of side-reaction coefficients $\alpha_{L(H)}$ for organic ligands

Ligand		pH 2.0	3.0	3.5	4.0	4.5	5.0	5.5
1. CH_2O_2	HL	1.56	0.66	0.33	0.13	0.05	0.02	0.00
2. $C_2H_2O_3$	HL	1.21	0.40	0.17	0.06	0.02	0.01	0.00
3. $C_2H_2O_4$	H_2L	1.87	0.89	0.49	0.22	0.08	0.03	0.01
4. $C_2H_3O_2Cl$	HL	0.76	0.17	0.06	0.02	0.01	0.00	0.00
5. $C_2H_4O_2$	HL	2.56	1.57	1.10	0.67	0.33	0.13	0.05
6. $C_2H_4O_2S$	H_2L	9.56	7.68	6.88	6.21	5.65	5.12	4.61
7. $C_2H_4O_3$	HL	1.64	0.72	0.37	0.15	0.05	0.02	0.01
8. $C_2H_5NO_2$	HL	8.09	6.66	6.10	5.58	5.07	4.57	4.07
9. $C_2H_6N_2O$	HL	5.93	4.93	4.43	3.93	3.43	2.93	2.43
10. C_2H_7NO	L	7.52	6.52	6.02	5.52	5.02	4.52	4.02
11. C_2H_7NS	HL	14.92	12.92	11.92	10.92	9.92	8.92	7.92
12. $C_2H_8N_2$	L	12.97	10.97	9.97	8.97	7.97	6.97	5.98
13. $C_3H_4N_2$	HL	17.43	15.43	14.43	13.43	12.43	11.43	10.44
14. $C_3H_4O_3$	HL	0.45	0.07	0.02	0.01	0.00	0.00	0.00
15. $C_3H_4O_4$	H_2L	4.02	2.44	1.84	1.32	0.85	0.46	0.21
16. $C_3H_6O_2$	HL	2.67	1.68	1.20	0.75	0.39	0.17	0.06
17. $C_3H_6O_3$	HL	1.67	0.75	0.39	0.16	0.06	0.02	0.01
18. $C_3H_7NO_2$	H_2L	8.17	6.77	6.22	5.70	5.19	4.69	4.19
19. $C_3H_7NO_2$	H_2L	9.64	7.74	6.92	6.23	5.64	5.11	4.60
20. $C_3H_7NO_2$	HL	8.40	7.05	6.51	6.00	5.49	4.99	4.49
21. $C_3H_7NO_2S$	H_2L	14.69	12.47	11.45	10.44	9.44	8.44	7.44
22. $C_3H_7NO_3$	L	7.43	6.11	5.58	5.07	4.56	4.06	3.56
23. $C_3H_7NO_3$	HL	7.13	6.13	5.63	5.13	4.63	4.13	3.63
24. $C_3H_8OS_2$	H_2L	15.26	13.26	12.26	11.26	10.26	9.26	8.26
25. $C_3H_8NO_6P$	H_3L	11.70	9.41	8.38	7.37	6.39	5.45	4.59
26. $C_3H_{10}N_2$	L	12.63	10.63	9.63	8.63	7.63	6.64	5.65
27. $C_3H_{10}N_2$	L	15.26	13.26	12.26	11.26	10.26	9.26	8.26
28. $C_3H_{10}N_2O$	L	13.56	11.56	10.56	9.56	8.56	7.56	6.56
29. $C_3H_{11}N_3$	L	15.27	12.34	10.96	9.72	8.61	7.56	6.55
30. $C_4H_4O_4$	H_2L	3.01	1.35	0.77	0.37	0.15	0.05	0.02
31. $C_4H_4O_4$	H_2L	4.02	2.85	2.34	1.84	1.35	0.89	0.50
32. $C_4H_6O_4$	H_2L	5.24	3.28	2.36	1.55	0.92	0.46	0.20
33. $C_4H_6O_5$	H_2L	3.97	2.15	1.42	0.85	0.43	0.18	0.07
34. $C_4H_6O_6$	H_2L	2.83	1.20	0.64	0.29	0.11	0.04	0.01
35. $C_4H_7NO_3$	H_2L	7.10	5.78	5.24	4.73	4.22	3.72	3.22
36. $C_4H_7NO_4$	H_2L	8.23	6.50	5.89	5.36	4.85	4.34	3.84
37. $C_4H_7NO_4$	H_2L	9.60	7.44	6.55	5.81	5.19	4.65	4.14
38. $C_4H_8N_2O_2$	H_2L	18.35	16.35	15.35	14.35	13.35	12.35	11.35

$\log \alpha_{L(H)}$

pH 6.0	6.5	7.0	7.5	8.0	8.5	9.0	9.5	10.0	10.5	11.0	12.0
0.00	0.00	0.00	0.00	0.00	0.00	0.00	0.00	0.00	0.00	0.00	0.00
0.00	0.00	0.00	0.00	0.00	0.00	0.00	0.00	0.00	0.00	0.00	0.00
0.00	0.00	0.00	0.00	0.00	0.00	0.00	0.00	0.00	0.00	0.00	0.00
0.00	0.00	0.00	0.00	0.00	0.00	0.00	0.00	0.00	0.00	0.00	0.00
0.02	0.00	0.00	0.00	0.00	0.00	0.00	0.00	0.00	0.00	0.00	0.00
4.11	3.61	3.11	2.61	2.11	1.62	1.14	0.71	0.36	0.15	0.05	0.01
0.00	0.00	0.00	0.00	0.00	0.00	0.00	0.00	0.00	0.00	0.00	0.00
3.57	3.07	2.57	2.07	1.58	1.11	0.67	0.34	0.14	0.05	0.02	0.00
1.94	1.45	0.98	0.57	0.27	0.10	0.04	0.01	0.00	0.00	0.00	0.00
3.52	3.02	2.52	2.02	1.53	1.06	0.63	0.31	0.12	0.04	0.01	0.00
6.92	5.93	4.95	4.00	3.13	2.39	1.78	1.26	0.79	0.42	0.18	0.02
5.00	4.07	3.23	2.53	1.94	1.42	0.95	0.54	0.25	0.10	0.03	0.00
9.47	8.55	7.74	7.06	6.48	5.95	5.44	4.94	4.44	3.94	3.44	2.44
0.00	0.00	0.00	0.00	0.00	0.00	0.00	0.00	0.00	0.00	0.00	0.00
0.08	0.03	0.01	0.00	0.00	0.00	0.00	0.00	0.00	0.00	0.00	0.00
0.02	0.01	0.00	0.00	0.00	0.00	0.00	0.00	0.00	0.00	0.00	0.00
0.00	0.00	0.00	0.00	0.00	0.00	0.00	0.00	0.00	0.00	0.00	0.00
3.69	3.19	2.69	2.19	1.70	1.22	0.77	0.41	0.17	0.06	0.02	0.00
4.10	3.60	3.10	2.60	2.10	1.61	1.13	0.70	0.35	0.15	0.05	0.01
3.99	3.49	2.99	2.49	1.99	1.50	1.03	0.61	0.30	0.12	0.04	0.00
6.44	5.45	4.47	3.53	2.67	1.96	1.37	0.87	0.47	0.21	0.08	0.01
3.06	2.56	2.06	1.57	1.10	0.67	0.33	0.13	0.05	0.02	0.00	0.00
3.13	2.63	2.13	1.64	1.16	0.72	0.37	0.15	0.05	0.02	0.01	0.00
7.26	6.26	5.27	4.29	3.36	2.52	1.83	1.25	0.78	0.40	0.17	0.02
3.87	3.27	2.73	2.22	1.72	1.24	0.79	0.42	0.18	0.07	0.02	0.00
4.69	3.79	3.01	2.37	1.82	1.31	0.85	0.46	0.20	0.08	0.03	0.00
7.26	6.26	5.27	4.28	3.33	2.46	1.72	1.12	0.65	0.32	0.12	0.01
5.56	4.57	3.60	2.69	1.88	1.22	0.71	0.35	0.14	0.05	0.02	0.00
5.55	4.56	3.59	2.67	1.87	1.22	0.72	0.36	0.14	0.05	0.02	0.00
0.01	0.00	0.00	0.00	0.00	0.00	0.00	0.00	0.00	0.00	0.00	0.00
0.22	0.08	0.03	0.01	0.00	0.00	0.00	0.00	0.00	0.00	0.00	0.00
0.07	0.02	0.01	0.00	0.00	0.00	0.00	0.00	0.00	0.00	0.00	0.00
0.02	0.01	0.00	0.00	0.00	0.00	0.00	0.00	0.00	0.00	0.00	0.00
0.00	0.00	0.00	0.00	0.00	0.00	0.00	0.00	0.00	0.00	0.00	0.00
2.72	2.22	1.73	1.25	0.80	0.42	0.18	0.07	0.02	0.01	0.00	0.00
3.34	2.84	2.34	1.85	1.36	0.90	0.50	0.23	0.09	0.03	0.01	0.00
3.63	3.13	2.63	2.13	1.64	1.16	0.72	0.37	0.15	0.05	0.02	0.00
10.35	9.35	8.35	7.35	6.35	5.35	4.37	3.40	2.48	1.69	1.05	0.26

Table 10.2 (Continued)

Ligand		pH 2.0	3.0	3.5	4.0	4.5	5.0	5.5
39. $C_4H_8O_2$	HL	2.63	1.64	1.16	0.72	0.37	0.15	0.05
40. $C_4H_8O_2$	HL	2.63	1.64	1.16	0.72	0.37	0.15	0.05
41. $C_4H_8N_2O_3$	HL	7.23	5.44	4.72	4.12	3.59	3.08	2.57
42. $C_4H_9NO_3$	HL	7.74	6.35	5.80	5.29	4.78	4.28	3.78
43. $C_4H_9NO_3$	HL	8.21	6.39	5.65	5.04	4.49	3.98	3.47
44. $C_4H_{10}N_2$	L	11.30	9.30	8.30	7.31	6.33	5.40	4.56
45. $C_4H_{11}NO_3$	L	6.09	5.09	4.59	4.09	3.59	3.09	2.59
46. $C_4H_{13}NO_6P_2$	H_4L	17.04	14.05	12.57	11.12	9.76	8.55	7.53
47. $C_4H_{13}N_3$	L	17.09	14.11	12.66	11.29	10.05	8.93	7.88
48. $C_5H_4N_4$	HL	7.21	5.76	5.19	4.67	4.16	3.66	3.16
49. C_5H_5N	L	3.24	2.24	1.75	1.26	0.81	0.44	0.19
50. $C_5H_5N_5$	HL	9.74	7.78	6.84	6.01	5.31	4.72	4.19
51. $C_5H_6N_2$	L	4.06	3.06	2.56	2.06	1.57	1.10	0.67
52. $C_5H_8O_2$	HL	6.80	5.80	5.30	4.80	4.30	3.80	3.30
53. $C_5H_8O_4$	H_2L	5.16	3.19	2.25	1.42	0.77	0.35	0.13
54. $C_5H_9NO_2$	HL	8.63	7.41	6.89	6.38	5.88	5.38	4.88
55. $C_5H_9NO_3$	HL	8.32	6.67	6.04	5.49	4.97	4.46	3.96
56. $C_5H_9NO_4$	H_2L	9.79	7.82	6.87	6.00	5.27	4.65	4.11
57. $C_5H_9NO_4$	H_2L	7.93	6.61	6.08	5.57	5.06	4.56	4.06
58. $C_5H_9N_3$	L	11.90	9.90	8.90	7.90	6.91	5.94	5.00
59. $C_5H_{10}N_2O_3$	HL	7.40	6.07	5.53	5.02	4.51	4.01	3.51
60. $C_5H_{11}N$	L	9.01	8.01	7.51	7.01	6.51	6.01	5.51
61. $C_5H_{11}NO_2$	HL	7.94	6.56	6.01	5.50	4.99	4.49	3.99
62. $C_5H_{11}NO_2$	HL	8.13	6.73	6.18	5.66	5.15	4.65	4.15
63. $C_5H_{11}NS_2$	HL	1.33	0.48	0.22	0.08	0.03	0.01	0.00
64. $C_6H_2O_4Cl_2$	H_2L	0.71	0.15	0.05	0.02	0.01	0.00	0.00
65. $C_6H_2O_6$	H_2L	3.05	1.19	0.52	0.17	0.05	0.02	0.01
66. $C_6H_5NO_2$	HL	3.25	2.22	1.72	1.24	0.79	0.42	0.18
67. $C_6H_5NO_2$	HL	3.05	1.76	1.24	0.78	0.41	0.18	0.06
68. C_6H_6O	HL	7.82	6.82	6.32	5.82	5.32	4.82	4.32
69. $C_6H_6O_2$	H_2L	18.23	16.23	15.23	14.23	13.23	12.23	11.23
70. $C_6H_6O_2$	H_2L	15.36	13.36	12.36	11.36	10.36	9.36	8.36
71. $C_6H_6O_4$	HL	5.66	4.66	4.16	3.66	3.16	2.66	2.16
72. $C_6H_6N_2O_2$	HL	2.16	1.19	0.75	0.39	0.16	0.06	0.02
73. C_6H_6S	HL	4.46	3.46	2.96	2.46	1.96	1.47	1.01
74. $C_6H_6O_2S_2$	H_4L	16.11	14.11	13.11	12.11	11.11	10.11	9.11
75. $C_6H_6O_3$	H_3L	28.02	25.02	23.52	22.02	20.52	19.02	17.52
76. $C_6H_7O_3As$	H_2L	7.66	5.79	5.00	4.35	3.78	3.26	2.75

Ch. 10] Side-Reaction Coefficients for Protonation Equilibria

$\log \alpha_{L(H)}$

pH 6.0	6.5	7.0	7.5	8.0	8.5	9.0	9.5	10.0	10.5	11.0	12.0
0.02	0.01	0.00	0.00	0.00	0.00	0.00	0.00	0.00	0.00	0.00	0.00
0.02	0.01	0.00	0.00	0.00	0.00	0.00	0.00	0.00	0.00	0.00	0.00
2.07	1.58	1.11	0.67	0.34	0.14	0.05	0.02	0.01	0.00	0.00	0.00
3.28	2.78	2.28	1.79	1.30	0.85	0.46	0.20	0.08	0.03	0.01	0.00
2.97	2.47	1.97	1.48	1.01	0.60	0.29	0.11	0.04	0.01	0.00	0.00
3.85	3.26	2.73	2.22	1.72	1.24	0.79	0.42	0.18	0.07	0.02	0.00
2.09	1.60	1.12	0.69	0.35	0.14	0.05	0.02	0.01	0.00	0.00	0.00
6.69	6.02	5.45	4.93	4.42	3.92	3.42	2.92	2.42	1.93	1.44	0.56
6.87	5.86	4.86	3.87	2.90	1.98	1.18	0.59	0.25	0.09	0.03	0.00
2.66	2.16	1.67	1.19	0.75	0.39	0.16	0.06	0.02	0.01	0.00	0.00
0.07	0.02	0.01	0.00	0.00	0.00	0.00	0.00	0.00	0.00	0.00	0.00
3.68	3.17	2.67	2.17	1.68	1.20	0.75	0.39	0.17	0.06	0.02	0.00
0.33	0.13	0.05	0.02	0.00	0.00	0.00	0.00	0.00	0.00	0.00	0.00
2.80	2.30	1.81	1.32	0.86	0.48	0.21	0.08	0.03	0.01	0.00	0.00
0.04	0.01	0.00	0.00	0.00	0.00	0.00	0.00	0.00	0.00	0.00	0.00
4.38	3.88	3.38	2.88	2.38	1.89	1.40	0.93	0.53	0.25	0.09	0.01
3.46	2.96	2.46	1.96	1.47	1.01	0.59	0.28	0.11	0.04	0.01	0.00
3.60	3.09	2.59	2.09	1.60	1.12	0.69	0.35	0.14	0.05	0.02	0.00
3.56	3.06	2.56	2.06	1.57	1.10	0.67	0.33	0.13	0.05	0.02	0.00
4.17	3.47	2.88	2.35	1.84	1.35	0.89	0.50	0.22	0.08	0.03	0.00
3.01	2.51	2.01	1.52	1.05	0.63	0.31	0.12	0.04	0.01	0.00	0.00
5.01	4.51	4.01	3.51	3.01	2.51	2.01	1.52	1.05	0.63	0.31	0.04
3.49	2.99	2.49	1.99	1.50	1.03	0.61	0.30	0.12	0.04	0.01	0.00
3.65	3.15	2.65	2.15	1.66	1.18	0.74	0.38	0.16	0.06	0.02	0.00
0.00	0.00	0.00	0.00	0.00	0.00	0.00	0.00	0.00	0.00	0.00	0.00
0.00	0.00	0.00	0.00	0.00	0.00	0.00	0.00	0.00	0.00	0.00	0.00
0.00	0.00	0.00	0.00	0.00	0.00	0.00	0.00	0.00	0.00	0.00	0.00
0.07	0.02	0.01	0.00	0.00	0.00	0.00	0.00	0.00	0.00	0.00	0.00
0.02	0.01	0.00	0.00	0.00	0.00	0.00	0.00	0.00	0.00	0.00	0.00
3.82	3.32	2.82	2.32	1.83	1.34	0.88	0.49	0.22	0.08	0.03	0.00
10.23	9.23	8.23	7.24	6.25	5.30	4.43	3.69	3.07	2.52	2.01	1.04
7.36	6.37	5.38	4.42	3.54	2.77	2.14	1.60	1.10	0.67	0.33	0.05
1.67	1.19	0.75	0.39	0.16	0.06	0.02	0.01	0.00	0.00	0.00	0.00
0.01	0.00	0.00	0.00	0.00	0.00	0.00	0.00	0.00	0.00	0.00	0.00
0.59	0.28	0.11	0.04	0.01	0.00	0.00	0.00	0.00	0.00	0.00	0.00
8.12	7.14	6.21	5.36	4.65	4.05	3.52	3.01	2.50	2.00	1.51	0.62
16.02	14.52	13.03	11.54	10.07	8.65	7.35	6.19	5.15	4.19	3.35	2.05
2.25	1.76	1.27	0.82	0.44	0.19	0.07	0.02	0.01	0.00	0.00	0.00

Table 10.2 (Continued)

Ligand		pH 2.0	3.0	3.5	4.0	4.5	5.0	5.5
77. C_6H_7N	L	4.04	3.04	2.54	2.04	1.55	1.08	0.65
78. $C_6H_8O_6$	H_2L	11.37	9.41	8.48	7.66	6.97	6.38	5.85
79. $C_6H_8O_7$	H_3L	6.97	4.29	3.18	2.22	1.44	0.85	0.42
80. $C_6H_8O_7$*	H_3L	3.28	1.60	0.99	0.53	0.25	0.16	0.23
81. $C_6H_8O_7$**	H_3L	0.93	0.25	0.14	0.18	0.40	0.81	1.38
82. $C_6H_9NO_6$	H_3L	8.48	6.83	6.25	5.72	5.21	4.71	4.21
83. $C_6H_9N_3O_2$	HL	11.04	9.04	8.04	7.04	6.05	5.08	4.15
84. $C_6H_{10}N_2O_2$	H_2L	18.67	16.67	15.67	14.67	13.67	12.67	11.67
85. $C_6H_{10}O_4$	H_2L	5.29	3.31	2.36	1.49	0.80	0.36	0.13
86. $C_6H_{11}NO_4S$	H_2L	14.96	12.96	11.96	10.96	9.96	8.96	7.96
87. $C_6H_{11}NO_5$	H_2L	7.07	5.72	5.18	4.67	4.16	3.66	3.16
88. $C_6H_{11}N_3O_4$	HL	7.78	5.37	4.58	3.96	3.41	2.90	2.39
89. $C_6H_{12}N_2O_4$	H_2L	12.12	10.12	9.12	8.12	7.12	6.13	5.16
90. $C_6H_{13}NO_2$	HL	8.08	6.66	6.10	5.58	5.07	4.57	4.07
91. $C_6H_{13}NO_2$	HL	8.06	6.69	6.14	5.63	5.12	4.62	4.12
92. $C_6H_{14}N_2O_2$	HL	15.77	13.77	12.77	11.77	10.77	9.77	8.77
93. $C_6H_{14}N_4O_2$	HL	7.34	6.06	5.53	5.01	4.51	4.01	3.51
94. $C_6H_{15}NO_3$	L	5.80	4.80	4.30	3.80	3.30	2.80	2.30
95. $C_6H_{18}N_4$	L	20.65	16.82	15.07	13.45	11.91	10.40	8.92
96. $C_6H_{18}N_4$	L	21.98	18.98	17.48	15.98	14.48	12.98	11.48
97. $C_7H_5NO_5$	H_3L	10.11	8.25	7.53	6.93	6.40	5.89	5.38
98. $C_7H_5NO_4$	H_2L	3.03	1.74	1.22	0.77	0.40	0.17	0.06
99. $C_7H_5NS_2$	HL	4.93	3.93	3.43	2.93	2.43	1.94	1.45
100. $C_7H_6O_2$	HL	6.13	5.13	4.63	4.13	3.63	3.13	2.63
101. $C_7H_6O_2$	HL	4.70	3.70	3.20	2.70	2.20	1.71	1.23
102. $C_7H_6O_3$	H_2L	12.27	10.62	9.98	9.43	8.91	8.40	7.90
103. $C_7H_6O_5$	H_3L	18.41	15.43	13.98	12.60	11.35	10.22	9.17
104. $C_7H_6O_5$	H_3L	8.96	6.98	6.03	5.15	4.40	3.77	3.22
105. $C_7H_6O_6S$	H_3L	10.33	8.84	8.26	7.73	7.22	6.72	6.22
106. $C_7H_7NO_2$	HL	3.09	1.84	1.32	0.86	0.47	0.21	0.08
107. $C_7H_{12}O_4$	H_2L	5.39	3.41	2.45	1.57	0.86	0.39	0.15
108. $C_8H_5O_2SF_3$	HL	4.53	3.53	3.03	2.53	2.03	1.54	1.07
109. $C_8H_5N_5O_6$	H_3L	16.10	14.10	13.10	12.10	11.10	10.10	9.10
110. $C_8H_6O_4$	H_2L	3.75	2.13	1.51	1.00	0.57	0.27	0.10
111. $C_8H_8O_3$	HL	1.22	0.41	0.17	0.06	0.02	0.01	0.00
112. $C_8H_9NO_2$	HL	2.68	1.44	0.95	0.53	0.25	0.09	0.03
113. $C_8H_{14}N_4O_5$	HL	7.08	5.27	4.54	3.93	3.39	2.88	2.37
114. $C_8H_{16}N_2O_4$	H_2L	12.47	10.47	9.47	8.47	7.47	6.48	5.49

Ch. 10]   **Side-Reaction Coefficients for Protonation Equilibria**   261

$$\log \alpha_{L(H)}$$

pH 6.0	6.5	7.0	7.5	8.0	8.5	9.0	9.5	10.0	10.5	11.0	12.0
0.32	0.13	0.05	0.01	0.00	0.00	0.00	0.00	0.00	0.00	0.00	0.00
5.34	4.84	4.34	3.84	3.34	2.84	2.34	1.85	1.36	0.90	0.50	0.09
0.18	0.06	0.02	0.01	0.00	0.00	0.00	0.00	0.00	0.00	0.00	0.00
0.49	0.87	1.33	1.82	2.31	2.81	3.31	3.81	4.31	4.81	5.31	6.31
2.14	3.02	3.98	4.97	5.96	6.96	7.96	8.96	9.96	10.96	11.96	13.96
3.71	3.21	2.71	2.21	1.72	1.24	0.79	0.42	0.18	0.07	0.02	0.00
3.33	2.65	2.07	1.55	1.06	0.64	0.31	0.12	0.04	0.01	0.00	0.00
10.67	9.67	8.67	7.67	6.67	5.67	4.68	3.71	2.77	1.94	1.26	0.36
0.04	0.01	0.00	0.00	0.00	0.00	0.00	0.00	0.00	0.00	0.00	0.00
6.96	5.97	4.99	4.04	3.18	2.46	1.86	1.33	0.86	0.47	0.21	0.03
2.66	2.16	1.67	1.19	0.75	0.39	0.16	0.06	0.02	0.01	0.00	0.00
1.90	1.41	0.94	0.54	0.25	0.10	0.03	0.01	0.00	0.00	0.00	0.00
4.23	3.41	2.72	2.14	1.62	1.13	0.69	0.35	0.14	0.05	0.02	0.00
3.57	3.07	2.57	2.07	1.58	1.11	0.67	0.34	0.14	0.05	0.02	0.00
3.62	3.12	2.62	2.12	1.63	1.15	0.71	0.37	0.15	0.05	0.02	0.00
7.77	6.77	5.77	4.78	3.80	2.87	2.04	1.35	0.81	0.42	0.17	0.02
3.01	2.51	2.01	1.52	1.05	0.63	0.31	0.12	0.04	0.01	0.00	0.00
1.81	1.32	0.86	0.48	0.21	0.08	0.03	0.01	0.00	0.00	0.00	0.00
7.49	6.15	4.96	3.88	2.87	1.93	1.12	0.53	0.21	0.07	0.02	0.00
9.98	8.49	7.00	5.53	4.13	2.86	1.79	0.97	0.44	0.17	0.06	0.01
4.88	4.38	3.88	3.38	2.88	2.38	1.89	1.40	0.93	0.53	0.25	0.03
0.02	0.01	0.00	0.00	0.00	0.00	0.00	0.00	0.00	0.00	0.00	0.00
0.98	0.57	0.27	0.10	0.04	0.01	0.00	0.00	0.00	0.00	0.00	0.00
2.13	1.64	1.16	0.72	0.37	0.15	0.05	0.02	0.01	0.00	0.00	0.00
0.78	0.41	0.18	0.06	0.02	0.01	0.00	0.00	0.00	0.00	0.00	0.00
7.40	6.90	6.40	5.90	5.40	4.90	4.40	3.90	3.40	2.90	2.40	1.42
8.16	7.16	6.16	5.18	4.23	3.36	2.63	2.02	1.49	1.00	0.58	0.11
2.71	2.21	1.71	1.23	0.78	0.41	0.18	0.07	0.04	0.05	0.13	0.66
5.72	5.22	4.72	4.22	3.72	3.22	2.72	2.22	1.73	1.25	0.80	0.18
0.03	0.01	0.00	0.00	0.00	0.00	0.00	0.00	0.00	0.00	0.00	0.00
0.05	0.02	0.01	0.00	0.00	0.00	0.00	0.00	0.00	0.00	0.00	0.00
0.64	0.32	0.13	0.04	0.01	0.00	0.00	0.00	0.00	0.00	0.00	0.00
8.10	7.10	6.10	5.11	4.13	3.18	2.31	1.59	1.01	0.56	0.26	0.03
0.04	0.01	0.00	0.00	0.00	0.00	0.00	0.00	0.00	0.00	0.00	0.00
0.00	0.00	0.00	0.00	0.00	0.00	0.00	0.00	0.00	0.00	0.00	0.00
0.01	0.00	0.00	0.00	0.00	0.00	0.00	0.00	0.00	0.00	0.00	0.00
1.88	1.39	0.93	0.52	0.24	0.09	0.03	0.01	0.00	0.00	0.00	0.00
4.53	3.62	2.84	2.19	1.64	1.14	0.70	0.35	0.15	0.05	0.02	0.00

Table 10.2 (Continued)

Ligand		pH 2.0	3.0	3.5	4.0	4.5	5.0	5.5
115. $C_8H_{23}N_5$	L	23.59	19.60	17.62	15.67	13.80	12.07	10.46
116. $C_9H_6INO_4S$	H_2L	5.65	4.18	3.62	3.09	2.58	2.08	1.59
117. C_9H_7NO	HL	10.65	8.65	7.66	6.69	5.77	4.96	4.28
118. $C_9H_7NO_4S$	H_2L	8.36	6.40	5.49	4.69	4.02	3.46	2.93
119. $C_9H_7N_3O_2S$	H_2L	11.71	9.67	8.67	7.67	6.68	5.69	4.74
120. $C_9H_{11}NO_3$	H_2L	15.02	13.02	12.02	11.02	10.02	9.02	8.02
121. $C_9H_{11}NO_3$	H_2L	15.18	13.18	12.18	11.18	10.18	9.18	8.18
122. $C_{10}H_7NO_2$	HL	5.25	4.25	3.75	3.25	2.75	2.25	1.76
123. $C_{10}H_7NO_2$	HL	5.65	4.65	4.15	3.65	3.15	2.65	2.15
124. $C_{10}H_7NO_2$	HL	2.96	1.96	1.47	1.01	0.59	0.28	0.11
125. $C_{10}H_8N_2$	L	2.54	1.45	0.97	0.56	0.26	0.10	0.03
126. $C_{10}H_8N_2O_2$	H_2L	16.83	14.83	13.83	12.83	11.83	10.83	9.83
127. $C_{10}H_8O_8S_2$	H_3L	3.37	2.35	1.86	1.37	0.91	0.51	0.23
128. $C_{10}H_{10}O_2$	HL	6.89	5.89	5.39	4.89	4.39	3.89	3.39
129. $C_{10}H_{13}N_5O_4$	HL	11.86	9.97	9.15	8.47	7.89	7.36	6.85
130. $C_{10}H_{14}N_5O_7P$	H_3L	5.75	3.82	2.95	2.21	1.60	1.08	0.64
131. $C_{10}H_{14}N_5O_7P$	H_3L	5.49	3.57	2.71	1.99	1.40	0.91	0.50
132. $C_{10}H_{14}N_5O_7P$	H_3L	15.73	13.73	12.73	11.73	10.73	9.74	8.76
133. $C_{10}H_{15}N_5O_{10}P_2$	H_4L	6.36	4.41	3.49	2.68	2.01	1.45	0.96
134. $C_{10}H_{16}N_2O_8$	H_4L	13.36	10.46	9.34	8.30	7.30	6.31	5.38
135. $C_{10}H_{16}N_2O_8$*	H_4L	5.19	3.29	2.67	2.13	1.63	1.14	0.71
136. $C_{10}H_{16}N_2O_8$	H_4L	15.06	11.65	10.19	8.89	7.74	6.69	5.68
137. $C_{10}H_{16}N_5O_{13}P_3$	H_5L	6.57	4.61	3.68	2.84	2.15	1.57	1.06
138. $C_{10}H_{17}NO_4$	H_2L	9.19	7.87	7.33	6.82	6.31	5.81	5.31
139. $C_{10}H_{18}N_2O_7$	H_3L	11.88	9.33	8.24	7.21	6.24	5.34	4.55
140. $C_{10}H_{18}N_2O_7$*	H_3L	4.07	2.52	1.93	1.40	0.93	0.53	0.24
141. $C_{11}H_9N_3O_2$	H_2L	14.58	11.98	10.88	9.84	8.86	7.93	7.11
142. $C_{11}H_{12}N_2O_2$	HL	7.84	6.42	5.86	5.34	4.83	4.33	3.83
143. $C_{11}H_{18}N_2O_8$	H_4L	14.19	11.36	10.22	9.17	8.16	7.17	6.22
144. $C_{11}H_{18}N_2O_8$	H_4L	15.36	12.63	11.53	10.50	9.49	8.48	7.48
145. $C_{11}H_{18}N_2O_8$*	H_4L	6.90	5.17	4.57	4.04	3.53	3.02	2.52
146. $C_{12}H_7N_2Cl$	L	2.19	1.21	0.76	0.40	0.17	0.06	0.02
147. $C_{12}H_7N_2Cl$	L	2.07	1.11	0.67	0.34	0.14	0.05	0.02
148. $C_{12}H_7N_3O_2$	L	1.25	0.42	0.18	0.07	0.02	0.01	0.00
149. $C_{12}H_8N_2$	L	3.18	1.97	1.46	0.98	0.57	0.27	0.10
150. $C_{12}H_{20}N_2O_8$	H_4L	15.82	11.99	10.27	8.76	7.47	6.35	5.31
151. $C_{12}H_{20}N_2O_8$*	H_4L	8.36	5.60	4.59	3.89	3.34	2.82	2.31
152. $C_{12}H_{20}N_2O_8S$	H_4L	14.70	12.02	10.93	9.90	8.89	7.89	6.89

Ch. 10]   **Side-Reaction Coefficients for Protonation Equilibria**   263

$$\log \alpha_{L(H)}$$

pH 6.0	6.5	7.0	7.5	8.0	8.5	9.0	9.5	10.0	10.5	11.0	12.0
8.91	7.41	5.93	4.50	3.18	2.04	1.14	0.52	0.20	0.07	0.02	0.00
1.11	0.68	0.34	0.14	0.05	0.02	0.01	0.00	0.00	0.00	0.00	0.00
3.70	3.17	2.67	2.16	1.67	1.19	0.75	0.39	0.16	0.06	0.02	0.00
2.43	1.93	1.44	0.97	0.56	0.26	0.10	0.03	0.01	0.00	0.00	0.00
3.87	3.13	2.51	1.97	1.46	0.99	0.58	0.27	0.11	0.04	0.01	0.00
7.02	6.02	5.03	4.06	3.14	2.33	1.66	1.11	0.66	0.32	0.13	0.01
7.18	6.18	5.18	4.19	3.22	2.29	1.48	0,84	0.40	0.16	0.06	0.01
1.27	0.82	0.44	0.19	0.07	0.02	0.01	0.00	0.00	0.00	0.00	0.00
1.66	1.18	0.74	0.38	0.16	0.06	0.02	0.01	0.00	0.00	0.00	0.00
0.04	0.01	0.00	0.00	0.00	0.00	0.00	0.00	0.00	0.00	0.00	0.00
0.01	0.00	0.00	0.00	0.00	0.00	0.00	0.00	0.00	0.00	0.00	0.00
8.83	7.83	6.83	5.83	4.84	3.85	2.90	2.04	1.31	0.75	0.37	0.05
0.09	0.03	0.01	0.00	0.00	0.00	0.00	0.00	0.00	0.00	0.00	0.00
2.89	2.39	1.90	1.41	0.94	0.54	0.25	0.10	0.03	0.01	0.00	0.00
6.35	5.85	5.35	4.85	4.35	3.85	3.35	2.85	2.35	1.86	1.37	0.51
0.31	0.12	0.04	0.01	0.00	0.00	0.00	0.00	0.00	0.00	0.00	0.00
0.23	0.08	0.03	0.01	0.00	0.00	0.00	0.00	0.00	0.00	0.00	0.00
7.81	6.95	6.23	5.62	5.08	4.57	4.06	3.56	3.06	2.56	2.06	1.10
0.55	0.25	0.10	0.03	0.01	0.00	0.00	0.00	0.00	0.00	0.00	0.00
4.53	3.82	3.22	2.69	2.18	1.68	1.20	0.75	0.39	0.17	0.06	0.01
0.36	0.15	0.05	0.02	0.01	0.01	0.03	0.08	0.22	0.50	0.89	1.84
4.71	3.82	3.04	2.41	1.85	1.35	0.88	0.49	0.22	0.08	0.03	0.00
0.63	0.31	0.12	0.04	0.01	0.00	0.00	0.00	0.00	0.00	0.00	0.00
4.81	4.31	3.81	3.31	2.81	2.31	1.82	1.33	0.87	0.48	0.22	0.03
3.90	3.34	2.82	2.32	1.82	1.33	0.87	0.48	0.22	0.08	0.03	0.00
0.09	0.03	0.01	0.01	0.01	0.02	0.06	0.17	0.41	0.77	1.22	2.19
6.43	5.85	5.32	4.81	4.31	3.81	3.31	2.81	2.31	1.82	1.33	0.48
3.33	2.83	2.33	1.84	1.35	0.89	0.50	0.22	0.08	0.03	0.01	0.00
5.35	4.60	3.99	3.44	2.93	2.42	1.93	1.44	0,97	0.56	0.26	0.03
6.48	5.49	4.52	3.59	2.77	2.09	1.52	1.02	0.59	0.28	0.11	0.01
2.02	1.53	1.06	0.63	0.31	0.13	0.06	0.06	0.13	0.32	0.65	1.55
0.01	0.00	0.00	0.00	0.00	0.00	0.00	0.00	0.00	0.00	0.00	0.00
0.01	0.00	0.00	0.00	0.00	0.00	0.00	0.00	0.00	0.00	0.00	0.00
0.00	0.00	0.00	0.00	0.00	0.00	0.00	0.00	0.00	0.00	0.00	0.00
0.04	0.01	0.00	0.00	0.00	0.00	0.00	0.00	0.00	0.00	0.00	0.00
4.34	3.44	2.68	2.04	1.50	1.01	0.59	0.28	0.11	0.04	0.01	0.00
1.82	1.33	0.87	0.48	0.23	0.12	0.15	0.33	0.65	1.08	1.55	2.54
5.89	4.89	3.90	2.93	2.02	1.23	0.64	0.28	0.10	0.04	0.01	0.00

Table 10.2 (Continued)

Ligand		pH 2.0	3.0	3.5	4.0	4.5	5.0	5.5
153. $C_{12}H_{20}N_2O_8S$*	H_4L	7.87	5.66	5.02	4.49	3.97	3.47	2.97
154. $C_{12}H_{20}N_2O_9$	H_4L	15.33	12.52	11.38	10.33	9.32	8.31	7.31
155. $C_{12}H_{20}N_2O_9$*	H_4L	7.86	6.05	5.41	4.86	4.35	3.84	3.34
156. $C_{13}H_{10}N_2$	L	3.27	2.27	1.78	1.29	0.84	0.46	0.20
157. $C_{13}H_9N_3OS$	HL	7.13	6.10	5.60	5.10	4.60	4.10	3.60
158. $C_{13}H_{20}N_2O_8$	H_4L	14.33	11.68	10.61	9.58	8.57	7.57	6.57
159. $C_{14}H_{12}N_2$	L	3.85	2.85	2.35	1.86	1.37	0.91	0.51
160. $C_{14}H_{12}N_2$	L	3.95	2.95	2.45	1.95	1.47	1.00	0.58
161. $C_{14}H_{12}N_2$	L	3.60	2.60	2.10	1.61	1.13	0.70	0.35
162. $C_{14}H_{22}N_2O_8$	H_4L	16.64	13.27	11.89	10.68	9.60	8.59	7.64
163. $C_{14}H_{22}N_2O_8$*	H_4L	6.24	3.87	2.99	2.28	1.70	1.19	0.74
164. $C_{14}H_{22}N_4O_{10}$	H_4L	10.48	7.66	6.54	5.50	4.50	3.51	2.56
165. $C_{14}H_{22}N_4O_{10}$*	H_4L	5.18	3.36	2.74	2.20	1.70	1.21	0.76
166. $C_{14}H_{23}N_3O_{10}$	H_5L	18.19	14.44	12.88	11.46	10.19	9.06	8.01
167. $C_{14}H_{23}N_3O_{10}$*	H_5L	9.74	6.99	5.93	5.01	4.24	3.61	3.06
168. $C_{14}H_{24}N_2O_{10}$	H_4L	15.33	12.50	11.38	10.34	9.33	8.32	7.32
169. $C_{14}H_{24}N_2O_{10}$*	H_4L	7.86	6.03	5.41	4.87	4.36	3.85	3.35
170. $C_{15}H_{11}N_3O$	HL	10.15	8.45	7.80	7.23	6.71	6.20	5.70
171. $C_{15}H_{11}N_3O$	HL	8.14	6.40	5.72	5.14	4.61	4.10	3.60
172. $C_{16}H_{12}N_2O_5S$	H_3L	15.64	13.64	12.64	11.64	10.64	9.65	8.66
173. $C_{16}H_{13}N_2O_{11}S_2As$	H_6L	15.87	12.09	10.81	9.72	8.69	7.68	6.68
174. $C_{18}H_{12}N_2$	L	2.72	1.73	1.25	0.80	0.42	0.18	0.07
175. $C_{18}H_{20}N_2O_6$	H_4L	28.88	24.88	22.88	20.88	18.89	16.90	14.94
176. $C_{18}H_{30}N_4O_{12}$	H_6L	23.45	18.41	16.34	14.46	12.75	11.18	9.70
177. $C_{18}H_{30}N_4O_{12}$*	H_6L	14.95	10.91	9.34	7.96	6.75	5.68	4.70
178. $C_{18}H_{30}N_4O_{12}$**	H_6L	7.51	4.47	3.40	2.52	1.81	1.24	0.76
179. $C_{19}H_{14}O_7S$	H_4L	24.39	21.36	19.86	18.36	16.86	15.36	13.86
180. $C_{20}H_{13}N_3O_7S$	H_3L	14.19	12.19	11.19	10.19	9.19	8.20	7.21
181. $C_{20}H_{24}N_2O_6$	H_4L	28.42	24.43	22.45	20.51	18.66	16.94	15.34
182. $C_{20}H_{24}N_2O_6$*	H_4L	17.96	14.97	13.49	12.05	10.70	9.48	8.38
183. $C_{20}H_{24}N_2O_6$**	H_4L	8.96	6.97	5.99	5.05	4.20	3.48	2.88
184. $C_{21}H_{18}O_5S$	H_2L	5.99	4.95	4.45	3.95	3.45	2.95	2.45
185. $C_{22}H_{18}N_4O_{14}S_2As_2$	H_8L	20.17	16.28	14.70	13.19	11.72	10.32	9.05
186. $C_{23}H_{18}O_9S$	H_4L	13.79	11.60	10.58	9.58	8.60	7.64	6.77
187. $C_{24}H_{16}N_2$	L	2.80	1.81	1.32	0.86	0.48	0.21	0.08
188. $C_{26}H_{24}N_9O_9S$	H_6L	15.14	12.49	11.39	10.36	9.35	8.34	7.35
189. $C_{26}H_{24}N_9O_9S$*	H_6L	6.24	4.59	3.99	3.46	2.95	2.44	1.95
190. $C_{26}H_{24}N_9O_9S$**	H_6L	0.80	0.15	0.05	0.02	0.01	0.00	0.01

Ch. 10] Side-Reaction Coefficients for Protonation Equilibria 265

$$\log \alpha_{L(H)}$$

pH 6.0	6.5	7.0	7.5	8.0	8.5	9.0	9.5	10.0	10.5	11.0	12.0
2.47	1.97	1.48	1.01	0.60	0.31	0.22	0.36	0.68	1.11	1.59	2.58
6.31	5.31	4.32	3.33	2.37	1.49	0.78	0.33	0.12	0.04	0.01	0.00
2.84	2.34	1.85	1.36	0.90	0.52	0.31	0.36	0.65	1.07	1.54	2.53
0.07	0.02	0.01	0.00	0.00	0.00	0.00	0.00	0.00	0.00	0.00	0.00
3.10	2.60	2.10	1.61	1.13	0.70	0.35	0.15	0.05	0.02	0.01	0.00
5.58	4.61	3.69	2.88	2.21	1.64	1.14	0.69	0.35	0.14	0.05	0.01
0.23	0.09	0.03	0.01	0.00	0.00	0.00	0.00	0.00	0.00	0.00	0.00
0.28	0.11	0.04	0.01	0.00	0.00	0.00	0.00	0.00	0.00	0.00	0.00
0.15	0.05	0.02	0.01	0.00	0.00	0.00	0.00	0.00	0.00	0.00	0.00
6.78	6.06	5.46	4.92	4.41	3.90	3.40	2.90	2.40	1.91	1.42	0.55
0.38	0.16	0.06	0.02	0.01	0.00	0.00	0.00	0.00	0.01	0.02	0.15
1.71	1.01	0.52	0.22	0.08	0.03	0.01	0.00	0.00	0.00	0.00	0.00
0.41	0.21	0.22	0.42	0.78	1.23	1.71	2.20	2.70	3.20	3.70	4.70
6.99	5.99	4.99	4.02	3.09	2.27	1.59	1.04	0.59	0.28	0.11	0.01
2.54	2.04	1.54	1.07	0.64	0.32	0.14	0.09	0.14	0.33	0.66	1.56
6.32	5.32	4.33	3.34	2.38	1.49	0.78	0.33	0.12	0.04	0.01	0.00
2.85	2.35	1.86	1.37	0.91	0.52	0.31	0.36	0.65	1.07	1.54	2.53
5.20	4.70	4.20	3.70	3.20	2.70	2.20	1.71	1.23	0.78	0.41	0.06
3.10	2.60	2.10	1.61	1.13	0.70	0.35	0.15	0.05	0.02	0.01	0.00
7.70	6.80	6.02	5.37	4.81	4.29	3.78	3.28	2.78	2.28	1.79	0.85
5.69	4.71	3.78	2.93	2.22	1.63	1.12	0.68	0.34	0.14	0.05	0.01
0.02	0.01	0.00	0.00	0.00	0.00	0.00	0.00	0.00	0.00	0.00	0.00
13.05	11.28	9.65	8.12	6.66	5.30	4.09	3.04	2.13	1.39	0.82	0.17
8.33	7.10	6.00	4.96	3.96	2.99	2.08	1.29	0.70	0.32	0.12	0.01
3.83	3.10	2.50	1.96	1.46	0.99	0.58	0.29	0.20	0.32	0.62	1.51
0.39	0.16	0.06	0.02	0.02	0.05	0.14	0.35	0.76	1.38	2.18	4.07
12.37	10.88	9.42	8.04	6.78	5.66	4.65	3.76	3.00	2.37	1.83	0.87
6.25	5.37	4.60	3.97	3.42	2.90	2.39	1.90	1.41	0.94	0.54	0.10
13.80	12.29	10.80	9.34	7.95	6.68	5.55	4.50	3.51	2.58	1.77	0.62
7.34	6.33	5.34	4.38	3.49	2.72	2.09	1.54	1.05	0.62	0.31	0.16
2.34	1.83	1.34	0.88	0.49	0.22	0.09	0.04	0.05	0.12	0.31	1.16
1.95	1.47	1.00	0.58	0.28	0.11	0.04	0.01	0.00	0.00	0.00	0.00
7.91	6.88	5.94	5.12	4.44	3.87	3.34	2.83	2.33	1.84	1.35	0.50
6.02	5.40	4.85	4.34	3.83	3.33	2.83	2.33	1.84	1.35	0.89	0.22
0.03	0.01	0.00	0.00	0.00	0.00	0.00	0.00	0.00	0.00	0.00	0.00
6.36	5.39	4.47	3.67	3.01	2.44	1.92	1.42	0.95	0.55	0.25	0.03
1.46	0.99	0.57	0.27	0.11	0.04	0.02	0.02	0.05	0.15	0.35	1.13
0.02	0.05	0.13	0.33	0.67	1.10	1.58	2.08	2.61	3.21	3.91	5.69

Table 10.2 (Continued)

Ligand		pH						
		2.0	3.0	3.5	4.0	4.5	5.0	5.5
191. $C_{30}H_{18}N_6O_{21}S_6$	H_9L	14.60	12.60	11.60	10.60	9.60	8.60	7.61
192. $C_{31}H_{32}N_2O_{13}S$	H_6L	25.03	20.63	18.81	17.18	15.64	14.14	12.67
193. $C_{31}H_{32}N_2O_{13}S*$	H_6L	14.75	11.35	10.03	8.90	7.86	6.86	5.89
194. $C_{31}H_{32}N_2O_{13}S**$	H_6L	6.29	3.89	3.07	2.44	1.90	1.40	0.93
195. $C_{32}H_{32}N_2O_{12}$	H_6L	31.49	26.44	24.26	22.19	20.17	18.17	16.18
196. $C_{32}H_{32}N_2O_{12}*$	H_6L	21.48	17.38	15.67	14.09	12.57	11.06	9.57
197. $C_{37}H_{44}N_2O_{13}S$	H_6L	30.09	24.46	22.12	20.20	18.53	16.98	15.46
198. $C_{37}H_{44}N_2O_{13}S**$	H_6L	9.54	5.91	4.57	3.65	2.98	2.43	1.91
199. $C_{37}H_{44}N_2O_{13}S****$	H_6L	2.34	0.71	0.37	0.45	0.78	1.23	1.71
200. $C_{38}H_{44}N_2O_{12}$	H_6L	15.60	13.60	12.60	11.60	10.60	9.60	8.61

List of organic ligands
(abbreviations and/or synonyms used in this work are given in parentheses)

1. Formic acid HL
2. Glyoxylic (oxoacetic) acid HL
3. Oxalic acid (ox) H_2L
4. Monochloroacetic acid (Cl-ac) HL
5. Acetic acid (ac) HL
6. Mercaptoacetic (thioglycollic) acid (TGA) HL
7. Hydroxyacetic acid HL
8. Aminoacetic acid (glycine) (glyc) HL
9. Glycinamide HL
10. 2-Aminoethanol L
11. 2-Aminoethanethiol HL
12. Ethylenediamine (en) L
13. Imidazole HL [$I = 0$]
14. Pyruvic (2-oxopropanoic) acid HL
15. Malonic acid H_2L
16. Propanoic acid HL
17. Lactic acid HL
18. Alanine HL
19. β-Alanine HL
20. Sarcosine HL
21. Cysteine (cyst) H_2L
22. Serine HL
23. Isoserine HL
24. 2,3-Dimercaptopropan-1-ol (BAL) H_2L
25. L-Phosphoserine H_3L
26. 1,2-Propylenediamine (pn) L
27. Trimethylenediamine L
28. 1,3-Diamino-2-propanol L

Side-Reaction Coefficients for Protonation Equilibria

$\log \alpha_{L(H)}$

pH 6.0	6.5	7.0	7.5	8.0	8.5	9.0	9.5	10.0	10.5	11.0	12.0
6.63	5.70	4.85	4.15	3.55	3.02	2.51	2.01	1.51	1.04	0.62	0.12
11.26	9.98	8.83	7.77	6.75	5.75	4.76	3.79	2.87	2.07	1.41	0.47
4.98	4.20	3.55	2.99	2.47	1.97	1.48	1.01	0.59	0.29	0.13	0.19
0.52	0.24	0.09	0.03	0.01	0.01	0.02	0.05	0.13	0.33	0.67	1.73
14.21	12.29	10.51	8.94	7.60	6.45	5.39	4.38	3.38	2.42	1.53	0.35
8.10	6.68	5.41	4.36	3.55	2.92	2.37	1.86	1.37	0.91	0.52	0.34
13.97	12.50	11.10	9.80	8.65	7.58	6.56	5.56	4.58	3.64	2.78	1.47
1.42	0.95	0.55	0.25	0.10	0.03	0.01	0.01	0.03	0.09	0.23	0.92
2.22	2.75	3.35	4.05	4.90	5.83	6.81	7.81	8.83	9.89	11.03	13.72
7.62	6.66	5.76	4.98	4.34	3.78	3.26	2.75	2.25	1.76	1.27	0.44

29. 1,2,3-Triaminopropanol L
30. Maleic acid H$_2$L
31. Fumaric acid H$_2$L
32. Succinic acid H$_2$L
33. Malic acid H$_2$L
34. D-Tartaric acid (tart) H$_2$L
35. Asparagine HL
36. Iminodiacetic acid (IDA) H$_2$L
37. Aspartic acid H$_2$L
38. Butane-2,3-dione dioxime (biacetyl dioxime, dimethylglyoxime) H$_2$L
39. Butyric (butanoic) acid HL
40. Isobutyric (2-methylpropanoic) acid HL
41. Diglycine HL
42. Homoserine HL
43. Threonine HL
44. Piperazine L
45. Tris(hydroxymethyl)aminomethane (THAM) L
46. N-Ethyliminodimethylenediphosphonic acid H$_4$L
47. Iminobis(2-ethylamine) (diethylenetriamine) (dien) L
48. Purine HL
49. Pyridine (py) L
50. Adenine HL
51. 3-Aminopyridine L
52. Pentane-2,4-dione (acetylacetone) (acac) HL
53. Glutaric acid H$_2$L
54. Proline HL
55. Hydroxyproline HL
56. Glutamic acid (glutam) H$_2$L
57. N-Methyliminodiacetic acid (MIDA) H$_2$L
58. Histamine (histam) L
59. Glutamine HL

Table 10.2 (Continued)

60. Piperidine L
61. Valine HL
62. Norvaline HL
63. N,N-Diethyldithiocarbamic acid (DDC) HL
64. Chloranilic acid H_2L
65. Rhodizonic acid ($I = 3.0$) H_2L
66. Picolinic acid HL
67. Nicotinic acid ($I = 0.5$) HL
68. Phenol HL
69. Pyrocatechol H_2L
70. Resorcinol H_2L
71. Kojic acid HL
72. Cupferron (N-nitrosophenylhydroxylamine) HL
73. Benzenethiol HL
74. Tiron (1,2-dihydroxybenzene-3,5-disulphonic acid) H_4L
75. Pyrogallol H_3L
76. Benzenearsonic acid H_2L
77. 4-Picoline L
78. Ascorbic acid (asco) H_2L
79. Citric acid (citr) H_3L
80. Citric acid (citr) $H_2 \cdot HL$
81. Citric acid (citr) $H \cdot H_2L$
82. Nitrilotriacetic acid (NTA) H_3L
83. Histidine (histid) HL
84. Cyclohexane-1,2-dione dioxime (nioxime) H_2L
85. Adipic acid H_2L
86. N-(2-Mercaptoethyl)iminodiacetic acid H_3L (20°C)
87. N-(2-Hydroxyethyl)iminodiacetic acid H_2L
88. Triglycine HL
89. Ethylenediamine-N,N'-diacetic acid (EDDA) H_2L
90. Leucine HL
91. Isoleucine HL
92. Lysine HL
93. Arginine HL
94. Nitrilotri-2-ethanol (triethanolamine) (TEA) L
95. Triethylenetetramine (trien) L
96. Nitrilotris(2-ethylamine) (tren) L
97. 4-Hydroxypyridine-2,6-dicarboxylic acid (chelidamic acid) H_3L
98. Dipicolinic acid H_2L
99. 2-Mercaptobenzo-1,3-thiazole HL
100. Salicylaldehyde HL
101. Tropolone HL
102. Salicylic acid (sal) H_2L
103. Gallic acid H_3L
104. Gallic acid $H_2 \cdot HL$
105. Sulphosalicylic acid (sulfosal) H_3L
106. Anthranilic acid HL

Table 10.2 (Continued)

107. Pimelic acid H_2L
108. Thenoyltrifluoroacetone (TTA) HL
109. Purpuric acid H_3L
110. Phthalic acid H_2L
111. Mandelic acid HL
112. Phenylglycine HL
113. Tetraglycine HL
114. Ethylenediamine-N,N'-di-3-propanoic acid (EDDP) H_2L
115. Tetraethylenepentamine (tetren) L
116. 7-Iodo-8-hydroxyquinoline-5-sulphonic acid (ferron) H_2L
117. 8-Hydroxyquinoline, 8-quinolinol (oxine) HL
118. 8-Hydroxyquinoline-5-sulphonic acid (sulfoxine) H_2L
119. Thiazolylazoresorcinol (TAR) ($20°C$) H_2L
120. o-Tyrosine H_2L
121. Tyrosine H_2L
122. 2-Nitroso-1-naphthol HL
123. 1-Nitroso-2-naphthol HL
124. Quinoline-2-carboxylic acid (quinaldic acid) HL
125. 2,2'-Bipyridyl (bipy) L
126. α-Furil dioxime H_2L
127. Chromotropic acid H_3L
128. 1-Phenylbutane-1,3-dione (benzoylacetone) HL
129. Adenosine HL
130. Adenosine-2'-monophosphoric acid, dihydrogen salt (AMP-2) H_3L
131. Adenosine-3'-monophosphoric acid, dihydrogen salt (AMP-3) H_3L
132. Adenosine-5'-monophosphoric acid, dihydrogen salt (AMP-5) H_3L
133. Adenosine-5'-diphosphoric acid, trihydrogen salt (ADP) H_4L
134. Ethylenedinitrilotetra-acetic acid (EDTA) H_4L
135. Ethylenedinitrilotetra-acetic acid (EDTA) $H_3 \cdot HL$
136. Ethylenedi-iminodibutanedioic acid (EDDS) H_4L
137. Adenosine-5'-triphosphoric acid, tetrahydrogen salt (ATP) H_5L
138. N-(Cyclohexyl)iminodiacetic acid H_2L
139. N-(2-Hydroxyethyl)ethylenedinitrilo-N',N'-triacetic acid (HEDTA) H_3L
140. N-(2-Hydroxyethyl)ethylenedinitrilo-N',N'-triacetic acid (HEDTA) $H_2 \cdot HL$
141. Pyridylazoresorcinol (PAR) H_2L
142. Tryptophan (tryp)
143. DL-(Methylene)dinitrilotetra-acetic acid (PDTA) ($20°C$) H_4L
144. Trimethylenedinitrilotetra-acetic acid (TMDTA) ($20°C$) H_4L
145. Trimethylenedinitrilotetra-acetic acid (TMDTA) ($20°C$) $H_3 \cdot HL$
146. 2-Chloro-1,10-phenanthroline L
147. 5-Chloro-1,10-phenanthroline L
148. 5-Nitro-1,10-phenanthroline L
149. 1,10-Phenanthroline L
150. Ethylenedi-iminodi-2-pentanedioic acid (EDDG) H_4L
151. Ethylenedi-iminodi-2-pentanedioic acid (EDDG) $H_3 \cdot HL$
152. Thiobis(ethylenenitrilo)tetra-acetic acid (TEDTA) ($20°C$) H_4L
153. Thiobis(ethylenenitrilo)tetra-acetic acid (TEDTA) ($20°C$) $H_3 \cdot HL$

Table 10.2 (Continued)

154. Oxybis(ethylenenitrilo)tetra-acetic acid (EEDTA) (20° C) H_4L
155. Oxybis(ethylenenitrilo)tetra-acetic acid (EEDTA) (20° C) $H_3 \cdot HL$
156. 1-(1′,3′-Thiazol-2′-ylazo)-2-naphthol (TAN) ($I = 0.05$; 20°C) HL
157. 5-Methyl-1,10-phenanthroline L
158. *trans*-1,2-Cyclopentylenedinitrilotetra-acetic acid (CPDTA) (20°C) H_4L
159. 2,9-Dimethyl-1,10-phenanthroline (neocuproine) L
160. 4,7-Dimethyl-1,10-phenanthroline L
161. 5,6-Dimethyl-1,10-phenanthroline L
162. *trans*-1,2-Cyclohexylenedinitrilotetra-acetic acid (DCTA) (20°C) H_4L
163. *trans*-1,2-Cyclohexylenedinitrilotetra-acetic acid (DCTA) (20°C) $H_3 \cdot HL$
164. Ethylenebis(imino-1-oxoethylenenitrilo)tetra-acetic acid (DGENTA) H_4L
165. Ethylenebis(imino-1-oxoethylenenitrilo)tetra-acetic acid (DGENTA) $H_3 \cdot HL$
166. Diethylenetrinitrilopenta-acetic acid (DTPA) H_5L
167. Diethylenetrinitrilopenta-acetic acid (DTPA) $H_4 \cdot HL$
168. Ethylenebis(oxyethylenenitrilo)tetra-acetic acid (EGTA) (20°C) H_4L
169. Ethylenebis(oxyethylenenitrilo)tetra-acetic acid (EGTA) (20°C) $H_3 \cdot HL$
170. 1-(2′-pyridylazo)-2-naphthol (PAN) (30°C) HL
171. 1-(2′-pyridylazo)-4-naphthol (*p*-PAN) (30°C) HL
172. Solochrome Violet R H_3L
173. 2-(2′-Arsonophenylazo)chromotropic acid (Arsenazo I) H_6L
174. 5-Phenyl-1,10-phenanthroline L
175. Ethylenedi-iminobis[(2-hydroxyphenyl)acetic] acid (EHPG) H_4L
176. Triethylenetetranitrilohexa-acetic acid (TTHA) H_6L
177. Triethylenetetranitrilohexa-acetic acid (TTHA) $H_5 \cdot HL$
178. Triethylenetetranitrilohexa-acetic acid (TTHA) $H_4 \cdot H_2L$
179. Pyrocatechol Violet H_4L
180. Eriochrome Black T H_3L
181. N,N'-Bis(2-hydroxybenzyl)ethylenedinitrilo-N,N'-diacetic acid (HBED) H_4L
182. N,N'-Bis(2-hydroxybenzyl)ethylenedinitrilo-N,N'-diacetic acid (HBED) $H_3 \cdot HL$
183. N,N'-Bis(2-hydroxybenzyl)ethylenedinitrilo-N,N'-diacetic acid (HBED) $H_2 \cdot H_2L$
184. Cresol Red H_2L
185. 2,7-Bis(2′-arsonophenylazo)chromotropic acid (Arsenazo III) ($I = 0.2$) H_8L
186. Eriochrome Cyanine R H_4L
187. 4,7-Diphenyl-1,10-phenanthroline (bathophenanthroline) ($I = 0.3$, dioxane 50%) L
188. Semixylenol Orange H_6L
189. Semixylenol Orange $H_5 \cdot HL$
190. Semixylenol Orange $H_4 \cdot H_2L$
191. Calcichrome ($I = 0.2$; 20°C) H_9L
192. Xylenol Orange ($I = 0.2$; 20°C) H_6L
193. Xylenol Orange ($I = 0.2$; 20°C) $H_5 \cdot HL$
194. Xylenol Orange ($I = 0.2$; 20°C) $H_4 \cdot H_2L$
195. Phthaleincomplexone ($I = 0.2$; 20°C) H_6L
196. Phthaleincomplexone ($I = 0.2$; 20°C) $H_5 \cdot HL$
197. Methylthymol Blue ($I = 0.2$; $T = $?) H_6L
198. Methylthymol Blue ($I = 0.2$; $T = $?) $H_4 \cdot H_2L$
199. Methylthymol Blue ($I = 0.2$; $T = $?) $H_2 \cdot H_4L$
200. Thymolphthalexone ($I = 0.2$; 20°C) H_6L

Table 10.2 (Continued)

Alphabetical list of organic ligands
[ligands marked with asterisks are partially protonated, to the extent denoted by the number of asterisks]

Ligand	No. in Table
ac	5
acac	52
acetylacetone	52
acetic acid	5
adenine	50
adenosine	129
-5′-diphosphoric acid, trihydrogen salt	133
-2′-monophosphoric acid, dihydrogen salt	130
-3′-monophosphoric acid, dihydrogen salt	131
-5′-monophosphoric acid, dihydrogen salt	132
-5′-triphosphoric acid, tetrahydrogen salt	137
adipic acid	85
ADP	133
alanine	18
β-alanine	19
aminoacetic acid	8
2-aminoethanethiol	11
2-aminoethanol	10
3-aminopyridine	51
AMP-2	130
AMP-3	131
AMP-5	132
anthranilic acid	106
arginine	93
Arsenazo I	173
Arsenazo III	185
2-(2′arsonophenylazo)chromotropic acid (Arsenazo I)	173
asco	78
ascorbic acid	78
asparagine	35
aspartic acid	37
ATP	137
BAL	24
bathophenanthroline	187
benzenearsonic acid	76
benzenethiol	73
benzoylacetone	128
biacetyl dioxime	38
bipy	125
2,2′-bipyridyl	125

Table 10.2 (Continued)

Ligand	No. in Table
2,7-bis(2'-arsonophenylazo)chromotropic acid (Arsenazo III)	185
N,N'-bis(2-hydroxybenzyl)ethylenedinitrilo-N,N'-diacetic acid ($\alpha_{L(H)}$; $\alpha_{HL(H)}$; $\alpha_{H_2L(H)}$)	181; 182*; 183**
butane-2,3-dione dioxime	38
butanoic acid	39
butyric acid	39
Calcichrome	191
chelidamic acid	97
chloranilic acid	64
chloroacetic acid (Cl-ac)	4
2-chloro-1,10-phenanthroline	146
5-chloro-1,10-phenanthroline	147
chromotropic acid	127
citr (see citric acid)	
citric acid (citr) ($\alpha_{L(H)}$; $\alpha_{HL(H)}$; $\alpha_{H_2L(H)}$)	79; 80*; 81**
Cl-ac	4
CPDTA	158
Cresol Red	184
cupferron	72
cyclohexane-1,2-dione dioxime	84
trans-1,2-cyclohexylenedinitrilotetra-acetic acid ($\alpha_{L(H)}$; $\alpha_{HL(H)}$)	162; 163*
N-(cyclohexyl)iminodiacetic acid	138
trans-1,2-cyclopentylenedinitrilotetra-acetic acid	158
cyst	21
cysteine	21
DCTA ($\alpha_{L(H)}$; $\alpha_{HL(H)}$)	162; 163*
DDC	63
DGENTA ($\alpha_{L(H)}$; $\alpha_{HL(H)}$)	164; 165*
1,3-diamino-2-propanol	28
dien	47
N,N-diethyldithiocarbamic acid	63
diethylenetriamine	47
diethylenetrinitrilopenta-acetic acid ($\alpha_{L(H)}$; $\alpha_{HL(H)}$)	166; 167*
diglycine	41
1,2-dihydroxybenzene-3,5-disulphonic acid	74
2,3-dimercaptopropan-1-ol	24
dimethylglyoxime	38
2,9-dimethyl-1,10-phenanthroline	159
4,7-dimethyl-1,10-phenanthroline	160
5,6-dimethyl-1,10-phenanthroline	161
4,7-diphenyl-1,10-phenanthroline	187
dipicolinic acid	98
DTPA ($\alpha_{L(H)}$; $\alpha_{HL(H)}$)	166; 167*

Table 10.2 (Continued)

Ligand	No. in Table
EDDA	89
EDDG ($\alpha_{L(H)}$; $\alpha_{HL(H)}$)	150; 151*
EDDP	114
EDDS	136
EDTA ($\alpha_{L(H)}$; $\alpha_{HL(H)}$)	134; 135*
EEDTA ($\alpha_{L(H)}$; $\alpha_{HL(H)}$)	154; 155*
EGTA ($\alpha_{L(H)}$; $\alpha_{HL(H)}$)	168; 169*
EHPG	175
en	12
Eriochrome Black T	180
Eriochrome Cyanine R	186
ethylenebis(imino-1-oxoethylenenitrilo)tetra-acetic acid ($\alpha_{L(H)}$; $\alpha_{HL(H)}$)	164; 165*
ethylenebis(oxyethylenenitrilo)tetra-acetic acid ($\alpha_{L(H)}$; $\alpha_{HL(H)}$)	168; 169*
ethylenediamine	12
ethylenediamine-N,N'-diacetic acid	89
ethylenediamine-N,N'-di-3-propanoic acid	114
ethylenedi-iminobis[(2-hydroxyphenyl)acetic] acid	175
ethylenedi-iminodibutanedioic acid	136
ethylenedi-iminodi-2-pentanedioic acid ($\alpha_{L(H)}$; $\alpha_{HL(H)}$)	150; 151*
ethylenedinitrilotetra-acetic acid ($\alpha_{L(H)}$; $\alpha_{HL(H)}$)	134; 135*
N-ethyliminodimethylenediphosphonic acid	46
ferron	116
formic acid	1
fumaric acid	31
α-furil dioxime	126
gallic acid ($\alpha_{L(H)}$; $\alpha_{HL(H)}$)	103; 104*
glutam	56
glutamic acid	56
glutamine	59
glutaric acid	53
glyc	8
glycinamide	9
glycine	8
glyoxylic acid	2
HBED ($\alpha_{L(H)}$; $\alpha_{HL(H)}$; $\alpha_{H_2L(H)}$)	181; 182*; 183**
HEDTA ($\alpha_{L(H)}$; $\alpha_{HL(H)}$)	139; 140*
histam	58
histamine	58
histid	83
histidine	83
homoserine	42
hydroxyacetic acid	7
N-(2-hydroxyethyl)ethylenedinitrilo-N,N',N'-triacetic acid ($\alpha_{L(H)}$; $\alpha_{HL(H)}$)	139; 140*

Table 10.2 (Continued)

Ligand	No. in Table
hydroxyproline	55
4-hydroxypyridine-2,6-dicarboxylic acid	97
8-hydroxyquinoline	117
8-hydroxyquinoline-5-sulphonic acid	118
IDA	36
imidazole	13
iminobis(2-ethylamine)	47
iminodiacetic acid	36
7-iodo-8-hydroxyquinoline-5-sulphonic acid	116
isobutyric acid	40
isoleucine	91
isoserine	23
kojic acid	71
lactic acid	17
leucine	90
lysine	92
maleic acid	30
malic acid	33
malonic acid	15
mandelic acid	111
mercaptoacetic acid	6
2-mercaptobenzo-1,3-thiazole	99
N-(2-mercaptoethyl)iminodiacetic acid	86
DL-(methylene)dinitrilotetra-acetic acid	143
N-methyliminodiacetic acid	57
5-methyl-1,10-phenanthroline	157
2-methylpropanoic acid	40
Methylthymol Blue ($\alpha_{L(H)}$; $\alpha_{H_2L(H)}$; $\alpha_{H_4L(H)}$)	197; 198**; 199****
MIDA	57
monochloroacetic acid	4
neocuproine	159
nicotinic acid	67
nioxime	84
nitrilotriacetic acid	82
nitrilotri-2-ethanol	94
nitrilotris(2-ethylamine)	96
5-nitro-1,10-phenanthroline	148
N-nitrosophenylhydroxylamine	72
1-nitroso-2-naphthol	123
2-nitroso-1-naphthol	122
norvaline	62
NTA	82
ox	3

Table 10.2 (Continued)

Ligand	No. in Table
oxalic acid	3
oxine	117
oxoacetic acid	2
2-oxopropanoic acid	14
oxybis(ethylenenitrilo)tetra-acetic acid ($\alpha_{L(H)}$; $\alpha_{HL(H)}$)	154; 155*
PAN	170
p-PAN	171
PAR	141
PDTA	143
pentane-2,4-dione	52
1,10-phenanthroline	149
phenol	68
1-phenylbutane-1,3-dione	128
phenylglycine	112
5-phenyl-1,10-phenanthroline	174
L-phosphoserine	25
phthaleincomplexone ($\alpha_{L(H)}$; $\alpha_{HL(H)}$)	195; 196*
phthalic acid	110
piperazine	44
piperidine	60
4-picoline	77
picolinic acid	66
pimelic acid	107
pn	26
proline	54
propanoic acid	16
1,2-propylenediamine	26
purine	48
purpuric acid	109
py	49
pyridine	49
1-(2'-pyridylazo)-2-naphthol	170
1-(2'-pyridylazo)-4-naphthol	171
pyridylazoresorcinol	141
pyrocatechol	69
Pyrocatechol Violet	179
pyrogallol	75
pyruvic acid	14
quinaldic acid	124
quinoline-2-carboxylic acid	124
resorcinol	70
rhodizonic acid	65
sal	102

Table 10.2 (Continued)

Ligand	No. in Table
salicylaldehyde	100
salicylic acid	102
sarcosine	20
Semixylenol Orange ($\alpha_{L(H)}$; $\alpha_{HL(H)}$; $\alpha_{H_2L(H)}$)	188; 189*; 190**
serine	22
Solochrome Violet R	172
succinic acid	32
sulfoxine	118
sulfosal	105
sulphosalicylic acid	105
TAN	156
TAR	119
tart	34
D-tartaric acid	34
TEA	94
TEDTA ($\alpha_{L(H)}$; $\alpha_{HL(H)}$)	152; 153*
tetraethylenepentamine	115
tetraglycine	113
tetren	115
THAM	45
thenoyltrifluoroacetone	108
1-(1',3'-thiazol-2'-ylazo)-2-naphthol	156
thiazolylazoresorcinol	119
thiobis(ethylenenitrilo)tetra-acetic acid ($\alpha_{L(H)}$; $\alpha_{HL(H)}$)	152; 153*
thioglycollic acid	6
threonine	43
thymolphthalexone	200
tiron	74
TMDTA ($\alpha_{L(H)}$; $\alpha_{HL(H)}$)	144; 145*
tren	96
1,2,3-triaminopropanol	29
trien	95
triethanolamine	94
triethylenetetramine	95
triethylenetetranitrilohexa-acetic acid ($\alpha_{L(H)}$; $\alpha_{HL(H)}$; $\alpha_{H_2L(H)}$)	176; 177*; 178**
triglycine	88
trimethylenediamine	27
trimethylenedinitrilotetra-acetic acid ($\alpha_{L(H)}$; $\alpha_{HL(H)}$)	144; 145*
tris(hydroxymethyl)aminomethane	45
tropolone	101
tryp	142
tryptophan	142
TTA	108

Table 10.2 (Continued)

Ligand	No. in Table
TTHA ($\alpha_{L(H)}$; $\alpha_{HL(H)}$; $\alpha_{H_2L(H)}$)	176; 177*; 178**
tyrosine	121
o-tyrosine	120
valine	61
Xylenol Orange ($\alpha_{L(H)}$; $\alpha_{HL(H)}$; $\alpha_{H_2L(H)}$)	192; 193*; 194**

CHAPTER 11

Side-reaction coefficients for complexation equilibria of metal ions

Formation of various complex species of the metal ion with interfering ligands affects the solution equilibria to an extent similar to that for protonation of ligands. In aqueous medium formation of hydroxo-complexes is the most frequently encountered case, but it is also necessary to consider formation of complexes with components of a buffer or with a masking agent.

The extent of each such competing reaction can be expressed by the side-reaction coefficient for the reaction of the metal M with a ligand X. The value of this coefficient $\alpha_{M(X)}$ is calculated as indicated by the expression in Eq. (3.6), p. 45, or by Eq. (3.19) if simultaneous formation of complexes with several secondary ligands is to be expected.

For the calculation of $\alpha_{M(X)}$ it is necessary to know the stability constants for all the resulting complex species and the equilibrium concentration of the ligand; however, only the concentration of hydroxide ion is easily estimated, on the basis of pH measurement. With other complex-forming compounds only their total concentration in solution is known, which can be corrected, if necessary, for the amount bound in complexes. The equilibrium concentration of a ligand species, which is subject to protonation, is a function of pH. These factors make the calculation of the coefficient somewhat difficult. Further complication may be caused by the existence of polynuclear hydroxo-complexes; then the value of $\alpha_{M(X)}$ depends also on the equilibrium concentration of the free metal ion. Such problems can be dealt with by use of the program HALTAFALL [1]. Simpler cases have been discussed by Nagypál and Beck [2].

Tabulations of side-reaction coefficients allow easy and rapid estimation of conditional values of equilibrium constants for various types of reactions in aqueous solutions. In his pioneer monograph Ringbom [3] presented a table of $\log \alpha_{M(X)}$ for 23 metal ions, with various numbers of ligands (ranging, according to the importance of the metal ion, up to over 20). A very extensive collection of graphs presenting the dependence of $\log \alpha_{M(X)}$ on pH is available in the book by Kragten [4]. Table 11.1 presents values of $\log \alpha_{M(OH)}$ as a function of pH for three different metal concentrations: 1×10^{-1}, 1×10^{-5} and $1 \times 10^{-10} M$. The computation of these coefficients was based on stability constants of hydroxo-complexes calculated for $I \sim 1$ and $25°C$ from the data given by Baes and Mesmer [5]. Some examples of the pH-dependence of $\log \alpha_{M(OH)}$ are given in Figs. 11.1–11.4. In Tables 11.2 and 11.3 the values of $\log \alpha_{M(L,OH)}$ are listed for selected inorganic and organic ligands. The metal ions are arranged in alphabetical order, inorganic ligands in the sequence of subgroups of the periodic system. The arrangement of the organic ligands and the abbreviations used are given in Table 11.3. The total concentration of the free ligand in the system is taken to be 0.01, 0.1 and 1.0M, as stated in the second column of the table. Depending on the importance of the ligand the coefficients were calculated for all three or only some of these concentrations. For computation of the data in Tables 11.2 and 11.3 the stability constants from Chapters 6 and 7 were used. Because some values of these constants were not available for the standard conditions chosen, the listed side-reaction coefficients may be taken as valid approximations for the range of ionic strength between 0.1 and 1 and a temperature of 20–25°C. The programs for computation of the Tables in Chapters 10 and 11 were written by Dr. Suchánek, University of Chemical Technology, Prague.

The pH values for the compilation were chosen in the same way as for Chapter 10 (cf. p. 252), the total pH range being 2–12, with the main region between pH 3 and 11 covered at intervals of 0.5 pH unit to give better interpolation.

REFERENCES

[1] N. Ingri, W. Kakołowicz, L. G. Sillén and B. Warnquist, *Talanta*, 1967, **14**, 1261 (*Errata: Talanta*, 1968, **15**, No. 3, xi).
[2] I. Nagypál and M. T. Beck, *Talanta*, 1982, **29**, 473.
[3] A. Ringbom, *Complexation in Analytical Chemistry*, pp. 352–360, Interscience, New York, 1963.
[4] J. Kragten, *Atlas of Metal-Ligand Equilibria in Aqueous Solution*, Horwood, Chichester, 1978.
[5] C. F. Baes, Jr. and R. E. Mesmer, *The Hydrolysis of Cations*, Wiley, New York, 1976.

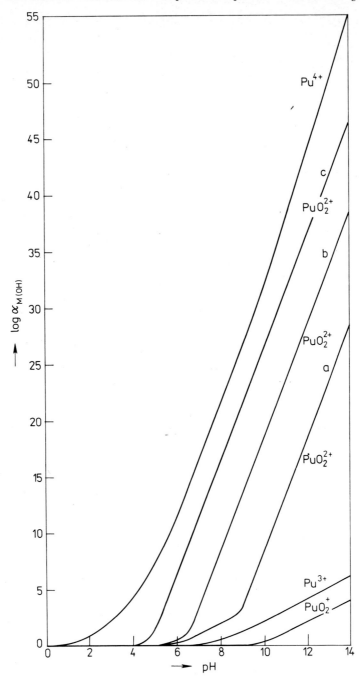

Fig. 11.1 — Side-reaction coefficients $\alpha_{M(OH)}$ of Pu^{3+}, Pu^{4+}, PuO_2^+, and PuO_2^{2+} as a function of pH ($c_{PuO_2^{2+}}$: a — $10^{-10}M$; b — $10^{-5}M$; c — $10^{-1}M$; no polynuclear complexes were considered for Pu^{3+}, Pu^{4+} and PuO_2^+).

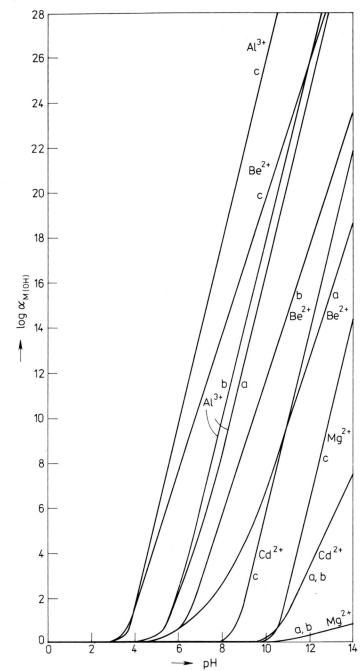

Fig. 11.2 — Side-reaction coefficients $\alpha_{M(OH)}$ of Be^{2+}, Mg^{2+}, Cd^{2+}, and Al^{3+} as a function of pH (c_M: a — $10^{-10}M$; b — $10^{-5}M$; c — $10^{-1}M$).

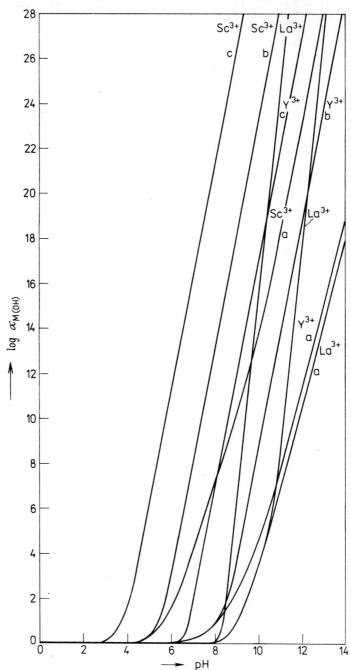

Fig. 11.3 — Side-reaction coefficients $\alpha_{M(OH)}$ of Sc^{3+}, Y^{3+}, La^{3+} as a function of pH (c_M: a — $10^{-10}M$; b — $10^{-5}M$; c — $10^{-1}M$).

Ch. 11] Side-Reaction Coefficients for Complexation Equilibria of Metal Ions 283

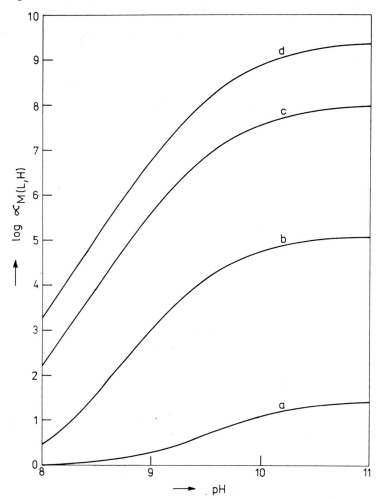

Fig. 11.4 — Side-reaction coefficient for the formation of zinc-ammine complexes as a function of pH (presence of hydroxo-complexes is not considered in this case; $\log K_H(NH_3) = 9.3$, see Table 3.1, p. 46, for the stability constants). The total concentration of free ammonia is constant ($[NH_3']$: a — 0.01; b — 0.1; c — 0.5; d — $1M$).

Table 11.1
Values of side-reaction coefficients $\alpha_{M(OH)}$

Ion	Concentration [M], M	pH 2.0	3.0	3.5	4.0	4.5	5.0	5.5	6.0
Ag^+		0.00	0.00	0.00	0.00	0.00	0.00	0.00	0.00
Al^{3+}	10^{-10}	0.00	0.00	0.00	0.02	0.06	0.28	0.98	2.13
	10^{-5}	0.00	0.00	0.00	0.02	0.06	0.28	1.01	2.29
	10^{-1}	0.00	0.01	0.22	1.79	3.78	5.78	7.78	9.78
Am^{3+}		0.00	0.00	0.00	0.01	0.02	0.06	0.18	0.41
Au^{3+}		5.15	8.11	9.60	11.10	12.60	14.10	15.60	17.10
Be^{2+}	10^{-10}	0.00	0.00	0.00	0.01	0.03	0.08	0.22	0.48
	10^{-5}	0.00	0.00	0.00	0.01	0.03	0.08	0.22	0.54
	10^{-1}	0.00	0.05	0.42	1.66	3.15	4.65	6.15	7.65
Bi^{3+}		0.62	1.61	2.33	3.36	4.66	6.10	7.58	9.07
Cd^{2+}	10^{-5}	0.00	0.00	0.00	0.00	0.00	0.00	0.00	0.00
	10^{-1}	0.00	0.00	0.00	0.00	0.00	0.00	0.00	0.00
Ce^{3+}	10^{-10}	0.00	0.00	0.00	0.00	0.00	0.00	0.00	0.00
	10^{-5}	0.00	0.00	0.00	0.00	0.00	0.00	0.00	0.00
	10^{-1}	0.00	0.00	0.00	0.00	0.00	0.00	0.00	0.00
Ce^{4+}	10^{-10}	4.33	6.38	7.78	9.62	11.60	13.60	15.60	17.60
	10^{-5}	6.63	10.60	12.60	14.60	16.60	18.60	20.60	22.60
	10^{-1}	10.63	14.60	16.60	18.60	20.60	22.60	24.60	26.60
Co^{2+}	10^{-5}	0.00	0.00	0.00	0.00	0.00	0.00	0.00	0.00
	10^{-1}	0.00	0.00	0.00	0.00	0.00	0.00	0.00	0.00
Cr^{3+}	10^{-10}	0.00	0.01	0.03	0.10	0.26	0.56	1.02	1.62
	10^{-5}	0.00	0.01	0.03	0.12	0.77	2.58	4.58	6.58
	10^{-1}	0.03	2.58	4.58	6.58	8.58	10.58	12.58	14.58
Cu^{2+}	10^{-10}	0.00	0.00	0.00	0.00	0.00	0.00	0.00	0.00
	10^{-5}	0.00	0.00	0.00	0.00	0.00	0.00	0.00	0.00
	10^{-1}	0.00	0.00	0.00	0.00	0.00	0.01	0.12	0.63
Es^{3+}		0.00	0.00	0.00	0.01	0.02	0.06	0.18	0.41
Fe^{2+}		0.00	0.00	0.00	0.00	0.00	0.00	0.00	0.00
Fe^{3+}	10^{-10}	0.07	0.49	1.01	1.75	2.64	3.62	4.65	5.76
	10^{-5}	0.07	0.49	1.03	1.82	2.98	4.72	6.68	8.68
	10^{-1}	1.03	4.68	6.68	8.68	10.68	12.68	14.68	16.68
Ga^{3+}		0.03	0.24	0.68	1.63	2.96	4.42	5.92	7.45
Hf^{4+}		1.25	3.05	4.09	5.24	6.60	8.25	10.14	12.19
Hg_2^{2+}		0.00	0.00	0.01	0.04	0.12	0.30	0.62	1.04
Hg^{2+}	10^{-5}	0.01	0.23	0.83	1.73	2.71	3.70	4.70	5.70
	10^{-1}	0.02	1.12	2.57	4.06	5.56	7.06	8.56	10.06
In^{3+}	10^{-10}	0.00	0.02	0.08	0.28	0.90	2.01	3.37	4.82

Ch. 11] Side-Reaction Coefficients for Complexation Equilibria of Metal Ions

$\log \alpha_{M(OH)}$

pH 6.5	7.0	7.5	8.0	8.5	9.0	9.5	10.0	10.5	11.0	12.0
0.00	0.00	0.00	0.00	0.00	0.00	0.00	0.00	0.01	0.02	0.34
3.51	5.02	6.66	8.45	10.35	12.32	14.31	16.30	18.30	20.30	24.30
3.96	5.85	7.81	9.80	11.79	13.79	15.79	17.79	19.79	21.79	25.79
11.78	13.78	15.78	17.78	19.78	21.78	23.78	25.78	27.78	29.78	33.78
0.78	1.23	1.71	2.20	2.70	3.20	3.70	4.20	4.70	5.20	6.20
18.60	20.10	21.60	23.10	24.60	26.10	27.60	29.11	30.64	32.21	35.75
0.88	1.35	1.89	2.52	3.30	4.27	5.45	6.78	8.22	9.70	12.69
1.33	2.67	4.15	5.65	7.15	8.65	10.15	11.65	13.15	14.65	17.65
9.15	10.65	12.15	13.65	15.15	16.65	18.15	19.65	21.15	22.65	25.65
10.57	12.07	13.57	15.07	16.57	18.07	19.57	21.07	22.57	24.07	27.08
0.00	0.00	0.00	0.00	0.01	0.02	0.06	0.19	0.59	1.35	3.36
0.00	0.00	0.00	0.01	0.22	1.77	3.76	5.76	7.76	9.76	13.76
0.00	0.00	0.01	0.04	0.12	0.30	0.62	1.04	1.51	2.02	5.48
0.00	0.00	0.01	0.04	0.12	0.70	2.98	5.48	7.98	10.48	15.48
0.00	0.02	1.02	3.48	5.98	8.48	10.98	13.48	15.98	18.48	23.48
19.60	21.60	23.60	25.60	27.60	29.60	31.60	33.60	35.60	37.60	41.60
24.60	26.60	28.60	30.60	32.60	34.60	36.60	38.60	40.60	42.60	46.60
28.60	30.60	32.60	34.60	36.60	38.60	40.60	42.60	44.60	46.60	50.60
0.00	0.00	0.00	0.00	0.02	0.06	0.26	0.85	1.74	2.73	4.92
0.00	0.00	0.01	0.35	2.10	4.09	6.09	8.09	10.09	12.09	16.09
2.38	3.28	4.27	5.40	6.87	8.69	10.65	12.64	14.63	16.63	20.63
8.58	10.58	12.58	14.58	16.58	18.58	20.58	22.58	24.58	26.58	30.58
16.58	18.58	20.58	22.58	24.58	26.58	28.58	30.58	32.58	34.58	38.58
0.01	0.02	0.06	0.18	0.45	0.94	1.68	2.70	4.07	5.76	9.62
0.01	0.03	0.17	0.68	1.55	2.52	3.52	4.52	5.53	6.58	9.65
1.52	2.51	3.51	4.51	5.51	6.51	7.51	8.51	9.51	10.51	12.51
0.78	1.23	1.71	2.20	2.70	3.20	3.70	4.20	4.70	5.20	6.20
0.00	0.00	0.00	0.01	0.02	0.06	0.18	0.44	0.95	1.95	4.81
7.01	8.38	9.83	11.33	12.88	14.51	16.28	18.16	20.12	22.11	26.10
10.68	12.68	14.68	16.68	18.68	20.68	22.68	24.68	26.68	28.68	32.68
18.68	20.68	22.68	24.68	26.68	28.68	30.68	32.68	34.68	36.68	40.68
9.05	10.75	12.60	14.53	16.51	18.50	20.50	22.50	24.50	26.50	30.50
14.42	16.78	19.23	21.71	24.20	26.70	29.20	31.70	34.20	36.70	41.70
1.51	2.00	2.50	3.00	3.50	4.00	4.50	5.00	5.50	6.00	7.00
6.70	7.70	8.70	9.71	10.73	11.79	12.94	14.22	15.61	17.08	20.06
11.56	13.06	14.56	16.06	17.56	19.06	20.56	22.06	23.56	25.06	28.06
6.31	7.80	9.31	10.82	12.35	13.95	15.66	17.51	19.44	21.42	25.41

Table 11.1 (Continued)

Ion	Concentration [M], M	pH 2.0	3.0	3.5	4.0	4.5	5.0	5.5	6.0
In^{3+}	10^{-5}	0.00	0.02	0.13	1.22	3.17	5.17	7.17	9.17
	10^{-1}	1.20	5.17	7.17	9.17	11.17	13.17	15.17	17.17
La^{3+}	10^{-10}	0.00	0.00	0.00	0.00	0.00	0.00	0.00	0.00
	10^{-5}	0.00	0.00	0.00	0.00	0.00	0.00	0.00	0.00
	10^{-1}	0.00	0.00	0.00	0.00	0.00	0.00	0.00	0.00
Ln^{3+} *	10^{-10}	0.00	0.00	0.00	0.00	0.00	0.00	0.00	0.00
*	10^{-5}	0.00	0.00	0.00	0.00	0.00	0.00	0.00	0.00
*	10^{-1}	0.00	0.00	0.00	0.00	0.00	0.00	0.00	0.00
Mg^{2+}	10^{-5}	0.00	0.00	0.00	0.00	0.00	0.00	0.00	0.00
	10^{-1}	0.00	0.00	0.00	0.00	0.00	0.00	0.00	0.00
Mn^{2+}	10^{-10}	0.00	0.00	0.00	0.00	0.00	0.00	0.00	0.00
	10^{-5}	0.00	0.00	0.00	0.00	0.00	0.00	0.00	0.00
	10^{-1}	0.00	0.00	0.00	0.00	0.00	0.00	0.00	0.00
Ni^{2+}	10^{-5}	0.00	0.00	0.00	0.00	0.00	0.00	0.00	0.00
	10^{-1}	0.00	0.00	0.00	0.00	0.00	0.00	0.00	0.00
Np^{4+}		0.15	0.70	1.13	1.61	2.10	2.60	3.10	3.60
NpO_2^+		0.00	0.00	0.00	0.00	0.00	0.00	0.00	0.00
NpO_2^{2+}	10^{-10}	0.00	0.00	0.00	0.01	0.04	0.11	0.27	0.57
	10^{-5}	0.00	0.00	0.00	0.01	0.04	0.12	0.46	2.29
	10^{-1}	0.00	0.02	0.16	0.86	2.80	5.27	7.77	10.27
Pa^{4+}		3.52	5.89	7.27	8.72	10.21	11.70	13.20	14.70
PaO_2^+		0.01	0.01	0.04	0.12	0.30	0.62	1.04	1.51
Pb^{2+}	10^{-10}	0.00	0.00	0.00	0.00	0.00	0.00	0.00	0.01
	10^{-5}	0.00	0.00	0.00	0.00	0.00	0.00	0.00	0.01
	10^{-1}	0.00	0.00	0.00	0.00	0.00	0.01	0.39	2.30
Pd^{2+}		0.48	2.05	3.02	4.05	5.30	7.04	9.00	11.00
Pu^{3+}		0.00	0.00	0.00	0.00	0.00	0.00	0.00	0.01
Pu^{4+}		1.16	3.12	4.31	5.74	7.45	9.43	11.62	13.98
PuO_2^+		0.00	0.00	0.00	0.00	0.00	0.00	0.00	0.00
PuO_2^{2+}	10^{-10}	0.00	0.00	0.00	0.00	0.01	0.04	0.12	0.30
	10^{-5}	0.00	0.00	0.00	0.00	0.01	0.04	0.12	0.32
	10^{-1}	0.00	0.00	0.00	0.03	0.24	1.49	3.88	6.38
Rh^{3+}		0.03	0.25	0.55	0.95	1.43	2.16	3.82	5.80
Sb^{3+}		0.00	0.70	2.04	3.51	5.00	6.50	8.00	9.50
Sc^{3+}	10^{-10}	0.00	0.00	0.01	0.03	0.09	0.27	0.73	1.60
	10^{-5}	0.00	0.00	0.01	0.03	0.09	0.30	1.05	3.14
	10^{-1}	0.00	0.07	0.42	1.48	3.64	6.13	8.63	11.13

Ch. 11] Side-Reaction Coefficients for Complexation Equilibria of Metal Ions

$$\log \alpha_{M(OH)}$$

pH	6.5	7.0	7.5	8.0	8.5	9.0	9.5	10.0	10.5	11.0	12.0
	11.17	13.17	15.17	17.17	19.17	21.17	23.17	25.17	27.17	29.17	33.17
	19.17	21.17	23.17	25.17	27.17	29.17	31.17	33.17	35.17	37.17	41.17
	0.00	0.00	0.01	0.04	0.20	0.91	2.20	3.67	5.23	6.90	10.64
	0.00	0.00	0.01	0.04	0.20	0.91	2.20	3.67	5.25	8.51	17.50
	0.00	0.00	0.01	0.06	2.07	6.50	11.00	15.50	20.00	24.50	33.50
	0.00	0.01	0.03	0.08	0.21	0.48	0.86	1.34	2.56	4.98	9.98
	0.00	0.01	0.03	0.33	2.48	4.98	7.48	9.98	12.48	14.98	19.98
	0.61	2.98	5.48	7.98	10.48	12.98	15.48	17.98	20.48	22.98	27.98
	0.00	0.00	0.00	0.00	0.00	0.00	0.00	0.00	0.01	0.04	0.31
	0.00	0.00	0.00	0.00	0.00	0.00	0.00	0.02	0.59	2.46	6.46
	0.00	0.00	0.00	0.00	0.00	0.00	0.02	0.05	0.14	0.37	1.84
	0.00	0.00	0.00	0.00	0.00	0.01	0.05	0.56	1.91	3.40	6.40
	0.00	0.00	0.00	0.01	0.26	1.42	2.90	4.40	5.90	7.40	10.40
	0.00	0.00	0.00	0.00	0.01	0.04	0.18	0.74	1.77	3.14	6.69
	0.00	0.02	0.72	2.62	4.62	6.62	8.62	10.62	12.62	14.62	18.62
	4.10	4.60	5.10	5.60	6.10	6.60	7.10	7.60	8.10	8.60	9.60
	0.00	0.01	0.02	0.06	0.16	0.38	0.74	1.18	1.66	2.15	3.15
	0.99	1.46	1.95	2.66	4.77	7.27	9.77	12.27	14.77	17.27	22.27
	4.77	7.27	9.77	12.27	14.77	17.27	19.77	22.27	24.77	27.27	32.27
	12.77	15.27	17.77	20.27	22.77	25.27	27.77	30.27	32.77	35.27	40.27
	16.20	17.70	19.20	20.70	22.20	23.70	25.20	26.70	28.20	29.70	32.70
	2.00	2.50	3.00	3.50	4.00	4.50	5.00	5.50	6.00	6.50	7.50
	0.02	0.07	0.20	0.45	0.86	1.39	2.08	2.93	3.96	5.17	7.97
	0.02	0.07	0.20	0.53	2.07	5.76	9.76	13.76	17.76	21.76	29.76
	5.77	9.76	13.76	17.76	21.76	25.76	29.76	33.76	37.76	41.76	49.76
	13.00	15.00	17.00	19.00	21.00	23.00	25.00	27.00	29.00	31.00	35.00
	0.02	0.06	0.18	0.41	0.78	1.23	1.71	2.20	2.70	3.20	4.20
	16.43	18.91	21.40	23.90	26.40	28.90	31.40	33.90	36.40	38.90	43.90
	0.00	0.00	0.00	0.01	0.02	0.05	0.15	0.35	0.70	1.13	2.10
	0.62	1.04	1.51	2.00	2.51	3.53	5.88	8.38	10.88	13.38	18.38
	1.09	3.38	5.88	8.38	10.88	13.38	15.88	18.38	20.88	23.38	28.38
	8.88	11.38	13.88	16.38	18.88	21.38	23.88	26.38	28.88	31.38	36.38
	7.80	9.80	11.80	13.80	15.80	17.80	19.80	21.80	23.80	25.80	29.80
	11.00	12.50	14.00	15.50	17.00	18.50	20.00	21.51	23.03	24.60	28.05
	2.79	4.17	5.62	7.11	8.62	10.16	11.78	13.62	15.79	18.18	23.13
	5.63	8.13	10.63	13.13	15.63	18.13	20.63	23.13	25.63	28.13	33.13
	13.63	16.13	18.63	21.13	23.63	26.13	28.63	31.13	33.63	36.13	41.13

Table 11.1 (Continued)

Ion	Concentration [M], M	pH 2.0	3.0	3.5	4.0	4.5	5.0	5.5	6.0
Sn^{2+}	10^{-10}	0.01	0.07	0.23	0.71	1.51	2.45	3.44	4.43
	10^{-5}	0.01	0.07	0.24	0.72	1.64	3.13	5.04	7.03
	10^{-1}	0.06	3.03	5.03	7.03	9.03	11.03	13.03	15.03
Th^{4+}	10^{-5}	0.00	0.05	0.19	0.70	1.64	3.02	4.84	6.81
	10^{-1}	0.02	0.79	1.71	2.71	3.70	4.71	5.76	7.06
Ti^{3+}	10^{-10}	0.06	0.41	0.78	1.23	1.71	2.20	2.70	3.20
	10^{-5}	0.06	0.41	0.78	1.24	1.75	2.32	3.00	3.82
	10^{-1}	0.22	1.72	2.71	3.70	4.70	5.70	6.70	7.70
TiO^{2+}		0.10	0.90	1.76	2.72	3.71	4.70	5.70	6.70
Tl^{3+}		2.09	4.83	6.31	7.80	9.30	10.80	12.30	13.80
U^{4+}		1.00	2.88	3.99	5.25	6.72	8.49	10.61	12.97
UO_2^{2+}	10^{-10}	0.00	0.00	0.00	0.01	0.02	0.06	0.18	0.41
	10^{-5}	0.00	0.00	0.00	0.01	0.03	0.19	1.71	4.18
	10^{-1}	0.00	0.10	0.61	2.25	4.68	7.18	9.68	12.18
V^{3+}	10^{-10}	0.06	0.39	0.75	1.19	1.67	2.16	2.66	3.17
	10^{-5}	0.06	0.39	0.75	1.26	2.11	3.43	4.90	6.40
	10^{-1}	0.15	1.72	3.02	4.44	5.91	7.41	8.90	10.40
VO^{2+}	10^{-5}	0.00	0.00	0.00	0.03	0.19	0.80	1.72	2.71
	10^{-1}	0.00	0.01	0.08	0.46	1.30	2.28	3.27	4.27
VO_2^+		0.02	0.20	0.55	1.20	2.07	3.02	4.01	5.01
Y^{3+}	10^{-10}	0.00	0.00	0.00	0.00	0.00	0.00	0.01	0.02
	10^{-5}	0.00	0.00	0.00	0.00	0.00	0.00	0.01	0.02
	10^{-1}	0.00	0.00	0.00	0.00	0.00	0.00	0.01	0.02
Zn^{2+}	10^{-10}	0.00	0.00	0.00	0.00	0.00	0.00	0.00	0.00
	10^{-5}	0.00	0.00	0.00	0.00	0.00	0.00	0.00	0.00
	10^{-1}	0.00	0.00	0.00	0.00	0.00	0.00	0.00	0.00
Zr^{4+}	10^{-10}	1.89	3.76	4.83	6.04	7.49	9.21	11.15	13.27
	10^{-5}	3.01	6.98	8.98	10.98	12.98	14.98	16.98	18.98
	10^{-1}	10.98	14.98	16.98	18.98	20.98	22.98	24.98	26.98

Ch. 11]   **Side-Reaction Coefficients for Complexation Equilibria of Metal Ions**   289

$$\log \alpha_{M(OH)}$$

pH										
6.5	7.0	7.5	8.0	8.5	9.0	9.5	10.0	10.5	11.0	12.0
5.43	6.43	7.44	8.46	9.52	10.67	11.99	13.51	15.23	17.10	21.03
9.03	11.03	13.03	15.03	17.03	19.03	21.03	23.03	25.03	27.03	31.03
17.03	19.03	21.03	23.03	25.03	27.03	29.03	31.03	33.03	35.03	39.03
8.80	10.80	12.80	14.80	16.80	18.80	20.80	22.80	24.80	26.80	30.80
8.84	10.80	12.80	14.80	16.80	18.80	20.80	22.80	24.80	26.80	30.80
3.70	4.20	4.70	5.20	5.70	6.21	6.74	7.32	8.00	8.82	10.71
4.74	5.71	6.71	7.70	8.70	9.70	10.70	11.70	12.70	13.70	15.70
8.70	9.70	10.70	11.70	12.70	13.70	14.70	15.70	16.70	17.70	19.70
7.70	8.70	9.70	10.70	11.70	12.70	13.70	14.70	15.70	16.70	18.70
15.30	16.80	18.30	19.80	21.30	22.80	24.31	25.82	27.35	28.95	32.50
15.42	17.91	20.40	22.90	25.40	27.90	30.40	32.90	35.40	37.90	42.90
0.78	1.23	2.00	4.18	6.68	9.18	11.68	14.18	16.68	19.18	24.18
6.68	9.18	11.68	14.18	16.68	19.18	21.68	24.18	26.68	29.18	34.18
14.68	17.18	19.68	22.18	24.68	27.18	29.68	32.18	34.68	37.18	42.18
3.73	4.60	5.93	7.40	8.90	10.40	11.90	13.40	14.90	16.40	19.40
7.90	9.40	10.90	12.40	13.90	15.40	16.90	18.40	19.90	21.40	24.40
11.90	13.40	14.90	16.40	17.90	19.40	20.90	22.40	23.90	25.40	28.40
3.71	4.71	5.71	6.71	7.71	8.71	9.71	10.71	11.71	12.71	14.71
5.27	6.27	7.27	8.27	9.27	10.27	11.27	12.27	13.27	14.27	16.27
6.03	7.08	8.21	9.48	10.86	12.32	13.81	15.30	16.80	18.30	21.33
0.06	0.18	0.44	0.88	1.53	2.38	3.44	4.70	6.13	7.73	11.37
0.06	0.18	0.44	0.89	1.76	3.89	6.38	8.88	11.38	13.88	18.88
0.15	1.89	4.38	6.88	9.38	11.88	14.38	16.88	19.38	21.88	26.88
0.00	0.00	0.01	0.05	0.27	0.91	1.83	2.83	3.87	4.99	7.89
0.00	0.00	0.01	0.05	0.27	0.91	1.83	2.83	3.87	5.04	10.00
0.00	0.00	0.02	0.06	0.29	0.92	1.83	2.89	5.03	8.00	14.00
15.55	17.95	20.42	22.91	25.40	27.90	30.40	32.90	35.40	37.90	42.90
20.98	22.98	24.98	26.98	28.98	30.98	32.98	34.98	36.99	39.01	43.24
28.98	30.98	32.98	34.98	36.98	38.98	40.98	42.98	44.98	46.98	50.98

Table 11.2

Values of side-reaction coefficients $\alpha_{M(L,OH)}$ for inorganic ligands

Ligand	Concentration [L], M	pH 2.0	3.0	3.5	4.0	4.5	5.0	5.5	6.0
		Ag^+							
CN^-	0.01	2.10	4.10	5.10	6.10	7.10	8.10	9.10	10.10
	0.10	4.10	6.10	7.10	8.10	9.10	10.10	11.10	12.10
	1.00	6.10	8.10	9.10	10.10	11.10	12.10	13.10	14.10
NH_3	0.01	0.00	0.00	0.00	0.00	0.00	0.00	0.00	0.00
	0.10	0.00	0.00	0.00	0.00	0.00	0.00	0.01	0.03
	1.00	0.00	0.00	0.00	0.00	0.01	0.03	0.13	0.51
HS^-	0.10	23.37	23.37	23.37	23.37	23.37	23.36	23.34	23.27
SO_3^{2-}	0.10	0.29	1.09	1.58	2.11	2.69	3.38	4.18	5.00
$S_2O_3^{2-}$	0.01	20.14	20.26	20.27	20.28	20.28	20.28	20.28	20.28
	0.10	25.14	25.26	25.27	25.28	25.28	25.28	25.28	25.28
		Al^{3+}							
F^-	0.01	6.41	9.05	9.71	9.99	10.09	10.12	10.13	10.13
	0.10	10.22	13.34	14.12	14.44	14.56	14.60	14.61	14.62
	1.00	14.73	18.45	19.36	19.75	19.88	19.93	19.94	19.95
		Am^{3+}							
SO_4^{2-}	0.10	0.90	0.96	0.97	0.97	0.97	0.98	0.99	1.04
	1.00	2.60	2.68	2.69	2.69	2.69	2.69	2.70	2.70
F^-	0.10	3.05	5.17	5.68	5.89	5.97	5.99	6.00	6.00
	1.00	5.99	8.16	8.67	8.89	8.96	8.99	9.00	9.00
		Au^+							
CN^-	0.01	20.40	22.40	23.40	24.40	25.40	26.40	27.40	28.40
	0.10	22.40	24.40	25.40	26.40	27.40	28.40	29.40	30.40
NH_3	0.10	10.20	12.20	13.20	14.20	15.20	16.20	17.20	18.20
		Au^{3+}							
NH_3	0.01	5.15	8.11	9.60	11.10	12.60	14.10	15.60	17.10
	0.10	5.15	8.11	9.60	11.10	12.60	14.10	15.60	17.10
	1.00	5.15	8.11	9.60	11.10	12.60	14.11	15.63	17.18
Cl^-	0.10	13.50	13.50	13.51	13.53	13.64	14.27	15.61	17.10
	1.00	17.50	17.50	17.50	17.50	17.51	17.53	17.60	17.85
		Ba^{2+}							
$P_3O_{10}^{5-}$	1.00	0.00	0.06	0.19	0.46	0.83	1.22	1.55	1.80
		Be^{2+}							
SO_4^{2-}	1.00	2.14	2.26	2.27	2.27	2.27	2.27	2.27	2.28
F^-	0.10	5.75	8.18	8.81	9.08	9.17	9.21	9.22	9.22
	1.00	9.20	12.00	12.68	12.96	13.06	13.10	13.11	13.11
		Bi^{3+}							
SCN^-	1.00	4.02	4.18	4.20	4.26	4.79	6.10	7.58	9.07
Cl^-	1.00	7.02	7.02	7.02	7.02	7.02	7.07	7.69	9.08
Br^-	1.00	9.68	9.68	9.68	9.68	9.68	9.68	9.69	9.78
I^-	1.00	18.80	18.80	18.80	18.80	18.80	18.80	18.80	18.80

Side-Reaction Coefficients for Complexation Equilibria of Metal Ions

$$\log \alpha_{M(L,OH)}$$

pH 6.5	7.0	7.5	8.0	8.5	9.0	9.5	10.0	10.5	11.0	12.0
Ag^+										
11.10	12.09	13.07	14.01	14.84	15.45	15.79	15.93	15.98	16.00	16.01
13.10	14.09	15.07	16.02	16.86	17.50	17.86	18.02	18.07	18.09	18.10
15.10	16.10	17.10	18.11	19.13	20.03	20.59	20.84	20.93	20.95	20.97
0.01	0.03	0.13	0.50	1.21	2.00	2.62	2.97	3.12	3.18	3.20
0.13	0.51	1.28	2.21	3.14	3.98	4.61	4.97	5.12	5.17	5.20
1.29	2.23	3.22	4.19	5.14	5.98	6.61	6.97	7.12	7.17	7.20
23.12	22.82	22.42	21.95	21.46	20.97	20.47	19.97	19.47	18.97	17.96
5.65	6.00	6.14	6.19	6.21	6.22	6.22	6.22	6.22	6.22	6.22
20.28	20.28	20.28	20.28	20.28	20.28	20.28	20.28	20.28	20.28	20.28
25.28	25.28	25.28	25.28	25.28	25.28	25.28	25.28	25.28	25.28	25.28
Al^{3+}										
10.13	10.13	10.14	10.30	11.80	13.79	15.79	17.79	19.79	21.79	25.79
14.62	14.62	14.62	14.62	14.62	14.68	15.82	17.79	19.79	21.79	25.79
19.95	19.95	19.95	19.95	19.95	19.95	19.95	19.95	20.18	21.80	25.79
Am^{3+}										
1.16	1.40	1.77	2.22	2.71	3.20	3.70	4.20	4.70	5.20	6.20
2.70	2.71	2.74	2.82	3.00	3.32	3.74	4.21	4.70	5.20	6.20
6.01	6.01	6.01	6.01	6.01	6.01	6.01	6.01	6.03	6.07	6.41
9.00	9.00	9.00	9.00	9.00	9.00	9.00	9.00	9.00	9.00	9.00
Au^+										
29.40	30.39	31.37	32.31	33.14	33.75	34.08	34.23	34.28	34.29	34.30
31.40	32.39	33.37	34.31	35.14	35.75	36.08	36.23	36.28	36.29	36.30
19.20	20.20	21.19	22.17	23.10	23.91	24.49	24.81	24.93	24.98	25.00
Au^{3+}										
18.60	20.10	21.60	23.10	24.20	26.10	27.60	29.11	30.63	32.20	35.73
18.60	20.10	21.60	23.10	24.20	26.10	27.60	29.11	30.63	32.20	35.73
18.81	20.57	22.45	24.36	26.20	27.83	29.00	29.73	30.70	32.20	35.73
18.60	20.10	21.60	23.10	24.20	26.10	27.60	29.11	30.63	32.20	35.73
18.72	20.12	21.60	23.10	24.20	26.10	27.60	29.11	30.63	32.20	35.73
Ba^{2+}										
2.05	2.37	2.72	3.01	3.19	3.26	3.29	3.30	3.30	3.30	3.30
Be^{2+}										
2.32	2.81	4.16	5.65	7.15	8.65	10.15	11.65	13.15	14.65	17.65
9.22	9.22	9.22	9.22	9.22	9.32	10.20	11.65	13.15	14.65	17.65
13.11	13.11	13.11	13.11	13.11	13.11	13.11	13.13	13.43	14.66	17.65
Bi^{3+}										
10.57	12.07	13.57	15.07	16.57	18.07	19.57	21.07	22.57	24.07	27.08
10.57	12.07	13.57	15.07	16.57	18.07	19.57	21.07	22.57	24.07	27.08
10.62	12.07	13.57	15.07	16.57	18.07	19.57	21.07	22.57	24.07	27.08
18.80	18.80	18.80	18.80	18.81	18.88	19.64	21.07	22.57	24.07	27.08

Table 11.2 (Continued)

Ligand	Concentration [L], M	pH 2.0	3.0	3.5	4.0	4.5	5.0	5.5	6.0
		Ca^{2+}							
$P_2O_7^{4-}$	0.10	0.00	0.01	0.03	0.08	0.22	0.48	0.85	1.27
	1.00	0.01	0.08	0.22	0.49	0.88	1.33	1.79	2.25
$P_3O_{10}^{5-}$	0.10	0.00	0.06	0.17	0.41	0.77	1.19	1.60	1.98
	1.00	0.03	0.39	0.77	1.23	1.70	2.16	2.59	2.98
$P_4O_{13}^{6-}$	0.10	0.00	0.03	0.09	0.26	0.55	0.97	1.44	1.97
	1.00	0.01	0.23	0.54	0.95	1.43	1.92	2.43	2.96
F^-	1.00	0.14	0.48	0.60	0.65	0.67	0.68	0.68	0.68
		Cd^{2+}							
CN^-	0.01	0.00	0.00	0.00	0.01	0.02	0.05	0.16	0.45
	0.10	0.00	0.01	0.02	0.05	0.16	0.45	1.14	2.29
	1.00	0.01	0.05	0.16	0.45	1.14	2.29	3.73	5.38
SCN^-	1.00	2.38	2.47	2.47	2.47	2.48	2.48	2.48	2.48
NH_3	0.01	0.00	0.00	0.00	0.00	0.00	0.00	0.00	0.00
	0.10	0.00	0.00	0.00	0.00	0.00	0.00	0.00	0.01
	1.00	0.00	0.00	0.00	0.00	0.00	0.01	0.02	0.07
HS^-	0.10	0.01	0.10	0.29	0.71	1.39	2.27	3.22	4.21
SO_4^{2-}	1.00	2.32	2.48	2.49	2.49	2.49	2.49	2.49	2.49
$S_2O_3^{2-}$	1.00	7.17	7.25	7.26	7.26	7.26	7.26	7.26	7.26
		Ce^{3+}							
$P_2O_7^{4-}$	0.10	5.25	7.41	8.42	9.42	10.42	11.39	12.31	13.12
	1.00	6.25	8.41	9.42	10.42	11.42	12.39	13.31	14.12
SO_4^{2-}	1.00	3.02	3.15	3.16	3.16	3.17	3.17	3.17	3.17
F^-	1.00	1.81	2.53	2.70	2.77	2.80	2.81	2.81	2.81
		Ce^{4+}							
$P_2O_7^{4-}$	0.10	6.87	10.61	12.60	14.60	16.60	18.60	20.60	22.60
	1.00	7.56	10.65	12.61	14.60	16.60	18.60	20.60	22.60
SO_4^{2-}	0.10	7.38	10.60	12.60	14.60	16.60	18.60	20.60	22.60
	1.00	10.28	10.81	12.60	14.60	16.60	18.60	20.60	22.60
		Co^{2+}							
NH_3	0.01	0.00	0.00	0.00	0.00	0.00	0.00	0.00	0.00
	0.10	0.00	0.00	0.00	0.00	0.00	0.00	0.00	0.00
	1.00	0.00	0.00	0.00	0.00	0.00	0.00	0.01	0.02
$P_2O_7^{4-}$	0.10	0.03	0.33	0.68	1.12	1.63	2.18	2.81	3.47
	1.00	0.25	1.10	1.59	2.09	2.62	3.18	3.81	4.47
$P_3O_{10}^{5-}$	0.10	0.14	1.01	1.52	2.03	2.54	3.07	3.63	4.20
	1.00	0.69	1.97	2.51	3.03	3.54	4.07	4.63	5.20
		Co^{3+}							
NH_3	0.01	0.00	0.03	0.08	0.24	0.63	1.36	2.51	4.12
	0.10	0.03	0.24	0.63	1.36	2.51	4.12	6.20	8.67
	1.00	0.24	1.36	2.51	4.12	6.20	8.67	11.41	14.31

Ch. 11] Side-Reaction Coefficients for Complexation Equilibria of Metal Ions

$$\log \alpha_{M(L,OH)}$$

pH	6.5	7.0	7.5	8.0	8.5	9.0	9.5	10.0	10.5	11.0	12.0
Ca^{2+}											
	1.71	2.17	2.64	3.09	3.46	3.71	3.83	3.89	3.92	3.98	4.38
	2.70	3.17	3.64	4.09	4.46	4.71	4.83	4.89	4.92	4.98	5.38
	2.37	2.81	3.27	3.70	4.04	4.25	4.35	4.40	4.44	4.52	5.02
	3.37	3.81	4.27	4.70	5.04	5.25	5.35	5.40	5.44	5.52	6.02
	2.52	3.06	3.56	3.97	4.24	4.38	4.44	4.46	4.47	4.47	4.47
	3.51	4.06	4.56	4.97	5.24	5.38	5.44	5.46	5.47	5.47	5.47
	0.68	0.68	0.68	0.68	0.68	0.68	0.68	0.68	0.68	0.68	0.68
Cd^{2+}											
	1.14	2.29	3.72	5.33	7.03	8.64	9.86	10.56	10.86	10.96	11.01
	3.73	5.38	7.18	9.06	10.92	12.59	13.84	14.54	14.84	14.95	15.00
	7.19	9.11	11.07	13.03	14.91	16.58	17.84	18.54	18.84	18.95	18.99
	2.48	2.48	2.48	2.48	2.48	2.48	2.48	2.48	2.48	2.51	3.41
	0.00	0.01	0.02	0.07	0.20	0.45	0.79	1.05	1.23	1.57	3.36
	0.02	0.07	0.21	0.54	1.16	2.00	2.82	3.36	3.60	3.69	3.89
	0.21	0.56	1.23	2.26	3.63	5.11	6.34	7.08	7.40	7.51	7.56
	5.20	6.20	7.21	8.22	9.25	10.31	11.41	12.38	13.05	13.37	13.51
	2.49	2.49	2.49	2.49	2.49	2.49	2.49	2.50	2.50	2.52	3.42
	7.26	7.26	7.26	7.26	7.26	7.26	7.26	7.26	7.26	7.26	7.26
Ce^{3+}											
	13.79	14.36	14.87	15.33	15.71	15.96	16.08	16.13	16.14	16.15	16.23
	14.79	15.36	15.87	16.33	16.71	16.96	17.08	17.13	17.14	17.15	17.16
	3.17	3.17	3.17	3.17	3.17	3.17	3.38	5.48	7.98	10.48	15.48
	2.81	2.81	2.81	2.81	2.81	2.81	3.20	5.48	7.98	10.48	15.48
Ce^{4+}											
	24.60	26.60	28.60	30.60	32.60	34.60	36.60	38.60	40.60	42.60	46.60
	24.60	26.60	28.60	30.60	32.60	34.60	36.60	38.60	40.60	42.60	46.60
	24.60	26.60	28.60	30.60	32.60	34.60	36.60	38.60	40.60	42.60	46.60
	24.60	26.60	28.60	30.60	32.60	34.60	36.60	38.60	40.60	42.60	46.60
Co^{2+}											
	0.00	0.00	0.01	0.02	0.06	0.17	0.41	0.92	1.75	2.73	4.92
	0.01	0.02	0.06	0.16	0.42	0.90	1.48	1.92	2.25	2.83	4.92
	0.05	0.16	0.45	1.07	2.12	3.47	4.72	5.50	5.85	5.98	6.07
	4.08	4.62	5.13	5.58	5.96	6.21	6.33	6.38	6.39	6.40	6.41
	5.08	5.62	6.13	6.58	6.96	7.21	7.33	7.38	7.39	7.40	7.40
	4.76	5.28	5.77	6.22	6.59	6.82	6.94	6.98	6.99	7.00	7.00
	5.76	6.28	6.77	7.22	7.59	7.82	7.94	7.98	7.99	8.00	8.00
Co^{3+}											
	6.20	8.66	11.39	14.23	17.02	19.54	21.43	22.50	22.96	23.12	23.19
	11.41	14.30	17.25	20.18	23.01	25.53	27.42	28.50	28.96	29.12	29.19
	17.27	20.26	23.23	26.18	29.01	31.53	33.42	34.50	34.96	35.12	35.19

Table 11.2 (Continued)

Ligand	Concentration [L], M	pH 2.0	3.0	3.5	4.0	4.5	5.0	5.5	6.0
		Co^{3+}							
H_2O_2	0.01	2.30	3.30	3.80	4.30	4.80	5.30	5.80	6.30
	0.10	3.30	4.30	4.80	5.30	5.80	6.30	6.80	7.30
	1.00	4.30	5.30	5.80	6.30	6.80	7.30	7.80	8.30
		Cr^{3+}							
SCN^-	0.10	2.03	2.08	2.09	2.09	2.11	2.70	4.58	6.58
F^-	0.10	4.38	6.42	6.91	7.11	7.18	7.20	7.21	7.30
	1.00	7.28	9.40	9.89	10.10	10.17	10.19	10.20	10.20
		Cu^+							
CN^-	0.01	0.00	0.23	0.90	1.85	2.86	3.89	4.99	6.21
	0.10	0.23	1.85	2.86	3.89	4.99	6.21	7.56	9.00
NH_3	0.01	0.00	0.00	0.00	0.01	0.04	0.11	0.28	0.62
	0.10	0.00	0.01	0.04	0.11	0.28	0.62	1.14	1.83
	1.00	0.01	0.11	0.28	0.62	1.14	1.83	2.69	3.63
$S_2O_3^{2-}$	0.10	10.79	10.85	10.86	10.86	10.86	10.86	10.86	10.86
SO_3^{2-}	0.10	2.42	3.50	4.01	4.51	5.01	5.50	5.99	6.45
		Cu^{2+}							
SCN^-	0.10	0.95	0.99	0.99	0.99	0.99	0.99	0.99	0.99
NH_3	0.01	0.00	0.00	0.00	0.00	0.00	0.00	0.01	0.02
	0.10	0.00	0.00	0.00	0.00	0.01	0.02	0.07	0.20
	1.00	0.00	0.00	0.01	0.02	0.07	0.20	0.53	1.18
NO_2^-	0.10	0.06	0.27	0.37	0.42	0.44	0.45	0.45	0.45
HPO_4^{2-}	0.10	0.00	0.00	0.00	0.00	0.00	0.03	0.20	0.81
	1.00	0.00	0.00	0.00	0.00	0.03	0.21	0.83	1.75
$P_2O_7^{4-}$	0.10	4.43	5.00	5.37	5.91	6.64	7.52	8.47	9.42
	1.00	6.43	7.00	7.37	7.91	8.64	9.52	10.47	11.42
$P_3O_{10}^{5-}$	0.10	0.83	2.14	2.68	3.20	3.73	4.28	4.89	5.51
	1.00	1.77	3.14	3.68	4.20	4.73	5.28	5.89	6.51
$P_4O_{13}^{6-}$	0.10	2.01	2.61	2.92	3.36	3.92	4.62	5.43	6.27
	1.00	3.03	3.95	4.46	5.02	5.68	6.47	7.34	8.20
Cl^-	1.00	1.30	1.30	1.30	1.30	1.30	1.30	1.30	1.30
		Fe^{2+}							
CN^-	0.01	0.00	0.00	0.00	0.00	0.00	0.01	1.17	4.14
	0.10	0.00	0.00	0.00	0.01	1.17	4.14	7.14	10.14
SCN^-	1.00	0.84	0.86	0.86	0.86	0.86	0.86	0.86	0.86
HPO_4^{2-}	0.10	0.00	0.00	0.00	0.00	0.00	0.00	0.00	0.00
	1.00	0.00	0.00	0.00	0.00	0.00	0.00	0.00	0.01
$P_3O_{10}^{5-}$	0.10	1.21	1.37	1.40	1.39	1.35	1.25	1.03	0.74
	1.00	2.18	2.35	2.38	2.38	2.34	2.23	1.99	1.66
		Fe^{3+}							
CN^-	0.01	0.07	0.49	1.01	1.80	3.49	6.35	9.34	12.34
	0.10	0.07	0.72	3.34	6.34	9.34	12.34	15.34	18.34

Ch. 11] Side-Reaction Coefficients for Complexation Equilibria of Metal Ions   295

$$\log \alpha_{M(L,OH)}$$

pH 6.5	7.0	7.5	8.0	8.5	9.0	9.5	10.0	10.5	11.0	12.0
Co^{3+}										
6.80	7.30	7.80	8.30	8.80	9.30	9.80	10.29	10.77	11.20	11.75
7.80	8.30	8.80	9.30	9.80	10.30	10.80	11.29	11.77	12.20	12.75
8.80	9.30	9.80	10.30	10.80	11.30	11.80	12.29	12.77	13.20	13.75
Cr^{3+}										
8.58	10.58	12.58	14.58	16.58	18.58	20.58	22.58	24.58	26.58	30.58
8.60	10.58	12.58	14.58	16.58	18.58	20.58	22.58	24.58	26.58	30.58
10.21	10.73	12.58	14.58	16.58	18.58	20.58	22.58	24.58	26.58	30.58
Cu^{+}										
7.56	8.99	10.46	11.90	13.26	14.39	15.14	15.51	15.65	15.70	15.72
10.48	11.97	13.47	14.97	16.42	17.69	18.55	18.97	19.14	19.19	19.22
1.14	1.83	2.68	3.60	4.53	5.36	5.99	6.35	6.50	6.55	6.58
2.69	3.63	4.60	5.58	6.52	7.36	7.99	8.35	8.50	8.55	8.58
4.61	5.60	6.59	7.57	8.52	9.36	9.99	10.35	10.50	10.55	10.58
10.86	10.86	10.86	10.86	10.86	10.86	10.86	10.86	10.86	10.86	10.86
6.81	7.02	7.12	7.15	7.16	7.16	7.16	7.16	7.16	7.16	7.16
Cu^{2+}										
0.99	0.99	1.01	1.13	1.64	2.53	3.52	4.52	5.53	6.58	9.65
0.08	0.22	0.59	1.25	2.19	3.24	4.15	4.84	5.60	6.59	9.65
0.53	1.18	2.19	3.53	5.05	6.53	7.65	8.26	8.51	8.61	9.69
2.20	3.57	5.22	7.02	8.84	10.46	11.65	12.29	12.56	12.65	12.70
0.45	0.46	0.52	0.82	1.57	2.52	3.52	4.52	5.53	6.58	9.65
1.62	2.08	2.19	2.21	2.29	2.69	3.54	4.52	5.53	6.58	9.65
2.61	3.08	3.18	3.20	3.21	3.28	3.69	4.54	5.53	6.58	9.65
10.33	11.23	12.15	13.04	13.78	14.27	14.51	14.60	14.64	14.65	14.65
12.33	13.23	14.15	15.04	15.78	16.27	16.51	16.60	16.64	16.65	16.65
6.10	6.63	7.13	7.58	7.95	8.18	8.30	8.34	8.35	8.37	9.67
7.10	7.63	8.13	8.58	8.95	9.18	9.30	9.34	9.35	9.36	9.83
7.01	7.57	8.01	8.37	8.62	8.77	8.86	8.99	9.22	9.58	10.57
8.93	9.47	9.86	10.19	10.43	10.56	10.61	10.65	10.72	10.87	11.56
1.30	1.30	1.31	1.38	1.73	2.55	3.52	4.52	5.53	6.58	9.65
Fe^{2+}										
7.13	10.12	13.09	15.98	18.68	20.89	22.32	23.01	23.27	23.36	23.40
13.13	16.12	19.09	21.98	24.68	26.89	28.32	29.01	29.27	29.36	29.40
0.86	0.86	0.86	0.86	0.87	0.87	0.89	0.96	1.19	1.98	4.81
0.01	0.06	0.34	1.07	1.91	2.42	2.57	2.59	2.60	2.67	4.81
0.07	0.37	1.10	2.03	2.90	3.42	3.56	3.59	3.59	3.59	4.83
0.57	0.72	1.04	1.31	1.46	1.52	1.55	1.57	1.64	2.10	4.81
1.45	1.64	2.00	2.29	2.84	2.51	2.53	2.54	2.55	2.64	4.81
Fe^{3+}										
15.33	18.32	21.29	24.18	26.88	29.09	30.52	31.21	31.47	31.56	32.71
21.33	24.32	27.29	30.18	32.88	35.09	36.52	37.21	37.47	37.56	37.60

Table 11.2 (Continued)

Ligand	Concentration [L], M	pH 2.0	3.0	3.5	4.0	4.5	5.0	5.5	6.0
		Fe^{3+}							
SCN^-	0.10	2.47	2.57	2.58	2.64	3.11	4.72	6.68	8.68
	1.00	6.54	6.68	6.69	6.69	6.69	6.70	6.99	8.68
HPO_4^{2-}	0.10	2.71	3.73	4.23	4.73	5.23	5.76	6.80	8.68
	1.00	3.71	4.73	5.23	5.73	6.23	6.72	7.31	8.71
$P_2O_7^{4-}$	0.10	0.07	0.49	1.01	1.89	4.63	8.11	11.60	15.07
	1.00	0.07	0.49	1.07	3.24	6.62	10.11	13.60	17.07
$P_3O_{10}^{5-}$	0.10	8.39	11.55	12.74	13.79	14.76	15.61	16.25	16.60
	1.00	10.39	13.55	14.74	15.79	16.76	17.61	18.25	18.60
H_2O_2	0.10	0.09	0.56	1.07	1.82	2.97	4.71	6.68	8.68
SO_4^{2-}	1.00	2.32	2.34	2.35	2.44	3.05	4.72	6.68	8.68
$S_2O_3^{2-}$	1.00	1.19	1.26	1.40	1.88	2.97	4.71	6.68	8.68
F^-	0.10	6.12	8.26	8.78	8.99	9.07	9.09	9.10	9.24
	1.00	9.09	11.26	11.77	11.99	12.06	12.09	12.10	12.10
Cl^-	1.00	1.05	1.12	1.31	1.85	2.97	4.71	6.68	8.68
		Ga^{3+}							
SCN^-	1.00	1.18	1.22	1.30	1.76	2.97	4.42	5.92	7.45
F^-	0.10	4.68	6.72	7.21	7.41	7.48	7.50	7.52	7.78
	1.00	7.58	9.70	10.19	10.40	10.47	10.49	10.50	10.50
Cl^-	1.00	0.37	0.48	0.78	1.64	2.96	4.42	5.92	7.45
		$Ge(OH)_4$							
F^-	0.10	11.80	14.43	14.89	14.99	15.00	15.00	15.00	15.00
	1.00	16.72	19.28	19.62	19.53	19.30	19.13	19.05	19.02
		Hf^{4+}							
SO_4^{2-}	0.10	3.38	3.60	4.18	5.25	6.60	8.25	10.14	12.19
	1.00	5.36	5.43	5.46	5.65	6.63	8.25	10.14	12.19
F^-	0.10	22.93	28.29	30.19	31.28	31.75	31.92	31.97	31.99
	1.00	28.92	34.28	36.19	37.28	37.75	37.92	37.97	37.99
Cl^-	1.00	1.34	3.06	4.09	5.24	6.60	8.25	10.14	12.19
		Hg_2^{2+}							
$P_2O_7^{4-}$	0.10	0.00	0.00	0.05	0.82	2.63	4.43	6.00	7.30
	1.00	0.00	0.03	0.79	2.67	4.60	6.41	8.00	9.30
$P_3O_{10}^{5-}$	0.10	0.00	0.00	0.01	0.07	0.61	2.14	3.70	5.04
	1.00	0.00	0.00	0.02	0.48	2.18	4.01	5.63	6.99
SO_4^{2-}	1.00	4.68	4.82	4.83	4.84	4.84	4.84	4.84	4.84
		Hg^{2+}							
CN^-	0.01	16.43	18.43	19.43	20.43	21.43	22.43	23.44	24.45
	0.10	18.43	20.43	21.43	22.43	23.44	24.45	25.49	26.61
	1.00	20.43	22.43	23.44	24.45	25.49	26.61	27.89	29.39
SCN^-	0.10	17.61	17.73	17.74	17.74	17.74	17.74	17.74	17.74
	1.00	21.57	21.69	21.70	21.70	21.70	21.70	21.70	21.70

Ch. 11]   Side-Reaction Coefficients for Complexation Equilibria of Metal Ions   297

$$\log \alpha_{M(L,OH)}$$

pH 6.5	7.0	7.5	8.0	8.5	9.0	9.5	10.0	10.5	11.0	12.0
Fe^{3+}										
10.68	12.68	14.68	16.68	18.68	20.68	22.68	24.68	26.68	28.68	32.68
10.68	12.68	14.68	16.68	18.68	20.68	22.68	24.68	26.68	28.68	32.68
10.68	12.68	14.68	16.68	18.68	20.68	22.68	24.68	26.68	28.68	32.68
10.68	12.68	14.68	16.68	18.68	20.68	22.68	24.68	26.68	28.68	32.68
18.18	19.43	19.59	19.42	19.21	20.68	22.68	24.68	26.68	28.68	32.68
20.18	21.43	21.59	21.42	21.06	20.94	22.68	24.68	26.68	28.68	32.68
16.74	16.74	16.61	16.81	18.68	20.68	22.68	24.68	26.68	28.68	32.68
18.74	18.74	18.60	18.26	18.71	20.68	22.68	24.68	26.68	28.68	32.68
10.68	12.68	14.68	16.68	18.68	20.68	22.68	24.68	26.68	28.68	32.68
10.68	12.68	14.68	16.68	18.68	20.68	22.68	24.68	26.68	28.68	32.68
10.68	12.68	14.68	16.68	18.68	20.68	22.68	24.68	26.68	28.68	32.68
10.69	12.68	14.68	16.68	18.68	20.68	22.68	24.68	26.68	28.68	32.68
12.12	12.78	14.68	16.68	18.68	20.68	22.68	24.68	26.68	28.68	32.68
10.68	12.68	14.68	16.68	18.68	20.68	22.68	24.68	26.68	28.68	32.68
Ga^{3+}										
9.05	10.75	12.60	14.53	16.51	18.50	20.50	22.50	24.50	26.50	30.50
9.06	10.75	12.60	14.53	16.51	18.50	20.50	22.50	24.50	26.50	30.50
10.52	10.95	12.60	14.53	16.51	18.50	20.50	22.50	24.50	26.50	30.50
9.05	10.75	12.60	14.53	16.51	18.50	20.50	22.50	24.50	26.50	30.50
$Ge(OH)_4$										
15.00	15.00	15.00	15.00	15.00	15.00	15.00	15.00	15.00	15.00	15.00
19.00	19.00	19.00	19.00	19.00	19.00	19.00	19.00	19.00	19.00	19.00
Hf^{4+}										
14.42	16.78	19.23	21.71	24.20	26.70	29.20	31.70	34.20	36.70	41.70
14.42	16.78	19.23	21.71	24.20	26.70	29.20	31.70	34.20	36.70	41.70
32.00	32.00	32.00	32.00	32.00	32.00	32.00	32.00	32.18	34.20	36.70	41.70
38.00	38.00	38.00	38.00	38.00	38.00	38.00	38.00	38.00	38.02	41.70
14.42	16.78	19.23	21.71	24.20	26.70	29.20	31.70	34.20	36.70	41.70
Hg_2^{2+}										
8.37	9.23	9.85	10.19	10.35	10.47	10.66	10.97	11.38	11.85	12.84
10.36	11.23	11.84	12.17	12.31	12.37	12.41	12.49	12.66	12.97	13.85
6.17	7.17	8.04	8.70	9.13	9.45	9.81	10.24	10.71	11.20	12.20
8.13	9.13	9.99	10.63	11.01	11.19	11.32	11.51	11.82	12.24	13.20
4.84	4.84	4.84	4.85	4.86	4.90	5.00	5.23	5.59	6.03	7.00
Hg^{2+}										
25.49	26.60	27.86	29.29	30.80	32.10	32.94	33.33	33.47	33.52	33.54
27.89	29.38	31.08	32.88	34.51	36.00	36.88	37.28	37.43	37.48	37.50
31.12	32.99	34.92	36.83	38.58	39.99	40.87	41.28	41.43	41.48	41.50
17.74	17.74	17.74	17.74	17.74	17.74	17.74	17.74	17.75	17.83	20.06
21.70	21.70	21.70	21.70	21.70	21.70	21.70	21.70	21.70	21.70	21.71

Table 11.2 (Continued)

Ligand	Concentration [L], M	pH 2.0	3.0	3.5	4.0	4.5	5.0	5.5	6.0
		Hg^{2+}							
NH_3	0.01	0.10	0.81	1.60	2.53	3.51	4.50	5.50	6.50
	0.10	0.76	2.46	3.43	4.42	5.42	6.42	7.42	8.42
	1.00	2.45	4.42	5.42	6.42	7.42	8.42	9.42	10.42
$P_2O_7^{4-}$	0.10	0.01	0.25	0.94	2.06	3.37	4.71	5.98	7.12
	1.00	0.01	0.37	1.43	2.83	4.27	5.67	6.96	8.11
S^{2-}	0.10	25.96	27.96	28.96	29.96	30.97	31.98	33.01	34.07
SO_3^{2-}	0.10	17.63	18.86	19.38	19.89	20.39	20.91	21.44	22.03
SO_4^{2-}	1.00	2.28	2.42	2.44	2.51	2.89	3.72	4.70	5.70
$S_2O_3^{2-}$	0.10	31.21	31.26	31.26	31.26	31.26	31.26	31.26	31.26
Cl^-	0.01	9.27	9.27	9.27	9.27	9.27	9.27	9.27	9.27
	0.10	11.72	11.72	11.72	11.72	11.72	11.72	11.72	11.72
Br^-	0.01	13.68	13.68	13.68	13.68	13.68	13.68	13.68	13.68
	0.10	17.10	17.10	17.10	17.10	17.10	17.10	17.10	17.10
		In^{3+}							
SO_4^{2-}	1.00	3.03	3.15	3.16	3.17	3.47	5.17	7.17	9.17
F^-	0.10	2.87	5.05	5.63	5.89	5.98	6.07	7.20	9.17
	1.00	6.00	8.72	9.39	9.68	9.78	9.81	9.82	9.91
Cl^-	0.10	2.10	2.14	2.22	2.42	3.30	5.17	7.17	9.17
	1.00	4.36	4.37	4.37	4.39	4.46	5.26	7.17	9.17
		K^+							
$P_2O_7^{4-}$	1.00	0.00	0.00	0.00	0.00	0.00	0.00	0.00	0.00
$P_3O_{10}^{5-}$	1.00	0.00	0.00	0.00	0.00	0.00	0.00	0.00	0.02
$P_4O_{13}^{6-}$	1.00	0.00	0.00	0.00	0.01	0.04	0.11	0.28	0.54
		La^{3+}							
$Fe(CN)_6^{3-}$ *	1.00	3.70	3.70	3.70	3.70	3.70	3.70	3.70	3.70
$P_2O_7^{4-}$	0.10	3.20	5.56	6.60	7.62	8.62	9.62	10.60	11.55
	1.00	4.20	6.56	7.60	8.62	9.62	10.62	11.60	12.55
SO_4^{2-}	1.00	2.40	2.49	2.50	2.50	2.50	2.50	2.50	2.50
F^-	1.00	1.67	2.39	2.56	2.63	2.66	2.67	2.67	2.67
		Li^+							
$P_2O_7^{4-}$	1.00	0.00	0.00	0.01	0.05	0.13	0.31	0.57	0.83
$P_3O_{10}^{5-}$	1.00	0.00	0.06	0.17	0.40	0.75	1.14	1.49	1.72
$P_4O_{13}^{6-}$	1.00	0.00	0.00	0.01	0.04	0.11	0.28	0.57	0.94
		Ln^{3+}							
$Fe(CN)_6^{3-}$ *	1.00	3.70	3.70	3.70	3.70	3.70	3.70	3.70	3.70
$P_2O_7^{4-}$	0.10	7.47	9.84	10.88	11.89	12.90	13.89	14.87	15.82
	1.00	8.47	10.84	11.88	12.89	13.90	14.89	15.87	16.82
SO_4^{2-}	1.00	1.78	1.84	1.85	1.85	1.85	1.85	1.85	1.85
F^-	1.00	2.30	3.02	3.19	3.26	3.29	3.30	3.30	3.30
		Mg^{2+}							
PO_4^{3-}	0.10	0.10	0.17	0.18	0.21	0.26	0.39	0.57	0.73
	1.00	0.55	0.75	0.79	0.85	0.97	1.18	1.44	1.64

Ch. 11] Side-Reaction Coefficients for Complexation Equilibria of Metal Ions

$\log \alpha_{M(L,OH)}$

pH	6.5	7.0	7.5	8.0	8.5	9.0	9.5	10.0	10.5	11.0	12.0
Hg^{2+}											
	7.50	8.49	9.49	10.48	11.43	12.33	13.19	14.26	15.62	17.08	20.06
	9.42	10.42	11.42	12.41	13.38	14.28	15.02	15.50	15.95	17.10	20.06
	11.42	12.43	13.46	14.53	15.81	16.99	18.08	18.74	19.03	19.14	20.11
	8.16	9.09	9.91	10.61	11.26	12.02	13.00	14.23	15.61	17.08	20.06
	9.14	10.08	10.88	11.55	12.13	12.70	13.35	14.32	15.63	17.08	20.06
	35.15	36.12	36.93	37.65	38.42	39.30	40.25	41.24	42.23	43.23	45.22
	22.68	23.28	23.70	23.90	23.97	24.00	24.01	24.01	24.01	24.01	24.01
	6.70	7.70	8.70	9.71	10.73	11.79	12.94	14.22	15.61	17.08	20.06
	31.26	31.26	31.26	31.26	31.26	31.26	31.26	31.26	31.26	31.26	31.26
	9.27	9.28	9.38	9.85	10.75	11.79	12.94	14.22	15.61	17.08	20.06
	11.72	11.72	11.72	11.73	11.76	12.06	12.96	14.22	15.61	17.08	20.06
	13.68	13.68	13.68	13.68	13.68	13.68	13.75	14.33	15.62	17.08	20.06
	17.10	17.10	17.10	17.10	17.10	17.10	17.10	17.10	17.12	17.39	20.06
In^{3+}											
	11.17	13.17	15.17	17.17	19.17	21.17	23.17	25.17	27.17	29.17	33.17
	11.17	13.17	15.17	17.17	19.17	21.17	23.17	25.17	27.17	29.17	33.17
	11.19	13.17	15.17	17.17	19.17	21.17	23.17	25.17	27.17	29.17	33.17
	11.17	13.17	15.17	17.17	19.17	21.17	23.17	25.17	27.17	29.17	33.17
	11.17	13.17	15.17	17.17	19.17	21.17	23.17	25.17	27.17	29.17	33.17
K^+											
	0.01	0.04	0.13	0.29	0.52	0.71	0.80	0.84	0.86	0.86	0.86
	0.07	0.19	0.44	0.76	1.07	1.27	1.36	1.39	1.40	1.41	1.41
	0.83	1.05	1.23	1.40	1.56	1.66	1.70	1.71	1.72	1.72	1.72
La^{3+}											
	3.70	3.70	3.70	3.70	3.70	3.70	3.71	3.98	5.27	8.51	17.50
	12.42	13.16	13.79	14.39	15.02	15.65	16.16	16.45	16.57	16.61	17.55
	13.43	14.22	15.00	15.84	16.71	17.50	18.07	18.38	18.51	18.55	18.61
	2.50	2.50	2.50	2.50	2.50	2.51	2.68	3.70	5.26	8.51	17.50
	2.67	2.67	2.67	2.67	2.67	2.68	2.80	3.71	5.26	8.51	17.50
Li^+											
	1.01	1.17	1.38	1.68	1.99	2.21	2.33	2.37	2.38	2.39	2.39
	1.85	1.95	2.09	2.32	2.57	2.74	2.82	2.86	2.87	2.87	2.87
	1.30	1.61	1.91	2.20	2.44	2.56	2.61	2.63	2.64	2.64	2.64
Ln^{3+}											
	3.70	3.70	3.70	3.70	3.73	5.00	7.48	9.98	12.48	14.98	19.98
	16.69	17.42	18.03	18.56	19.04	19.45	19.75	19.90	19.97	19.99	20.29
	17.69	18.42	19.03	19.56	20.04	20.45	20.75	20.90	20.97	20.99	21.04
	1.85	1.85	1.85	1.86	2.57	4.98	7.48	9.98	12.48	14.98	19.98
	3.30	3.30	3.30	3.30	3.36	4.99	7.48	9.98	12.48	14.98	19.98
Mg^{2+}											
	0.81	0.85	0.86	0.86	0.87	0.89	0.94	1.07	1.32	1.65	2.22
	1.75	1.79	1.80	1.81	1.82	1.84	1.89	2.03	2.30	2.64	3.22

Table 11.2 (Continued)

Ligand	Concentration [L], M	pH 2.0	3.0	3.5	4.0	4.5	5.0	5.5	6.0
		Mg^{2+}							
$P_2O_7^{4-}$	0.10	0.00	0.05	0.14	0.34	0.67	1.09	1.52	1.93
	1.00	0.03	0.33	0.68	1.11	1.58	2.06	2.51	2.95
$P_3O_{10}^{5-}$	0.10	0.01	0.15	0.38	0.74	1.17	1.62	2.04	2.41
	1.00	0.09	0.71	1.17	1.66	2.14	2.61	3.03	3.41
F^-	1.00	0.49	1.08	1.24	1.30	1.33	1.34	1.34	1.34
		Mn^{2+}							
SCN^-	1.00	0.71	0.73	0.74	0.74	0.74	0.74	0.74	0.74
NH_3	0.01	0.00	0.00	0.00	0.00	0.00	0.00	0.00	0.00
	0.10	0.00	0.00	0.00	0.00	0.00	0.00	0.00	0.00
	1.00	0.00	0.00	0.00	0.00	0.00	0.00	0.00	0.00
HPO_4^{2-}	0.10	0.00	0.03	0.09	0.23	0.50	0.85	1.18	1.41
	1.00	0.02	0.22	0.51	0.91	1.35	1.79	2.16	2.40
$P_3O_{10}^{5-}$	0.10	0.19	1.15	1.67	2.18	2.69	3.21	3.75	4.30
	1.00	0.81	2.12	2.66	3.18	3.69	4.21	4.75	5.30
SO_4^{2-}	1.00	1.97	2.22	2.25	2.26	2.26	2.26	2.26	2.26
		Mn^{3+}							
$P_2O_7^{4-}$	0.10	8.63	12.98	15.01	17.02	19.00	20.95	22.81	24.47
	1.00	10.88	14.98	17.01	19.02	21.00	22.95	24.81	26.47
		Na^+							
$P_2O_7^{4-}$	1.00	0.00	0.00	0.00	0.00	0.00	0.00	0.00	0.00
$P_3O_{10}^{5-}$	1.00	0.00	0.00	0.02	0.05	0.13	0.30	0.52	0.70
$P_4O_{13}^{6-}$	1.00	0.00	0.00	0.00	0.01	0.04	0.11	0.27	0.54
		Ni^{2+}							
CN^-	0.01	1.09	2.52	3.45	4.50	5.69	7.09	8.73	10.56
	0.10	5.06	6.52	7.45	8.50	9.69	11.09	12.73	14.56
SCN^-	1.00	4.13	4.21	4.22	4.22	4.22	4.22	4.22	4.22
NH_3	0.01	0.00	0.00	0.00	0.00	0.00	0.00	0.00	0.00
	0.10	0.00	0.00	0.00	0.00	0.00	0.00	0.00	0.01
	1.00	0.00	0.00	0.00	0.00	0.00	0.01	0.03	0.09
		Np^{4+}							
SO_4^{2-}	0.10	3.40	3.45	3.46	3.47	3.48	3.52	3.62	3.84
	1.00	5.36	5.42	5.42	5.42	5.42	5.42	5.43	5.43
F^-	0.10	15.05	18.62	19.90	20.62	20.93	21.05	21.08	21.09
	1.00	19.05	22.62	23.90	24.62	24.93	25.05	25.08	25.09
		NpO_2^{2+}							
SO_4^{2-}	1.00	2.59	2.67	2.68	2.77	3.56	5.00	6.50	8.00
F^-	0.10	2.99	4.41	4.75	4.90	4.96	5.28	6.51	8.00
	1.00	4.96	6.41	6.75	6.89	6.95	6.97	7.09	8.04
		Pb^{2+}							
$P_2O_7^{4-}$	0.10	0.00	0.00	0.02	0.14	0.67	1.55	2.46	3.30
	1.00	0.00	0.02	0.14	0.68	1.58	2.54	3.50	4.49

Ch. 11] Side-Reaction Coefficients for Complexation Equilibria of Metal Ions 301

$$\log \alpha_{M(L,OH)}$$

pH 6.5	7.0	7.5	8.0	8.5	9.0	9.5	10.0	10.5	11.0	12.0
Mg^{2+}										
2.34	2.85	3.51	4.27	4.97	5.45	5.68	5.77	5.80	5.82	5.85
3.51	4.35	5.28	6.18	6.93	7.42	7.66	7.76	7.79	7.80	7.80
2.80	3.24	3.69	4.12	4.46	4.67	4.77	4.82	4.86	4.96	5.50
3.80	4.24	4.69	5.12	5.46	5.67	5.77	5.82	5.86	5.96	6.50
1.34	1.34	1.34	1.34	1.34	1.34	1.34	1.34	1.34	1.34	1.36
Mn^{2+}										
0.74	0.74	0.74	0.74	0.74	0.74	0.75	0.91	1.93	3.40	6.40
0.00	0.00	0.00	0.00	0.01	0.02	0.07	0.57	1.91	3.40	6.40
0.00	0.00	0.00	0.01	0.04	0.11	0.23	0.66	1.91	3.40	6.40
0.00	0.01	0.04	0.13	0.36	0.80	1.36	1.78	2.23	3.42	6.40
1.53	1.57	1.58	1.59	1.59	1.59	1.59	1.61	2.07	3.41	6.40
2.51	2.56	2.57	2.58	2.58	2.58	2.58	2.58	2.64	3.45	6.40
4.84	5.36	5.85	6.30	6.67	6.90	7.02	7.06	7.07	7.08	7.16
5.84	6.36	6.85	7.30	7.67	7.90	8.02	8.06	8.07	8.08	8.09
2.26	2.26	2.26	2.26	2.26	2.26	2.26	2.27	2.42	3.43	6.40
Mn^{3+}										
25.84	26.99	27.97	28.79	29.37	29.67	29.79	29.83	29.84	29.85	29.85
27.84	28.99	29.97	30.79	31.37	31.67	31.79	31.83	31.84	31.85	31.85
Na^+										
0.02	0.07	0.18	0.40	0.67	0.87	0.98	1.02	1.03	1.04	1.04
0.80	0.87	0.98	1.15	1.37	1.53	1.61	1.64	1.65	1.65	1.65
0.83	1.06	1.25	1.45	1.63	1.73	1.77	1.79	1.79	1.80	1.80
Ni^{2+}										
12.49	14.45	16.41	18.30	19.99	21.28	22.01	22.33	22.44	22.48	22.50
16.49	18.45	20.41	22.30	23.99	25.28	26.01	26.33	26.44	26.48	26.50
4.22	4.22	4.22	4.22	4.22	4.22	4.22	4.22	4.23	4.26	6.69
0.00	0.01	0.03	0.09	0.34	0.54	0.95	1.32	1.91	3.15	6.69
0.03	0.09	0.25	0.65	1.38	2.39	3.38	4.04	4.33	4.46	6.69
0.25	0.66	1.46	2.71	4.37	6.20	7.76	8.71	9.12	9.27	9.33
Np^{4+}										
4.19	4.63	5.11	5.60	6.10	6.60	7.10	7.60	8.10	8.60	9.60
5.44	5.48	5.59	5.82	6.18	6.63	7.11	7.60	8.10	8.60	9.60
21.10	21.10	21.10	21.10	21.10	21.10	21.10	21.10	21.10	21.10	21.10
25.10	25.10	25.10	25.10	25.10	25.10	25.10	25.10	25.10	25.10	25.10
NpO_2^{2+}										
9.50	11.00	12.50	14.00	15.50	17.00	18.50	20.00	21.50	23.00	26.00
9.50	11.00	12.50	14.00	15.50	17.00	18.50	20.00	21.50	23.00	26.00
9.50	11.00	12.50	14.00	15.50	17.00	18.50	20.00	21.50	23.00	26.00
Pb^{2+}										
4.06	4.85	5.70	6.56	7.29	7.78	9.77	13.76	17.76	21.76	29.76
5.56	6.61	7.61	8.52	9.28	9.77	10.20	13.76	17.76	21.76	29.76

Table 11.2 (Continued)

Ligand	Concentration [L], M	pH 2.0	3.0	3.5	4.0	4.5	5.0	5.5	6.0
Pb^{2+}									
$S_2O_3^{2-}$	0.10	3.32	3.39	3.39	3.40	3.40	3.40	3.40	3.40
	1.00	6.42	6.50	6.51	6.51	6.51	6.51	6.51	6.51
F^-	1.00	0.85	2.04	2.36	2.50	2.55	2.57	2.57	2.57
Cl^-	1.00	1.81	1.81	1.81	1.81	1.81	1.81	1.81	1.81
Br^-	1.00	2.53	2.53	2.53	2.53	2.53	2.53	2.53	2.53
I^-	1.00	4.05	4.05	4.05	4.05	4.05	4.05	4.05	4.05
Pd^{2+}									
CN^-	0.01	5.56	9.56	11.56	13.56	15.56	17.56	19.56	21.56
	0.10	9.56	13.56	15.56	17.56	19.56	21.56	23.57	25.58
SCN^-	0.10	23.07	23.19	23.20	23.20	23.20	23.20	23.20	23.20
NH_3	0.01	0.71	2.26	3.26	4.38	5.78	7.53	9.45	11.42
	0.10	1.88	4.12	5.61	7.36	9.25	11.22	13.21	15.20
	1.00	4.12	7.36	9.25	11.22	13.21	15.20	17.20	19.20
Cl^-	0.10	7.68	7.68	7.68	7.68	7.68	7.77	9.02	11.00
	1.00	11.52	11.52	11.52	11.52	11.52	11.52	11.52	11.64
Br^-	0.10	10.93	10.93	10.93	10.93	10.93	10.93	10.93	11.27
I^-	0.10	20.50	20.50	20.50	20.50	20.50	20.50	20.50	20.50
Pu^{4+}									
SO_4^{2-}	0.10	2.62	3.25	4.32	5.74	7.45	9.43	11.62	13.98
F^-	0.10	4.76	5.49	5.68	6.04	7.46	9.43	11.62	13.98
Sb^{3+}									
F^-	0.10	9.85	11.40	12.33	13.31	14.30	15.30	16.30	17.30
	1.00	9.85	11.40	12.34	13.31	14.30	15.30	16.30	17.30
Cl^-	0.10	9.85	11.40	12.33	13.31	14.30	15.30	16.30	17.30
	1.00	9.85	11.40	12.33	13.31	14.30	15.30	16.30	17.30
Sc^{3+}									
SO_4^{2-}	0.10	1.98	2.10	2.11	2.12	2.12	2.12	2.15	3.18
	1.00	3.82	3.96	3.97	3.98	3.98	3.98	3.98	4.04
F^-	0.01	6.85	9.42	10.05	10.31	10.41	10.44	10.45	10.45
	1.00	14.49	17.33	17.99	18.26	18.35	18.39	18.40	18.40
Sn^{2+}									
F^-	0.10	3.56	5.67	6.18	6.39	6.47	6.50	6.52	7.14
	1.00	6.49	8.66	9.17	9.39	9.46	9.49	9.50	9.50
Cl^-	1.00	2.07	2.07	2.07	2.08	2.20	3.17	5.04	7.03
Br^-	1.00	1.49	1.49	1.50	1.55	1.86	3.14	5.04	7.03
I^-	1.00	2.94	2.94	2.94	2.94	2.96	3.35	5.04	7.03
Sn^{4+}									
F^-	0.10	11.92	16.71	18.04	18.66	18.93	19.95	21.90	23.90
Sr^{2+}									
$P_2O_7^{4-}$	0.10	0.00	0.00	0.00	0.00	0.00	0.00	0.02	0.14
$P_3O_{10}^{5-}$	0.10	0.01	0.14	0.38	0.74	1.17	1.63	2.04	2.38
$P_4O_{13}^{6-}$	0.10	0.00	0.03	0.09	0.23	0.51	0.91	1.36	1.82

Ch. 11]   Side-Reaction Coefficients for Complexation Equilibria of Metal Ions   303

$$\log \alpha_{M(L,OH)}$$

pH 6.5	7.0	7.5	8.0	8.5	9.0	9.5	10.0	10.5	11.0	12.0
Pb^{2+}										
3.40	3.40	3.40	3.40	3.42	5.76	9.76	13.76	17.76	21.76	29.76
6.51	6.51	6.51	6.51	6.51	6.58	9.76	13.76	17.76	21.76	29.76
2.57	2.57	2.57	2.58	2.69	5.76	9.76	13.76	17.76	21.76	29.76
1.81	1.81	1.81	1.82	2.25	5.76	9.76	13.76	17.76	21.76	29.76
2.53	2.53	2.53	2.53	2.65	5.76	9.76	13.76	17.76	21.76	29.76
4.05	4.05	4.05	4.05	4.05	5.77	9.76	13.76	17.76	21.76	29.76
Pd^{2+}										
23.56	25.57	27.59	29.62	31.61	33.33	34.48	35.03	35.25	35.32	35.51
27.62	29.72	31.93	34.21	36.40	38.22	39.41	39.98	40.20	40.27	40.30
23.20	23.20	23.20	23.20	23.20	23.41	25.01	27.00	29.00	31.00	35.00
13.41	15.41	17.40	19.37	21.30	23.15	25.03	27.00	29.00	31.00	35.00
17.20	19.19	21.18	23.13	24.99	26.62	27.79	28.43	29.17	31.00	35.00
21.20	23.19	25.18	27.13	28.99	30.62	31.78	32.41	32.67	32.76	35.00
13.00	15.00	17.00	19.00	21.00	23.00	25.00	27.00	29.00	31.00	35.00
31.01	15.00	17.00	19.00	21.00	23.00	25.00	27.00	29.00	31.00	35.00
13.00	15.00	17.00	19.00	21.00	23.00	25.00	27.00	29.00	31.00	35.00
20.50	20.50	20.50	20.51	21.12	23.00	25.00	27.00	29.00	31.00	35.00
Pu^{4+}										
16.43	18.91	21.40	23.90	26.40	28.90	31.40	33.90	36.40	38.90	43.90
16.43	18.91	21.40	23.90	26.40	28.90	31.40	33.90	36.40	38.90	43.90
Sb^{3+}										
18.30	19.30	20.30	21.30	22.30	23.30	24.30	25.31	26.33	27.40	29.85
18.30	19.30	20.30	21.30	22.30	23.30	24.30	25.31	26.33	27.40	29.85
18.30	19.30	20.30	21.30	22.30	23.30	24.30	25.31	26.33	27.40	29.85
18.30	19.30	20.30	21.30	22.30	23.30	24.30	25.31	26.33	27.40	29.85
Sc^{3+}										
5.63	8.13	10.63	13.13	15.63	18.13	20.63	23.13	25.63	28.13	33.13
5.64	8.13	10.63	13.13	15.63	18.13	20.63	23.13	25.63	28.13	33.13
10.45	10.45	10.85	13.13	15.63	18.13	20.63	23.13	25.63	28.13	33.13
18.40	18.40	18.40	18.40	18.40	18.59	20.63	23.13	25.63	28.13	33.13
Sn^{2+}										
9.03	11.03	13.03	15.03	17.03	19.03	21.03	23.03	25.03	27.03	31.03
9.63	11.04	13.03	15.03	17.03	19.03	21.03	23.03	25.03	27.03	31.03
9.03	11.03	13.03	15.03	17.03	19.03	21.03	23.03	25.03	27.03	31.03
9.03	11.03	13.03	15.03	17.03	19.03	21.03	23.03	25.03	27.03	31.03
9.03	11.03	13.03	15.03	17.03	19.03	21.03	23.03	25.03	27.03	31.03
Sn^{4+}										
25.90	27.90	29.90	31.90	33.90	35.90	37.90	39.90	41.90	43.90	47.90
Sr^{2+}										
0.45	0.89	1.34	1.74	2.02	2.17	2.23	2.25	2.26	2.26	2.26
2.65	2.96	3.34	3.74	4.09	4.33	4.44	4.48	4.49	4.50	4.50
2.24	2.63	3.01	3.37	3.62	3.76	3.81	3.83	3.84	3.84	3.84

Table 11.2 (Continued)

Ligand	Concentration [L], M	pH 2.0	3.0	3.5	4.0	4.5	5.0	5.5	6.0
		Th^{4+}							
$H_2PO_4^-$	0.10	5.02	5.43	5.47	5.48	5.47	5.45	5.47	6.82
	1.00	7.02	7.43	7.47	7.48	7.47	7.44	7.35	7.29
SO_4^{2-}	0.10	4.60	4.69	4.70	4.70	4.71	4.71	5.08	6.81
	1.00	7.32	7.43	7.44	7.44	7.44	7.44	7.44	7.53
F^-	0.10	12.33	14.99	15.63	15.89	15.99	16.02	16.03	16.03
IO_3^-	0.10	4.11	4.18	4.19	4.19	4.19	4.22	4.93	6.81
		TiO^{2+}							
H_2O_2	0.10	4.32	4.32	4.32	4.33	4.41	4.85	5.72	6.70
	1.00	6.30	6.30	6.30	6.30	6.30	6.31	6.40	6.85
SO_4^{2-}	1.00	2.12	2.17	2.30	2.82	3.72	4.70	5.70	6.70
F^-	0.10	8.13	11.50	12.55	13.08	13.29	13.37	13.39	13.40
		Tl^+							
HS^-	0.10	0.00	0.03	0.08	0.20	0.46	0.83	1.27	1.69
$S_2O_3^{2-}$	1.00	1.14	1.17	1.18	1.18	1.18	1.18	1.18	1.18
		Tl^{3+}							
CN^-	0.10	2.37	6.22	8.21	10.20	12.20	14.20	16.20	18.20
Cl^-	0.10	12.35	12.35	12.35	12.35	12.35	12.36	12.63	13.82
	1.00	16.31	16.31	16.31	16.31	16.31	16.31	16.31	16.31
		U^{4+}							
SCN^-	1.00	2.38	3.01	4.00	5.25	6.72	8.49	10.61	12.97
SO_4^{2-}	1.00	5.74	5.81	5.83	5.92	6.78	8.49	10.61	12.97
F^-	0.10	13.66	16.34	17.30	17.84	18.08	18.16	18.19	18.20
		UO_2^{2+}							
SCN^-	1.00	1.08	1.12	1.13	1.13	1.13	1.14	1.81	4.18
CO_3^{2-}	0.10	2.41	4.41	5.41	6.41	7.40	8.38	9.36	10.37
SO_4^{2-}	1.00	3.58	3.72	3.73	3.73	3.73	3.73	3.74	4.31
F^-	0.10	4.73	7.11	7.73	7.99	8.09	8.12	8.13	8.13
		VO^{2+}							
SCN^-	1.00	1.10	1.13	1.13	1.13	1.15	1.27	1.82	2.72
SO_4^{2-}	1.00	2.15	2.40	2.43	2.44	2.44	2.45	2.52	2.90
F^-	0.10	1.99	3.73	4.23	4.46	4.54	4.57	4.57	4.58
		VO_2^+							
H_2O_2	0.10	3.53	3.53	3.53	3.53	3.54	3.65	4.13	5.02
SO_4^{2-}	1.00	0.97	1.03	1.11	1.40	2.10	3.03	4.01	5.01
F^-	0.10	1.77	3.37	3.82	4.01	4.09	4.14	4.36	5.06
		Y^{3+}							
SO_4^{2-}	1.00	1.75	1.81	1.82	1.82	1.82	1.82	1.82	1.82
F^-	0.10	4.39	6.50	6.99	7.20	7.27	7.29	7.30	7.30
		Zn^{2+}							
CN^-	0.01	0.00	0.00	0.00	0.00	0.01	0.05	0.26	0.98
	0.10	0.00	0.00	0.01	0.05	0.26	0.98	2.21	3.85

Side-Reaction Coefficients for Complexation Equilibria of Metal Ions

$$\log \alpha_{M(L,OH)}$$

pH	6.5	7.0	7.5	8.0	8.5	9.0	9.5	10.0	10.5	11.0	12.0
Th^{4+}											
	8.80	10.80	12.80	14.80	16.80	18.80	20.80	22.80	24.80	26.80	30.80
	8.81	10.80	12.80	14.80	16.80	18.80	20.80	22.80	24.80	26.80	30.80
	8.80	10.80	12.80	14.80	16.80	18.80	20.80	22.80	24.80	26.80	30.80
	8.82	10.80	12.80	14.80	16.80	18.80	20.80	22.80	24.80	26.80	30.80
	16.03	16.03	16.03	16.06	16.87	18.80	20.80	22.80	24.80	26.80	30.80
	8.80	10.80	12.80	14.80	16.80	18.80	20.80	22.80	24.80	26.80	30.80
TiO^{2+}											
	7.70	8.70	9.70	10.70	11.70	12.70	13.70	14.70	15.70	16.70	18.70
	7.72	8.70	9.70	10.70	11.70	12.70	13.70	14.70	15.70	16.70	18.70
	7.70	8.70	9.70	10.70	11.70	12.70	13.70	14.70	15.70	16.70	18.70
	13.40	13.40	13.40	13.40	13.41	13.48	13.88	14.72	15.70	16.70	18.70
Tl^{+}											
	2.03	2.24	2.34	2.41	2.51	2.72	3.06	3.49	3.97	4.46	5.44
	1.18	1.18	1.18	1.18	1.18	1.18	1.18	1.18	1.18	1.18	1.18
Tl^{3+}											
	20.20	22.19	24.17	26.09	27.88	29.35	30.29	30.74	30.92	30.98	32.51
	15.30	16.80	18.30	19.80	21.30	22.80	24.31	25.82	27.35	28.95	32.50
	16.35	16.92	18.30	19.80	21.30	22.80	24.31	25.82	27.35	28.95	32.50
U^{4+}											
	15.42	17.91	20.40	22.90	25.40	27.90	30.40	32.90	35.40	37.90	42.90
	15.42	17.91	20.40	22.90	25.40	27.90	30.40	32.90	35.40	37.90	42.90
	18.20	18.38	20.40	22.90	25.40	27.90	30.40	32.90	35.40	37.90	42.90
UO_2^{2+}											
	6.68	9.18	11.68	14.18	16.68	19.18	21.68	24.18	26.68	29.18	34.18
	11.50	12.62	13.67	14.80	16.72	19.18	21.68	24.18	26.68	29.18	34.18
	6.68	9.18	11.68	14.18	16.68	19.18	21.68	24.18	26.68	29.18	34.18
	8.15	9.21	11.68	14.18	16.68	19.18	21.68	24.18	26.68	29.18	34.18
VO^{2+}											
	3.71	4.71	5.71	6.71	7.71	8.71	9.71	10.71	11.71	12.71	14.71
	3.73	4.71	5.71	6.71	7.71	8.71	9.71	10.71	11.71	12.71	14.71
	4.63	4.95	5.74	6.71	7.71	8.71	9.71	10.71	11.71	12.71	14.71
VO_2^{+}											
	6.03	7.08	8.21	9.48	10.86	12.32	13.81	15.30	16.80	18.30	21.33
	6.03	7.08	8.21	9.48	10.86	12.32	13.81	15.30	16.80	18.30	21.33
	6.03	7.08	8.21	9.48	10.86	12.32	13.81	15.30	16.80	18.30	21.33
Y^{3+}											
	1.82	1.82	1.83	1.86	2.09	3.89	6.38	8.88	11.38	13.88	18.88
	7.30	7.30	7.30	7.30	7.30	7.30	7.35	8.89	11.38	13.88	18.88
Zn^{2+}											
	2.21	3.85	5.72	7.64	9.51	11.19	12.44	13.14	13.44	13.55	13.60
	5.73	7.69	9.67	11.62	13.51	15.18	16.44	17.14	17.34	17.55	17.59

Table 11.2 (Continued)

Ligand	Concentration [L], M	pH 2.0	3.0	3.5	4.0	4.5	5.0	5.5	6.0
		Zn^{2+}							
NH_3	0.01	0.00	0.00	0.00	0.00	0.00	0.00	0.00	0.00
	0.10	0.00	0.00	0.00	0.00	0.00	0.00	0.00	0.00
	1.00	0.00	0.00	0.00	0.00	0.00	0.00	0.01	0.04
SCN^-	1.00	1.71	1.80	1.81	1.81	1.81	1.81	1.81	1.81
HPO_4^{2-}	1.00	0.01	0.16	0.39	0.75	1.18	1.61	1.98	2.22
$P_2O_7^{4-}$	0.10	0.00	0.00	0.02	0.14	0.69	1.60	2.57	3.52
$P_3O_{10}^{5-}$	0.10	0.21	1.20	1.72	2.24	2.76	3.32	3.94	4.57
HS^-	0.10	1.60	3.59	4.59	5.59	6.59	7.58	8.56	9.49
SO_4^{2-}	1.00	1.94	2.08	2.09	2.09	2.10	2.10	2.10	2.10
$S_2O_3^{2-}$	1.00	3.25	3.32	3.32	3.32	3.32	3.32	3.32	3.32
Cl^-	1.00	1.15	1.15	1.15	1.15	1.15	1.15	1.15	1.15
		Zr^{4+}							
SO_4^{2-}	0.10	4.61	6.98	8.98	10.98	12.98	14.98	16.98	18.98
F^-	0.10	19.45	23.02	24.30	25.02	25.33	25.45	25.48	25.49

Table 11.3

Values of side-reaction coefficients $\alpha_{M(L,OH)}$ for organic ligands

Ligand	Concentration [L], M	pH 2.0	3.0	3.5	4.0	4.5	5.0	5.5	6.0
		Ag^+							
DCTA	0.01	0.00	0.00	0.00	0.00	0.00	0.01	0.10	0.44
	0.10	0.00	0.00	0.00	0.00	0.01	0.10	0.54	1.27
DTPA	0.10	0.00	0.00	0.00	0.00	0.00	0.02	0.15	0.71
EDTA	0.01	0.00	0.00	0.01	0.03	0.10	0.27	0.59	1.04
	0.10	0.00	0.02	0.09	0.26	0.57	0.99	1.48	2.00
EGTA	0.10	0.00	0.00	0.01	0.05	0.14	0.34	0.68	1.12
NTA	0.10	0.00	0.00	0.00	0.01	0.04	0.11	0.28	0.58
bipy	0.10	0.23	1.81	2.73	3.55	4.15	4.47	4.60	4.65
en	0.10	0.00	0.00	0.00	0.01	0.02	0.05	0.15	0.36
phen	0.10	4.20	6.19	7.17	8.10	8.93	9.53	9.85	9.99
py	0.10	0.00	0.03	0.08	0.27	0.72	1.32	1.81	2.06
TEA	0.10	0.00	0.00	0.00	0.00	0.00	0.01	0.03	0.10
THAM	0.10	0.00	0.00	0.00	0.00	0.02	0.06	0.20	0.66
tren	0.10	0.00	0.00	0.00	0.00	0.01	0.02	0.07	0.20
trien	0.10	0.00	0.00	0.01	0.04	0.13	0.34	0.70	1.22
glyc	0.10	0.00	0.00	0.00	0.00	0.00	0.00	0.01	0.02
thiourea	0.10	9.00	9.84	9.93	9.96	9.97	9.98	9.98	9.98

Ch. 11] Side-Reaction Coefficients for Complexation Equilibria of Metal Ions

$\log \alpha_{M(L,OH)}$

pH 6.5	7.0	7.5	8.0	8.5	9.0	9.5	10.0	10.5	11.0	12.0
Zn^{2+}										
0.00	0.01	0.02	0.09	0.34	0.98	1.87	2.84	3.87	5.04	10.00
0.01	0.04	0.15	0.61	1.80	3.24	4.36	4.97	5.24	5.49	10.00
0.15	0.62	1.95	3.73	5.54	7.15	8.31	8.93	9.19	9.28	10.09
1.81	1.81	1.81	1.81	1.82	1.85	2.12	2.87	3.87	5.04	10.00
2.34	2.38	2.39	2.40	2.40	2.41	2.50	2.96	3.88	5.04	10.00
4.39	5.15	5.82	6.51	7.26	8.00	8.57	8.92	9.17	9.47	10.49
5.16	5.70	6.20	6.65	7.02	7.25	7.37	7.41	7.42	7.43	10.01
10.34	11.05	11.65	12.18	12.69	13.20	13.70	14.20	14.70	15.20	16.19
2.10	2.10	2.10	2.10	2.10	2.12	2.28	2.90	3.88	5.04	10.00
3.32	3.32	3.32	3.32	3.32	3.32	3.34	3.44	3.98	5.04	10.00
1.15	1.15	1.15	1.15	1.17	1.33	1.91	2.84	3.87	5.04	10.00
Zr^{4+}										
20.98	22.98	24.98	26.98	28.98	30.98	32.98	34.98	36.99	39.01	43.24
25.50	25.50	25.61	26.99	28.98	30.98	32.98	34.98	36.99	39.01	43.24

$\log \alpha_{M(L,OH)}$

pH 6.5	7.0	7.5	8.0	8.5	9.0	9.5	10.0	10.5	11.0	12.0
Ag^{+}										
1.01	1.58	2.11	2.62	3.13	3.63	4.13	4.63	5.12	5.61	6.48
1.97	2.57	3.11	3.62	4.13	4.63	5.13	5.63	6.12	6.61	7.48
1.63	2.62	3.59	4.52	5.34	6.02	6.57	7.02	7.33	7.50	7.60
1.55	2.07	2.57	3.08	3.57	4.06	4.51	4.88	5.13	5.25	5.31
2.54	3.06	3.57	4.08	4.57	5.06	5.51	5.88	6.13	6.25	6.31
1.62	2.18	2.85	3.62	4.43	5.11	5.55	5.76	5.84	5.87	5.88
1.00	1.46	1.95	2.44	2.92	3.37	3.74	3.98	4.09	4.14	4.16
4.66	4.67	4.67	4.67	4.67	4.67	4.67	4.67	4.67	4.67	4.67
0.68	1.06	1.58	2.45	3.44	4.40	5.26	5.92	6.31	6.48	6.56
10.04	10.05	10.06	10.06	10.06	10.06	10.06	10.06	10.06	10.06	10.06
2.16	2.20	2.21	2.21	2.21	2.21	2.21	2.21	2.21	2.21	2.22
0.26	0.58	1.01	1.41	1.65	1.76	1.79	1.80	1.81	1.81	1.82
1.44	2.34	3.20	3.88	4.29	4.47	4.54	4.56	4.57	4.57	4.57
0.49	0.98	1.71	2.65	3.73	4.76	5.63	6.24	6.58	6.72	6.79
1.83	2.46	3.11	3.82	4.44	5.54	6.30	6.60	6.65	6.65	6.65
0.07	0.24	0.74	1.56	2.46	3.31	3.96	4.34	4.51	4.57	4.60
9.98	9.98	9.98	9.98	9.98	9.98	9.98	9.98	9.98	9.98	9.98

Table 11.3 (Continued)

Ligand	Concentration [L], M	pH 2.0	3.0	3.5	4.0	4.5	5.0	5.5	6.0
		Al^{3+}							
ac	0.10	0.00	0.04	0.10	0.24	0.42	0.63	1.12	2.30
citr	0.01	0.07	2.03	3.32	4.60	5.80	6.87	7.80	8.59
	0.10	0.44	3.03	4.32	5.60	6.80	7.87	8.80	9.59
ox	0.10	7.32	10.14	11.14	11.72	11.98	12.07	12.11	12.12
tart	0.01	1.13	3.88	4.83	5.38	5.63	5.73	5.76	5.77
	0.10	2.81	5.86	6.83	7.38	7.63	7.72	7.76	7.77
sal	0.10	0.16	1.33	2.01	2.72	3.53	4.45	5.42	6.41
sulfosal	0.10	0.94	2.39	2.97	3.52	4.09	4.75	5.55	6.49
tiron	0.10	0.15	1.62	2.72	4.22	6.11	8.12	10.24	12.77
DCTA	0.01	1.06	4.27	5.63	6.82	7.90	8.91	9.86	10.73
	0.10	2.03	5.27	6.63	7.82	8.90	9.91	10.86	11.73
DTPA	0.10	1.82	4.58	5.69	6.72	7.73	8.73	9.75	10.81
EDTA	0.01	1.47	3.84	4.88	5.90	6.91	7.92	8.96	10.03
	0.10	2.46	4.84	5.88	6.90	7.91	8.92	9.96	11.03
EGTA	0.01	0.02	0.56	1.15	1.87	2.75	3.81	5.06	6.45
	0.10	0.13	1.43	2.12	2.86	3.75	4.81	6.06	7.44
NTA	0.10	2.17	3.61	4.17	4.71	5.28	5.94	6.74	7.64
acac	0.10	0.92	2.60	3.61	4.74	6.00	7.39	8.85	10.34
ferron	0.10	1.58	4.89	6.50	8.04	9.55	11.05	12.53	13.96
		Am^{3+} *							
ac	0.10	0.01	0.07	0.19	0.41	0.68	0.91	1.04	1.12
citr	0.10	0.23	2.51	3.66	4.86	6.18	7.33	8.16	8.65
ox	0.10	3.92	5.78	6.45	6.84	7.01	7.07	7.09	7.10
DCTA	0.10	2.03	5.52	6.91	8.12	9.20	10.21	11.16	12.02
DTPA	0.10	3.61	7.36	8.92	10.34	11.61	12.74	13.79	14.81
EDTA	0.01	2.64	5.54	6.66	7.70	8.70	9.69	10.62	11.47
	0.10	3.64	6.54	7.66	8.70	9.70	10.69	11.62	12.47
NTA	0.10	2.41	5.18	6.33	7.38	8.39	9.40	10.40	11.40
HTGA	0.10	0.05	0.30	0.47	0.58	0.64	0.66	0.70	0.79
		Ba^{2+}							
citr	0.10	1.74	2.41	2.52	2.48	2.27	1.94	1.72	1.76
tiron	0.10	0.00	0.00	0.00	0.00	0.00	0.01	0.03	0.10
DCTA	0.01	0.00	0.00	0.01	0.03	0.11	0.28	0.57	0.88
	0.10	0.00	0.01	0.06	0.24	0.58	1.01	1.44	1.83
DTPA	0.01	0.00	0.00	0.00	0.00	0.00	0.01	0.04	0.24
	0.10	0.00	0.00	0.00	0.00	0.01	0.07	0.30	0.93
EDTA	0.01	0.00	0.00	0.00	0.01	0.03	0.14	0.56	1.27
	0.10	0.00	0.00	0.01	0.05	0.21	0.67	1.44	2.25
EGTA	0.01	0.00	0.00	0.00	0.00	0.00	0.02	0.08	0.39
	0.10	0.00	0.00	0.00	0.01	0.04	0.14	0.48	1.20
NTA	0.10	0.00	0.00	0.00	0.01	0.02	0.06	0.16	0.38

Ch. 11] Side-Reaction Coefficients for Complexation Equilibria of Metal Ions

$\log \alpha_{M(L,OH)}$

pH 6.5	7.0	7.5	8.0	8.5	9.0	9.5	10.0	10.5	11.0	12.0
Al^{3+}										
3.96	5.85	7.81	9.80	11.79	13.79	15.79	17.79	19.79	21.79	25.79
9.31	10.09	10.97	11.92	12.91	13.94	15.17	16.88	18.83	20.83	24.83
10.31	11.09	11.97	12.92	13.91	14.91	15.94	17.17	18.88	20.83	24.83
12.12	12.12	12.12	12.12	12.29	13.80	15.79	17.79	19.79	21.79	25.79
5.78	6.11	7.81	9.80	11.79	13.79	15.79	17.79	19.79	21.79	25.79
7.77	7.77	8.09	9.80	11.79	13.79	15.79	17.79	19.79	21.79	25.79
7.42	8.46	9.58	10.82	12.20	13.69	15.29	17.03	18.90	20.85	24.83
7.53	8.70	10.02	11.44	12.92	14.41	15.93	17.49	19.13	20.92	24.83
15.61	18.41	20.94	23.06	24.85	26.45	27.98	29.49	30.99	32.46	35.14
11.47	12.14	12.83	13.64	14.55	15.52	16.51	17.58	18.99	20.85	24.83
12.47	13.14	13.83	14.64	15.55	16.52	17.51	18.51	19.58	20.99	24.83
11.97	13.25	14.63	16.02	17.34	18.51	19.56	20.51	21.32	22.02	24.84
11.09	12.13	13.15	14.16	15.17	16.18	17.22	18.30	19.47	20.97	24.83
12.09	13.13	14.15	15.16	16.17	17.18	18.22	19.29	20.37	21.51	24.85
7.90	9.40	10.91	12.46	14.05	15.60	16.99	18.19	19.39	20.93	24.83
8.90	10.40	11.91	13.46	15.05	16.60	17.98	19.17	20.26	21.40	24.84
8.61	9.62	10.65	11.76	12.98	14.31	15.68	17.13	18.88	20.84	24.83
11.83	13.32	14.79	16.20	17.46	18.41	18.95	19.20	20.01	21.93	25.93
15.26	16.27	16.88	17.16	17.29	17.38	17.55	17.88	18.93	20.83	24.83
Am^{3+} *										
1.23	1.45	1.79	2.23	2.71	3.20	3.70	4.20	4.70	5.20	6.20
8.88	8.96	8.99	9.00	9.00	9.00	9.00	9.00	9.00	9.00	9.00
7.10	7.10	7.10	7.10	7.10	7.10	7.10	7.10	7.10	7.11	7.15
12.74	13.34	13.88	14.39	14.90	15.40	15.90	16.40	16.89	17.38	18.25
15.81	16.81	17.78	18.71	19.53	20.21	20.76	21.21	21.52	21.69	21.79
12.18	12.78	13.31	13.82	14.32	14.80	15.25	15.61	15.83	15.94	15.99
13.18	13.78	14.31	14.82	15.32	15.80	16.25	16.61	16.83	16.94	16.99
12.40	13.40	14.39	15.38	16.34	17.22	17.93	18.38	18.59	18.66	18.70
0.98	1.31	1.74	2.21	2.70	3.20	3.70	4.20	4.70	5.20	6.20
Ba^{2+}										
1.84	1.88	1.89	1.89	1.89	1.90	1.90	1.90	1.90	1.90	1.90
0.24	0.48	0.73	0.91	1.00	1.04	1.08	1.17	1.37	1.69	2.49
1.18	1.48	1.87	2.32	2.80	3.29	3.79	4.29	4.78	5.27	6.14
2.15	2.47	2.86	3.31	3.80	4.29	4.79	5.29	5.78	6.27	7.14
0.88	1.80	2.76	3.69	4.51	5.19	5.74	6.19	6.50	6.67	6.77
1.83	2.80	3.76	4.69	5.51	6.19	6.74	7.19	7.50	7.67	7.77
1.97	2.56	3.10	3.61	4.11	4.60	5.05	5.42	5.67	5.79	5.85
2.96	3.56	4.10	4.61	5.11	5.60	6.05	6.42	6.67	6.79	6.85
1.15	2.10	3.07	4.03	4.92	5.63	6.08	6.29	6.37	6.40	6.41
2.12	3.09	4.07	5.03	5.92	6.63	7.08	7.29	7.37	7.40	7.41
0.74	1.18	1.66	2.14	2.62	3.06	3.42	3.64	3.74	3.78	3.80

Table 11.3 (Continued)

Ligand	Concentration [L], M	pH 2.0	3.0	3.5	4.0	4.5	5.0	5.5	6.0
		Be^{2+}							
ac	0.10	0.01	0.05	0.13	0.31	0.54	0.74	0.86	0.98
citr	0.10	0.11	0.46	0.81	1.42	2.10	2.67	3.08	3.33
ox	0.10	1.05	2.30	2.86	3.21	3.37	3.43	3.45	3.46
sulfosal	0.10	0.37	1.67	2.33	3.07	3.92	4.86	5.84	6.83
tiron	0.10	0.00	0.02	0.08	0.31	0.93	1.84	2.91	4.29
DCTA	0.10	0.00	0.00	0.02	0.23	0.96	1.92	2.87	3.73
EDTA	0.10	0.00	0.00	0.03	0.21	0.85	1.78	2.71	3.57
NTA	0.10	0.00	0.08	0.24	0.54	0.95	1.42	1.91	2.41
histam	0.10	0.00	0.00	0.00	0.02	0.09	0.44	1.25	2.36
histid	0.10	0.00	0.00	0.00	0.01	0.08	0.42	1.23	2.38
acac	0.10	0.21	1.09	1.88	2.79	3.76	4.75	5.75	6.75
		Bi^{3+}							
DCTA	0.01	16.59	20.02	21.40	22.60	23.67	24.66	25.57	26.36
	0.10	17.59	21.02	22.40	23.60	24.67	25.66	26.57	27.36
DTPA	0.10	18.28	21.65	23.11	24.47	25.70	26.80	27.84	28.85
EDTA	0.01	11.81	14.41	15.46	16.47	17.47	18.46	19.41	20.29
	0.10	12.81	15.41	16.46	17.47	18.47	19.46	20.41	21.29
NTA	0.10	9.09	11.93	13.03	14.07	15.08	16.08	17.08	18.08
thiourea	0.10	4.09	5.65	5.82	5.88	5.91	6.13	7.12	8.53
		Ca^{2+}							
ac	0.10	0.00	0.00	0.01	0.02	0.04	0.07	0.09	0.10
citr	0.01	0.01	0.04	0.09	0.22	0.48	0.84	1.18	1.40
	0.10	0.06	0.29	0.52	0.88	1.33	1.78	2.15	2.38
tiron	0.10	0.00	0.00	0.00	0.00	0.01	0.02	0.05	0.14
DCTA	0.01	0.00	0.00	0.08	0.63	1.61	2.61	3.56	4.42
	0.10	0.00	0.03	0.49	1.53	2.60	3.61	4.56	5.42
DTPA	0.10	0.00	0.01	0.12	0.55	1.21	1.84	2.45	3.12
EDTA	0.01	0.00	0.01	0.08	0.47	1.30	2.26	3.20	4.06
	0.10	0.00	0.07	0.46	1.31	2.28	3.26	4.20	5.06
EGTA	0.10	0.00	0.01	0.05	0.23	0.80	1.68	2.66	3.65
NTA	0.10	0.00	0.02	0.06	0.19	0.43	0.81	1.26	1.75
glutam	0.10	0.00	0.00	0.00	0.00	0.00	0.00	0.00	0.00
		Cd^{2+}							
ac	0.10	0.00	0.02	0.05	0.13	0.26	0.39	0.47	0.50
citr	0.01	0.01	0.04	0.10	0.27	0.60	1.02	1.39	1.64
	0.10	0.05	0.29	0.57	0.98	1.50	2.05	2.56	2.92
sal	0.10	0.00	0.00	0.00	0.00	0.00	0.00	0.00	0.00
sulfosal	0.10	0.00	0.00	0.00	0.00	0.00	0.00	0.00	0.00
DCTA	0.01	2.20	4.95	6.16	7.29	8.34	9.34	10.29	11.15
	0.10	3.19	5.95	7.16	8.29	9.34	10.34	11.29	12.15
DTPA	0.01	2.33	4.22	5.07	5.99	6.99	8.01	9.02	10.02
	0.10	3.33	5.22	6.07	6.99	7.99	9.01	10.02	11.02

Ch. 11] Side-Reaction Coefficients for Complexation Equilibria of Metal Ions 311

$$\log \alpha_{M(L,OH)}$$

pH 6.5	7.0	7.5	8.0	8.5	9.0	9.5	10.0	10.5	11.0	12.0
Be^{2+}										
1.41	2.61	4.08	5.58	7.08	8.58	10.08	11.58	13.08	14.58	17.58
3.44	3.53	4.18	5.58	7.08	8.58	10.08	11.58	13.08	14.58	17.58
3.46	3.52	4.23	5.65	7.15	8.65	10.15	11.65	13.15	14.65	17.65
7.83	8.83	9.83	10.83	11.83	12.83	13.83	14.82	15.79	16.71	18.14
6.04	7.87	9.55	10.96	12.15	13.22	14.24	15.24	16.24	17.22	19.03
4.45	5.05	5.60	6.22	7.20	8.59	10.08	11.58	13.08	14.58	17.58
4.30	4.90	5.46	6.10	7.17	8.59	10.08	11.58	13.08	14.58	17.58
2.91	3.46	4.30	5.60	7.08	8.58	10.08	11.58	13.08	14.58	17.58
3.59	4.73	5.79	6.81	7.85	8.94	10.17	11.59	13.08	14.58	17.58
3.61	4.74	5.78	6.77	7.73	8.76	10.09	11.58	13.08	14.58	17.58
7.75	8.74	9.71	10.62	11.40	11.93	12.20	12.39	13.17	14.59	17.58
Bi^{3+}										
27.01	27.57	28.08	28.58	29.04	29.43	29.71	29.88	30.05	30.30	31.13
28.01	28.57	29.08	29.58	30.04	30.43	30.71	30.88	31.05	31.30	32.13
29.85	30.84	31.79	32.67	33.40	33.94	34.31	34.52	34.68	34.89	35.64
21.05	21.68	22.22	22.74	23.23	23.71	24.13	24.47	24.74	25.56	29.42
22.05	22.68	23.22	23.74	24.23	24.71	25.13	25.46	25.71	26.08	29.42
19.08	20.07	21.05	21.99	22.82	23.44	23.78	23.93	24.10	25.45	29.42
10.01	11.50	13.00	14.50	16.00	17.50	19.00	20.51	22.03	23.58	26.98
Ca^{2+}										
0.11	0.11	0.11	0.11	0.11	0.11	0.11	0.11	0.11	0.11	0.12
1.50	1.54	1.56	1.56	1.56	1.56	1.56	1.56	1.56	1.56	1.56
2.49	2.53	2.54	2.55	2.55	2.55	2.55	2.55	2.55	2.55	2.55
0.34	0.63	0.92	1.14	1.32	1.55	1.90	2.33	2.81	3.29	4.18
5.14	5.74	6.28	6.79	7.30	7.80	8.30	8.80	9.29	9.78	10.65
6.14	6.74	7.28	7.79	8.30	8.80	9.30	9.80	10.29	10.78	11.65
3.91	4.81	5.75	6.66	7.48	8.16	8.71	9.16	9.47	9.64	9.74
4.79	5.39	5.93	6.44	6.94	7.43	7.88	8.25	8.50	8.62	8.68
5.79	6.39	6.93	7.44	7.94	8.43	8.88	9.25	9.50	9.62	9.68
4.65	5.64	6.63	7.59	8.48	9.19	9.64	9.85	9.93	9.96	9.97
2.25	2.76	3.30	3.91	4.62	5.38	6.04	6.47	6.67	6.74	6.77
0.00	0.00	0.01	0.03	0.08	0.19	0.34	0.47	0.53	0.56	0.58
Cd^{2+}										
0.51	0.51	0.51	0.51	0.51	0.52	0.53	0.58	0.79	1.39	3.36
1.76	1.80	1.82	1.83	1.83	1.83	1.83	1.83	1.85	1.99	3.50
3.09	3.16	3.18	3.19	3.19	3.19	3.19	3.19	3.20	3.20	3.67
0.00	0.01	0.02	0.06	0.16	0.39	0.75	1.20	1.70	2.24	3.65
0.01	0.03	0.10	0.26	0.56	0.97	1.44	1.92	2.40	2.86	3.77
11.87	12.47	13.01	13.52	14.03	14.53	15.03	15.53	16.02	16.51	17.38
12.87	13.47	14.01	14.52	15.03	15.53	16.03	16.53	17.02	17.51	18.38
11.02	12.01	12.98	13.91	14.73	15.41	15.97	16.41	16.72	16.89	16.99
12.02	13.01	13.98	14.91	15.73	16.41	16.97	17.41	17.72	17.89	17.99

Table 11.3 (Continued)

Ligand	Concentration [L], M	pH 2.0	3.0	3.5	4.0	4.5	5.0	5.5	6.0
		Cd^{2+}							
EDTA	0.01	1.95	4.14	5.10	6.07	7.06	8.03	8.97	9.83
	0.10	2.95	5.14	6.10	7.07	8.06	9.03	9.97	10.83
EGTA	0.10	1.86	3.80	4.61	5.47	6.41	7.39	8.38	9.38
NTA	0.10	0.51	2.01	2.59	3.12	3.64	4.17	4.74	5.41
bipy	0.10	1.00	3.20	4.47	5.66	6.53	7.01	7.21	7.28
dien	0.10	0.00	0.00	0.00	0.00	0.00	0.00	0.03	0.25
en	0.10	0.00	0.00	0.00	0.00	0.00	0.00	0.00	0.04
phen	0.10	3.10	5.83	7.27	8.67	9.90	10.80	11.29	11.49
py	0.10	0.00	0.00	0.01	0.04	0.12	0.27	0.46	0.59
tren	0.10	0.00	0.00	0.00	0.00	0.00	0.00	0.09	0.92
glyc	0.10	0.00	0.00	0.00	0.00	0.00	0.01	0.03	0.10
histam	0.10	0.00	0.00	0.00	0.00	0.00	0.00	0.03	0.17
acac	0.10	0.00	0.00	0.00	0.00	0.01	0.03	0.09	0.24
		Ce^{3+}							
citr	0.10	0.11	2.11	3.24	4.37	5.62	6.74	7.56	8.05
ox	0.10	2.95	5.57	6.70	7.40	7.72	7.83	7.87	7.88
sulfosal	0.10	0.00	0.00	0.00	0.00	0.01	0.02	0.06	0.16
DCTA	0.10	0.17	3.18	4.57	5.78	6.86	7.87	8.82	9.68
EDTA	0.01	0.63	3.40	4.52	5.56	6.56	7.55	8.49	9.35
	0.10	1.53	4.40	5.52	6.56	7.56	8.55	9.49	10.35
DTPA	0.10	1.17	4.89	6.45	7.87	9.14	10.27	11.32	12.34
EGTA	0.10	0.19	2.56	3.68	4.72	5.73	6.74	7.74	8.74
NTA	0.10	1.38	3.59	4.63	5.65	6.65	7.66	8.66	9.66
acac	0.10	0.00	0.01	0.03	0.09	0.23	0.52	0.94	1.48
		Co^{2+}							
ac	0.10	0.00	0.01	0.02	0.06	0.11	0.17	0.20	0.21
citr	0.10	0.10	0.71	1.29	1.97	2.62	3.18	3.58	3.83
sal	0.10	0.00	0.00	0.00	0.00	0.00	0.00	0.00	0.01
sulfosal	0.10	0.00	0.00	0.00	0.00	0.00	0.01	0.03	0.10
tiron	0.10	0.00	0.00	0.01	0.04	0.12	0.30	0.63	1.08
DCTA	0.01	1.80	4.60	5.83	6.97	8.03	9.03	9.98	10.84
	0.10	2.80	5.60	6.83	7.97	9.03	10.03	10.98	11.84
DTPA	0.01	3.15	5.07	5.91	6.75	7.57	8.39	9.29	10.24
	0.10	4.15	6.07	6.91	7.75	8.57	9.39	10.29	11.24
EDTA	0.01	1.89	4.03	4.97	5.93	6.91	7.88	8.82	9.68
	0.10	2.89	5.03	5.97	6.93	7.91	8.88	9.82	10.68
EGTA	0.10	0.06	1.05	1.64	2.20	2.77	3.44	4.23	5.14
HEDTA	0.10	1.63	4.17	5.26	6.29	7.26	8.16	8.95	9.60
NTA	0.10	1.00	2.61	3.19	3.72	4.23	4.74	5.26	5.81
bipy	0.10	5.31	8.55	9.98	11.22	12.11	12.60	12.80	12.87

Ch. 11] Side-Reaction Coefficients for Complexation Equilibria of Metal Ions 313

$\log \alpha_{M(L,OH)}$

pH 6.5	7.0	7.5	8.0	8.5	9.0	9.50	10.0	10.50	11.0	12.0
Cd^{2+}										
10.56	11.16	11.70	12.21	12.71	13.20	13.65	14.02	14.27	14.39	14.45
11.56	12.16	12.70	13.21	13.71	14.20	14.65	15.02	15.27	15.39	15.45
10.38	11.37	12.36	13.32	14.21	14.92	15.37	15.58	15.66	15.69	15.70
6.22	7.13	8.10	9.08	10.03	10.92	11.63	12.07	12.28	12.35	12.39
7.30	7.31	7.31	7.31	7.31	7.31	7.31	7.31	7.31	7.31	7.31
0.96	2.06	3.61	5.49	7.37	9.06	10.37	11.20	11.60	11.76	11.83
0.29	1.12	2.39	3.68	4.85	6.01	7.17	8.18	8.84	9.16	9.31
11.57	11.59	11.60	11.60	11.60	11.60	11.60	11.60	11.60	11.60	11.60
0.65	0.67	0.68	0.68	0.68	0.68	0.69	0.73	0.89	1.42	3.36
2.36	3.85	5.32	6.75	8.07	9.21	10.11	10.73	11.07	11.22	11.29
0.28	0.68	1.34	2.25	3.31	4.45	5.49	6.21	6.56	6.70	6.76
0.59	1.21	1.96	2.82	3.74	4.62	5.35	5.84	6.07	6.16	6.20
0.56	1.09	1.82	2.65	3.45	4.07	4.42	4.58	4.63	4.65	4.77
Ce^{3+}										
8.28	8.36	8.39	8.40	8.40	8.40	8.40	8.41	8.93	11.28	16.28
7.88	7.89	7.89	7.89	7.89	7.89	7.89	7.89	8.24	10.48	15.48
0.39	0.82	1.44	2.24	3.15	4.12	5.11	6.17	8.02	10.48	15.48
10.40	11.00	11.54	12.05	12.56	13.06	13.56	14.06	14.55	15.04	16.43
10.08	10.68	11.22	11.73	12.23	12.72	13.17	13.54	13.79	13.91	16.28
11.08	11.68	12.22	12.73	13.23	13.72	14.17	14.54	14.79	14.91	16.30
13.34	14.34	15.31	16.24	17.06	17.74	18.29	18.74	19.05	19.22	19.32
9.74	10.73	11.72	12.68	13.57	14.28	14.73	14.94	15.02	15.05	16.30
10.66	11.65	12.65	13.64	14.61	15.53	16.32	16.96	17.49	18.00	19.00
2.20	3.20	4.45	5.75	6.89	7.67	8.07	8.23	8.46	10.48	15.48
Co^{2+}										
0.21	0.22	0.22	0.22	0.23	0.25	0.39	0.88	1.74	2.73	4.92
3.94	3.98	3.99	4.00	4.00	4.00	4.00	4.00	4.00	4.04	5.16
0.03	0.08	0.22	0.50	0.91	1.42	2.03	2.79	3.68	4.63	6.58
0.26	0.55	0.96	1.43	1.92	2.41	2.91	3.41	3.90	4.39	5.40
1.62	2.40	3.54	4.70	5.77	6.80	7.80	8.80	9.77	10.68	11.99
11.56	12.16	12.70	13.21	13.72	14.22	14.72	15.22	15.71	16.20	17.07
12.56	13.16	13.70	14.21	14.72	15.22	15.72	16.22	16.71	17.20	18.07
11.22	12.21	13.18	14.10	14.93	15.61	16.16	16.60	16.92	17.09	17.18
12.22	13.21	14.18	15.10	15.93	16.61	17.16	17.60	17.92	18.09	18.18
10.41	11.01	11.55	12.06	12.56	13.05	13.50	13.87	14.12	14.24	14.30
11.41	12.01	12.55	13.06	13.56	14.05	14.50	14.87	15.12	15.24	15.30
6.10	7.09	8.07	9.03	9.91	10.63	11.08	11.29	11.37	11.39	11.41
10.16	10.68	11.18	11.68	12.17	12.63	13.02	13.28	13.42	13.47	13.50
6.44	7.20	8.09	9.03	9.98	10.86	11.57	12.01	12.22	12.29	12.33
12.89	12.90	12.90	12.90	12.90	12.90	12.90	12.90	12.90	12.90	12.90

Table 11.3 (Continued)

Ligand	Concentration [L], M	pH 2.0	3.0	3.5	4.0	4.5	5.0	5.5	6.0
		Co^{2+}							
dien	0.10	0.00	0.00	0.00	0.00	0.00	0.01	0.05	0.37
en	0.10	0.00	0.00	0.00	0.00	0.00	0.00	0.01	0.08
phen	0.10	7.30	10.28	11.74	13.15	14.38	15.28	15.77	15.97
py	0.10	0.00	0.00	0.01	0.03	0.09	0.19	0.33	0.42
TEA	0.10	0.00	0.00	0.00	0.00	0.00	0.00	0.01	0.03
trien	0.10	0.00	0.00	0.00	0.00	0.04	0.33	1.21	2.39
cyst	0.10	0.00	0.00	0.00	0.00	0.01	0.06	0.40	1.33
glutam	0.10	0.00	0.00	0.00	0.00	0.01	0.03	0.11	0.29
glyc	0.10	0.00	0.00	0.00	0.00	0.01	0.04	0.13	0.35
histam	0.10	0.00	0.00	0.00	0.00	0.00	0.01	0.08	0.41
histid	0.10	0.00	0.00	0.00	0.03	0.21	0.90	2.12	3.59
tryp	0.10	0.00	0.00	0.00	0.00	0.01	0.02	0.06	0.20
TGA	0.10	0.00	0.00	0.00	0.02	0.09	0.37	1.04	1.96
acac	0.10	0.00	0.01	0.03	0.10	0.27	0.59	1.11	1.80
		Co^{3+}							
EDTA	0.10	28.30	30.25	31.18	32.14	33.12	34.09	35.03	35.87
HEDTA	0.10	24.52	27.07	28.16	29.19	30.16	31.06	31.85	32.50
cyst	0.10	0.57	4.60	6.63	8.64	10.64	12.64	14.64	16.64
		Cr^{3+}							
ac	0.10	1.20	2.56	3.64	4.80	5.72	6.26	6.50	6.88
sulfosal	0.10	0.01	0.20	0.49	0.91	1.45	2.66	4.58	6.58
EDTA	0.10	9.21	11.86	12.95	13.98	14.99	15.97	16.92	17.79
NTA	0.10	0.63	2.17	2.75	3.28	3.79	4.32	5.05	6.61
		Cu^{+}							
bipy	0.10	9.41	10.50	10.98	11.39	11.69	11.85	11.92	11.94
en	0.10	0.00	0.00	0.00	0.00	0.00	0.01	0.06	0.37
glyc	0.10	1.08	2.47	3.03	3.55	4.06	4.56	5.06	5.56
histam	0.10	3.80	4.80	5.30	5.81	6.32	6.85	7.41	7.90
histid	0.10	4.55	4.86	4.89	4.89	4.88	4.82	4.67	4.32
phen	0.10	7.58	9.57	10.56	11.52	12.39	13.09	13.52	13.71
py	0.10	0.59	1.55	2.18	2.94	3.78	4.59	5.20	5.54
		Cu^{2+}							
ac	0.10	0.01	0.06	0.16	0.37	0.66	0.90	1.03	1.08
citr	0.01	0.09	0.49	1.04	1.90	2.86	3.80	4.66	5.39
	0.10	0.52	1.34	2.01	2.89	3.86	4.80	5.66	6.39
tart	0.01	0.01	0.28	0.57	0.80	0.96	1.27	2.02	2.99
	0.10	0.10	1.01	1.51	1.88	2.29	3.04	3.99	4.98
sal	0.10	0.00	0.04	0.15	0.40	0.78	1.24	1.74	2.31
sulfosal	0.10	0.01	0.14	0.40	0.78	1.25	1.77	2.37	3.09
tiron	0.10	0.04	0.31	0.68	1.24	2.16	3.78	5.72	7.68

Ch. 11] Side-Reaction Coefficients for Complexation Equilibria of Metal Ions   315

$$\log \alpha_{M(L,OH)}$$

pH 6.5	7.0	7.5	8.0	8.5	9.0	9.5	10.0	10.5	11.0	12.0
Co^{2+}										
1.21	2.46	4.19	6.10	7.94	9.54	10.72	11.41	11.72	11.84	11.89
0.51	1.56	3.02	4.53	6.03	7.48	8.82	9.90	10.59	10.92	11.08
16.05	16.07	16.08	16.08	16.08	16.08	16.08	16.08	16.08	16.08	16.08
0.46	0.48	0.48	0.49	0.49	0.50	0.59	0.96	1.76	2.73	4.92
0.08	0.20	0.40	0.60	0.73	0.79	0.85	1.09	1.78	2.73	4.92
3.64	4.82	5.91	6.93	7.88	8.74	9.40	9.80	9.98	10.05	10.09
2.83	4.71	6.63	8.42	9.98	11.27	12.36	13.31	14.08	14.60	14.95
0.64	1.18	1.89	2.74	3.64	4.49	5.17	5.58	5.76	5.83	5.93
0.78	1.47	2.42	3.58	4.86	6.09	7.05	7.62	7.87	7.96	8.00
1.06	1.84	2.73	3.71	4.76	5.84	6.83	7.50	7.83	7.96	8.02
4.94	6.09	7.13	8.10	8.97	9.65	10.06	10.24	10.31	10.33	10.34
0.57	1.30	2.35	3.62	4.99	6.36	7.58	8.46	8.95	9.15	9.24
2.94	3.93	4.93	5.92	6.91	7.86	8.74	9.43	9.85	10.04	10.14
2.65	3.58	4.54	5.48	6.32	6.95	7.31	7.46	7.51	7.53	7.55
Co^{3+}										
36.58	37.18	37.71	38.22	38.72	39.20	39.65	40.01	40.23	40.34	40.39
33.06	33.58	34.08	34.58	35.07	35.53	35.92	36.18	36.32	36.37	36.40
18.63	20.60	22.53	24.32	25.88	27.17	28.26	29.21	29.98	30.50	30.85
Cr^{3+}										
8.58	10.58	12.58	14.58	16.58	18.58	20.58	22.58	24.58	26.58	30.58
8.58	10.58	12.58	14.58	16.58	18.58	20.58	22.58	24.58	26.58	30.58
18.55	19.25	19.99	20.85	21.78	22.75	23.69	24.57	25.38	26.67	30.58
8.58	10.58	12.58	14.58	16.58	18.58	20.58	22.58	24.58	26.58	30.58
Cu^{+}										
11.95	11.95	11.95	11.95	11.95	11.95	11.95	11.95	11.95	11.95	11.95
1.12	1.98	2.74	3.38	3.92	4.42	4.85	5.18	5.37	5.46	5.50
6.06	6.56	7.06	7.55	8.02	8.45	8.78	8.97	9.05	9.09	9.10
8.23	8.38	8.44	8.45	8.43	8.36	8.18	7.82	7.30	6.75	5.71
3.71	2.93	2.31	2.08	1.98	1.82	1.52	1.13	0.72	0.38	0.06
13.78	13.81	13.82	13.82	13.82	13.82	13.82	13.82	13.82	13.82	13.82
5.69	5.74	5.76	5.76	5.76	5.76	5.76	5.76	5.76	5.76	5.76
Cu^{2+}										
1.09	1.10	1.12	1.21	1.67	2.54	3.52	4.52	5.53	6.58	9.65
6.00	6.54	7.05	7.56	8.06	8.56	9.06	9.56	10.06	10.56	11.56
7.00	7.54	8.05	8.56	9.06	9.56	10.06	10.56	11.06	11.56	12.56
3.98	4.98	5.98	6.98	7.98	8.98	9.98	10.98	11.98	12.98	14.98
5.98	6.98	7.98	8.98	9.98	10.98	11.98	12.98	13.98	14.98	16.98
2.98	3.78	4.70	5.67	6.65	7.65	8.65	9.65	10.65	11.65	13.62
3.95	4.89	5.87	6.86	7.86	8.86	9.85	10.84	11.81	12.71	13.93
9.57	11.31	12.78	14.00	15.09	16.11	17.12	18.11	19.09	20.00	21.31

Table 11.3 (Continued)

Ligand	Concentration [L], M	pH 2.0	3.0	3.5	4.0	4.5	5.0	5.5	6.0
		Cu^{2+}							
DCTA	0.01	4.36	7.08	8.26	9.37	10.41	11.41	12.36	13.22
	0.10	5.36	8.08	9.26	10.37	11.41	12.41	13.36	14.22
DTPA	0.01	5.09	7.35	8.68	10.04	11.29	12.42	13.47	14.49
	0.10	6.09	8.35	9.68	11.04	12.29	13.42	14.47	15.49
EDTA	0.01	4.38	6.52	7.46	8.42	9.40	10.37	11.31	12.17
	0.10	5.38	7.52	8.46	9.42	10.40	11.37	12.31	13.17
EGTA	0.01	2.53	4.38	5.06	5.75	6.55	7.45	8.42	9.42
	0.10	3.53	5.38	6.06	6.75	7.55	8.45	9.42	10.42
HEDTA	0.10	5.18	7.27	8.30	9.30	10.26	11.17	11.95	12.60
NTA	0.10	3.51	5.17	5.75	6.28	6.79	7.32	7.87	8.51
bipy	0.10	15.69	19.87	21.82	23.65	25.24	26.57	27.70	28.75
dien	0.10	0.04	1.22	2.41	3.67	4.87	5.98	7.02	8.03
en	0.10	0.00	0.01	0.06	0.42	1.40	3.01	4.94	6.91
phen	0.10	16.94	20.93	22.91	24.84	26.67	28.27	29.59	30.73
py	0.10	0.01	0.07	0.21	0.53	1.09	1.81	2.45	2.82
TEA	0.10	0.00	0.01	0.02	0.05	0.15	0.36	0.72	1.20
tetren	0.10	2.21	4.98	6.26	7.53	8.84	10.16	11.52	12.95
THAM	0.10	0.00	0.00	0.01	0.03	0.10	0.34	1.00	2.25
tren	0.10	0.00	0.01	0.24	1.38	2.86	4.36	5.86	7.36
trien	0.10	0.21	2.74	4.19	5.65	7.12	8.61	10.08	11.53
cyst	0.10	3.89	6.14	7.17	8.18	9.18	10.18	11.18	12.18
glutam	0.10	0.00	0.04	0.30	0.99	1.92	2.94	3.97	4.97
glyc	0.10	0.05	0.68	1.29	2.06	2.95	3.91	4.90	5.89
histam	0.10	0.00	0.02	0.16	0.73	1.66	2.74	4.15	5.79
histid	2.00	1.38	3.46	4.72	6.12	7.58	9.09	10.60	12.05
	0.10	0.24	1.28	2.25	3.54	4.99	6.49	8.00	9.45
tryp	0.10	0.08	0.92	1.68	2.56	3.52	4.51	5.50	6.50
thiourea	0.10	14.10	14.36	14.39	14.40	14.40	14.40	14.40	14.40
acac	0.10	0.58	1.64	2.39	3.28	4.23	5.22	6.21	7.21
		Fe^{2+}							
ac	0.10	0.00	0.01	0.02	0.04	0.09	0.14	0.16	0.17
citr	0.01	0.00	0.05	0.21	0.58	1.10	1.60	1.99	2.23
	0.10	0.01	0.35	0.86	1.46	2.07	2.59	2.99	3.23
asco	0.10	0.00	0.00	0.01	0.02	0.04	0.05	0.06	0.06
sal	0.10	0.00	0.00	0.00	0.00	0.00	0.00	0.00	0.01
sulfosal	0.10	0.00	0.00	0.00	0.00	0.00	0.01	0.02	0.06
DCTA	0.10	2.20	4.96	6.22	7.38	8.44	9.44	10.38	11.23
DTPA	0.10	0.63	3.26	4.33	5.27	6.08	6.82	7.61	8.49

Ch. 11] Side-Reaction Coefficients for Complexation Equilibria of Metal Ions 317

$\log \alpha_{M(L,OH)}$

pH 6.5	7.0	7.5	8.0	8.5	9.0	9.5	10.0	10.5	11.0	12.0
Cu^{2+}										
13.94	14.54	15.08	15.59	16.10	16.60	17.10	17.60	18.09	18.58	19.45
14.94	15.54	16.08	16.59	17.10	17.60	18.10	18.60	19.09	19.58	20.45
15.49	16.48	17.46	18.38	19.21	19.89	20.44	20.88	21.20	21.37	21.46
16.49	17.48	18.46	19.38	20.21	20.89	21.44	21.88	22.20	22.37	22.46
12.90	13.50	14.04	14.55	15.05	15.54	15.99	16.38	16.67	16.91	17.57
13.90	14.50	15.04	15.55	16.05	16.54	16.99	17.38	17.67	17.91	18.57
10.45	11.55	12.79	14.28	15.93	17.56	18.98	20.19	21.26	22.29	24.30
11.45	12.55	13.79	15.28	16.93	18.56	19.98	21.19	22.26	23.29	25.30
13.16	13.68	14.18	14.68	15.17	15.63	16.02	16.28	16.42	16.47	16.50
9.29	10.18	11.13	12.11	13.06	13.95	14.66	15.10	15.31	15.39	15.51
29.76	30.77	31.77	32.77	33.77	34.77	35.77	36.77	37.77	38.77	40.77
9.04	10.09	11.40	13.13	14.95	16.54	17.72	18.41	18.72	18.84	18.92
8.85	10.68	12.29	13.62	14.75	15.76	16.67	17.40	17.86	18.08	18.19
31.78	32.79	33.80	34.80	35.80	36.80	37.80	38.80	39.80	40.80	42.80
2.98	3.03	3.05	3.05	3.07	3.17	3.65	4.53	5.53	6.58	9.65
1.79	2.47	3.14	3.63	3.88	4.01	4.19	4.80	5.73	6.73	9.17
14.41	15.88	17.30	18.62	19.76	20.66	21.28	21.60	21.73	21.78	21.80
3.89	5.67	7.37	8.73	9.57	10.09	10.77	11.71	12.70	13.71	15.71
8.86	10.35	11.83	13.29	14.67	15.96	17.15	18.18	18.99	19.63	20.69
12.89	14.11	15.21	16.24	17.20	18.07	18.78	19.32	19.77	20.23	21.20
13.17	14.14	15.08	15.91	16.60	17.16	17.61	17.92	18.09	18.16	18.20
5.98	6.98	7.97	8.96	9.91	10.78	11.46	11.87	12.06	12.13	12.16
6.89	7.89	8.88	9.87	10.81	11.67	12.33	12.71	12.88	12.94	12.97
7.27	8.51	9.60	10.62	11.59	12.50	13.24	13.72	13.96	14.05	14.09
13.34	14.46	15.50	16.47	17.34	18.02	18.42	18.60	18.67	18.69	18.70
10.74	11.86	12.89	13.86	14.74	15.41	15.82	16.00	16.07	16.09	16.10
7.50	8.50	9.49	10.47	11.42	12.26	12.89	13.25	13.40	13.45	13.48
14.40	14.40	14.40	14.40	14.40	14.40	14.40	14.40	14.40	14.40	14.40
8.21	9.20	10.17	11.08	11.86	12.39	12.65	12.76	12.79	12.80	12.81
Fe^{2+}										
0.17	0.18	0.18	0.18	0.19	0.22	0.30	0.51	0.98	1.96	4.81
2.34	2.38	2.40	2.40	2.40	2.40	2.40	2.41	2.42	2.59	5.01
3.34	3.38	3.39	3.40	3.40	3.40	3.40	3.40	3.40	3.42	5.02
0.07	0.07	0.07	0.07	0.09	0.13	0.26	0.52	1.02	1.97	4.81
0.02	0.06	0.16	0.39	0.76	1.25	1.85	2.60	3.47	4.42	6.39
0.17	0.41	0.78	1.28	1.89	2.64	3.52	4.47	5.42	6.31	7.54
11.95	12.55	13.08	13.59	14.10	14.60	15.10	15.60	16.09	16.58	17.42
9.44	10.42	11.40	12.37	13.33	14.32	15.45	16.71	17.95	19.10	21.19

Table 11.3 (Continued)

Ligand	Concentration [L], M	pH 2.0	3.0	3.5	4.0	4.5	5.0	5.5	6.0
		Fe^{2+}							
EDTA	0.01	0.17	1.94	2.93	3.92	4.91	5.89	6.83	7.69
	0.10	0.76	2.94	3.93	4.92	5.91	6.89	7.83	8.69
EGTA	0.10	0.00	0.17	0.51	1.05	1.76	2.63	3.58	4.56
HEDTA	0.10	0.35	2.05	3.03	4.01	4.97	5.87	6.65	7.30
en	0.10	0.00	0.00	0.00	0.00	0.00	0.00	0.00	0.00
trien	0.10	0.00	0.00	0.00	0.00	0.00	0.00	0.00	0.04
cyst	0.10	0.00	0.00	0.00	0.00	0.00	0.00	0.00	0.01
glutam	0.10	0.00	0.00	0.00	0.00	0.00	0.00	0.01	0.04
glyc	0.10	0.00	0.00	0.00	0.00	0.00	0.02	0.05	0.14
histid	0.10	0.00	0.00	0.00	0.00	0.00	0.01	0.07	0.46
tryp	0.10	0.00	0.00	0.00	0.00	0.00	0.00	0.01	0.04
TGA	0.10	0.00	0.00	0.00	0.00	0.00	0.00	0.03	0.25
BAL	0.10	0.00	0.00	0.00	0.00	0.00	0.00	0.00	0.12
acac	0.10	0.00	0.01	0.02	0.05	0.14	0.35	0.72	1.25
		Fe^{3+}							
ac	0.10	0.34	1.57	2.53	3.56	4.45	5.11	6.47	8.43
Cl-ac	0.10	0.55	1.08	1.34	1.89	2.99	4.72	6.68	8.68
citr	0.01	2.97	10.12	13.34	16.25	18.82	21.01	22.85	24.35
	0.10	4.80	12.12	15.34	18.25	20.82	23.01	24.85	26.35
ox	0.10	10.68	13.51	14.51	15.09	15.35	15.44	15.48	15.49
tart	0.01	2.20	5.46	6.57	7.28	7.64	7.79	7.85	8.53
	0.10	4.20	7.46	8.57	9.28	9.64	9.79	9.84	9.87
sal	0.10	13.48	18.88	21.06	22.87	24.49	26.03	27.54	29.05
sulfosal	0.10	1.15	2.99	3.94	4.94	5.95	6.96	7.99	9.16
tiron	0.10	3.27	5.96	7.92	10.00	12.44	15.31	18.28	21.22
DCTA	0.10	12.46	15.83	17.21	18.42	19.50	20.51	21.46	22.32
DTPA	0.10	10.24	13.07	14.30	15.52	16.70	17.80	18.84	19.85
EDTA	0.01	9.72	12.53	13.64	14.68	15.69	16.67	17.62	18.49
	0.10	10.72	13.53	14.64	15.68	16.69	17.67	18.62	19.49
EGTA	0.10	4.31	7.14	8.26	9.30	10.31	11.32	12.32	13.32
HEDTA	0.10	6.93	9.53	10.71	11.95	13.27	14.62	15.88	17.02
NTA	0.10	6.45	8.75	9.84	10.87	11.88	12.89	13.89	14.89
TEA	0.10	0.10	0.62	1.19	1.97	3.00	4.53	6.45	8.45
glutam	0.10	0.87	3.21	4.22	5.11	5.86	6.49	7.19	8.71
glyc	0.10	0.82	2.17	2.73	3.26	3.81	4.84	6.69	8.68
histid	0.10	0.10	0.62	1.19	1.97	3.00	4.53	6.44	8.43
tryp	0.10	0.35	1.52	2.08	2.63	3.33	4.75	6.68	8.68
BAL	0.10	2.64	5.64	7.14	8.64	10.14	11.64	13.14	14.64
acac	0.10	2.95	5.37	6.78	8.25	9.74	11.23	12.73	14.23

Ch. 11] Side-Reaction Coefficients for Complexation Equilibria of Metal Ions 319

$\log \alpha_{M(L,OH)}$

pH 6.5	7.0	7.5	8.0	8.5	9.0	9.5	10.0	10.5	11.0	12.0
Fe^{2+}										
8.42	9.02	9.57	10.11	10.68	11.35	12.20	13.22	14.31	15.37	17.41
9.42	10.02	10.57	11.11	11.68	12.35	13.20	14.22	15.31	16.37	18.41
5.55	6.54	7.53	8.49	9.38	10.09	10.54	10.75	10.83	10.86	10.87
7.86	8.39	8.91	9.45	10.05	10.78	11.65	12.61	13.60	14.60	16.60
0.02	0.14	0.60	1.40	2.38	3.47	4.61	5.60	6.26	6.58	6.74
0.49	1.56	2.65	3.67	4.63	5.49	6.15	6.55	6.73	6.80	6.84
0.06	0.47	1.69	3.26	4.75	6.02	7.10	8.05	8.84	9.39	10.08
0.11	0.29	0.59	1.00	1.45	1.87	2.20	2.39	2.49	2.61	4.81
0.39	0.88	1.62	2.51	3.43	4.28	4.95	5.35	5.53	5.59	5.71
1.39	2.69	3.96	5.08	6.09	6.98	7.69	8.12	8.32	8.39	8.43
0.13	0.37	0.89	1.71	2.71	3.83	4.93	5.77	6.24	6.43	6.53
0.94	1.89	2.88	3.88	4.88	5.86	6.82	7.69	8.38	8.87	9.65
1.90	4.72	7.63	10.43	12.94	15.03	16.74	18.15	19.21	19.85	20.25
1.94	2.78	3.70	4.62	5.45	6.08	6.44	6.59	6.64	6.66	6.71
Fe^{3+}										
10.43	12.43	14.43	16.43	18.43	20.43	22.43	24.43	26.43	28.43	32.43
10.68	12.68	14.68	16.68	18.68	20.68	22.68	24.68	26.68	28.68	32.68
25.58	26.66	27.69	28.70	29.70	30.70	31.70	32.70	33.70	34.70	36.70
27.58	28.66	29.69	30.70	31.70	32.70	33.70	34.70	35.70	36.70	38.70
15.49	15.49	15.55	16.70	18.68	20.68	22.68	24.68	26.68	28.68	32.68
10.43	12.43	14.43	16.43	18.43	20.43	22.43	24.43	26.43	28.43	32.43
10.53	12.43	14.43	16.43	18.43	20.43	22.43	24.43	26.43	28.43	32.43
30.55	32.05	33.55	35.05	36.55	38.05	39.55	41.05	42.55	44.04	46.95
10.77	12.69	14.68	16.68	18.68	20.68	22.68	24.68	26.68	28.68	32.68
24.06	26.66	28.87	30.71	32.34	33.88	35.39	36.88	38.34	39.71	41.66
23.04	23.64	24.18	24.70	25.22	25.78	26.41	27.18	28.08	29.17	32.68
20.85	21.85	22.83	23.76	24.60	25.30	25.88	26.39	26.96	28.45	32.43
19.24	19.92	20.64	21.48	22.44	23.49	24.65	25.88	27.14	28.61	32.43
20.24	20.92	21.64	22.48	23.44	24.49	25.64	26.86	28.05	29.22	32.45
14.32	15.31	16.30	17.31	18.60	20.44	22.43	24.43	26.43	28.43	32.43
18.08	19.11	20.13	21.16	22.26	23.47	24.80	26.20	27.65	29.18	32.59
15.89	16.89	17.89	18.88	19.89	20.99	22.53	24.44	26.43	28.43	32.43
10.49	12.58	14.74	16.92	19.03	21.08	23.09	25.10	27.10	29.10	33.10
10.68	12.68	14.68	16.68	18.68	20.68	22.68	24.68	26.68	28.68	32.68
10.68	12.68	14.68	16.68	18.68	20.68	22.68	24.68	26.68	28.68	32.68
10.43	12.43	14.43	16.43	18.43	20.43	22.43	24.43	26.43	28.43	32.43
10.68	12.68	14.68	16.68	18.68	20.68	22.68	24.68	26.68	28.68	32.68
16.14	17.63	19.11	20.54	21.88	23.07	24.15	25.20	26.56	28.44	32.43
15.73	17.22	18.69	20.10	21.37	22.33	23.22	24.98	26.98	28.98	32.98

Table 11.3 (Continued)

Ligand	Concentration [L], M	pH 2.0	3.0	3.5	4.0	4.5	5.0	5.5	6.0
		Fe^{3+}							
ferron	0.10	5.28	9.65	11.35	12.92	14.45	15.95	17.43	18.86
sulfoxine	0.10	17.03	23.03	26.03	29.02	31.99	34.90	37.67	40.19
		Ga^{3+}							
citr	0.10	2.11	5.07	6.52	7.91	9.17	10.26	11.18	11.92
ox	0.10	10.04	12.88	13.88	14.46	14.72	14.81	14.85	14.86
tiron	0.10	2.14	4.14	5.14	6.14	7.14	8.14	9.14	10.13
DCTA	0.10	6.12	9.03	10.35	11.53	12.61	13.62	14.60	15.55
DTPA	0.10	8.52	11.28	12.39	13.42	14.42	15.41	16.41	17.42
EDTA	0.10	6.05	8.75	9.85	10.89	11.92	12.97	14.07	15.22
HEDTA	0.10	3.91	6.48	7.62	8.78	10.02	11.32	12.57	13.72
glyc	0.10	0.46	1.69	2.25	2.78	3.38	4.27	5.60	7.15
acac	0.10	1.97	3.87	4.92	6.08	7.37	8.78	10.25	11.73
ferron	0.10	10.70	14.16	15.79	17.33	18.85	20.35	21.83	23.26
		GeO_2							
tiron	0.10	4.70	6.70	7.70	8.70	9.70	10.69	11.67	12.62
EDTA	0.10	3.58	3.09	2.66	2.19	1.69	1.20	0.70	0.28
HEDTA	0.10	1.36	0.25	0.04	0.00	0.00	0.00	0.00	0.00
ferron	0.10	3.65	4.57	4.71	4.76	4.77	4.77	4.76	4.71
sulfoxine	0.10	0.00	0.04	0.27	0.98	1.91	2.82	3.59	4.11
		Hg_2^{2+}							
ox	0.10	1.25	3.24	4.02	4.56	4.82	4.93	4.98	5.02
		Hg^{2+}							
ac	0.10	3.08	6.76	8.65	10.37	11.71	12.51	12.87	13.00
citr	0.10	2.94	5.61	6.72	7.68	8.46	9.05	9.48	9.72
ox	0.10	6.79	7.77	8.17	8.44	8.58	8.63	8.65	8.66
tart	0.10	3.17	4.80	5.36	5.71	5.89	5.97	6.01	6.17
tiron	0.10	2.76	4.76	5.76	6.76	7.76	8.76	9.76	10.75
DCTA	0.10	8.49	11.08	12.26	13.37	14.41	15.41	16.36	17.22
DTPA	0.10	0.01	0.25	0.83	1.73	2.71	3.70	4.70	5.70
EDTA	0.10	8.37	10.48	11.38	12.33	13.30	14.27	15.22	16.07
EGTA	0.10	7.93	10.01	10.94	11.90	12.89	13.88	14.88	15.88
HEDTA	0.10	7.17	9.72	10.81	11.84	12.81	13.71	14.50	15.15
NTA	0.10	5.17	6.83	7.41	7.94	8.45	8.95	9.45	9.95
dien	0.10	7.17	10.14	11.59	12.95	14.17	15.28	16.32	17.33
en	0.10	0.89	2.85	3.90	5.04	6.32	7.81	9.50	11.31
py	0.10	1.49	3.39	4.37	5.34	6.27	7.06	7.62	7.92
tren	0.10	2.36	5.36	6.86	8.36	9.86	11.36	12.86	14.36
glyc	0.10	1.49	3.96	5.07	6.10	7.11	8.12	9.12	10.12
TGA	0.10	33.46	35.34	36.14	36.80	37.37	37.89	38.40	38.90
thiourea	0.10	22.72	23.45	23.53	23.55	23.56	23.56	23.56	23.57
acac	0.10	13.70	14.70	15.20	15.70	16.20	16.70	17.20	17.70

Ch. 11] Side-Reaction Coefficients for Complexation Equilibria of Metal Ions 321

$$\log \alpha_{M(L,OH)}$$

pH										
6.5	7.0	7.5	8.0	8.5	9.0	9.5	10.0	10.5	11.0	12.0

Fe^{3+}

| 20.16 | 21.17 | 21.78 | 22.05 | 22.15 | 22.19 | 22.63 | 24.43 | 26.43 | 28.43 | 32.43 |
| 42.45 | 44.53 | 46.50 | 48.33 | 49.92 | 51.25 | 52.38 | 53.43 | 54.44 | 55.45 | 57.45 |

Ga^{3+}

12.54	13.08	13.64	14.73	16.61	18.60	20.60	22.60	24.60	26.60	30.60
14.86	14.86	14.86	15.05	16.61	18.60	20.60	22.60	24.60	26.60	30.60
11.11	12.06	13.08	14.65	16.60	18.60	20.60	22.60	24.60	26.60	30.60
16.46	17.39	18.36	19.35	20.34	21.34	22.35	23.41	24.79	26.62	30.60
18.46	19.57	20.79	22.10	23.39	24.57	25.63	26.59	27.43	28.14	30.62
16.35	17.42	18.45	19.45	20.45	21.44	22.39	23.35	24.70	26.60	30.60
14.78	15.80	16.80	17.80	18.80	19.79	20.93	22.63	24.60	26.60	30.60
8.86	10.70	12.63	14.61	16.60	18.60	20.60	22.60	24.60	26.60	30.60
13.23	14.72	16.19	17.62	18.95	20.35	22.21	24.20	26.20	28.20	32.20
24.56	25.57	26.18	26.45	26.55	26.59	26.60	26.62	26.66	27.00	30.60

GeO_2

13.46	14.06	14.27	14.11	13.74	13.28	12.79	12.28	11.74	11.11	9.06
0.06	0.01	0.01	0.04	0.11	0.28	0.59	1.01	1.50	2.03	3.40
0.00	0.00	0.01	0.04	0.11	0.28	0.59	1.01	1.50	2.03	3.40
4.58	4.25	3.66	2.84	1.91	1.02	0.68	1.02	1.50	2.03	3.40
4.37	4.45	4.42	4.25	3.84	3.17	2.31	1.51	1.53	2.03	3.40

Hg_2^{2+}

| 5.11 | 5.31 | 5.65 | 6.08 | 6.55 | 7.04 | 7.54 | 8.04 | 8.54 | 9.04 | 10.04 |

Hg^{2+}

13.04	13.05	13.06	13.06	13.06	13.08	13.30	14.24	15.61	17.08	20.06
9.84	9.88	9.92	10.11	10.79	11.79	12.94	14.22	15.61	17.08	20.06
8.66	8.71	8.98	9.75	10.73	11.79	12.94	14.22	15.61	17.08	20.06
6.78	7.71	8.70	9.71	10.73	11.79	12.94	14.22	15.61	17.08	20.06
11.73	12.66	13.51	14.21	14.81	15.34	15.85	16.36	16.88	17.53	20.07
17.94	18.54	19.08	19.60	20.70	20.61	21.14	21.73	22.42	23.23	25.01
6.70	7.70	8.70	9.71	10.73	11.79	12.94	14.22	15.61	17.08	20.06
16.80	17.40	17.95	18.48	19.05	19.69	20.43	21.21	21.93	22.54	23.59
16.88	17.87	18.86	19.82	20.71	21.42	21.87	22.08	22.16	22.19	22.20
15.71	16.25	16.79	17.38	18.07	18.87	19.70	20.44	21.07	21.62	22.65
10.45	10.95	11.45	11.94	12.43	12.90	13.40	14.28	15.62	17.08	20.06
18.34	19.34	20.33	21.30	22.22	23.02	23.61	23.95	24.11	24.17	24.20
13.13	14.78	16.18	17.35	18.40	19.35	20.16	20.74	21.06	21.20	21.45
8.05	8.23	8.80	9.72	10.73	11.79	12.94	14.22	15.61	17.08	20.06
15.86	17.35	18.82	20.25	21.57	22.71	23.61	24.23	24.57	24.72	24.79
11.12	12.12	13.11	14.10	15.04	15.90	16.56	16.94	17.12	17.43	20.06
39.40	39.90	40.40	40.89	41.38	41.85	42.25	42.55	42.70	42.77	42.80
23.57	23.57	23.57	23.57	23.57	23.57	23.57	23.57	23.57	23.57	23.57
18.20	18.69	19.18	19.64	20.02	20.29	20.42	20.47	20.49	20.50	20.63

Table 11.3 (Continued)

Ligand	Concentration [L], M	pH 2.0	3.0	3.5	4.0	4.5	5.0	5.5	6.0
		In^{3+}							
ac	0.10	0.18	0.93	1.66	2.67	3.85	5.27	7.17	9.17
ox	0.10	5.31	7.20	7.86	8.26	8.43	8.49	8.53	9.25
tart	0.10	1.58	4.56	5.52	6.07	6.32	6.44	7.24	9.17
tiron	0.10	0.07	1.27	2.24	3.24	4.26	5.40	6.98	8.91
DCTA	0.10	11.16	14.53	15.93	17.16	18.32	19.51	20.78	22.06
DTPA	0.10	9.63	13.37	14.94	16.37	17.65	18.78	19.83	20.85
EDTA	0.10	10.66	13.44	14.54	15.58	16.58	17.57	18.51	19.37
HEDTA	0.10	7.20	9.75	10.84	11.87	12.84	13.76	14.55	15.21
NTA	0.10	7.41	9.07	9.65	10.18	10.69	11.19	11.69	12.19
acac	0.10	0.34	1.39	2.22	3.16	4.22	5.62	7.48	9.46
ferron	0.10	11.14	15.17	16.86	18.43	19.95	21.45	22.93	24.36
		La^{3+}							
ac	0.10	0.00	0.03	0.08	0.20	0.43	0.69	0.86	0.93
Cl-ac	0.10	0.10	0.35	0.43	0.47	0.48	0.48	0.48	0.48
citr	0.01	0.00	0.02	0.27	1.76	3.32	4.51	5.35	5.85
	0.10	0.00	0.26	1.86	3.75	5.31	6.51	7.35	7.85
ox	0.10	2.92	5.39	6.35	6.93	7.18	7.27	7.30	7.31
tart	0.01	0.16	0.79	1.31	1.79	2.08	2.20	2.24	2.26
	0.10	0.73	2.00	2.91	3.58	3.93	4.07	4.12	4.14
sulfosal	0.10	0.00	0.00	0.00	0.00	0.00	0.01	0.04	0.13
tiron	0.10	0.00	0.00	0.02	0.20	0.84	1.78	2.77	3.76
DCTA	0.10	0.20	2.76	4.10	5.29	6.36	7.37	8.32	9.18
DTPA	0.10	0.92	4.06	5.54	6.93	8.19	9.32	10.37	11.39
EDTA	0.01	0.50	2.96	4.05	5.08	6.09	7.07	8.01	8.87
	0.10	1.36	3.96	5.05	6.08	7.09	8.07	9.01	9.87
EGTA	0.10	0.12	2.35	3.46	4.50	5.51	6.52	7.52	8.52
HEDTA	0.10	0.71	3.16	4.25	5.28	6.25	7.17	7.96	8.62
NTA	0.10	1.08	2.84	3.67	4.58	5.55	6.54	7.53	8.53
HTGA	0.10	0.13	0.63	0.95	1.16	1.25	1.29	1.30	1.30
acac	0.10	0.00	0.01	0.02	0.06	0.16	0.39	0.78	1.33
		Lu^{3+*}							
ac	0.10	0.00	0.04	0.12	0.31	0.64	0.99	1.21	1.31
Cl-ac	0.10	0.12	0.41	0.51	0.55	0.56	0.57	0.57	0.57
citr	0.10	0.27	2.61	3.72	4.68	5.46	6.05	6.48	6.72
ox	0.10	4.32	7.13	8.22	8.89	9.19	9.30	9.34	9.35
tart	0.10	0.46	1.91	2.46	2.81	2.99	3.06	3.09	3.10
sulfosal	0.10	0.00	0.00	0.01	0.02	0.05	0.15	0.38	0.82
tiron	0.10	0.78	1.71	2.21	2.71	3.23	3.78	4.41	5.17

Ch. 11] Side-Reaction Coefficients for Complexation Equilibria of Metal Ions 323

$$\log \alpha_{M(L,OH)}$$

pH 6.5	7.0	7.5	8.0	8.5	9.0	9.5	10.0	10.5	11.0	12.0
In^{3+}										
11.17	13.17	15.17	17.17	19.17	21.17	23.17	25.17	27.17	29.17	33.17
11.17	13.17	15.17	17.17	19.17	21.17	23.17	25.17	27.17	29.17	33.17
11.17	13.17	15.17	17.17	19.17	21.17	23.17	25.17	27.17	29.17	33.17
10.90	12.90	14.90	16.90	18.90	20.90	22.90	24.90	26.90	28.90	32.90
23.25	24.35	25.38	26.39	27.40	28.40	29.40	30.40	31.39	32.38	34.27
21.85	22.85	23.83	24.76	25.60	26.30	26.87	27.34	27.76	28.94	32.90
20.10	20.71	21.27	21.85	22.50	23.28	24.16	25.24	26.92	28.90	32.90
15.78	16.30	16.81	17.45	18.93	20.90	22.90	24.90	26.90	28.90	32.90
12.70	13.37	14.92	16.90	18.90	20.90	22.90	24.90	26.90	28.90	32.90
11.46	13.46	15.46	17.46	19.46	21.46	23.46	25.46	27.46	29.46	33.46
25.66	26.67	27.28	27.55	27.65	27.69	27.72	27.77	27.94	28.99	32.90
La^{3+}										
0.96	0.97	0.97	0.98	1.00	1.22	2.22	3.67	5.25	8.51	17.50
0.48	0.49	0.49	0.51	0.65	1.37	2.68	4.16	5.74	8.54	17.50
6.07	6.16	6.19	6.20	6.20	6.20	6.20	6.20	6.33	8.54	17.50
8.07	8.16	8.19	8.20	8.20	8.20	8.20	8.20	8.20	8.70	17.50
7.32	7.32	7.32	7.32	7.32	7.32	7.32	7.32	7.32	8.54	17.50
2.26	2.26	2.26	2.26	2.27	2.31	2.82	4.17	5.74	8.54	17.50
4.14	4.14	4.14	4.15	4.15	4.15	4.16	4.45	5.75	8.54	17.50
0.33	0.71	1.30	2.08	2.98	3.94	4.93	5.90	6.84	8.56	17.50
4.74	5.67	6.52	7.22	7.82	8.35	8.86	9.37	9.87	10.36	17.50
9.90	10.50	11.04	11.55	12.06	12.56	13.06	13.56	14.05	14.54	17.50
12.39	13.39	14.37	15.30	16.14	16.84	17.41	17.87	18.21	18.41	18.56
9.60	10.20	10.74	11.25	11.75	12.24	12.69	13.06	13.31	13.43	17.50
10.60	11.20	11.74	12.25	12.75	13.24	13.69	14.06	14.31	14.43	17.50
9.52	10.51	11.50	12.46	13.35	14.06	14.51	14.72	14.80	14.83	17.50
9.19	9.71	10.22	10.72	11.21	11.69	12.13	12.52	12.90	13.32	17.50
9.53	10.53	11.52	12.51	13.47	14.35	15.06	15.51	15.72	15.79	17.51
1.30	1.30	1.30	1.30	1.32	1.58	2.69	4.16	5.74	8.54	17.50
2.07	3.03	4.17	5.40	6.51	7.29	7.68	7.84	7.89	8.61	17.50
Lu^{3+} *										
1.34	1.35	1.36	1.40	2.53	4.98	7.48	9.98	12.48	14.98	19.98
0.57	0.57	0.60	1.05	3.28	5.78	8.28	10.78	13.28	15.78	20.78
6.84	6.88	6.89	6.90	6.90	6.93	8.29	10.78	13.28	15.78	20.78
9.35	9.35	9.36	9.36	9.36	9.36	9.36	10.07	12.48	14.98	19.98
3.10	3.10	3.10	3.10	3.50	5.78	8.28	10.78	13.28	15.78	20.78
1.47	2.30	3.23	4.21	5.20	6.22	7.66	9.98	12.48	14.98	19.98
6.03	6.92	7.75	8.46	9.05	9.58	10.10	11.00	13.28	15.78	20.78

Table 11.3 (Continued)

Ligand	Concentration [L], M	pH 2.0	3.0	3.5	4.0	4.5	5.0	5.5	6.0
		Lu^{3+}*							
DCTA	0.10	2.66	5.77	7.13	8.32	9.40	10.41	11.36	12.22
DTPA	0.10	3.55	6.96	8.49	9.90	11.17	12.31	13.36	14.38
EDTA	0.01	2.84	5.46	6.55	7.58	8.59	9.57	10.51	11.37
	0.10	3.84	6.46	7.55	8.58	9.59	10.57	11.51	12.37
EGTA	0.10	1.30	4.10	5.22	6.26	7.27	8.28	9.28	10.28
HEDTA	0.10	9.00	12.55	14.14	15.67	17.14	18.56	19.85	21.01
NTA	0.10	2.34	5.18	6.32	7.38	8.39	9.40	10.40	11.40
HGTA	0.10	0.14	0.66	1.07	1.47	1.78	1.95	2.03	2.05
acac	0.10	0.00	0.03	0.10	0.26	0.58	1.09	1.79	2.72
		Mg^{2+}							
citr	0.10	0.04	0.20	0.38	0.70	1.15	1.62	2.00	2.23
ox	0.10	0.26	1.06	1.58	1.99	2.22	2.32	2.35	2.36
tiron	0.10	0.00	0.00	0.00	0.00	0.00	0.01	0.03	0.10
DCTA	0.10	0.00	0.00	0.01	0.09	0.56	1.44	2.38	3.24
DTPA	0.10	0.00	0.00	0.03	0.20	0.65	1.20	1.74	2.28
EDTA	0.01	0.00	0.00	0.00	0.02	0.11	0.54	1.33	2.17
	0.10	0.00	0.01	0.03	0.15	0.59	1.41	2.32	3.16
EGTA	0.01	0.00	0.00	0.00	0.00	0.00	0.00	0.00	0.01
	0.10	0.00	0.00	0.00	0.00	0.00	0.01	0.04	0.12
HEDTA	0.10	0.00	0.00	0.00	0.03	0.20	0.75	1.46	2.10
NTA	0.10	0.00	0.00	0.01	0.03	0.08	0.22	0.49	0.88
glutam	0.10	0.00	0.00	0.00	0.00	0.00	0.00	0.00	0.00
acac	0.10	0.00	0.00	0.00	0.00	0.01	0.02	0.06	0.17
		Mn^{2+}							
ac	0.10	0.00	0.01	0.02	0.04	0.09	0.13	0.16	0.17
citr	0.10	0.00	0.12	0.41	0.87	1.41	1.91	2.29	2.53
sal	0.10	0.00	0.00	0.00	0.00	0.00	0.00	0.00	0.00
sulfosal	0.10	0.00	0.00	0.00	0.00	0.00	0.00	0.00	0.01
tiron	0.10	0.00	0.00	0.00	0.02	0.05	0.13	0.33	0.65
DCTA	0.10	0.78	3.42	4.67	5.82	6.89	7.89	8.84	9.70
DTPA	0.10	0.03	1.63	2.72	3.70	4.62	5.54	6.49	7.47
EDTA	0.01	0.13	1.66	2.56	3.50	4.47	5.44	6.39	7.24
	0.10	0.65	2.65	3.55	4.50	5.47	6.44	7.39	8.24
EGTA	0.10	0.00	0.26	0.69	1.32	2.10	3.01	3.98	4.96
HEDTA	0.10	0.00	0.60	1.58	2.59	3.56	4.46	5.25	5.90
NTA	0.10	0.00	0.17	0.46	0.86	1.33	1.82	2.32	2.84
en	0.10	0.00	0.00	0.00	0.00	0.00	0.00	0.00	0.00
trien	0.10	0.00	0.00	0.00	0.00	0.00	0.00	0.00	0.00
glyc	0.10	0.00	0.00	0.00	0.00	0.00	0.00	0.00	0.01
TGA	0.10	0.00	0.00	0.00	0.00	0.00	0.01	0.02	0.07

Ch. 11] Side-Reaction Coefficients for Complexation Equilibria of Metal Ions   325

$$\log \alpha_{M(L,OH)}$$

pH 6.5	7.0	7.5	8.0	8.5	9.0	9.5	10.0	10.5	11.0	12.0
Lu^{3+}*										
12.94	13.54	14.08	14.59	15.10	15.60	16.10	16.60	17.09	17.59	20.78
15.38	16.37	17.35	18.29	19.13	19.83	20.39	20.86	21.20	21.39	21.58
12.10	12.70	13.24	13.75	14.25	14.74	15.19	15.56	15.81	16.16	20.78
13.10	13.70	14.24	14.75	15.25	15.74	16.19	16.56	16.81	16.96	20.78
11.28	12.27	13.26	14.22	15.11	15.82	16.27	16.48	16.56	16.65	20.78
22.08	23.10	24.10	25.10	26.09	27.06	27.96	28.75	29.40	29.97	31.00
12.40	13.40	14.39	15.38	16.34	17.22	17.93	18.38	18.59	18.66	20.78
2.06	2.07	2.07	2.09	3.31	5.78	8.28	10.78	13.28	15.78	20.78
3.86	5.18	6.57	7.92	9.08	9.87	10.27	10.94	13.28	15.78	20.78
Mg^{2+}										
2.34	2.38	2.40	2.40	2.40	2.40	2.40	2.40	2.40	2.40	2.41
2.36	2.37	2.37	2.37	2.37	2.37	2.37	2.37	2.37	2.37	2.37
0.25	0.54	0.94	1.39	1.87	2.36	2.86	3.36	3.86	4.35	5.24
3.96	4.56	5.10	5.61	6.12	6.62	7.12	7.62	8.11	8.60	9.47
2.84	3.50	4.27	5.11	5.92	6.60	7.17	7.63	7.97	8.17	8.28
2.89	3.49	4.03	4.54	5.04	5.53	5.98	6.35	6.60	6.72	6.78
3.89	4.49	5.03	5.54	6.04	6.53	6.98	7.35	7.60	7.72	7.78
0.05	0.14	0.43	1.03	1.78	2.45	2.89	3.09	3.17	3.20	3.21
0.32	0.70	1.26	1.99	2.77	3.45	3.88	4.09	4.17	4.20	4.21
2.66	3.18	3.68	4.18	4.67	5.13	5.52	5.78	5.92	5.97	6.00
1.34	1.83	2.32	2.81	3.29	3.73	4.09	4.31	4.41	4.45	4.47
0.00	0.01	0.03	0.08	0.20	0.42	0.66	0.83	0.91	0.95	1.04
0.42	0.87	1.51	2.29	3.07	3.68	4.03	4.18	4.23	4.25	4.26
Mn^{2+}										
0.17	0.17	0.17	0.17	0.17	0.18	0.21	0.61	1.91	3.40	6.40
2.64	2.68	2.69	2.70	2.70	2.70	2.70	2.71	2.88	3.93	6.90
0.00	0.01	0.04	0.12	0.30	0.63	1.09	1.71	2.66	3.97	6.91
0.04	0.12	0.31	0.64	1.08	1.61	2.22	2.97	3.84	4.74	6.94
1.07	1.54	2.19	3.14	4.20	5.32	6.57	7.93	9.34	10.70	12.64
10.42	11.02	11.56	12.07	12.58	13.08	13.58	14.08	14.57	15.06	15.93
8.46	9.45	10.43	11.36	12.20	12.90	13.47	13.93	14.27	14.47	14.59
7.97	8.57	9.11	9.62	10.12	10.61	11.06	11.43	11.68	11.80	11.86
8.97	9.57	10.11	10.62	11.12	11.61	12.06	12.43	12.68	12.80	12.86
5.96	6.95	7.94	8.90	9.79	10.50	10.95	11.16	11.24	11.27	11.28
6.46	6.98	7.48	7.98	8.47	8.93	9.32	9.58	9.72	9.77	9.80
3.39	4.03	4.80	5.68	6.60	7.47	8.18	8.62	8.83	8.90	8.94
0.00	0.00	0.03	0.12	0.38	0.87	1.56	2.27	2.82	3.55	6.40
0.00	0.00	0.00	0.00	0.00	0.01	0.12	0.96	2.40	3.90	6.90
0.02	0.05	0.16	0.42	0.92	1.56	2.14	2.50	2.81	3.83	6.80
0.21	0.49	0.96	1.61	2.43	3.31	4.16	4.85	5.27	5.47	6.92

Table 11.3 (Continued)

Ligand	Concentration [L], M	pH 2.0	3.0	3.5	4.0	4.5	5.0	5.5	6.0
		Mn^{3+}							
ox	0.10	10.85	13.12	14.14	14.87	15.25	15.40	15.45	15.47
DCTA	0.10	11.26	14.63	16.01	17.22	18.30	19.31	20.26	21.12
DTPA	0.10	13.08	17.00	18.56	19.96	21.19	22.30	23.34	24.35
EDTA	0.01	9.84	12.72	13.84	14.88	15.88	16.87	17.81	18.67
	0.10	10.84	13.72	14.84	15.88	16.88	17.87	18.81	19.67
EGTA	0.10	6.37	9.20	10.32	11.36	12.37	13.38	14.38	15.38
HEDTA	0.10	9.82	12.37	13.46	14.49	15.46	16.36	17.15	17.80
NTA	0.10	11.60	13.21	13.76	14.28	14.79	15.29	15.79	16.29
		Ni^{2+}							
ac	0.10	0.00	0.01	0.02	0.06	0.12	0.17	0.21	0.22
citr	0.01	0.03	0.29	0.71	1.35	2.01	2.57	2.98	3.23
	0.10	0.25	1.02	1.63	2.33	3.01	3.57	3.98	4.23
sal	0.10	0.00	0.00	0.00	0.00	0.00	0.00	0.00	0.02
sulfosal	0.10	0.00	0.00	0.00	0.00	0.01	0.02	0.06	0.18
tiron	0.10	0.00	0.00	0.01	0.04	0.12	0.31	0.65	1.14
DCTA	0.10	3.42	6.14	7.39	8.54	9.61	10.61	11.56	12.42
DTPA	0.10	5.15	8.74	10.29	11.71	12.97	14.11	15.16	16.17
EDTA	0.01	3.22	6.05	7.16	8.20	9.20	10.19	11.13	11.99
	0.10	4.22	7.05	8.16	9.20	10.20	11.19	12.13	12.99
EGTA	0.10	0.49	2.16	2.80	3.39	4.04	4.81	5.70	6.66
HEDTA	0.10	4.22	6.77	7.86	8.89	9.86	10.76	11.55	12.20
NTA	0.10	2.07	3.73	4.31	4.84	5.37	5.91	6.52	7.24
dien	0.10	0.00	0.00	0.00	0.01	0.11	0.68	1.69	3.07
en	0.10	0.00	0.00	0.00	0.00	0.00	0.04	0.29	1.18
py	0.10	0.00	0.02	0.05	0.14	0.35	0.70	1.05	1.27
TEA	0.10	0.00	0.00	0.00	0.00	0.00	0.01	0.03	0.09
tetren	0.10	0.00	0.02	0.26	1.27	2.77	4.39	5.96	7.49
THAM	0.10	0.00	0.00	0.00	0.00	0.00	0.01	0.05	0.14
tren	0.10	0.00	0.00	0.00	0.00	0.03	0.52	1.87	3.36
trien	0.10	0.00	0.00	0.02	0.28	1.26	2.57	3.98	5.61
cyst	0.10	0.00	0.00	0.00	0.01	0.14	1.27	3.19	5.18
glutam	0.10	0.00	0.00	0.00	0.02	0.09	0.28	0.65	1.17
glyc	0.10	0.00	0.00	0.02	0.05	0.15	0.40	0.89	1.63
histam	0.10	0.00	0.00	0.00	0.00	0.03	0.24	0.90	1.90
histid	0.10	0.00	0.02	0.14	0.69	1.76	3.33	5.12	6.76
tryp	0.10	0.00	0.01	0.02	0.06	0.20	0.54	1.18	2.03
TGA	0.10	0.00	0.00	0.00	0.02	0.19	0.88	2.19	4.81
BAL	0.10	0.00	0.00	0.00	0.01	0.56	2.47	5.13	8.53
acac	0.10	0.00	0.03	0.10	0.27	0.57	1.02	1.59	2.30

Ch. 11] Side-Reaction Coefficients for Complexation Equilibria of Metal Ions 327

$$\log \alpha_{M(L,OH)}$$

pH 6.5	7.0	7.5	8.0	8.5	9.0	9.5	10.0	10.5	11.0	12.0
Mn^{3+}										
15.47	15.48	15.48	15.48	15.48	15.48	15.48	15.48	15.48	15.48	15.48
21.84	22.44	22.98	23.49	24.00	24.50	25.00	25.50	25.99	26.48	27.35
25.35	26.34	27.29	28.16	28.89	29.44	29.80	29.98	30.06	30.09	30.10
19.40	20.00	20.54	21.05	21.55	22.04	22.49	22.86	23.11	23.23	23.29
20.40	21.00	21.54	22.05	22.55	23.04	23.49	23.86	24.11	24.23	24.29
16.38	17.37	18.36	19.32	20.20	20.91	21.36	21.58	21.66	21.69	21.70
18.36	18.88	19.38	19.88	20.37	20.83	21.22	21.48	21.62	21.67	21.70
16.79	17.29	17.78	18.24	18.66	18.97	19.14	19.21	19.24	19.25	19.25
Ni^{2+}										
0.22	0.22	0.22	0.23	0.23	0.25	0.34	0.79	1.77	3.14	6.69
3.34	3.38	3.39	3.40	3.40	3.40	3.40	3.40	3.42	3.64	6.49
4.34	4.38	4.39	4.40	4.40	4.40	4.40	4.40	4.40	4.43	6.49
0.05	0.13	0.33	0.67	1.12	1.65	2.29	3.06	3.96	4.93	7.02
0.41	0.78	1.24	1.76	2.35	3.05	3.89	4.82	5.76	6.65	7.89
1.80	2.77	4.01	5.19	6.26	7.29	8.29	9.29	10.26	11.17	12.48
13.14	13.74	14.28	14.79	15.30	15.80	16.30	16.80	17.29	17.78	18.65
17.17	18.17	19.14	20.07	20.89	21.57	22.12	22.57	22.88	23.05	23.15
12.72	13.32	13.86	14.37	14.87	15.36	15.81	16.19	16.44	16.59	16.90
13.72	14.32	14.86	15.37	15.87	16.36	16.81	17.19	17.44	17.59	17.90
7.65	8.64	9.62	10.58	11.47	12.18	12.63	12.84	12.92	12.95	12.96
12.76	13.28	13.78	14.28	14.77	15.23	15.62	15.88	16.02	16.07	16.10
8.10	9.05	10.02	11.00	11.96	12.84	13.56	14.00	14.21	14.28	14.32
4.90	6.87	8.85	10.80	12.64	14.24	15.42	16.11	16.42	16.54	16.59
2.76	4.67	6.70	8.60	10.27	11.79	13.14	14.24	14.93	15.26	15.42
1.37	1.40	1.41	1.42	1.42	1.42	1.43	1.49	1.92	3.15	6.69
0.24	0.52	0.87	1.19	1.38	1.47	1.50	1.56	1.95	3.15	6.69
8.99	10.47	11.90	13.22	14.36	15.26	15.88	16.20	16.33	16.38	16.40
1.79	5.71	9.51	13.03	16.15	18.93	21.53	24.06	26.57	29.08	34.08
4.86	6.35	7.82	9.25	10.57	11.71	12.61	13.23	13.57	13.72	13.79
9.39	13.23	16.67	19.79	22.69	25.26	27.23	28.39	28.91	29.11	29.19
7.17	9.13	11.01	12.72	14.16	15.34	16.33	17.12	17.65	17.92	18.05
1.85	2.68	3.61	4.57	5.52	6.38	7.06	7.47	7.66	7.73	7.78
2.58	3.71	5.00	6.39	7.78	9.06	10.04	10.61	10.87	10.96	11.00
3.10	4.31	5.53	6.83	8.19	9.50	10.60	11.33	11.67	11.81	11.87
8.13	9.29	10.33	11.30	12.17	12.85	13.26	13.44	13.51	13.53	13.54
2.98	3.96	4.95	5.93	6.88	7.72	8.35	8.71	8.86	8.91	8.94
7.79	10.79	13.79	16.77	19.73	22.60	25.22	27.29	28.56	29.14	29.42
12.02	15.50	18.93	22.23	25.24	27.83	30.04	31.95	33.51	34.65	36.05
3.15	4.08	5.03	5.94	6.71	7.24	7.50	7.61	7.64	7.66	7.70

Table 11.3 (Continued)

Ligand	Concentration [L], M	pH 2.0	3.0	3.5	4.0	4.5	5.0	5.5	6.0
		Np^{4+}							
DTPA	0.10	10.49	15.42	17.56	19.14	20.39	21.50	22.54	23.55
EDTA	0.01	9.55	12.29	13.35	14.37	15.37	16.36	17.31	18.19
	0.10	10.55	13.29	14.35	15.37	16.37	17.36	18.31	19.19
HEDTA	0.10	11.13	15.91	18.00	19.99	21.87	23.57	24.99	26.18
NTA	0.10	14.77	17.98	19.09	20.12	21.13	22.14	23.14	24.14
acac	0.10	1.73	3.95	5.41	7.16	9.06	11.03	13.02	15.02
		NpO_2^+							
EDTA	0.01	0.01	0.30	0.72	1.19	1.68	2.16	2.60	2.94
	0.10	0.05	1.05	1.64	2.17	2.67	3.16	3.60	3.94
HEDTA	0.10	0.04	0.64	1.15	1.66	2.13	2.53	2.82	3.00
		NpO_2^{2+}							
ac	0.10	0.02	0.22	0.62	1.35	2.16	2.69	2.94	3.10
Cl-ac	0.10	0.16	0.52	0.63	0.68	0.70	0.73	0.84	2.29
		Pb^{2+}							
ac	0.10	0.01	0.08	0.22	0.50	0.91	1.33	1.62	1.75
ox	0.10	0.91	2.29	2.90	3.27	3.44	3.50	3.52	3.52
citr	0.01	0.02	0.20	0.50	0.93	1.40	1.83	2.15	2.30
	0.10	0.20	1.01	1.55	2.16	2.75	3.36	3.81	4.02
tart	0.01	0.05	0.26	0.46	0.63	0.72	0.76	0.77	0.78
	0.10	0.33	1.15	1.61	1.92	2.06	2.12	2.14	2.15
tiron	0.10	0.00	0.00	0.04	0.28	1.00	1.95	2.94	3.93
DTPA	0.10	1.95	4.71	5.81	6.81	7.77	8.72	9.70	10.69
EDTA	0.01	3.44	5.67	6.66	7.64	8.63	9.61	10.55	11.41
	0.10	4.44	6.67	7.66	8.64	9.63	10.61	11.55	12.41
EGTA	0.10	1.56	3.38	4.01	4.58	5.19	5.90	6.74	7.68
HEDTA	0.10	3.62	6.16	7.22	8.15	8.95	9.60	10.16	10.68
NTA	0.10	2.04	3.70	4.28	4.81	5.32	5.82	6.32	6.82
dien	0.10	0.00	0.00	0.00	0.00	0.00	0.02	0.15	0.72
		Pd^{2+}							
EDTA	0.01	4.60	6.55	7.40	8.32	9.29	10.26	11.22	12.13
	0.10	5.60	7.55	8.40	9.32	10.29	11.26	12.21	13.10
NTA	0.10	10.67	12.26	12.81	13.33	13.84	14.34	14.84	15.34
en	0.10	0.48	2.05	3.02	4.05	5.30	7.04	9.00	11.00
glyc	0.10	8.98	11.82	12.94	13.98	14.99	16.00	17.00	18.00
acac	0.10	11.52	13.52	14.52	15.52	16.52	17.52	18.52	19.52
		Pu^{4+}							
acac	0.01	2.38	4.81	6.26	7.78	9.38	11.12	13.00	14.97
	0.10	4.81	7.77	9.38	11.11	12.98	14.93	16.91	18.90
		PuO_2^+							
ox	0.10	6.79	7.74	8.07	8.27	8.35	8.38	8.40	8.40

Ch. 11] Side-Reaction Coefficients for Complexation Equilibria of Metal Ions

$\log \alpha_{M(L,OH)}$

pH	6.5	7.0	7.5	8.0	8.5	9.0	9.5	10.0	10.5	11.0	12.0
Np^{4+}											
	24.55	25.54	26.49	27.36	28.09	28.64	29.00	29.18	29.26	29.29	29.30
	18.95	19.58	20.12	20.64	21.13	21.60	22.02	22.32	22.49	22.56	22.60
	19.95	20.58	21.12	21.64	22.13	22.60	23.02	23.32	23.49	23.56	23.60
	27.25	28.26	29.23	30.11	30.82	31.27	31.48	31.55	31.58	31.59	31.59
	25.14	26.13	27.11	28.05	28.88	29.50	29.84	29.98	30.04	30.05	30.06
	17.01	18.99	20.93	22.76	24.31	25.37	25.90	26.11	26.19	26.21	26.22
NpO_2^+											
	3.16	3.28	3.37	3.53	3.81	4.19	4.60	4.96	5.23	5.44	6.09
	4.16	4.28	4.37	4.53	4.81	5.19	5.60	5.96	6.23	6.44	7.09
	3.15	3.36	3.69	4.12	4.58	5.03	5.42	5.70	5.87	6.03	6.62
NpO_2^{2+}											
	4.78	7.27	9.77	12.27	14.77	17.27	19.77	22.27	24.77	27.27	32.27
	4.77	7.27	9.77	12.27	14.77	17.27	19.77	22.27	24.77	27.27	32.27
Pb^{2+}											
	1.80	1.81	1.82	1.84	2.26	5.76	9.76	13.76	17.76	21.76	29.76
	3.53	3.53	3.53	3.53	3.54	5.76	9.76	13.76	17.76	21.76	29.76
	2.36	2.38	2.39	2.40	2.56	5.76	9.76	13.76	17.76	21.76	29.76
	4.10	4.13	4.14	4.14	4.14	5.77	9.76	13.76	17.76	21.76	29.76
	0.78	0.79	0.82	0.92	2.09	5.76	9.76	13.76	17.76	21.76	29.76
	2.15	2.15	2.15	2.16	2.41	5.76	9.76	13.76	17.76	21.76	29.76
	4.94	6.10	7.38	8.57	9.64	10.67	11.68	13.79	17.76	21.76	29.76
	11.68	12.67	13.65	14.58	15.42	16.12	16.69	17.15	17.53	20.47	28.47
	12.14	12.74	13.28	13.79	14.29	14.78	15.23	15.60	16.56	20.47	28.47
	13.14	13.74	14.28	14.79	15.29	15.78	16.23	16.60	17.00	20.47	28.47
	8.65	9.64	10.63	11.59	12.57	13.18	13.63	13.86	16.47	20.47	28.47
	11.19	11.69	12.19	12.68	13.17	13.63	14.02	14.29	16.47	20.47	28.47
	7.32	7.82	8.32	8.81	9.29	9.73	10.10	12.47	16.47	20.47	28.47
	1.65	2.64	3.63	4.60	5.55	6.50	8.50	12.47	16.47	20.47	28.47
Pd^{2+}											
	13.23	15.01	17.00	19.00	21.00	23.00	25.00	27.00	29.00	31.00	35.00
	13.91	15.11	17.00	19.00	21.00	23.00	25.00	27.00	29.00	31.00	35.00
	15.84	16.35	17.22	19.01	21.00	23.00	25.00	27.00	29.00	31.00	35.00
	13.00	15.00	17.00	19.00	21.00	23.00	25.00	27.00	29.00	31.00	35.00
	19.00	20.00	21.00	21.99	22.96	23.91	25.15	27.01	29.00	31.00	35.00
	20.52	21.51	22.49	23.44	24.29	25.16	26.81	28.80	30.80	32.80	36.80
Pu^{4+}											
	17.03	19.19	21.50	23.92	26.40	28.90	31.40	33.90	36.40	38.90	43.90
	20.89	22.87	24.82	26.65	28.20	29.41	31.41	33.90	36.40	38.90	43.90
PuO_2^+											
	8.40	8.40	8.40	8.40	8.40	8.40	8.40	8.40	8.40	8.40	8.40

Table 11.3 (Continued)

Ligand	Concentration [L], M	pH 2.0	3.0	3.5	4.0	4.5	5.0	5.5	6.0
		PuO_2^+							
EDTA	0.01	0.15	2.51	3.63	4.67	5.67	6.66	7.59	8.44
	0.10	0.70	3.51	4.63	5.67	6.67	7.66	8.59	9.44
HEDTA	0.10	1.41	3.94	5.03	6.06	7.03	7.93	8.72	9.37
		PuO_2^{2+}							
ac	0.10	0.02	0.14	0.36	0.79	1.37	1.83	2.05	2.13
Cl-ac	0.10	0.11	0.35	0.42	0.46	0.47	0.48	0.51	0.61
		$Sb(OH)_3$							
tart	0.10	10.74	8.00	6.11	3.82	1.21	0.05	0.12	0.30
tiron	0.10	5.25	4.35	4.25	4.58	5.06	5.55	6.04	6.53
DCTA	0.10	25.01	25.38	25.26	24.97	24.55	24.06	23.51	22.87
DTPA	0.10	24.18	24.02	23.75	23.48	23.16	22.77	22.31	21.82
EDTA	0.01	28.98	28.84	28.46	28.00	27.50	26.99	26.42	25.77
	0.10	29.98	29.84	29.46	29.00	28.50	27.99	27.42	26.77
HEDTA	0.10	12.53	12.30	12.18	12.11	12.07	12.02	11.93	11.85
		Sc^{3+}							
ox	0.10	6.80	10.13	11.43	12.20	12.54	12.66	12.70	12.71
tart	0.10	8.67	10.30	10.86	11.21	11.39	11.46	11.49	11.50
tiron	0.10	2.51	3.99	4.91	5.87	6.86	7.86	8.86	9.85
DCTA	0.10	8.46	11.83	13.21	14.42	15.50	16.51	17.46	18.32
DTPA	0.10	5.13	8.87	10.44	11.87	13.15	14.28	15.33	16.35
EDTA	0.01	7.94	10.56	11.65	12.68	13.69	14.67	15.61	16.47
	0.10	8.94	11.56	12.65	13.68	14.69	15.67	16.61	17.47
EGTA	0.10	1.88	4.70	5.82	6.86	7.87	8.88	9.88	10.88
HEDTA	0.10	4.42	6.97	8.06	9.09	10.06	10.96	11.75	12.40
NTA	0.10	5.24	8.56	9.72	10.77	11.79	12.80	13.80	14.80
acac	0.10	0.34	1.44	2.30	3.25	4.23	5.22	6.22	7.22
		Sn^{2+}							
tart	0.10	2.30	5.51	6.62	7.33	7.69	7.84	7.89	7.97
DCTA	0.10	5.14	7.34	8.46	9.55	10.59	11.57	12.47	13.26
DTPA	0.10	5.40	7.85	8.89	9.92	10.94	11.95	12.96	13.96
EDTA	0.01	3.95	6.11	7.09	8.08	9.07	10.06	11.01	11.89
	0.10	4.95	7.11	8.09	9.08	10.07	11.06	12.01	12.89
EGTA	0.10	3.34	5.39	6.38	7.38	8.38	9.38	10.38	11.38
		Sr^{2+}							
citr	0.10	0.00	0.00	0.00	0.00	0.00	0.00	0.00	0.00
tiron	0.10	0.00	0.00	0.00	0.00	0.00	0.01	0.03	0.08
DCTA	0.10	0.00	0.00	0.00	0.03	0.29	1.04	1.95	2.81
DTPA	0.10	0.00	0.00	0.00	0.03	0.15	0.49	1.05	1.80
EDTA	0.01	0.00	0.00	0.00	0.01	0.02	0.08	0.22	0.52
	0.10	0.00	0.01	0.02	0.07	0.20	0.46	0.87	1.39
EGTA	0.10	0.00	0.00	0.00	0.01	0.05	0.17	0.55	1.29

Ch. 11] Side-Reaction Coefficients for Complexation Equilibria of Metal Ions

$\log \alpha_{M(L,OH)}$

pH 6.5	7.0	7.5	8.0	8.5	9.0	9.5	10.0	10.5	11.0	12.0
PuO_2^+										
9.15	9.75	10.28	10.79	11.29	11.77	12.22	12.58	12.80	12.91	12.96
10.15	10.75	11.28	11.79	12.29	12.77	13.22	13.58	13.80	13.91	13.96
9.93	10.45	10.95	11.45	11.94	12.40	12.79	13.05	13.19	13.24	13.27
PuO_2^{2+}										
2.19	3.41	5.88	8.38	10.88	13.38	15.88	18.38	20.88	23.38	28.38
1.15	3.38	5.88	8.38	10.88	13.38	15.88	18.38	20.88	23.38	28.38
$Sb(OH)_3$										
0.62	1.04	1.51	2.00	2.50	3.00	3.50	4.00	4.50	5.00	6.00
6.98	7.36	7.54	7.46	7.15	6.72	6.24	5.74	5.24	4.72	3.51
22.09	21.19	20.23	19.24	18.25	17.25	16.25	15.25	14.24	13.23	11.10
21.32	20.82	20.30	19.73	19.07	18.27	17.34	16.30	15.14	13.84	10.96
24.99	24.11	23.25	22.67	22.48	22.42	22.35	22.21	21.93	21.54	20.59
25.99	25.11	24.25	23.67	23.48	23.42	23.35	23.21	22.93	22.54	21.59
11.79	11.77	11.76	11.75	11.74	11.70	11.59	11.35	10.99	10.54	9.57
Sc^{3+}										
12.72	12.72	12.72	13.23	15.58	18.08	20.58	23.08	25.58	28.08	33.08
11.50	11.50	11.55	13.09	15.58	18.08	20.58	23.08	25.58	28.08	33.08
10.83	11.76	12.62	13.68	15.93	18.43	20.93	23.43	25.93	28.43	33.43
19.04	19.64	20.18	20.69	21.20	21.70	22.23	23.51	25.93	28.43	33.43
17.35	18.35	19.33	20.26	21.10	21.80	22.38	23.52	25.93	28.43	33.43
17.20	17.80	18.34	18.85	19.35	19.86	21.02	23.43	25.93	28.43	33.43
18.20	18.80	19.34	19.85	20.35	20.85	21.46	23.44	25.93	28.43	33.43
11.88	12.87	13.86	14.84	16.13	18.43	20.93	23.43	25.93	28.43	33.43
12.96	13.48	13.99	14.52	15.97	18.43	20.93	23.43	25.93	28.43	33.43
15.80	16.80	17.79	18.78	19.74	20.63	21.48	23.44	25.93	28.43	33.43
8.22	9.25	10.76	13.13	15.63	18.13	20.63	23.13	25.63	28.13	33.13
Sn^{2+}										
9.17	11.15	13.15	15.15	17.15	19.15	21.15	23.15	25.15	27.15	31.15
13.91	14.47	14.99	15.61	17.06	19.03	21.03	23.03	25.03	27.03	31.03
14.95	15.94	16.89	17.77	18.51	19.33	21.04	23.03	25.03	27.03	31.03
12.65	13.28	13.89	15.11	17.03	19.03	21.03	23.03	25.03	27.03	31.03
13.65	14.28	14.83	15.51	17.05	19.03	21.03	23.03	25.03	27.03	31.03
12.38	13.38	14.39	15.54	17.19	19.15	21.15	23.15	25.15	27.15	31.15
Sr^{2+}										
0.02	0.37	1.51	2.02	2.05	2.05	2.05	2.05	2.05	2.05	2.05
0.20	0.41	0.66	0.83	0.92	0.99	1.09	1.30	1.64	2.07	2.93
3.53	4.13	4.67	5.18	5.69	6.19	6.69	7.19	7.68	8.17	9.04
2.69	3.64	4.60	5.53	6.37	7.07	7.64	8.10	8.44	8.64	8.75
0.98	1.48	1.99	2.49	2.98	3.47	3.92	4.29	4.54	4.66	4.72
1.94	2.47	2.98	3.49	3.98	4.47	4.92	5.29	5.54	5.66	5.72
2.21	3.18	4.16	5.12	6.01	6.72	7.17	7.38	7.46	7.49	7.50

Table 11.3 (Continued)

Ligand	Concentration [L], M	pH 2.0	3.0	3.5	4.0	4.5	5.0	5.5	6.0
Sr^{2+}									
HEDTA	0.10	0.00	0.00	0.00	0.00	0.04	0.28	0.93	1.70
NTA	0.10	0.00	0.00	0.00	0.01	0.03	0.08	0.22	0.49
Th^{4+}									
ac	0.10	0.57	2.00	3.07	4.26	5.36	6.08	6.44	7.00
Cl-ac	0.10	1.51	2.79	3.08	3.19	3.24	3.44	4.85	6.81
ox	0.10	14.98	17.82	18.82	19.40	19.66	19.75	19.79	19.80
sal	0.10	4.70	6.92	7.42	7.61	7.68	7.70	7.71	7.81
sulfosal	0.10	1.43	2.82	3.37	3.89	4.40	4.90	5.51	6.86
sulfoxine	0.10	0.45	3.85	6.53	9.37	12.19	15.19	18.25	21.28
DCTA	0.10	8.58	13.12	16.89	20.30	23.45	26.47	29.38	32.09
DTPA	0.10	9.79	13.21	14.74	16.16	17.43	18.56	19.61	20.63
EDTA	0.01	8.21	12.56	15.78	18.86	21.87	24.84	27.73	30.44
	0.10	9.21	14.55	17.78	20.86	23.87	26.84	29.73	32.44
HEDTA	0.10	5.45	11.02	14.07	17.08	20.07	23.02	25.86	28.49
NTA	0.10	16.47	20.13	21.71	23.24	24.75	26.25	27.75	29.25
acac	0.10	0.30	1.50	2.49	3.70	5.14	6.83	8.68	10.62
Ti^{3+}									
EDTA	0.01	5.84	8.72	9.84	10.88	11.88	12.87	13.81	14.67
	0.10	6.84	9.72	10.84	11.88	12.88	13.87	14.81	15.67
acac	0.10	3.33	5.31	6.43	7.68	9.07	10.52	12.01	13.50
TiO^{2+}									
ox	0.10	4.77	6.59	7.25	7.64	7.81	7.87	7.90	7.93
asco	0.10	18.75	20.68	21.54	22.19	22.57	22.73	22.79	22.81
sulfosal	0.10	3.97	6.45	7.18	7.76	8.29	8.80	9.30	9.80
tiron	0.10	2.40	6.30	8.30	10.30	12.30	14.30	16.29	18.27
EDTA	0.01	1.84	4.72	5.84	6.88	7.88	8.87	9.81	10.67
	0.10	2.84	5.72	6.84	7.88	8.88	9.87	10.81	11.67
Tl^{+}									
citr	0.10	0.00	0.00	0.00	0.00	0.02	0.06	0.15	0.24
DCTA	0.10	0.00	0.00	0.00	0.00	0.00	0.00	0.00	0.03
DTPA	0.10	0.00	0.00	0.00	0.01	0.03	0.13	0.36	0.71
EDTA	0.01	0.00	0.00	0.00	0.00	0.01	0.04	0.12	0.36
	0.10	0.00	0.00	0.01	0.03	0.10	0.28	0.63	1.14
EGTA	0.10	0.00	0.00	0.00	0.00	0.01	0.04	0.12	0.30
NTA	0.10	0.00	0.00	0.00	0.00	0.01	0.05	0.13	0.32
Tl^{3+}									
ac	0.10	3.70	6.52	8.29	10.11	11.78	13.07	13.82	14.16
ferron	0.10	15.50	19.85	21.55	23.12	24.65	26.15	27.63	29.06
DCTA	0.10	23.92	27.41	28.80	30.00	31.07	32.06	32.97	33.76
DTPA	0.10	28.75	32.04	33.50	34.86	36.09	37.20	38.24	39.25
EDTA	0.01	22.72	25.48	26.55	27.57	28.58	29.60	30.63	31.69

Ch. 11] Side-Reaction Coefficients for Complexation Equilibria of Metal Ions

$\log \alpha_{M(L,OH)}$

pH	6.5	7.0	7.5	8.0	8.5	9.0	9.5	10.0	10.5	11.0	12.0
Sr^{2+}											
	2.37	2.95	3.47	3.98	4.47	4.93	5.32	5.58	5.72	5.77	5.80
	0.88	1.34	1.82	2.31	2.79	3.23	3.59	3.81	3.91	3.95	3.97
Th^{4+}											
	8.80	10.80	12.80	14.80	16.80	18.80	20.80	22.80	24.80	26.80	30.80
	8.80	10.80	12.80	14.80	16.80	18.80	20.80	22.80	24.80	26.80	30.80
	19.80	19.80	19.80	19.80	19.80	19.84	20.84	22.80	24.80	26.80	30.80
	9.12	11.10	13.10	15.10	17.10	19.10	21.10	23.10	25.10	27.10	31.10
	8.80	10.80	12.80	14.80	16.80	18.80	20.80	22.80	24.80	26.80	30.80
	24.28	27.23	30.10	32.74	34.93	36.58	37.84	38.94	39.97	40.98	42.98
	34.54	36.75	38.82	40.85	42.86	44.86	46.86	48.86	50.85	52.83	56.57
	21.64	22.63	23.62	24.59	25.53	26.43	27.34	28.24	29.06	29.75	31.30
	32.89	35.10	37.18	39.20	41.20	43.17	45.08	46.83	48.32	49.56	51.69
	34.89	37.10	39.18	41.20	43.20	45.17	47.08	48.83	50.32	51.56	53.69
	30.85	33.00	35.05	37.06	39.04	40.96	42.73	44.27	45.54	46.65	48.70
	30.75	32.25	33.75	35.24	36.72	38.16	39.52	40.74	41.84	42.88	44.90
	12.60	14.58	16.52	18.34	19.89	20.96	21.63	23.12	25.10	27.10	31.10
Ti^{3+}											
	15.40	16.00	16.54	17.05	17.55	18.04	18.49	18.86	19.11	19.23	19.29
	16.40	17.00	17.54	18.05	18.55	19.04	19.49	19.86	20.11	20.23	20.29
	14.99	16.48	17.94	19.31	20.47	21.26	21.66	21.82	21.87	21.89	21.90
TiO^{2+}											
	8.11	8.76	9.71	10.70	11.70	12.70	13.70	14.70	15.70	16.70	18.70
	22.82	22.82	22.82	22.82	22.82	22.82	22.81	22.78	22.70	22.49	21.33
	10.30	10.81	11.32	11.86	12.48	13.21	14.08	15.03	16.01	17.00	19.00
	20.20	22.01	23.54	24.66	25.45	26.05	26.58	27.09	27.59	28.06	28.74
	11.40	12.00	12.54	13.06	13.56	14.07	14.61	15.24	16.05	17.01	19.00
	12.40	13.00	13.54	14.05	14.55	15.04	15.50	15.92	16.36	17.07	19.00
Tl^{+}											
	0.29	0.31	0.32	0.32	0.32	0.32	0.32	0.32	0.32	0.32	0.33
	0.16	0.44	0.85	1.32	1.80	2.30	2.80	3.30	3.79	4.28	5.15
	1.16	1.64	2.12	2.60	3.05	3.48	3.92	4.33	4.65	4.84	4.96
	0.79	1.29	1.79	2.30	2.79	3.28	3.73	4.10	4.35	4.47	4.53
	1.72	2.27	2.79	3.29	3.79	4.28	4.73	5.10	5.35	5.47	5.53
	0.61	1.03	1.50	1.98	2.45	2.87	3.15	3.30	3.35	3.37	3.38
	0.65	1.08	1.55	2.04	2.52	2.96	3.33	3.57	3.68	3.73	3.75
Tl^{3+}											
	14.54	15.66	17.11	18.60	20.10	21.60	23.10	24.61	26.12	27.66	31.01
	30.36	31.37	31.98	32.25	32.35	32.38	32.40	32.40	32.40	32.40	32.75
	34.41	34.97	35.48	35.98	36.44	36.82	37.09	37.22	37.27	37.29	37.30
	40.25	41.24	42.19	43.06	43.79	44.34	44.70	44.88	44.96	44.99	45.00
	32.77	33.82	34.84	35.84	36.83	37.80	38.72	39.52	40.19	40.76	41.80

Table 11.3 (Continued)

Ligand	Concentration [L], M	pH 2.0	3.0	3.5	4.0	4.5	5.0	5.5	6.0
		Tl^{3+}							
EDTA	0.10	23.72	26.48	27.55	28.57	29.58	30.60	31.63	32.69
NTA	0.10	15.21	18.42	19.53	20.56	21.57	22.58	23.58	24.58
		U^{4+}							
DCTA	0.10	12.97	21.72	25.49	28.90	32.05	35.07	37.98	40.69
EDTA	0.10	15.48	23.27	26.51	29.59	32.61	35.57	38.45	41.14
acac	0.10	1.61	3.62	4.92	6.53	8.38	10.34	12.36	14.45
		UO_2^{2+}							
ac	0.10	0.03	0.29	0.78	1.65	2.52	3.08	3.35	4.25
Cl-ac	0.10	0.20	0.60	0.71	0.76	0.78	0.82	1.75	4.18
ox	0.10	5.43	7.34	8.03	8.44	8.63	8.70	8.72	8.73
asco	0.10	0.08	0.49	0.86	1.24	1.54	1.82	2.35	4.20
citr	0.01	0.99	6.29	8.51	10.42	11.98	13.18	14.02	14.52
	0.10	2.94	8.29	10.51	12.42	13.98	15.18	16.02	16.52
sal	0.10	0.51	1.14	1.43	1.79	2.24	2.78	3.42	4.49
sulfosal	0.10	0.22	1.33	1.91	2.49	3.14	3.92	4.82	5.79
tiron	0.10	0.23	1.13	1.92	2.84	3.81	4.80	5.80	6.79
DCTA	0.10	0.31	3.40	4.78	5.99	7.07	8.08	9.03	9.89
EDTA	0.01	0.60	2.14	2.78	3.41	4.13	4.95	5.83	6.65
	0.10	1.49	3.14	3.78	4.41	5.13	5.95	6.83	7.65
EGTA	0.10	0.66	2.39	3.00	3.55	4.07	4.61	5.20	5.91
NTA	0.10	0.34	1.74	2.31	2.84	3.35	3.85	4.35	4.93
sulfoxine	0.10	1.63	6.11	8.93	11.55	13.90	16.04	18.09	20.11
DDC	0.10	15.18	16.02	16.29	16.42	16.47	16.49	16.50	16.50
acac	0.10	0.28	1.07	1.68	2.45	3.34	4.30	5.28	6.28
		V^{3+}							
EDTA	0.10	11.54	14.42	15.54	16.58	17.58	18.57	19.51	20.37
NTA	0.10	4.32	7.43	8.59	9.64	10.66	11.67	12.67	13.76
		VO^{2+}							
ox	0.10	6.57	8.46	9.12	9.52	9.69	9.75	9.77	9.78
sal	0.10	0.10	1.13	1.75	2.35	3.00	3.77	4.66	5.62
sulfosal	0.10	0.53	1.92	2.60	3.36	4.24	5.22	6.47	8.20
tiron	0.10	0.17	2.64	4.59	6.59	8.59	10.59	12.60	14.72
DCTA	0.10	2.46	5.83	7.21	8.42	9.50	10.51	11.46	12.32
EDTA	0.01	3.38	6.52	7.96	9.42	10.90	12.37	13.81	15.17
	0.10	4.38	7.52	8.96	10.42	11.90	13.37	14.81	16.17
sulfoxine	0.10	0.68	2.58	3.58	4.79	7.36	10.27	13.04	15.56
		VO_2^+							
ac	0.10	0.04	0.32	0.69	1.26	2.02	2.93	3.91	4.91
asco	0.10	1.71	1.69	1.65	1.66	2.12	3.03	4.01	5.01
EDTA	0.10	4.99	5.91	6.23	6.69	7.37	8.20	9.09	9.93

Ch. 11]   Side-Reaction Coefficients for Complexation Equilibria of Metal Ions   335

$$\log \alpha_{M(L,OH)}$$

pH	6.5	7.0	7.5	8.0	8.5	9.0	9.5	10.0	10.5	11.0	12.0
Tl^{3+}											
	33.77	34.82	35.84	36.84	37.83	38.80	39.72	40.52	41.19	41.76	42.80
	25.58	26.57	27.55	28.49	29.32	29.94	30.28	30.42	30.48	30.50	32.50
U^{4+}											
	43.14	45.35	47.42	49.45	51.46	53.46	55.46	57.46	59.45	61.43	65.17
	43.56	45.76	47.82	49.84	51.84	53.80	55.69	57.41	58.87	60.08	62.19
	16.65	18.99	21.43	23.91	26.40	28.90	31.40	33.90	36.40	38.90	43.90
UO_2^{2+}											
	6.68	9.18	11.68	14.18	16.68	19.18	21.68	24.18	26.68	29.18	34.18
	6.68	9.18	11.68	14.18	16.68	19.18	21.68	24.18	26.68	29.18	34.18
	8.73	9.31	11.68	14.18	16.68	19.18	21.68	24.18	26.68	29.18	34.18
	6.68	9.18	11.68	14.18	16.68	19.18	21.68	24.18	26.68	29.18	34.18
	14.74	14.83	14.86	14.95	16.68	19.18	21.68	24.18	26.68	29.18	34.18
	16.74	16.83	16.86	16.87	17.08	19.18	21.68	24.18	26.68	29.18	34.18
	6.69	9.18	11.68	14.18	16.68	19.18	21.68	24.18	26.68	29.18	34.18
	7.02	9.19	11.68	14.18	16.68	19.18	21.68	24.18	26.68	29.18	34.18
	7.80	9.30	11.68	14.18	16.68	19.18	21.68	24.18	26.68	29.18	34.18
	10.61	11.22	12.02	14.18	16.68	19.18	21.68	24.18	26.68	29.18	34.18
	7.44	9.20	11.68	14.18	16.68	19.18	21.68	24.18	26.68	29.18	34.18
	8.36	9.38	11.68	14.18	16.68	19.18	21.68	24.18	26.68	29.18	34.18
	7.01	9.19	11.68	14.18	16.68	19.18	21.68	24.18	26.68	29.18	34.18
	6.70	9.18	11.68	14.18	16.68	19.18	21.68	24.18	26.68	29.18	34.18
	22.11	24.09	26.02	27.84	29.44	30.76	31.89	32.94	33.95	34.96	36.96
	16.50	16.50	16.50	16.50	16.90	19.18	21.68	24.18	26.68	29.18	34.18
	7.37	9.23	11.68	14.18	16.68	19.18	21.68	24.18	26.68	29.18	34.18
V^{3+}											
	21.10	21.70	22.24	22.76	23.29	23.85	24.47	25.15	25.81	26.41	27.45
	14.67	15.67	16.66	17.65	18.62	19.52	20.25	20.73	21.00	21.56	24.40
VO^{2+}											
	9.78	9.78	9.78	9.78	9.78	9.82	10.05	10.76	11.72	12.71	14.71
	6.61	7.60	8.60	9.60	10.60	11.60	12.60	13.60	14.60	15.60	17.57
	10.17	12.16	14.16	16.16	18.16	20.16	22.16	24.14	26.11	28.01	31.23
	17.23	20.02	22.69	25.11	27.30	29.37	31.39	33.39	35.39	37.37	41.16
	13.04	13.64	14.18	14.69	15.20	15.70	16.20	16.70	17.19	17.68	18.55
	16.40	17.50	18.54	19.55	20.55	21.54	22.49	23.36	24.11	24.73	25.79
	17.40	18.50	19.54	20.55	21.55	22.54	23.49	24.36	25.11	25.73	26.79
	17.82	19.90	21.87	23.70	25.29	26.62	27.75	28.80	29.81	30.82	32.82
VO_2^+											
	5.91	6.93	8.00	9.15	10.45	11.85	13.32	14.81	16.30	17.80	20.81
	6.03	7.08	8.21	9.48	10.86	12.32	13.81	15.30	16.80	18.30	21.33
	10.65	11.25	11.79	12.30	12.80	13.30	13.88	14.89	16.31	17.80	20.81

Table 11.3 (Continued)

Ligand	Concentration [L], M	pH 2.0	3.0	3.5	4.0	4.5	5.0	5.5	6.0
Y^{3+}									
ac	0.10	0.00	0.03	0.08	0.22	0.48	0.79	0.99	1.08
ox	0.10	3.52	5.45	6.23	6.77	7.05	7.16	7.19	7.21
citr	0.10	0.26	2.58	3.69	4.65	5.43	6.02	6.45	6.69
tiron	0.10	0.01	0.13	0.38	0.92	1.72	2.65	3.63	4.61
DCTA	0.10	2.61	5.64	6.99	8.17	9.25	10.26	11.21	12.07
DTPA	0.10	3.01	6.54	8.09	9.50	10.78	11.91	12.96	13.98
EDTA	0.10	3.63	6.51	7.63	8.67	9.67	10.66	11.60	12.46
acac	0.10	0.01	0.05	0.15	0.38	0.83	1.50	2.37	3.38
Zn^{2+}									
ac	0.10	0.00	0.00	0.01	0.03	0.08	0.16	0.24	0.29
citr	0.01	0.01	0.14	0.44	0.98	1.61	2.16	2.58	2.83
	0.10	0.10	0.69	1.27	1.94	2.61	3.20	3.68	4.00
sal	0.10	0.00	0.00	0.00	0.00	0.00	0.00	0.00	0.01
sulfosal	0.10	0.00	0.00	0.00	0.00	0.00	0.01	0.03	0.09
tiron	0.10	0.00	0.00	0.01	0.02	0.07	0.23	0.66	1.44
DCTA	0.10	2.68	5.35	6.58	7.72	8.78	9.78	10.73	11.59
DTPA	0.10	3.80	5.78	6.65	7.50	8.27	8.98	9.74	10.60
EDTA	0.01	2.11	4.92	6.46	7.98	9.49	10.97	12.41	13.77
	0.10	3.11	5.92	7.46	8.98	10.49	11.97	13.41	14.77
HEDTA	0.10	1.73	4.27	5.36	6.39	7.36	8.26	9.05	9.70
NTA	0.10	1.25	2.89	3.47	4.00	4.51	5.01	5.52	6.05
en	0.10	0.00	0.00	0.00	0.00	0.00	0.00	0.01	0.12
py	0.10	0.00	0.00	0.01	0.02	0.06	0.13	0.23	0.30
tetren	0.10	0.00	0.00	0.03	0.22	0.89	2.13	3.66	5.19
trien	0.10	0.00	0.00	0.00	0.01	0.12	0.77	1.97	3.32
glutam	0.10	0.00	0.00	0.00	0.00	0.01	0.04	0.12	0.34
glyc	0.10	0.00	0.00	0.00	0.01	0.03	0.09	0.25	0.62
histam	0.10	0.00	0.00	0.00	0.00	0.00	0.03	0.20	0.82
histid	0.10	0.00	0.01	0.54	2.40	4.38	6.33	8.18	9.83
tryp	0.10	0.00	0.00	0.01	0.02	0.07	0.20	0.53	1.17
BAL	0.10	0.00	0.08	0.48	1.34	2.54	5.14	8.13	11.13
TGA	0.10	0.00	0.06	0.30	0.92	1.83	3.00	4.61	6.53
acac	0.10	0.00	0.00	0.01	0.03	0.09	0.25	0.56	1.03
Zr^{4+}									
DCTA	0.10	15.52	19.01	20.40	21.60	22.67	23.66	24.57	25.36
DTPA	0.10	18.88	22.80	24.36	25.76	27.02	28.18	29.35	30.63
EDTA	0.01	14.04	16.94	18.06	19.10	20.11	21.11	22.10	23.09
	0.10	15.04	17.94	19.06	20.10	21.11	22.11	23.10	24.11

Ch. 11] Side-Reaction Coefficients for Complexation Equilibria of Metal Ions

$\log \alpha_{M(L,OH)}$

pH	6.5	7.0	7.5	8.0	8.5	9.0	9.5	10.0	10.5	11.0	12.0
Y^{3+}											
	1.12	1.14	1.18	1.30	1.84	3.89	6.38	8.88	11.38	13.88	18.88
	7.21	7.21	7.21	7.21	7.21	7.21	7.50	9.68	12.18	14.68	19.68
	6.81	6.85	6.86	6.87	6.87	6.87	7.35	9.68	12.18	14.68	19.68
	5.59	6.52	7.37	8.07	8.67	9.20	9.72	10.33	12.19	14.68	19.68
	12.79	13.39	13.93	14.44	14.95	15.45	15.95	16.45	16.94	17.43	19.69
	14.98	15.98	16.96	17.89	18.73	19.43	20.00	20.46	20.80	21.00	21.13
	13.19	13.79	14.33	14.84	15.34	15.83	16.28	16.65	16.90	17.02	19.68
	4.53	5.81	7.18	8.53	9.68	10.47	10.87	11.04	12.21	14.68	19.68
Zn^{2+}											
	0.31	0.32	0.32	0.33	0.36	0.53	1.11	2.03	3.03	4.09	7.64
	2.95	2.99	3.01	3.01	3.01	3.02	3.05	3.28	4.03	5.14	9.76
	4.15	4.21	4.23	4.24	4.24	4.24	4.24	4.26	4.43	5.19	9.76
	0.04	0.11	0.28	0.60	1.04	1.58	2.24	3.05	4.02	5.15	9.76
	0.23	0.52	0.98	1.61	2.40	3.31	4.27	5.25	6.21	7.11	9.77
	2.58	4.18	5.82	7.23	8.42	9.49	10.51	11.51	12.51	13.49	15.28
	12.31	12.91	13.45	13.96	14.47	14.97	15.47	15.97	16.46	16.95	17.82
	11.54	12.51	13.48	14.40	15.23	15.91	16.46	16.90	17.22	17.39	17.48
	15.00	16.10	17.14	18.15	19.15	20.14	21.09	21.96	22.71	23.33	24.39
	16.00	17.10	18.14	19.15	20.15	21.14	22.09	22.96	23.71	24.33	25.39
	10.26	10.78	11.28	11.78	12.27	12.73	13.12	13.38	13.52	13.57	13.60
	6.61	7.28	8.07	8.97	9.90	10.77	11.48	11.92	12.13	12.21	12.33
	0.72	2.03	3.48	4.76	5.93	7.09	8.22	9.14	9.69	9.93	10.33
	0.33	0.35	0.35	0.38	0.49	0.97	1.84	2.83	3.87	5.04	10.00
	6.69	8.17	9.60	10.92	12.06	12.96	13.58	13.90	14.03	14.08	14.10
	4.64	5.86	6.95	7.97	8.93	9.79	10.45	10.85	11.03	11.10	11.15
	0.76	1.40	2.23	3.18	4.18	5.17	6.01	6.54	6.78	6.87	10.00
	1.25	2.10	3.11	4.24	5.47	6.67	7.62	8.19	8.44	8.53	10.02
	1.86	2.98	4.07	5.17	6.33	7.53	8.58	9.28	9.62	9.76	10.22
	11.20	12.36	13.40	14.37	15.24	15.92	16.33	16.51	16.58	16.60	16.61
	2.11	3.27	4.61	6.02	7.41	8.67	9.61	10.15	10.38	10.46	10.57
	14.12	17.10	20.03	22.83	25.34	27.43	29.14	30.55	31.61	32.25	32.65
	8.51	10.51	12.50	14.49	16.46	18.38	20.13	21.51	22.35	22.74	22.92
	1.66	2.44	3.33	4.21	4.97	5.50	5.77	5.87	5.91	5.98	9.76
Zr^{4+}											
	26.01	26.57	27.09	27.67	29.02	30.98	32.98	34.98	36.99	39.01	43.24
	32.01	33.46	34.90	36.27	37.50	38.54	39.40	40.08	40.66	41.19	43.28
	24.14	25.67	27.64	29.65	31.64	33.60	35.49	37.21	38.67	39.89	43.32
	25.54	27.57	29.63	31.64	33.64	35.60	37.49	39.21	40.67	41.88	44.07

Table 11.3
Side-reaction coefficients of metal ions with organic ligands in the presence of hydroxide ions (temperature 20–25°C, $I = 0.1$–1.0).

The order of ligands in this table is as follows:

Order no.	Abbreviation	Name of the ligand
1	ac	acetic acid
2	Cl-ac	monochloroacetic acid
3	citr	citric acid
4	ox	oxalic acid
5	asco	ascorbic acid
6	tart	D-tartaric acid
7	sal	salicylic acid
8	sulfosal	sulphosalicylic acid
9	tiron	tiron
10	DCTA	trans-1,2-cyclohexylenedinitrilotetra-acetic acid
11	DTPA	diethylenetrinitrilopenta-acetic acid
12	EDTA	ethylenedinitrilotetra-acetic acid
13	HEDTA	N-(2-hydroxyethyl)-ethylenedinitrilo-N,N',N'-triacetic acid
14	EGTA	ethylenebis(oxyethylenenitrilo)tetra-acetic acid
15	NTA	nitrilotriacetic acid
16	bipy	2,2'-bipyridyl
17	dien	diethylenetriamine
18	en	ethylenediamine
19	1,10-phen	1,10-phenanthroline
20	py	pyridine
21	TEA	triethanolamine
22	tetren	tetraethylenepentamine
23	THAM	tris(hydroxymethyl)aminomethane
24	tren	nitrilotris(2-ethylamine)
25	trien	triethylenetetramine
26	cyst	cysteine
27	glutam	glutamine
28	glyc	glycine
29	histam	histamine
30	histid	histidine
31	tryp	tryptophan
32	TGA	thioglycollic acid
33	BAL	2,3-dimercaptopropan-1-ol
34	thiourea	thiourea
35	acac	acetylacetone
36	sulfoxine	8-hydroxyquinoline-5-sulphonic acid
37	ferron	7-iodo-8-hydroxyquinoline-5-sulphonic acid

CHAPTER 12

Diagrams of precipitation equilibria

Calculation of the solubility of a sparingly soluble electrolyte M_mA_n in water is based on the expression for the solubility product K_s [cf. Section 2.4.6, Eqs. (2.77) and (2.78)]:

$$c_{M_mA_n} = [K_s/(m^m + n^n)]^{\frac{1}{m+n}}. \tag{12.1}$$

In presence of excess of the precipitant ion, for example A^{m-}, the solubility is given by the residual concentration of the ion M^{n+} being precipitated, calculated from

$$c_{M_mA_n} = [M]/m = \frac{1}{m}(K_s/[A]^n)^{\frac{1}{m}}. \tag{12.2}$$

12.1 GRAPHS OF CONDITIONAL SOLUBILITY PRODUCTS

When problems involving precipitation equilibria are solved, it should always be kept in mind that precipitation is often considerably influenced by various side-reactions, namely by formation of complexes between the ions of the precipitate itself, by complexation with other ligands present in the solution and, last but not least, by various protolytic reactions in which the ions H^+ and/or OH^- participate. So only very seldom can the stoichiometric solubility product K_s be directly used for calculation of solubility. As explained in Section 3.1.2 [cf. Eq. (3.22)], a solubility affected by side-reactions can easily be calculated by means of expressions similar to those given in Eqs. (12.1) and (12.2): it is only necessary to substitute a conditional solubility product K'_s in place of K_s. The required values of the conditional solubility products for some common sparingly soluble compounds (electrolytes) can be read from the graphs in Figs. 12.1–12.5. They represent plots of logarithms of conditional

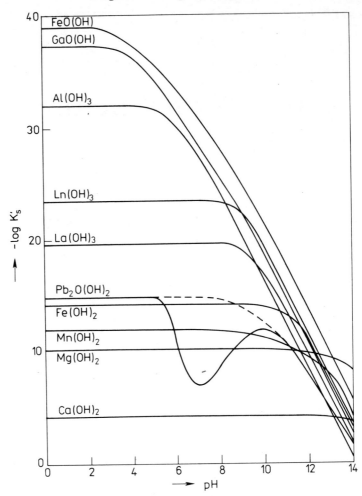

Fig. 12.1 — Conditional solubility products of various metal hydroxides influenced by hydroxo-complexes, as functions of pH (polynuclear hydroxo-complexes are not present, except for Pb(II), if $c_M \leq 1M$). In the broken part of the curve for Pb(II) polynuclear complexes are not considered.

solubility products as a function of pH, which allow readings with an error in the second decimal. In most cases this corresponds to the reliability with which the solubility products can be determined and is better than that with which a certain conditional value can be applied.

The equilibrium concentration of the free precipitant [cf. the ion A^{m-} in Eq. (12.2)], necessary for calculation of the solubility, is often not known. It can be estimated from the total concentration of the precipitant

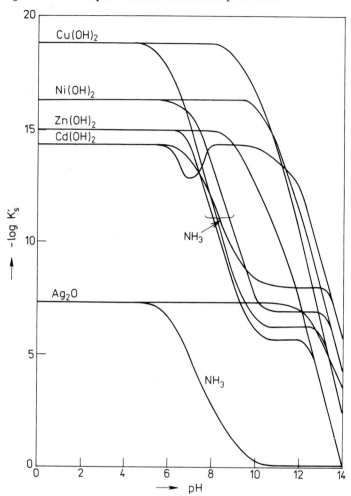

Fig. 12.2 — Conditional solubility products of various metal hydroxides influenced by hydroxo-complexes and also by ammine-complexes (at $c_{NH_3} = 1.0M$) as functions of pH (polynuclear complexes are not present, except for Cd(II), if $c_M \leqq 1M$). Curves calculated for the presence of ammonia are marked NH_3.

with the aid of a graphical representation of the concentrations of all species taking part in the precipitation equilibrium, usually presented in the form of a logarithmic concentration diagram (see Section 12.2). These diagrams give an overall view of various effects on the solubility and concentration levels of individual species of the system, in dependence on the reaction conditions and controlled variables, namely, the pH value, concentration of precipitant or complex-forming ions, etc. Predominance-

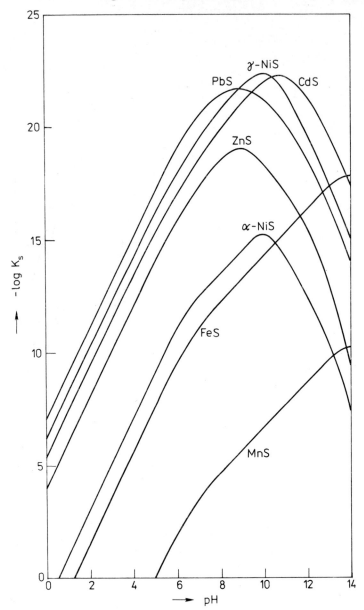

Fig. 12.3 — Conditional solubility products of various metal sulphides, as functions of pH.

area diagrams (discussed in Section 12.3), which can be used for similar purposes, allow representation of the precipitation equilibrium as a function of two controlled variables.

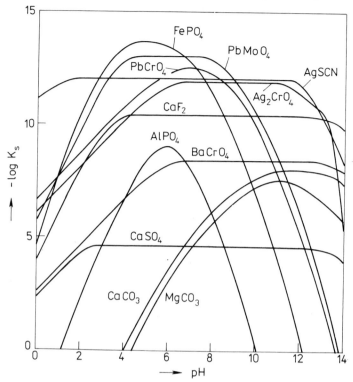

Fig. 12.4 — Conditional solubility products of various sparingly soluble compounds important in analytical chemistry, as functions of pH.

12.2 LOGARITHMIC CONCENTRATION DIAGRAMS FOR PRECIPITATION EQUILIBRIA

Construction of a logarithmic diagram for a complexation system in equilibrium with a solid phase has already been discussed in Section 4.2.2. Here some other examples of these diagrams and their applications will be given, but without details of the construction.

12.2.1 pH-Dependence solubility diagram for boric acid

Boric acid dissolves in water to a greater extent than many sparingly soluble compounds: 6 g of boric acid will dissolve in 100 ml of water at 25°C to give a saturated solution of about $1M$ concentration. In aqueous solution boric acid is very slightly dissociated (pK_{a1} = 9.24 at

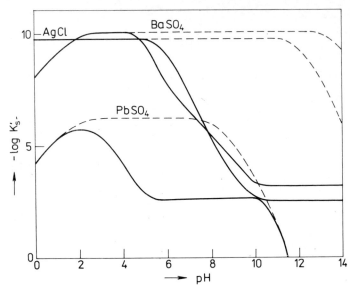

Fig. 12.5 — Conditional solubility products in the presence of masking agents, as functions of pH.
AgCl — the solubility product in the presence of ammonia ($c_{NH_3} = 1.0M$).
$PbSO_4$ — the solubility product in the presence of acetate ($c_{Ac} = 1.0M$).
$BaSO_4$ — the solubility product in the presence of EDTA ($c_L = 0.1M$).
Broken curves — only protonation of the anion and hydroxo-complexes of the cation were considered.

$I = 0$ and $25°C$). On neutralization, polyborates are formed; therefore, the amount of boric acid dissolved increases considerably as the pH increases. In addition to monoborate, di-, tri-, and tetraborates are the most important species [1]; there is also some evidence that pentaborate is formed [2].

Boric acid and borates may also be formulated as hydroxo-complexes; their mutual equilibria are then expressed by the following reaction schemes [3]:

$$\begin{aligned} B(OH)_3 + OH^- &\rightleftharpoons B(OH)_4^- \\ 2\,B(OH)_3 + OH^- &\rightleftharpoons B_2O(OH)_5^- + H_2O \\ 3\,B(OH)_3 + OH^- &\rightleftharpoons B_3O_3(OH)_4^- + 3\,H_2O \\ 4\,B(OH)_3 + 2\,OH^- &\rightleftharpoons B_4O_5(OH)_4^{2-} + 5\,H_2O. \end{aligned} \quad (12.3)$$

From the expressions for the stability constants of these hydroxo-complexes and the ionic product of water (see tables in Chapters 5 and 6) the pH-dependence of the concentration of the individual anions is obtained in the form of straight lines, making the diagram in Fig. 12.6, which holds for a saturated solution of boric acid ($[B(OH)_3] \sim 1M$) and ionic strength 1 (or 0). The relevant equations are given in the caption to Fig. 12.6.

The diagram for boric acid and the borates indicates that the solubility of boric acid is constant and independent of pH up to a pH value of about 6.7 (or 7). At higher pH values it rapidly increases, first owing to formation of triborate (within the pH range 7–8) and then (in more alkaline solution) to the presence of tetraborate. The contribution of mono- and diborate to the overall solubility is negligible for such a saturated solution. A different situation occurs in a dilute solution of boric acid, as can be seen from a distribution diagram for boric acid.

The solubility diagram for boric acid and the borates represents an example of a logarithmic diagram showing a system of complexes which are in equilibrium with a sparingly dissociated and moderately soluble compound.

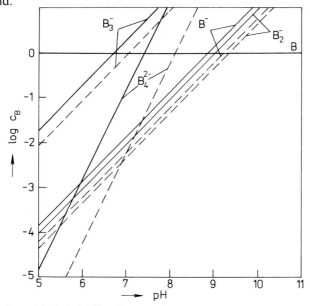

Fig. 12.6 — Logarithmic solubility diagram of boric acid in water as a function of pH [temperature 25°C; $I = 1$ (full lines) or 0 (broken lines)].
Lines in the diagram:
B^-: $\log[B(OH)_4^-]$ = pH − 8.88 ($I = 1$) or pH − 9.24 ($I = 0$)
B: $\log[B(OH)_3]$ = 0 (independent of pH and I)
B_2^-: $\log[B_2O(OH)_5^-]$ = pH − 9.05 ($I = 1$) or pH − 9.36 ($I = 0$)
B_3^-: $\log[B_3O_3(OH)_4^-]$ = pH − 6.75 ($I = 1$) or pH − 7.03 ($I = 0$)
B_4^{2-}: $\log[B_4O_5(OH)_4^{2-}]$ = 2 pH − 14.9 ($I = 1$) or pH − 16.3 ($I = 0$)

12.2.2 Logarithmic solubility diagram for iron(III) hydroxide, showing pH-dependence

Iron(III) hydroxide (FeOOH) in solution is in equilibrium with the ion Fe(III), mononuclear hydroxo-complexes $Fe(OH)_n^{(3-n)}$ (where n is 1–4), and the polynuclear complexes $Fe_2(OH)_2^{4+}$ and $Fe_3(OH)_4^{5+}$ (e.g. see ref. [4]). Thus the construction of a logarithmic solubility diagram is based on the equilibria between the solid hydroxide phase and particular complex ions in solution, which are characterized by the solubility product K_s and the constants K_{snm} to be calculated from the expression $\log K_{snm} = \log K_s + \log \beta_{nm}$ (see Sections 2.4.6 and 4.2.2). The necessary values of the constants are given in Chapters 6 and 8.

On rearrangement of the expressions for the constants K_s and K_{snm} to pH-relationships for the concentration of Fe(III) and particular hydroxo-complexes, the linear equations given in the caption to Fig. 12.7 are obtained. The logarithmic concentration diagram illustrates the change of composition of a solution saturated with FeOOH, as a function of pH. The full line represents the overall solubility of iron(III) hydroxide. The ion Fe(III) is the principal iron species in solution in a sufficiently acidic medium; then at pH > 2.2 hydroxo-complexes begin to predominate in solution. First, it is the species $FeOH^{2+}$, and then $Fe(OH)_2^+$. Between pH 6.3 and 9.6, where the solubility of iron(III) hydroxide reaches a minimum, the complex $Fe(OH)_3$ predominates. Its concentration does not depend on pH. Finally, at pH > 9.6 the solubility increases again owing to an increasing contribution from $Fe(OH)_4^-$. The minimum solubility of FeOOH can be located at the intersection of the straight lines for the complexes $Fe(OH)_2^+$ and $Fe(OH)_4^-$, which lies outside the diagram in Fig. 12.7. Point 1 indicates the corresponding pH value of 7.95. Of course, the minimum solubility is given as the sum of the concentrations of the complex $Fe(OH)_3$, which significantly predominates, and of the complex species $Fe(OH)_2^+$ and $Fe(OH)_4^-$, as read at pH 7.95:

$$c_{min} = 10^{-11.5} + 2 \times 10^{-13.15} = 3.3 \times 10^{-12} M$$

the contributions of other equilibrium components being entirely negligible. This result is in full agreement with the value obtained by calculation based on the following expressions (e.g. see ref. [5]):

$$c_{min} = K_s(2\beta_2^{1/2}\beta_4^{1/2} + \beta_3) = 10^{-12.85} + 10^{-11.5} = 3.3 \times 10^{-12} M$$

and

$$[OH^-]_{min} = (K_{s2}/K_{s4})^{1/2} = 10^{-6.05}.$$

Sec. 12.2] Logarithmic Concentration Diagrams for Precipitation Equilibria 347

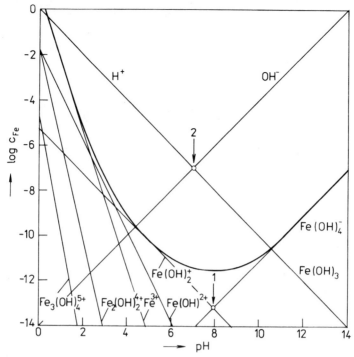

Fig. 12.7 — Logarithmic solubility diagram of ferric hydroxide (α-FeOOH), as a function of pH (temperature 25°C, $I = 0$). The heavy line represents the solubility dependence on pH, and is the sum of concentrations of all iron species in solution. The solubility minimum is found at pH = 7.95 (point 1); the solubility in pure water is found at point 2 (pH = 7.0). Lines in the diagram:

$\log [\text{Fe}^{3+}] = -3\,\text{pH} + 0.5$
$\log [\text{FeOH}^{2+}] = -2\,\text{pH} - 1.7$
$\log [\text{Fe(OH)}_2^+] = -\text{pH} - 5.2$
$\log [\text{Fe(OH)}_3] = -11.5$

$\log [\text{Fe(OH)}_4^-] = \text{pH} - 21.1$
$\log [\text{Fe}_2(\text{OH})_2^{4+}] = -4\,\text{pH} - 2.0$
$\log [\text{Fe}_3(\text{OH})_4^{5+}] = -5\,\text{pH} - 4.8$

The solubility of the hydroxide FeOOH in water, without any pH adjustment, depends on the pH which results in the equilibrium of a pure precipitate of iron(III) hydroxide with water. This pH value can be calculated on the basis of ligand balance for the ion OH$^-$, which holds for a saturated solution of FeOOH. In this rather complicated expression, which involves the concentrations of all electrically charged species in solution and thus agrees with the electroneutrality condition, the concentrations of the ions H$^+$ and OH$^-$ clearly predominate: this can be seen from the location of the lines $\log [\text{H}^+] = -\text{pH}$ and $\log [\text{OH}^-] = -14 + \text{pH}$, which are plotted as auxiliary straight lines in Fig. 12.7. An equal concentration of these ions is obviously attained at pH 7.0; the very sparingly soluble FeOOH has no influence on the resulting pH of an aqueous solution. This is always so with all sparingly soluble salts which yield such low concentrations of ions that their hydrolysis cannot noticeably affect the pH value of pure water.

The solubility of iron (III) hydroxide at pH 7.0 (cf. point 2 in Fig. 12.7) is $4.0 \times 10^{-12} M$ [which represents the sum of the concentrations of the predominating complexes $Fe(OH)_2^+$ and $Fe(OH)_3$]. Calculation based on the corresponding value of the conditional solubility product (cf. Fig. 12.1) gives the result

$$c_{Fe} = K_s'[OH^-]^{-3} = 10^{-11.4} M$$

which is in agreement with the value read from the diagram (cf. Section 4.2.2.1, p. 76).

12.2.3 Logarithmic solubility diagram of indium(III) hydroxide, showing dependence on pH and concentration of chloride

Salts of indium(III) are hydrolysed in aqueous solution, to form, according to Baes and Mesmer [6], mononuclear hydroxo-complexes of composition $In(OH)_n^{(3-n)}$ (where n is 1–4), besides polynuclear species. The formation of hydroxo-complexes already becomes apparent at pH $\geq$ 2; even at an In(III) concentration as low as $10^{-5} M$ the contribution of polynuclear hydroxo-complexes, e.g. $In_3(OH)_4^{5+}$, predominate. The solubility of indium(III) hydroxide, e.g. $In_2O_3 \cdot xH_2O$, in excess of hydroxide is relatively small.

When chloride is present, the chloro-complexes $InCl_p^{(3-p)}$ (where p is 1–4) and mixed hydroxo-chloro-complexes $InOHCl^+$ and $In_2(OH)_2Cl_2^{2+}$ exist in the equilibrium; thus the hydrolysis may be suppressed to a certain extent.

The data on the stability of hydroxo- and chloro-complexes (cf. Tables 6.12 and 6.18) are thus necessary for construction of a logarithmic solubility diagram for indium(III) hydroxide, showing the dependence on pH at a constant chloride concentration (Fig. 12.8) or the dependence on chloride concentration at a selected pH value (e.g. Fig. 12.9).

The equations for the line segments composing the diagram in Fig. 12.8 can be derived as indicated above from the expressions for the equilibrium constants K_{snpm} [cf. Eq. (2.80), p. 41] which characterize the equilibria between solid indium(III) hydroxide and the relevant species in solution:

$$K_{snpm} = K_s^m \beta_{npm} \tag{12.4}$$

Sec. 12.2]   Logarithmic Concentration Diagrams for Precipitation Equilibria

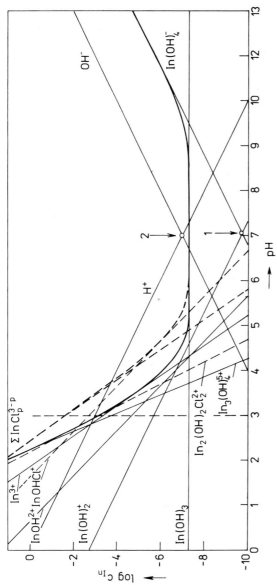

Fig. 12.8 — Logarithmic solubility diagram of $In_2O_3 \cdot xH_2O$ as a function of pH (full lines) and in the presence of chloride ($[Cl^-] = 0.1M$; broken lines). Temperature 25°C, $I = 0.1$. The heavy line represents the solubility dependence on pH in the absence of Cl^- and is the sum of all chloride-free species in solution. The heavy broken line is the concentration sum of the chloro-complexes $InCl_p^{(3-p)}$. Point 1 gives the pH for the lowest solubility. The solubility in pure water is found at pH = 7 (point 2). Equations for lines in the diagram:

$\log [In^{3+}]$ = $-3\,\text{pH} + 5.60$;
$\log [InOH^{2+}]$ = $-2\,\text{pH} + 1.30$;
$\log [In(OH)_2^+]$ = $-\text{pH} - 2.72$;
$\log [InOHCl^+]$ = $-2\,\text{pH} + \log [Cl^-] + 4.25$
$\log [In(OH)_2Cl_2^{2+}]$ = $-4\,\text{pH} + 2\log [Cl^-] + 10.94$
$\log [InCl^{2+}]$ = $-3\,\text{pH} + \log [Cl^-] + 7.83$
$\log [InCl_2^+]$ = $-3\,\text{pH} + 2\log [Cl^-] + 9.04$
$\log [InCl_3]$ = $-3\,\text{pH} + 3\log [Cl^-] + 9.45$.

$\log [In(OH)_3]$ = -7.33
$\log [In(OH)_4^-]$ = $\text{pH} - 16.78$
$\log [In_3(OH)_4^{5+}]$ = $-5\,\text{pH} + 11.23$

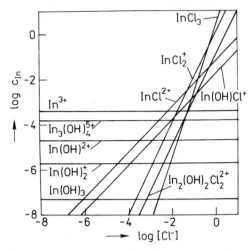

Fig. 12.9 — Logarithmic solubility diagram for $In_2O_3 \cdot xH_2O$ at pH 3.0 for varied $[Cl^-]$. Equations of the lines, given in the text to Fig. 12.8, were written for pH 3.0.

where the constants β_{npm} define the stability of the hydroxo- (subscript n), chloro- (subscript p), and mixed hydroxo-chloro-complexes of indium(III). The data used for construction of Fig. 12.8 refer to 25°C and ionic strength $I = 0.1$ (NaCl). To allow easier appreciation of the influence of chloride on the solubility of indium(III) hydroxide (at constant $0.1 M$ Cl^- concentration) broken lines are used to plot the relationships for complexes containing chloride. The total solubility of indium(III) hydroxide is indicated by a full line, which has a dashed section in the region where the solubility is influenced by chloride. Concentrations of individual chloro-complexes $InCl_p^{(3-p)}$ are not plotted separately, since the diagram would then not be clear; instead of that a summation line for $\sum_{p=1}^{3} [InCl_p^{(3-p)}]$ is plotted.

Figure 12.8 shows that in the absence of chloride, polynuclear hydroxo-complexes and the free ion In(III) predominate in acidic solution. The solubility of indium(III) hydroxide decreases sharply at pH > 2 to reach a minimum at pH 7.05 (read from the diagram; calculation gives $[OH^-]_{min} = (K_{s201}/K_{s401})^{1/2} = 10^{-6.75}$, i.e. $pH_{min} = 7.03$). The minimum solubility is actually determined by the stability of the complex $In(OH)_3$, and is $4.7 \times 10^{-8} M$. At pH > 9 the solubility increases again.

The solubility of $In_2O_3 \cdot xH_2O$ in pure water is given by the location of point 2 in Fig. 12.8, which corresponds to pH 7.0. As the solubility of indium(III) hydroxide is very low, the pH of the aqueous solution is not noticeably influenced by hydrolysis. This can be seen by setting up a balance for hydroxide ion: the amount of OH^- released cannot significantly influence that resulting from the autoprotolysis. As can be seen from the diagram, the solubility of indium(III) hydroxide in pure water is nearly identical to its minimum solubility.

In presence of chloride ($[Cl^-] = 0.1M$) the solubility of indium(III) hydroxide is somewhat higher. The greatest effect is due to the single-ligand chloro-complexes which begin to have an influence even at pH about 2. The mixed complex $InOHCl^+$ contributes most to the increase in solubility between pH 4 and 6. At higher pH the formation of chloro-complexes has no further influence. Thus between pH 3 and 5 the presence of chloride (at the given $0.1M$ concentration) increases the solubility of $In_2O_3 \cdot xH_2O$ by about one order of magnitude.

The influence of varying the concentration of chloride on the solubility of indium(III) hydroxide at pH 3.0 can be appraised from the family of curves in Fig. 12.9. The concentrations of the ion In(III) and its hydroxo-complexes naturally do not depend on the concentration of chloride: for a given pH they are constant. Free indium(III) ion and the polynuclear hydroxo-complex are the predominating species. At $[Cl^-] = 6 \times 10^{-3} M$ the chloro-complexes begin to predominate; first, it is the complex $InCl^{2+}$, then (with increasing $[Cl^-]$) $InCl_2^+$ and, finally, at $\log[Cl^-] > -0.5$ the complex $InCl_3$ becomes a species of importance. In the case under consideration, mixed complexes are always present, and though their concentration levels are lower than those of the chloro-complexes themselves, their contribution is by no means negligible. When chloride reaches about $1M$ concentration, the solubility of indium(III) hydroxide is increased by 4 orders of magnitude.

12.2.4 Predominance:area diagram for the equilibrium system of indium(III) hydroxo- and chloro-complexes

The diagram in Fig. 12.10 illustrates the effect of varied pH and $\log[Cl^-]$ on the composition of a solution saturated with indium(III) hydroxide, in which complexes of the type $In_m(OH)_n Cl_p^{(3m-n-p)}$ are present. Whenever such a diagram is used or constructed (see Section 4.3.2, p. 84) it is necessary to bear in mind that the boundary lines between the regions of predominance of relevant species correspond to equal concentrations (or activities if considered) of these species. The relationships, already used for construction of logarithmic diagrams for saturated solutions of

352   **Diagrams of Precipitation Equilibria**   [Ch. 12

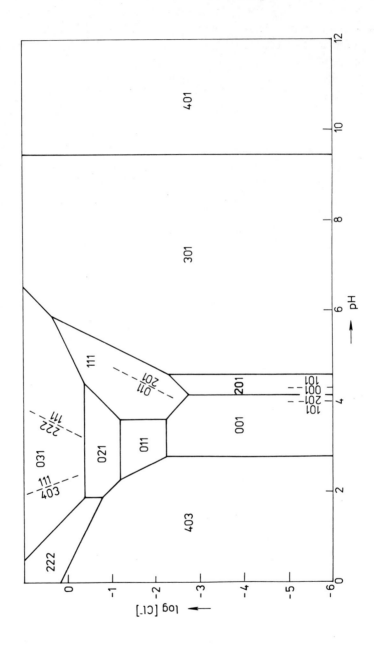

Sec. 12.2]   **Logarithmic Concentration Diagrams for Precipitation Equilibria**   353

indium(III) hydroxide can also be used for the construction of a predominance-area diagram if it is possible to neglect the influence of changes in ionic strength on the values of the equilibrium constants.

The distribution surfaces for 11 equilibrium species present in the saturated solution under consideration intersect in 55 intersection lines. Their equations for the diagram relationships of $\log[Cl^-] = f(pH)$ are obtained by combining the expressions for the relevant species of the system. However, only a fraction of the intersection lines corresponding to the boundaries between the areas of predominance is shown in this diagram. As a guide, the following logarithmic diagrams, already discussed

---

Fig. 12.10 − Predominance-area diagram for various forms of indium(III) in aqueous solution saturated with indium(III) hydroxide and in presence of chloride. Areas of predominant existence are denoted by the coefficients *npm* which characterize the complex species $In_m(OH)_nCl_p$. The broken lines indicate boundaries between non-predominating species. Boundaries and equations for the straight lines (lines not shown in the diagram are in parentheses):

403/001   pH = 2.82;
(001/101   pH = 4.30;
101/201   pH = 4.02; under the surface for the species 201)
001/201   pH = 4.16;
201/301   pH = 4.61;
301/401   pH = 9.45;
001/011 $\log[Cl^-] = -2.23$;
011/021 $\log[Cl^-] = -1.21$;
021/031 $\log[Cl^-] = -0.41$
403/011 $\log[Cl^-] = -2\,pH + 3.40$;
403/021 $\log[Cl^-] = -pH + 1.10$
(403/031 $\log[Cl^-] = -\frac{2}{3}pH + 0.59$, under surface for species 222)
(403/111 $\log[Cl^-] = -3\,pH + 6.98$, under surface for species 031)
403/222 $\log[Cl^-] = -\frac{1}{2}pH + 0.15$;
(222/111 $\log[Cl^-] = 2\,pH - 6.69$, under surface for species 031)
(222/011 $\log[Cl^-] = pH - 3.11$, under surface for species 403)
222/031 $\log[Cl^-] = -pH + 1.49$;
222/021   pH = 1.90;
(011/201 $\log[Cl^-] = 2\,pH - 10.55$;
011/301 $\log[Cl^-] = 3\,pH - 15.16$;
021/301 $\log[Cl^-] = \frac{3}{2}pH - 8.19$, located below surface for species 111)
031/301 $\log[Cl^-] = pH - 5.59$;
001/111 $\log[Cl^-] = -pH + 1.35$;
011/111   pH = 3.58;
021/111 $\log[Cl^-] = pH - 4.79$;
031/111 $\log[Cl^-] = \frac{1}{2}pH - 2.60$;
111/201 $\log[Cl^-] = pH - 6.97$;
111/301 $\log[Cl^-] = 2\,pH - 11.58$.

above, may be used: Fig. 12.8 (which shows, however, the pH-dependence only for a single chloride concentration) and Fig. 12.9 (which is valid only for pH 3). It is thus necessary to find some further boundary lines and to decide on their validity; this can be done only in the course of the diagram construction.

First, the equations for the boundaries of hydroxo-complexes and the free In(III) ion can be taken into consideration. The boundary lines will be denoted with a combination of the stoichiometric indices *npm* which characterize the composition of the relevant complexes. The relationships so obtained are given in the caption to Fig. 12.10. The boundary lines for the hydroxo-complexes are parallel to the axis carrying the $\log[Cl^-]$ scale, as the formation of these complexes is not influenced by concentration of chloride. The boundary line 403/001 appears at the lowest pH, and line 001/201 comes next, as the boundary 001/101 corresponds to a higher pH value than the previous one. The complex $InOH^{2+}$ does not predominate and so its boundary line is not shown in the diagram (cf. the location of the straight line for $InOH^{2+}$ in Fig. 12.8). The boundary lines 201/301 and 301/401 follow the sequence of number of coordinated OH^- ions.

Further simple expressions are written for the boundary lines between the areas for individual chloro-complexes: all these lines run parallel to the pH axis, as their location is not influenced by the pH value. The lines 001/011, 011/021, and 021/031 come in sequence with increasing concentration of chloride.

The polynuclear hydroxo-complex, which predominates in a strongly acidic medium, is converted into chloro-complexes at $[Cl^-] = 6 \times 10^{-3} M$. The relevant boundary lines depend on both pH and $\log[Cl^-]$. The line 403/031 and a part of the line 403/021 are hidden below the predominance area for the complex $In_2(OH)_2Cl_2^{2+}$ which is converted into chloro-complexes at higher pH values. This binuclear complex is transformed into the species $InOHCl^+$ at low concentration levels and so the boundary line 222/111 is covered by the surfaces for the chloro-complexes. However, at pH > 3.6 the contribution of the complex $InOHCl^+$ becomes greater and its distribution surface covers the intersection lines for the surfaces of the chloro- and hydroxo-indates; thus the lines 011/201, 021/301, and also a part of the boundary 031/301 are not shown in the diagram.

The predominance-area diagram for all these species, which are found in equilibrium with solid indium(III) hydroxide, illustrates the fact that in strongly acidic medium the binuclear hydroxo-complex predominates, and is converted into free indium(III) ion, which exists at pH 3–4. Mononuclear hydroxo-complexes are then formed as pH 7 is approached. The complex $In(OH)_3$ has the largest predominance area. It has already been realized from the logarithmic solubility diagram in Fig. 12.8 that this is the region for the lowest solubility of this compound.

The presence of chloride affects the composition of the equilibrium system more significantly in a mildly acidic medium at $[Cl^-] > 2 \times 10^{-3} M$, where the mixed-ligand mononuclear complex predominates. With further increase in chloride concentration the area for predominance of chloro-complexes becomes broader, and at $\log[Cl^-] = 0$ extends over nearly the whole region below pH 6. The formation of chloro-complexes thus contributes to an increase in solubility of indium(III) hydroxide; generally, the hydrolysis of indium(III) is thus suppressed.

It is useful to compare the information on indium compounds presented in this chapter, with the interpretation of the diagrams for the redox equilibria of indium given in Chapter 13 (cf. Figs. 13.6 and 13.7).

REFERENCES

[1] C. F. Baes, Jr. and R. E. Mesmer, *The Hydrolysis of Cations*, p. 111, Wiley, New York 1976.
[2] M. Barrés, *Rev. Chim. Miner.*, 1967, **4**, 803.
[3] F. A. Cotton and G. Wilkinson, *Advanced Inorganic Chemistry*, 2nd Ed., p. 263, Wiley, New York 1966.
[4] See ref. [1], p. 235.
[5] L. Šůcha and S. Kotrlý, *Solution Equilibria in Analytical Chemistry*, p. 237†. Van Nostrand Reinhold, London 1972.
[6] See ref. [1], p. 327.

---

† The expression should be rewritten for a precipitate of the composition MA_3.

CHAPTER 13

Diagrams for oxidation–reduction equilibria

Many oxidation–reduction (redox) reactions should be treated in a general sense as stepwise equilibrium systems if the element considered can exist in several oxidation states. The relevant redox potentials are the principal constants characterizing such redox systems (see Chapter 9), the composition of which can easily be illustrated by logarithmic activity (concentration) ratio diagrams (Figs.13.1–13.5). The redox equilibria are considerably affected by reaction conditions which control the extent of various side-reactions. The important factors are the pH value if protonation equilibria are present, and the concentrations of substances taking part in complexation or precipitation reactions in which some species of the redox system are involved. In such cases it is convenient to introduce the conditional redox potential, $E^{\circ\prime}$, as the main dependent variable of the equilibrium system (cf. p. 53). The dependence of the composition of a redox system on the values of the conditional redox potential and a further variable (pH, log [L]) controlling the equilibrium can be represented by a predominance-area diagram (Figs. 13.6–13.9). The principle of construction and the modes of application of these diagrams have already been presented in Chapter 4, p. 79–86.

13.1 LOGARITHMIC ACTIVITY-RATIO DIAGRAMS OF SOME REDOX SYSTEMS

The relative contributions of individual redox species in a system with varied oxidation–reduction potential can easily be presented as a graph if the effects of side-reactions can be neglected or if all other variables are kept constant. As an example the redox system of iron in aqueous solution will be treated here in some detail.

13.1.1 Logarithmic activity-ratio diagram of redox forms of iron in aqueous solution

Iron can exist in aqueous solution in form of iron(III) and iron(II) ions and can be reduced to the metal. Iron(II) undergoes hydrolysis at pH > 7, whereas iron(III) is already hydrolysed at pH about 1. The hydroxo-complexes formed are mononuclear for iron(II), and more likely polynuclear for iron(III) (at $c_{Fe(III)} > 10^{-5} M$). When the solubility product is reached, iron(II) is precipitated as a metastable hydroxide $Fe(OH)_2$; the stable solid phase existing in equilibrium with both the Fe(II) ion and its hydroxo-complexes is, according to thermodynamic calculations, magnetite, Fe_3O_4. Iron(III) is precipitated as hydrated iron(III) oxide, $Fe_2O_3 \cdot xH_2O$; the actual composition depends on the temperature of the medium from which it is precipitated. After long aging (up to several years) a compound corresponding to the formula FeOOH is obtained. Amorphous iron(III) hydroxide of this composition is somewhat more soluble than its stable crystalline α-FeOOH form.

In a strongly basic medium it is also possible to consider oxidation of iron(III) to yield the ion FeO_4^{2-} with iron as Fe(VI).

13.1.2 Simplified logarithmic activity-ratio diagram for redox forms of iron

This type of diagram can be used to illustrate the equilibria of the redox system Fe(VI)–Fe(III)–Fe(II)–Fe if it is assumed that no hydrolysis takes place. The following data valid for 25°C and zero ionic strength (cf. Table 9.1, p. 229) will be taken into consideration: $E^\circ_{Fe^{3+},Fe^{2+}} = E^\circ_{3,2} = 0.770$ V; $E^\circ_{Fe^{2+},Fe} = E^\circ_{2,0} = -0.473$ V; $E^{\circ\prime}_{FeO_4^{2-},FeOOH(s)} = E^{\circ\prime}_{6,3} = 0.71$ V (for $a_{OH^-} = 1$).

If the activity of the Fe(II) ion is taken as unity as a basis for construction of the diagram, the graphical representation of the logarithms of the activity ratios for individual components of the system is obtained in the form of straight lines; their equations can be obtained from the Nernst–Peters expressions written for the following reaction schemes (see Section 4.2.3, p. 77):

(1) $Fe^{2+} + 2e \rightleftharpoons Fe_{(s)}$

$pe = pe^\circ_{2,0} + \frac{1}{2}\log(a_{Fe^{2+}}/a_{Fe}) = -8.02 + \frac{1}{2}\log(a_2/a_0)$

and hence

line Fe: $\log(a_0/a_2) = -2pe - 16.04$.

(2) Taking unit activity for the ion Fe(II) gives

line Fe^{2+}: $\log(a_2/a_2) = 0$.

(3) $Fe^{3+} + e \rightleftharpoons Fe^{2+}$

$pe = pe^{\circ}_{3,2} + \log(a_{Fe^{3+}}/a_{Fe^{2+}}) = 13.05 + \log(a_3/a_2)$

line Fe^{3+}: $\log(a_3/a_2) = pe - 13.05$.

(4) $FeO_4^{2-} + 4e + 8H^+ \rightleftharpoons Fe^{2+} + 4H_2O$; $K_{6,2}$

$pe = pe^{\circ\prime}_{6,2} + \frac{1}{4}\log(a_{FeO_4^{2-}}/a_{Fe^{2+}}) + \frac{1}{4}\log a_{H^+}^8$.

The value of $pe^{\circ\prime}_{6,2}$ has to be calculated from given data. The reaction scheme corresponding to the equilibrium constant $K_{6,2}$ can be obtained by summing the following equilibria:

$FeO_4^{2-} + 3H_2O + 3e \rightleftharpoons FeOOH_{(s)} + 5OH^-$; K_A

$FeOOH_{(s)} + H_2O \rightleftharpoons Fe^{3+} + 3OH^-$; K_B

$Fe^{3+} + e \rightleftharpoons Fe^{2+}$; K_C

$8H^+ + 8OH^- \rightleftharpoons 8H_2O$. K_D.

The equilibrium constant $K_{6,2}$ for the redox equilibrium (4) is thus given by the product of the constants for all these partial reactions

$K_{6,2} = K_A K_B K_C K_D$

where the constant $K_B = K_{s(FeOOH)} = 10^{-41.5}$ represents the solubility product of iron(III) hydroxide (in the form of α-FeOOH) and $K_D = K_w^{-8} = 10^{112.0}$. The constants K_A and K_C are calculated from the values of the conditional redox potential $E^{\circ\prime}_{6,3}$ and the standard redox potential $E^{\circ}_{3,2}$, respectively [cf. Eq. (2.59), p. 37]:

$\log K_A = 3E^{\circ\prime}_{6,3}/0.059 = 36.10$; $\quad \log K_C = E^{\circ}_{3,2}/0.059 = 13.05$.

Hence, on insertion of the numerical values, $\log K_{6,2} = 119.65$, and the redox potential for oxidation of Fe(II) to FeO_4^{2-} is obtained: $E^{\circ\prime}_{6,2} = (0.059 \log K_{6,2})/4 = 1.76 \sim 1.8$ V.

The relative activity of FeO_4^{2-} is represented in the diagram by a straight line, but its position depends on the pH of the solution. For $a_{H^+} = 1$ the following equation holds:

$pe = 29.9 + \frac{1}{4}\log(a_6/a_2)$

and hence we have

line (dashed) FeO_4^{2-}: $\log(a_6/a_2) = 4pe - 119$.

Sec. 13.1]   Logarithmic Activity-Ratio Diagrams of Some Redox Systems

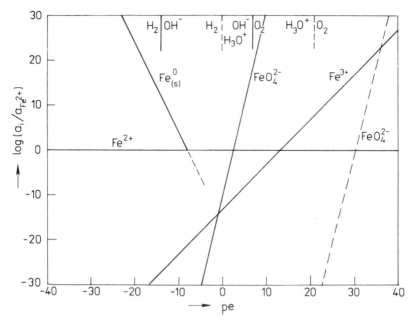

Fig. 13.1 — Logarithmic diagram of the activity ratios in the system Fe–Fe^{2+}–Fe^{3+}–FeO_4^{2-}, related to $a_{Fe^{2+}}$ (temperature 25°C, pH = 0; broken line for FeO_4^{2-} applies to pH = 14). The values of the (conditional) standard potentials of the redox pairs OH^-, H_2 and O_2, OH^- at pH = 14, and redox pairs H_3O^+, H_2 and O_2, H_3O^+ at pH = 1, are included in the diagram.

Analogously at pH 14 ($a_{H^+} = 10^{-14}$) it holds that

$$pe = 29.9 + \tfrac{1}{4}\log(a_6/a_2) - 28$$

and hence

line (full) FeO_4^{2-}: $\log(a_6/a_2) = 4pe - 7.6$.

The diagram, which is shown in Fig. 13.1, is then completed by plotting the values of $pe^{o'}_{OH^-,H_2}$, $pe^{o'}_{H_3O^+,H_2}$, $pe^{o'}_{O_2,OH^-}$, and $pe^{o'}_{O_2,H_3O^+}$, which correspond to reduction and oxidation of water at pH 0 or 14, respectively.

13.1.3 Logarithmic activity-ratio diagram for the redox system of iron in presence of hydrolysis products

The composition of the redox system Fe(VI)–Fe(III)–Fe(II)–Fe is considerably influenced by pH when hydrolysis products of both Fe(III) and Fe(II) are present in aqueous solution; the diagram is therefore constructed for certain chosen pH values, e.g. pH 0, 7, and 14. In analogy to the diagram in Fig. 13.1, the logarithms of the activity ratios of individual redox species are taken with respect to unit activity of the iron(II) ion.

In addition to the redox potentials given above the following data (valid for $25°C$ and $I = 0$) will be used:

conditional redox potential for the redox pair $Fe_3O_{4(s)}$, Fe(II) (obtained by calculation similar to that given above in Section 13.1.2): $E^{o'}_{Fe_3O_{4(s)}, Fe^{2+}} = 1.230$ V (for $a_{H^+} = 1$);

solubility products of iron(II) and iron(III) hydroxides:

$\log K_{s \, Fe(OH)_2} = -15.15,$

$\log K_{s(FeOOH)} = -41.5$ (identical with the value of $\log K_B$ used in Section 13.1.2);

stability constants for hydroxo-complexes of iron(II):

$\log \beta_1 = 4.5, \quad \log \beta_2 = 7.4, \quad \log \beta_3 = 11, \quad \log \beta_4 = 10;$

stability constants for hydroxo-complexes of iron(III):

$\log \beta_1 = 11.81, \quad \log \beta_2 = 22.33, \quad \log \beta_3 \sim 30, \quad \log \beta_4 = 34.4,$
$\log \beta_{22} = 25.05, \quad \log \beta_{43} = 49.7.$

From these data it is necessary to evaluate first the values of the conditional redox potentials for the redox pairs corresponding to conversion of particular forms of iron into the ion Fe(II), the activity of which is chosen as the reference value for the diagram.

The straight lines corresponding to the ratio of the activity of Fe, Fe(II) and Fe(III) to that of Fe(II) have the same locations (independent of pH) as the analogous lines in Fig. 13.1.

The activity ratio for the anion FeO_4^{2-} is represented by straight lines (denoted by the formula of the ion) which are dependent on the pH values chosen. The equations for pH 0 and 14 are given above; for pH 7 the following equation is analogously obtained:

line FeO_4^{2-}: $\log(a_6/a_2) = 4pe - 64.$

The equations for further straight lines can similarly be derived as indicated above for the case of the species FeO_4^{2-}. The sequence of redox reactions of Section 13.1.2 will be continued here in order to stress the relationship to the diagram in Fig. 13.1 and to avoid some misunderstanding.

(5) $Fe_3O_{4(s)} + 2e + 8H^+ \rightleftharpoons 3Fe^{2+} + 4H_2O$

$pe = pe^{o'}_{3,2,8H^+} + \frac{1}{2}\log(a_{Fe_3O_4}/a_{Fe^{2+}}) + \frac{1}{2}\log a^8_{H^+} - \frac{1}{2}\log a^2_{Fe^{2+}}$
$= 20.85 + \frac{1}{2}\log(a_{Fe_3O_4}/a_{Fe^{2+}}) + 4\log a_{H^+} - \log a_{Fe^{2+}}.$

Sec. 13.1] Logarithmic Activity-Ratio Diagrams of Some Redox Systems

The activity ratios for the oxide Fe_3O_4 and also for polynuclear complexes are dependent on the concentrations of the relevant free iron ions in solution; therefore, it is necessary to choose a certain concentration of the free ion Fe(II) (the activity of this form represents a reference value for the diagram) and treat it as a constant for the diagram. The diagrams for different pH values will thus be constructed for a concentration corresponding to $a_{Fe^{2+}} = 1M$ (Figs. 13.2–13.4) and the influence of a change in the concentration of Fe(II) will be shown by a diagram constructed for pH 7 and a concentration of iron(II) reduced so that $a_{Fe^{2+}} = 10^{-3} M$ (Fig. 13.5).

For the redox equilibrium (5) the following pH relationships are obtained for unit activity of Fe(II):

lines $Fe_3O_{4(s)}$: $\log(a_{Fe_3O_4}/a_{Fe^{2+}}) =$

at pH 0, $2pe - 41.70$
at pH 7, $2pe + 14.30$
at pH 14, $2pe + 70.30$
at pH 7 $2pe + 8.30$
and $a_{Fe^{2+}} = 10^{-3} M$.

(6) $Fe(OH)_{2(s)} \rightleftharpoons Fe^{2+} + 2 OH^-$

The precipitation equilibrium, controlled by the solubility product, does not depend on the redox potential. By convention, $a_{Fe(OH)_{2(s)}} = 1$ and hence $a_{Fe^{2+}} = K_s/a_{OH}^2$: thus the activity ratios of the precipitate $Fe(OH)_2$ and the ion Fe(II) are given by the following lines parallel to the abscissa:

lines $Fe(OH)_{2(s)}$: $\log(a_{Fe(OH)_{2(s)}}/a_{Fe^{2+}}) =$

at pH 0, -12.85
at pH 7, 1.15
at pH 14, 15.15

The other complex-formation equilibria of iron(II) do not depend on the redox potential either, but they are influenced by the pH and the stability constants of iron(II) hydroxo-complexes.

(7) $FeOH^+ \rightleftharpoons Fe^{2+} + OH^-$

The expression for $\beta_{Fe^{II}OH}$ can be rewritten as

$\log(a_{FeOH^+}/a_{Fe^{2+}}) = \log \beta_{Fe^{II}OH} + \log a_{OH^-}$

and hence the straight line equations are obtained as

lines $FeOH^+$: $\log a_{FeOH^+}/a_{Fe^{2+}} =$
at pH 0, -9.5
at pH 7, -2.5
at pH 14, 4.5.

The other three equilibria of the iron(II) hydroxo-complexes are treated in a similar way.

(8) $Fe(OH)_2 \rightleftharpoons Fe^{2+} + 2\,OH^-$

$\log(a_{Fe(OH)_2}/a_{Fe^{2+}}) = \log \beta_{Fe^{II}(OH)_2} + 2\log a_{OH^-}$

and hence

lines $Fe(OH)_2$: $\log(a_{Fe(OH)_2}/a_{Fe^{2+}}) =$
at pH 0, -20.6
at pH 7, -6.6
at pH 14, 7.4.

(9) $Fe(OH)_3^- \rightleftharpoons Fe^{2+} + 3\,OH^-$

$\log(a_{Fe(OH)_3^-}/a_{Fe^{2+}}) = \log \beta_{Fe^{II}(OH)_3} + 3\log a_{OH^-}$

so the pH relationships for the species $Fe(OH)_3^-$ are

lines $Fe(OH)_3^-$: $\log(a_{Fe(OH)_3^-}/a_{Fe^{2+}}) =$
at pH 0, -31†
at pH 7, -10
at pH 14, 11.

(10) $Fe(OH)_4^{2-} \rightleftharpoons Fe^{2+} + 4\,OH^-$

$\log(a_{Fe(OH)_4^{2-}}/a_{Fe^{2+}}) = \log \beta_{Fe^{II}(OH)_4} + 4\log a_{OH^-}$

so the straight lines for the species $Fe(OH)_4^{2-}$ are

lines $Fe(OH)_4^{2-}$: $\log(a_{Fe(OH)_4^{2-}}/a_{Fe^{2+}}) =$
at pH 0, -46†
at pH 7, -18
at pH 14, 10.

---

† Not plotted in Fig. 13.2.

Sec. 13.1]   Logarithmic Activity-Ratio Diagrams of Some Redox Systems

The activity ratios of various forms of iron(III) with respect to the activity of iron(II) depend not only on the pH but also on the redox potential; for example,

(11) $FeOOH_{(s)} + e + 3H^+ \rightleftharpoons Fe^{2+} + 2H_2O$; $K_{s,3,2}$

$$pe = pe^{\circ\prime}_{s,3,2} + \log(a_{FeOOH}/a_{Fe^{2+}}) + \log a_{H^+}^3.$$

The value of $pe^{\circ\prime}_{s,3,2}$ is evaluated as in case (4) from the equilibrium constant $K_{s,3,2}$, which represents the product of the constants for the three reactions involved in the redox equilibrium (11), namely those for the dissociation of FeOOH, the reduction of Fe^{3+} to Fe^{2+}, and the dissociation of water:

$$K_{s,3,2} = K_{s(FeOOH)} K_{3,2}/K_w^3.$$

On inserting numerical values, $pe^{\circ\prime}_{s,3,2} = \log K_{s,3,2} = 13.55$, and the activity-ratio expression is

$$\log(a_{FeOOH}/a_{Fe^{2+}}) = pe - 13.55 - 3\log a_{H^+}$$

to give the straight-line relationships for $FeOOH_{(s)}$ in Figs. 13.2–13.5

lines $FeOOH_{(s)}$: $\log(a_{FeOOH}/a_{Fe^{2+}}) =$
at pH 0, $pe - 13.55$
at pH 7, $pe + 7.45$
at pH 14, $pe + 28.45.$

A similar approach is used to derive the values and expressions necessary to obtain the relative activity of the other mononuclear hydroxo-complexes of iron(III).

(12) $FeOH^{2+} + e + H^+ \rightleftharpoons Fe^{2+} + H_2O$; $K_{3(11),2}$

$$pe = pe^{\circ\prime}_{3(11),2} + \log(a_{FeOH^{2+}}/a_{Fe^{2+}}) + \log a_{H^+}$$

$$K_{3(11),2} = K_{3,2}/K_w \beta_{Fe^{III}OH}$$

$$\log K_{3(11),2} = pe^{\circ\prime}_{3(11),2} = 15.24$$

so the following equation holds

$$\log(a_{FeOH^{2+}}/a_{Fe^{2+}}) = pe - 15.24 - \log a_{H^+}$$

and for the three pH values considered the following expressions are obtained:

lines $FeOH^{2+}$: $\log(a_{FeOH^{2+}}/a_{Fe^{2+}}) =$

at pH 0, $pe - 15.24$
at pH 7, $pe - 8.24$
at pH 14, $pe - 1.24$.

(13) $Fe(OH)_2^+ + e + 2H^+ \rightleftharpoons Fe^{2+} + 2H_2O; \quad K_{3(21),2}$

$$pe = pe^{\circ\prime}_{3(21),2} + \log(a_{Fe(OH)_2^+}/a_{Fe^{2+}}) + \log a_{H^+}^2$$

$$K_{3(21),2} = K_{3,2}/K_w^2 \beta_{Fe^{III}(OH)_2}$$

$$\log K_{3(21),2} = pe^{\circ\prime}_{3(21),2} = 18.72.$$

Hence the activity ratio obtained is:

$$\log(a_{Fe(OH)_2^+}/a_{Fe^{2+}}) = pe - 18.72 - 2\log a_{H^+}$$

and the three straight-line equations are

lines $Fe(OH)_2^+$: $\log(a_{Fe(OH)_2^+}/a_{Fe^{2+}}) =$

at pH 0, $pe - 18.72$
at pH 7, $pe - 4.72$
at pH 14, $pe + 9.28$.

(14) $Fe(OH)_3 + e + 3H^+ \rightleftharpoons Fe^{2+} + 3H_2O; \quad K_{3(31),2}$

$$pe = pe^{\circ\prime}_{3(31),2} + \log(a_{Fe(OH)_3}/a_{Fe^{2+}}) + \log a_{H^+}^3$$

$$K_{3(31),2} = K_{3,2}/K_w^3 \beta_{Fe(OH)_3}$$

$$\log K_{3(31),2} = pe^{\circ\prime}_{3(31),2} = 25.$$

The activity ratio is thus given by the expression

$$\log(a_{Fe(OH)_3}/a_{Fe^{2+}}) = pe - 25 - 3\log a_{H^+},$$

which gives the straight-line relationships for $Fe(OH)_3$ at the pH values considered:

lines $Fe(OH)_3$: $\log(a_{Fe(OH)_3}/a_{Fe^{2+}}) =$

at pH 0, $pe - 25$
at pH 7, $pe - 4$
at pH 14, $pe + 17$.

(15) $Fe(OH)_4^- + e + 4H^+ \rightleftharpoons Fe^{2+} + 4H_2O; \quad K_{3(14),2}$

$$pe = pe^{\circ\prime}_{3(41),2} + \log(a_{Fe(OH)_4^-}/a_{Fe^{2+}}) + \log a_{H^+}^4$$

Sec. 13.1]   Logarithmic Activity-Ratio Diagrams of Some Redox Systems   365

$$K_{3(41),2} = K_{3,2}/K_w^4 \beta_{Fe^{III}(OH)_4}$$

$$\log K_{3(41),2} = pe^{\circ\prime}_{3(41),2} = 34.65.$$

The activity ratio is thus given by

$$\log(a_{Fe(OH)_4^-}/a_{Fe^{2+}}) = pe - 34.65 - 4 \log a_{H^+}$$

from which the relationships for $Fe(OH)_4^-$ are obtained as

lines $Fe(OH)_4^-$: $\log(a_{Fe(OH)_4^-}/a_{Fe^{2+}}) =$
at pH 0, $pe - 34.65$
at pH 7, $pe - 6.65$
at pH 14, $pe - 21.35.$

The redox equilibria of polynuclear complexes of iron(III) are represented by the reaction schemes (16) and (17).

(16) $Fe_2(OH)_2^{4+} + 2e + 2H^+ \rightleftharpoons 2 Fe^{2+} + 2 H_2O;$ $K_{3(22),2}$

$$pe = pe^{\circ\prime}_{3(22),2} + \tfrac{1}{2}\log(a_{Fe_2(OH)_2^{4+}}/a_{Fe^{2+}}) + \tfrac{1}{2}\log a_{H^+}^2 - \tfrac{1}{2}\log a_{Fe^{2+}}$$

$$K_{3(22),2} = K_{3,2}^2/K_w^2 \beta_{Fe_2^{III}(OH)_2}$$

$$\log K_{3(22),2} = 2pe^{\circ\prime}_{3(22),2} = 29.05.$$

Hence, the activity ratio is

$$\log(a_{Fe_2(OH)_2^{4+}}/a_{Fe^{2+}}) = 2pe - 29.05 - 2\log a_{H^+} + \log a_{Fe^{2+}}$$

and the straight-line relationships for the species $Fe_2(OH)_2^{4+}$ written for the condition that $a_{Fe^{2+}} = 1M$ are

lines $Fe_2(OH)_2^{4+}$: $\log(a_{Fe_2(OH)_2^{4+}}/a_{Fe^{2+}}) =$
at pH 0, $2pe - 29.05$
at pH 7, $2pe - 15.05$
at pH 14, $2pe - 1.05$

If $a_{Fe^{2+}} = 10^{-3}M$ and pH = 7, the following equation is obtained:

$$\log(a_{Fe_2(OH)_2^{4+}}/a_{Fe^{2+}}) = 2pe - 18.05.$$

(17) $Fe_3(OH)_4^{5+} + 3e + 4H^+ \rightleftharpoons 3 Fe^{2+} + 4 H_2O;$ $K_{3(43),2}$

$$pe = pe^{\circ\prime}_{3(43),2} + \tfrac{1}{3}\log(a_{Fe_3(OH)_4^{5+}}/a_{Fe^{2+}}) + \tfrac{1}{3}\log a_{H^+}^4 - \tfrac{1}{3}\log a_{Fe^{2+}}^2$$

$$K_{3(43),2} = K_{3,2}^3/K_w^4 \beta_{Fe_3(OH)_4},$$

$$\log K_{3(43),2} = 3pe^{\circ\prime}_{3(43),2} = 45.45.$$

Hence the activity ratio is

$$\log(a_{Fe_3(OH)_4^{5+}}/a_{Fe^{2+}}) = 3pe - 45.45 - 4\log a_{H^+} - 2\log a_{Fe^{2+}}$$

and the straight-line relationships for $a_{Fe^{2+}} = 1$ and the pH values being considered are

lines $Fe_3(OH)_4^{5+}$: $\log(a_{Fe_3(OH)_4^{5+}}/a_{Fe^{2+}}) =$

at pH 0, $3pe - 45.45$
at pH 7, $3pe - 17.45$
at pH 14, $3pe + 10.55$.

If $a_{Fe^{2+}} = 10^{-3} M$ and pH $= 7$, the following equation results:

$$\log(a_{Fe_3(OH)_4^{5+}}/a_{Fe^{2+}}) = 3pe - 23.45.$$

The logarithmic activity-ratio diagrams for the redox system Fe(VI)–Fe(III)–Fe(II)–Fe in aqueous medium and the conditions being considered are illustrated in Figs. 13.2–13.5. These diagrams make easier the comprehension of the behaviour of iron and its compounds in aqueous medium, as a function of both the redox potential and pH value; they

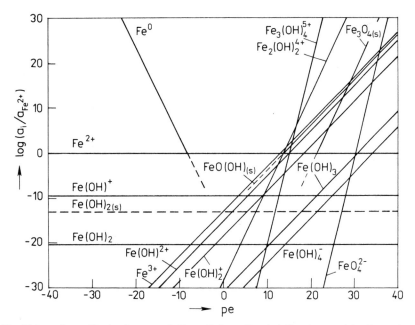

Fig. 13.2 — Logarithmic diagram of the activity ratios (relative to $a_{Fe^{2+}}$) in the system Fe–Fe(II)–Fe(III)–Fe(VI) with formation of hydroxo-complexes and solid compounds taken into account (temperature $25°C$, pH $= 0$, $a_{Fe^{2+}} = 1M$).

Sec. 13.1] Logarithmic Activity-Ratio Diagrams of Some Redox Systems 367

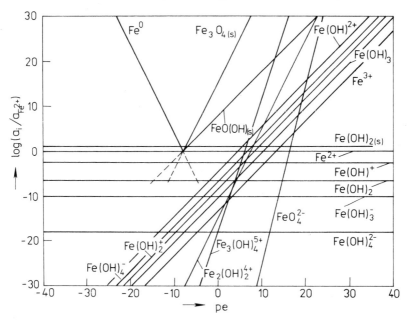

Fig. 13.3 — Logarithmic diagram of the activity ratios as in Fig. 13.2 (pH = 7).

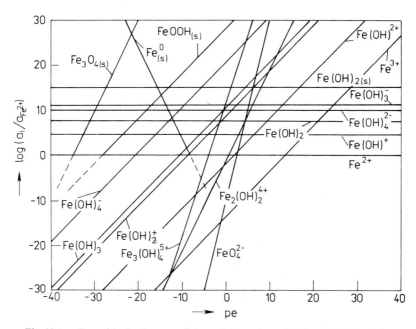

Fig. 13.4 — Logarithmic diagram of the activity ratios as in Fig. 13.2 (pH = 14).

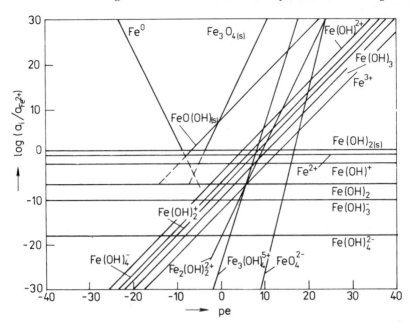

Fig. 13.5 — Logarithmic diagram of the activity ratios as in Fig. 13.2 (pH = 7, $a_{Fe^{2+}} = 1 \times 10^{-3} M$).

also give information on the influence of the total concentration of iron on the composition of the solution.

The simple diagram in Fig. 13.1 provides illustration of the known fact that in strongly acidic aqueous solution (pH 0) elemental iron is first oxidized to Fe(II) and then to Fe(III). In the absence of oxidizing agents other than protons the dissolution of metallic iron is accompanied by evolution of hydrogen. The oxidation to iron(VI) would necessitate a redox potential higher than that corresponding to oxidation of water to free oxygen. Thus under the given conditions such an oxidation process is not possible; the effect of oxygen overpotential (about 0.5 V, cf. ref. [1]) will shift the line H_3O^+/O_2 approximately 8.5pe units higher with respect to the $pe^{o'}_{O_2, H_3O^+}$ value in Fig. 13.1, but even at pe = 30 it would still result that $a_{FeO_4^{2-}} \ll a_{Fe^{2+}}$.

In contrast, in a strongly basic medium elemental iron is in equilibrium with a very small concentration of Fe(III) ion at the $pe^{o'}_{OH^-, H_2}$ value corresponding to pH 14; however, this holds only for the simplifying assumption that the hydrolysis products of iron(II) can be neglected. The diagram in Fig. 13.1 indicates that by increasing the redox potential it is possible to oxidize iron(II) directly to FeO_4^{2-} without formation of an appreciable amount of Fe(III).

The broken part of the line for Fe^0 corresponds either to pure iron ($a_{Fe(s)} = 1$) in equilibrium with species in solution, for which $a_i > 1$, or to an alloy (or generally to a solid mixture with other compounds where $a_{Fe(s)} < 1$). The activity ratio for a solid is plotted in the same way in the other diagrams for similar situations.

If the logarithmic activity-ratio diagram for iron involves all the forms which can exist in aqueous solution at varied pH and redox potential, only the region of predominant existence of Fe(III) is changed with respect to the simplified case discussed above, for a strongly acidic medium and high concentrations of solutes (pH = 0, $a_{Fe^{2+}} = 1M$, cf. Fig. 13.2). The ion Fe(III) predominates in aqueous solution up to an activity corresponding to a concentration of Fe(II) reaching about $10M$, as indicated by the intersection points of the relevant lines in Fig. 13.2. At higher concentrations polynuclear complexes would already predominate: first the complex $Fe_2(OH)_2^{4+}$ and then $Fe_3(OH)_4^{5+}$. The solubility of the hydroxide α-FeOOH corresponds to about $a_{Fe^{3+}} = 3M$, which is not a high value in view of the low pH. However, it is necessary to take into account that the solubility of freshly precipitated iron(III) hydroxide is much higher than that of its most stable, aged alpha, modification.

The composition of the equilibrium system of iron undergoes a significant change if the pH of the aqueous solution is increased to 7 ($a_{Fe^{2+}} = 1M$), as shown in Fig. 13.3. Metallic iron is oxidized under these conditions to form iron(II) hydroxide which is in equilibrium with an appreciable amount of Fe(II) and hydroxo-complexes of the type $Fe(OH)_n^{(2-n)+}$; the highest concentration is reached by the complex $Fe(OH)^+$. The activity ratios of the iron(II) compounds are not influenced by the total concentration of iron in the system, as can be seen from comparison of Figs. 13.3 and 13.5.

Oxidation of iron(II) hydroxide yields other solid compounds, namely $Fe_3O_{4(s)}$ and $FeO(OH)_{(s)}$. For a high concentration level, i.e. if $a_{Fe(II)} = 10M$, the compound $Fe_3O_{4(s)}$ represents, according to the diagram, the most stable form of iron when the redox potential is increased above $pe = -6.5$ (i.e. above approximately -0.4 V). Iron(III) hydroxide may be formed by oxidation of $Fe(OH)_{2(s)}$ only in a more dilute solution, as indicated by Fig. 13.5 which is constructed for $a_{Fe^{2+}} = 10^{-3}M$. It seems that the tabulated value of the redox potential for the pair $Fe_3O_{4(s)}$, Fe(II) [cf. reaction (5)] is rather low; thus magnetite may be formed by the oxidation of $Fe(OH)_{2(s)}$ apparently more easily than the hydroxide

FeOOH$_{(s)}$. The activity of the polynuclear iron(III) hydroxo-complexes is also influenced by the concentration level. Among the mononuclear hydroxo-complexes the complex species Fe(OH)$^{2+}$ predominates at pH 7.

At pH 14 and a high concentration level (see Fig. 13.4) elemental iron should be oxidized mainly to Fe$_3$O$_{4(s)}$. At a lower concentration the hydroxides Fe(OH)$_{2(s)}$ and then FeOOH$_{(s)}$ are formed from metallic iron. Hydroxo-complexes are formed in equilibrium with these solid phases; the iron(II) species Fe(OH)$_3^-$ and Fe(OH)$_4^{2-}$ and the iron(III) species Fe(OH)$_4^-$ and Fe(OH)$_3$ predominate in strongly basic medium.

13.2 PREDOMINANCE-AREA DIAGRAMS

A redox system influenced by another chemical process, for example, by complexation or precipitation equilibria, can only be represented with the aid of a three-dimensional graph. A predominance-area diagram is a useful simplification: at the cost of some loss of information the equilibrium system controlled by two variables is represented as a plane diagram. Examples of such diagrams for oxidation–reduction equilibria will be given in the next sections.

13.2.1 Diagram of redox, complexation and precipitation equilibria of indium in aqueous solution, showing the dependence on pH

Indium(III) can easily hydrolyse in aqueous solution to form mononuclear hydroxo-complexes In(OH)$_n^{(3-n)+}$ (where n is 1–4) together with a polynuclear species which most probably has the composition In$_3$(OH)$_4^{5+}$ (cf. ref. [2]). If an indium concentration above the value given by the solubility product is reached, a precipitate of In(OH)$_3$ (or In$_2$O$_3 \cdot x$H$_2$O) is formed. Indium(III) and its hydroxo-complexes are reduced in aqueous solution to the metal. Indium(I) is unstable in aqueous medium and disproportionates to In(III) and the metal.

In the predominance-area diagram represented in Fig. 13.6 the areas corresponding to particular forms of indium are separated by line segments which indicate equal activities of interacting species. For redox reactions these straight lines represent the dependence of the relevant conditional redox potential on pH. If one of the reacting substances is present in the solid state, the line separating the areas corresponds to (hypothetical) unit activity of the species in solution, which then further increases within that area. Therefore no areas can be shown in the diagram for solutes which have activities lower than 1. It is thus useful to construct for such systems another diagram representing the composition of a saturated solution but without showing the presence of a solid phase [which is indium(III) hydroxide in this case]. In such a diagram the regions of existence will also be shown for the species which predominate in saturated solution, but at activities lower than 1.

The predominance-area diagrams for indium, given here as examples for treatment of similar oxidation–reduction systems, will be discussed for both situations. The diagram showing regions of predominant existence of individual forms of indium in presence of a precipitate of indium(III) hydroxide, as a function of the redox potential and the pH, is illustrated in Fig. 13.6. The predominance-area diagram for the absence of solid indium(III) hydroxide is shown in Fig. 13.7. The equilibria of indium will be characterized by constants which are valid for a $3M$ $NaClO_4$ solution and $25°C$ (for the data see the tables in Chapters 5 and 9). At high ionic strength of the medium the activity coefficients of all the ions present can be considered as constant, over a large range of concentrations. The regions of predominant existence are separated by straight lines which correspond to the following equilibria and relevant equations.

(1) $In^{3+} + 3e \rightleftharpoons In_{(s)}$; $E_1^{\circ\prime} = -0.343$ V

$$E = -0.343 + \frac{0.059}{3} \log [In^{3+}]$$

so the boundary line (broken in Fig. 13.6 and full in Fig. 13.7) for $In^{3+}/In_{(s)}$, where $[In^{3+}] = 1M$, is written as

line 1: $E = -0.343$ V.

(2) $In^{3+} + 2e \rightleftharpoons In^+$; $E_2^{\circ\prime} = -0.425$ V

and the boundary In^{3+}/In^+, where the activities of the two ions are equal is shown by (broken line in Fig. 13.6)

line 2: $E = -0.425$ V.

(3) $In^+ + e \rightleftharpoons In_{(s)}$; $E_3^{\circ\prime} = 3E_1^{\circ\prime} - 2E_2^{\circ\prime} = -0.179$ V

and the boundary $In^+/In_{(s)}$ (broken line in Fig. 13.6) is

line 3: $E = -0.179$ V.

(4) $In(OH)_{3(s)} \rightleftharpoons In^{3+} + 3 OH^-$; $\log K_s = -36.22$
$\log K_w = -14.18$

hence the boundary $In^{3+}/In(OH)_{3(s)}$ for $[In^{3+}] = 1M$ is represented by

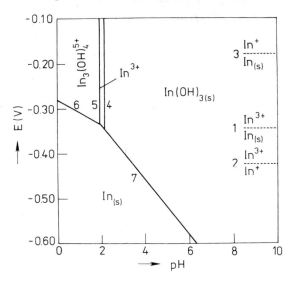

Fig. 13.6 — Predominance-area diagram for the system In–In(I)–In(III) (temperature 25°C, $I = 3.0$).

(full line in Fig. 13.6),

line 4: $\mathrm{pH} = 14.2 + \frac{1}{3}\log K_s = 2.1$.

Whether to represent hydroxo-complexes of indium(III) in the diagram (Fig. 13.6) when solid indium(III) hydroxide is present, can be deduced from the relevant pH values at which particular hydroxo-complexes attain unit concentration (cf. Section 4.3.2, p. 84). It is useful to summarize the values of the constants, the equations for calculation of the concentrations, and the resulting pH values corresponding to the concentration of $1M$, as follows:

$\log K_s = -36.22;$ $\log[\mathrm{In}^{3+}] = \log K_s - 3\log K_w - 3\,\mathrm{pH}$
$\phantom{\log K_s = -36.22;\quad \log[\mathrm{In}^{3+}]} = 6.32 - 3\,\mathrm{pH};$
$\phantom{\log K_s = -36.22;\quad \log[\mathrm{In}^{3+}]\;} \mathrm{pH} = 2.11;$

$\log K_{s1} = -26.44;$ $\log[\mathrm{InOH}^{2+}] = \log K_{s1} - 2\log K_w - 2\,\mathrm{pH}$
$\phantom{\log K_{s1} = -26.44;\quad \log[\mathrm{InOH}^{2+}]} = 1.92 - 2\,\mathrm{pH};$
$\phantom{\log K_{s1} = -26.44;\quad \log[\mathrm{InOH}^{2+}]\;} \mathrm{pH} = 0.96;$

$\log K_{s2} = -16.54;$ $\log[\mathrm{In(OH)}_2^+] = \log K_{s2} - \log K_w - \mathrm{pH}$
$\phantom{\log K_{s2} = -16.54;\quad \log[\mathrm{In(OH)}_2^+]} = -2.36 - \mathrm{pH};$
$\phantom{\log K_{s2} = -16.54;\quad \log[\mathrm{In(OH)}_2^+]\;} \mathrm{pH} = -2.36;$

$\log K_{s3} = -7.33;$ $\log [\text{In(OH)}_3] = \log K_{s3} = -7.33;$
                                        independent of pH

$\log K_{s4} = -2.78;$ $\log [\text{In(OH)}_4^-] = \log K_{s4} + \log K_w + \text{pH}$
                                        $= -16.96 + \text{pH};$
                                        $\text{pH} = 16.96;$

$\log K_{s43} = -61.23;$ $\log [\text{In}_3(\text{OH})_4^{5+}] = \log K_{s43} - 5 \log K_w - 5\,\text{pH}$
                                        $= 9.67 - 5\,\text{pH};$
                                        $\text{pH} = 1.93.$

It can be seen that sparingly soluble indium(III) hydroxide predominates at pH > 2.11, when the concentration of In(III) reaches the value of $1M$; in a more acidic solution the ion In(III) covers the area for solid hydroxide. In agreement with the calculation given on p. 372, this boundary is represented by line 4. At pH = 1.93 the ternuclear complex reaches unit concentration and thus the boundary line $\text{In}^{3+}/\text{In}_3(\text{OH})_4^{5+}$, which corresponds to equal concentration of indium(III) in the two species, is obtained on consideration of the above-given expressions, as

$$[\text{In}^{3+}] = 3[\text{In}_3(\text{OH})_4^{5+}]; \quad 6.32 - 3\,\text{pH} = 9.67 - 5\,\text{pH} + 0.48$$

and hence

line 5: pH = 1.91.

At this pH indium(III) begins to predominate in the form of the polynuclear complex in the system. The presence of the other hydroxo-complexes is not shown at all in the diagram in Fig. 13.6, though they exist in the solution in equilibrium with solid indium(III) hydroxide. However, at pH > 2.11, only solid phases predominate in the equilibrium.

The boundary $\text{In}^{3+}/\text{In}_{(s)}$ should lie between pH 2.11 and 1.91 as indicated in Fig. 13.6: the equilibrium limiting concentration of indium(III) is increased rapidly from $1M$ at pH 2.11 to about $4M$ at pH 1.91. This can easily be realized by calculation from the solubility product.

The remaining two boundaries correspond to the equilibria (6) and (7).

(6) $\text{In}_3(\text{OH})_4^{5+} + 9e + 4\,\text{H}^+ \rightleftharpoons 3\,\text{In}_{(s)} + 4\,\text{H}_2\text{O}$

$$E_6^{\circ\prime} = E_1^{\circ\prime} - \frac{0.059}{9} \log \beta_{43} K_w^4 = -0.282\ \text{V}$$

$$E = E_6^{\circ\prime} - \frac{4}{9} 0.059\ \text{pH} + \frac{0.059}{9} \log \frac{1}{3}.$$

hence the boundary $\text{In}_3(\text{OH})_4^{5+}/\text{In}_{(s)}$ for $0.333M$ concentration of the polynuclear complex is defined as

line 6: $E = -0.282 - 0.003 - 0.026$ pH.

(7) $\text{In(OH)}_{3(s)} + 3\text{H}^+ + 3e \rightleftharpoons \text{In}_{(s)} + 3\text{H}_2\text{O}$

$E^{\circ\prime} = E_1^{\circ\prime} + \dfrac{0.059}{3} \log K_s/K_w^3 = -0.219$ V

$E = E_7^{\circ\prime} - 0.059$ pH,

and the boundary $\text{In(OH)}_{3(s)}/\text{In}_{(s)}$ is found as

line 7: $E = -0.219 - 0.059$ pH.

The diagram in Fig. 13.7 shows the indium(III) hydroxo-complexes prevailing in particular ranges of pH in a solution saturated with indium(III) hydroxide. It is further necessary to define the boundaries which correspond to the redox equilibria of these complexes with metallic indium when they attain $1M$ concentration. In a saturated solution of indium(III) hydroxide such a concentration can be realized only for the

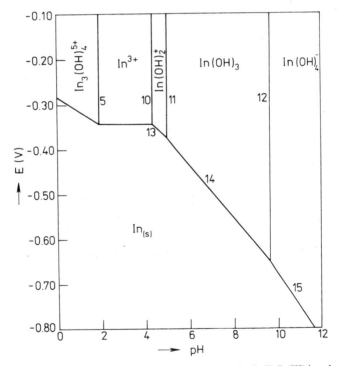

Fig. 13.7 — Predominance-area diagram for the system In–In(I)–In(III) in solution saturated with In(OH)_3 (temperature $25°$C, $I = 3.0$).

ion In(III) and the polynuclear hydroxo-complex, but these boundaries are useful to show the pH-dependence of the redox potential. From the boundaries in Fig. 13.6 only the lines 1, 5, and 6 remain. The hypothetical boundaries 2 and 3 will be omitted and line 4 not considered.

Further new boundaries are given by the lines which correspond to the following equilibria.

(8) $In^{3+} + OH^- \rightleftharpoons In(OH)^{2+}$; $\quad \log K_1 = 9.78$

line 8: $pH = 14.18 - \log K_1 = 4.40$.

(9) $In(OH)^{2+} + OH^- \rightleftharpoons In(OH)_2^+$; $\quad \log K_2 = 9.90$

line 9: $pH = 14.18 - \log K_2 = 4.28$.

The boundary 9 comes before line 8 in the diagram, as the indium(III) dihydroxo-complex is somewhat more stable than the monohydroxo-complex. Thus the following boundary will appear in the diagram:

(10) $In^{3+} + 2\,OH^- \rightleftharpoons In(OH)_2^+$; $\quad \log K_1 K_2 = 19.68$

line 10: $pH = 14.18 - \frac{1}{2}\log K_1 K_2 = 4.34$.

Subsequent equilibria are written as follows

(11) $In(OH)_2^+ + OH^- \rightleftharpoons In(OH)_3$; $\quad \log K_3 = 9.21$

line 11: $pH = 14.18 - \log K_3 = 4.97$.

(12) $In(OH)_3 + OH^- \rightleftharpoons In(OH)_4^-$; $\quad \log K_4 = 4.55$

line 12: $pH = 14.18 - \log K_4 = 9.63$.

(13) $In(OH)_2^+ + 3e + 2\,H^+ \rightleftharpoons In_{(s)} + 2\,H_2O$

$$E_{13}^{\circ\prime} = E_1^{\circ\prime} - \frac{0.059}{3}\log K_1 K_2 K_w^2 = -0.172\ V$$

line 13: $E = -0.172 - 0.039\ pH$.

(14) $In(OH)_3 + 3e + 3\,H^+ \rightleftharpoons In_{(s)} + 3\,H_2O$

$$E_{14}^{\circ\prime} = E_1^{\circ\prime} - \frac{0.059}{3}\log K_1 K_2 K_3 K_w^3 = -0.075\ V$$

line 14: $E = -0.075 - 0.059\ pH$.

(15) $In(OH)_4^- + 3e + 4\,H^+ \rightleftharpoons In_{(s)} + 4\,H_2O$

$$E_{15}^{\circ\prime} = E_1^{\circ\prime} - \frac{0.059}{3}\log K_1 K_2 K_3 K_4 K_w^4 = 0.115\ V$$

line 15: $E = -0.115 - 0.079\ pH$.

The diagrams in Figs. 13.6 and 13.7 illustrate some points of importance for the chemistry of indium in aqueous solution. First, the thermodynamic instability of the ion In(I) in aqueous solution is evident; secondly, the oxidation of metallic indium proceeds more easily in basic medium, yielding indium(III) hydroxo-complexes and sparingly soluble indium(III) hydroxide. In an acidic solution of an indium(III) salt and in the absence of chloride (cf. Section 12.2.4, p. 351) a polynuclear indium(III) hydroxo-complex, probably $In_3(OH)_4^{5+}$, represents a predominating species at pH < 2. Free indium(III) ion has only a narrow predominance region at about pH 2; between pH 2 and 3 indium(III) at concentrations commonly encountered is hydrolysed to form a precipitate of indium(III) hydroxide. With increasing pH the solubility of this hydroxide increases only negligibly; it has to be considered only in a strongly basic medium. For example, at pH 11 the concentration of the indium(III) tetrahydroxo-complex reaches a value of only $10^{-6} M$.

13.2.2 Diagram for redox equilibria in a system of oxygen, hydrogen peroxide and water, showing pH-dependence

Depending on the reaction conditions, oxygen is reduced either to hydrogen peroxide or to water. In aqueous solution hydrogen peroxide can be protonated to form $H_3O_2^+$ or undergo dissociation to HO_2^- and H^+.

The predominance-area diagram in Fig. 13.8 illustrates the following relationships (for 25°C and $I \to 0$).

(1) $E°(\frac{1}{2}O_{2(g)} + 2 H^+ + 2e \rightleftharpoons H_2O) = 1.229$ V

line 1: $E = E_1° + \frac{0.059}{2} \log a_{H^+}^2$

$E = 1.229 - 0.059$ pH.

(2) $E°(O_{2(g)} + 2 H^+ + 2e \rightleftharpoons H_2O_2) = 0.682$ V

line 2: $E = E_2° + \frac{0.059}{2} \log a_{H^+}^2$

$E = 0.682 - 0.059$ pH.

(3) $E°(H_2O_2 + 2 H^+ + 2e \rightleftharpoons 2 H_2O) = 1.77$ V

line 3: $E = E_3° + \frac{0.059}{2} \log a_{H^+}^2$

$E = 1.77 - 0.059$ pH.

Predominance-Area Diagrams

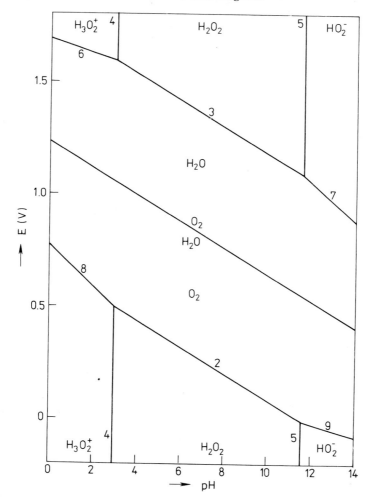

Fig. 13.8 — Predominance-area diagram for the system O_2–H_2O_2–H_2O, with acid-base forms of hydrogen peroxide taken into account (temperature 25°C).

(4) $\log K_{a1}(H_3O_2^+ \rightleftharpoons H_2O_2 + H^+) = -3$
line 4: pH = 3.

(5) $\log K_{a2}(H_2O_2 \rightleftharpoons HO_2^- + H^+) = -11.65$
line 5: pH = 11.6.

(6) $E^{\circ\prime}(H_3O_2^+ + H^+ + 2e \rightleftharpoons 2H_2O) = E_3^\circ + \dfrac{0.059}{2}\log K_{a1} = 1.68$ V

line 6: $E = E_6^{\circ\prime} + \dfrac{0.059}{2} \log a_{H^+}$

$E = 1.68 - 0.030 \text{ pH}$.

(7) $E^{\circ\prime}(HO_2^- + H_2O + 2e \rightleftharpoons 3\,OH^-)$

which can be rewritten as

$E^{\circ\prime}(HO_2^- + 3H^+ + 2e \rightleftharpoons 2H_2O) = E_3^\circ + \dfrac{0.059}{2} \log(1/K_{a2}) = 2.11 \text{ V}$

line 7: $E = E_7^{\circ\prime} + \dfrac{0.059}{2} \log a_{H^+}^3$

$E = 2.11 - 0.089 \text{ pH}$.

(8) $E_8^\circ{}'(O_{2(g)} + 3H^+ + 2e \rightleftharpoons H_3O_2^+) = E_2^\circ - \dfrac{0.059}{2} \log K_{a1} = 0.770 \text{ V}$

line 8: $E = E_8^{\circ\prime} + \dfrac{0.059}{2} \log a_{H^+}^3$,

$E = 0.770 - 0.089 \text{ pH}$.

(9) $E^{\circ\prime}(O_{2(g)} + H_2O + 2e \rightleftharpoons HO_2^- + OH^-)$

which can be rewritten as

$E^{\circ\prime}(O_{2(g)} + H^+ + 2e \rightleftharpoons HO_2^-) = E_2^\circ - \dfrac{0.059}{2} \log(1/K_{a2}) = 0.339 \text{ V}$

line 9: $E = E_9^{\circ\prime} + \dfrac{0.059}{2} \log a_{H^+}$

$E = 0.339 - 0.030 \text{ pH}$.

The diagram illustrates both the thermodynamic instability of hydrogen peroxide (formation of which is strongly endothermic) and its oxidizing power. It is evident that hydrogen peroxide is formed more easily by the reduction of oxygen than by the oxidation of water. This fact explains the presence of hydrogen peroxide in some reductors (cf. refs. [3, 4]). This reduction proceeds more easily in a basic medium, where hydrogen peroxide can also be reduced readily to water.

13.2.3 Dependence of redox equilibria of cadmium in aqueous solution on the concentration of chloride ions

Cadmium(II) forms $CdCl_n^{(2-n)+}$ chloro-complexes (n being 1–4) which are reduced to metallic cadmium at various potentials. The predominance-area diagram in Fig. 13.9 shows the boundary lines defined by the

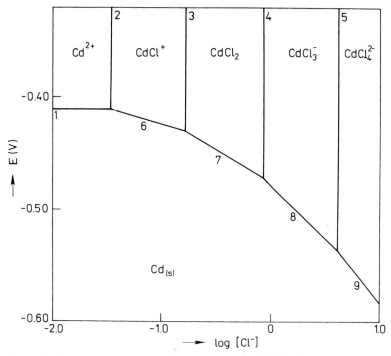

Fig. 13.9 — Predominance-area diagram for the system Cd–Cd(II) in the presence of chloride ions in acidic solution (temperature 25°C, $I = 3.0$).

following constants and equations [for 25°C and ionic strength 3 ($NaClO_4$)]. An acidic medium is assumed (pH < 7) and so hydrolysis of cadmium(II) need not be considered.

(1) $E°(Cd^{2+} + 2e \rightleftharpoons Cd_{(s)}) = E_1° = -0.4111$ V
   line 1: $E = -0.411$.

(2) $\log K_1(Cd^{2+} + Cl^- \rightleftharpoons CdCl^+) = 1.46$
   line 2: $\log [Cl^-] = -\log K_1 = -1.46$.

(3) $\log K_2(CdCl^+ + Cl^- \rightleftharpoons CdCl_2) = 0.78$
   line 3: $\log [Cl^-] = -\log K_2 = -0.78$.

(4) $\log K_3(CdCl_2 + Cl^- \rightleftharpoons CdCl_3^-) = 0.07$
   line 4: $\log [Cl^-] = -\log K_3 = -0.07$.

(5) $\log K_4(CdCl_3^- + Cl^- \rightleftharpoons CdCl_4^{2-}) = -0.61$
   line 5: $\log [Cl^-] = -\log K_4 = 0.61$.

(6) $E^{\circ\prime}(CdCl^+ + 2e \rightleftharpoons Cd_{(s)} + Cl^-) = E_1^\circ - \dfrac{0.059}{2}\log K_1 = -0.454\ V$

line 6: $E = E_6^{\circ\prime} - \dfrac{0.059}{2}\log[Cl^-]$

$E = -0.454 - 0.030\log[Cl^-]$.

(7) $E^{\circ\prime}(CdCl_2 + 2e \rightleftharpoons Cd_{(s)} + 2Cl^-) = E_1^\circ - \dfrac{0.059}{2}\log K_1 K_2$
$= -0.477\ V$

line 7: $E = E_7^{\circ\prime} - \dfrac{0.059}{2}\log[Cl^-]^2$

$E = -0.477 - 0.059\log[Cl^-]$.

(8) $E^{\circ\prime}(CdCl_3^- + 2e \rightleftharpoons Cd_{(s)} + 3Cl^-) = E_1^\circ - \dfrac{0.059}{2}\log K_1 K_2 K_3$
$= -0.479\ V$

line 8: $E = E_8^{\circ\prime} - \dfrac{0.059}{2}\log[Cl^-]^3$

$E = -0.479 - 0.089\log[Cl^-]$.

(9) $E^{\circ\prime}(CdCl_4^{2-} + 2e \rightleftharpoons Cd_{(s)} + 4Cl^-) = E_1^\circ - \dfrac{0.059}{2}\log K_1 K_2 K_3 K_4$
$= -0.461\ V$

line 9: $E = E_9^{\circ\prime} - \dfrac{0.059}{2}\log[Cl^-]^4$

$E = -0.461 - 0.118\log[Cl^-]$.

As illustrated in Fig. 13.9, the predominance of anionic complexes cannot be expected to occur before the concentration of chloride reaches 1M; this fact should be considered when cadmium is separated in the form of chloro-complexes from other metal ions on an anion-exchanger.

The presence of chloride considerably lowers the formal redox potential of the Cd(II)/Cd couple, making the oxidation of metallic cadmium easy and enhancing the corrosion of cadmium-plated surfaces.

REFERENCES

[1] W. Stumm and J. J. Morgan, *Aquatic Chemistry*, Wiley–Interscience, New York, 1970; 2nd Ed., Wiley, New York, 1981.

[2] C. F. Baes, Jr. and E. R. Mesmer, *The Hydrolysis of Cations*, p. 327. Wiley, New York, 1976.

[3] H. A. Laitinen and W. E. Harris, *Chemical Analysis*, 2nd Ed., p. 314. McGraw-Hill, New York, 1975.

[4] R. A. Chalmers, D. A. Edmond and W. Moser, *Anal. Chim. Acta*, 1966, **35**, 404.

CHAPTER 14

Examples of applications of tables and graphs

In this last chapter we give some examples that illustrate various possible uses of data on equilibria in solution. The ways in which side-reaction coefficients and other auxiliary quantities are used for solving problems are illustrated. These may be useful for the similar situations that arise in analytical chemistry, chemical technology, biochemistry, and medicine.

14.1 CALCULATION OF THE pH IN AN ISOPOLYACID EQUILIBRIUM SYSTEM IN AQUEOUS SOLUTION

The pH of an aqueous solution in which acid–base equilibrium of a polyprotic acid is attained can be found, for example, by use of the logarithmic diagram for the acid, as shown in Section 4.2.1.2, p. 72. For a more complicated system, in which higher aggregates of the anion are formed, diagram construction is difficult, as the shape of the curves depends not only on pH but also on the concentration of the anions. Calculation of the pH in such a system is possible only if the total concentration, given by the sum of all the equilibrium species of the isopolyacid, and the equilibrium constants for all the processes involved, are known. The principle of calculation by subsequent approximations with the use of a computer will be exemplified by the equilibrium system of boric acid and polyborates. All the acid–base equilibria of this system are listed in Section 12.2.1, p. 344; the necessary equilibrium constants can be found in Table 5.1.

For the calculation it is necessary to write the following equilibria and balance equations.

$$\beta_{11} = [B(OH)_4^-]/[H_3BO_3][OH^-] \qquad (14.1)$$

$$\beta_{12} = [B_2O(OH)_5^-]/[H_3BO_3]^2[OH^-] \qquad (14.2)$$

Sec. 14.1] Calculation of the pH in an Isopolyacid Equilibrium System

$$\beta_{13} = [B_3O_3(OH)_4^-]/[H_3BO_3]^3 [OH^-] \tag{14.3}$$

$$\beta_{24} = [B_4O_5(OH)_4^{2-}]/[H_3BO_3]^4 [OH^-]^2 \tag{14.4}$$

$$K_w = [H^+][OH^-] \tag{14.5}$$

$$c_B = [H_3BO_3] + [B(OH)_4^-] + 2[B_2O(OH)_5^-]$$
$$+ 3[B_3O_3(OH)_4^-] + 4[B_4O_5(OH)_4^{2-}] \tag{14.6}$$

$$[Na^+] + [H^+] = [B(OH)_4^-] + [B_2O(OH)_5^-]$$
$$+ [B_3O_3(OH)_4^-] + 2[B_4O_5(OH)_4^{2-}] + [OH^-] \tag{14.7}$$

The calculation then proceeds according to the following algorithm, which is a modification of the iterative method of Perrin and Sayce [1].
(1) A preliminary value of $[OH^-]_{calc}$ is calculated from Eq. (14.1), on the assumption that no polyborates are present in solution. Then the electroneutrality condition in Eq. (14.7) can be written as

$$[Na^+] = [B(OH)_4^-] + [OH^-] - [H^+]$$

and Eq. (14.6) is simplified to

$$c_B = [H_3BO_3] + [B(OH)_4^-].$$

In this step the problem is actually simplified to the calculation of the pH of a weak monoprotic acid in the presence of its salt with a strong base.
(2) The preliminary value of $[OH^-]_{calc}$ is then used for calculation of concentrations of individual forms of boric acid in solution by means of Eqs. (14.1)–(14.5), with the assumption that $[H_3BO_3]_{calc} = c_B - [Na^+]$.
(3) Equation (14.6) is now used for the calculation of $c_{B,calc}$ so as to introduce a correction for the equilibrium concentration of H_3BO_3:

$$[H_3BO_3]_{corr} = [H_3BO_3]_{calc} \, c_{B,calc}/c_B. \tag{14.8}$$

(4) From Eq. (14.7) the value of $[Na^+]_{calc}$ is found and used for correction of the preliminary value of $[OH^-]_{calc}$, by the expression

$$[OH^-]_{corr} = [OH^-]_{calc} [Na^+]_{calc}/[Na^+]. \tag{14.9}$$

(5) The corrected values for the concentrations of hydroxide ion and boric acid are used again to repeat the calculations of steps (2) and (3) of the algorithm. The resulting values of the concentrations $[Na^+]_{calc}$ and $c_{B,calc}$ are repeatedly inserted, together with the corrected value of $[OH^-]_{calc}$, into Eqs. (14.8) and (14.9). The calculations in steps (2)–(4) are repeated until the differences between c_B and $c_{B,calc}$ and simultaneously $[Na^+]$ and $[Na^+]_{calc}$ are less than 1% and 0.1%, respectively.

The accuracy and scope of application of the approach outlined was tested by calculation of the pH value given in the literature [2] for the temperature-dependence of the NBS standard buffer, 0.01m borax solution. The preliminary value of $[OH^-]_{calc}$ was in this case obtained from the pH (9.225) given for the borax buffer at 20°C. For the calculation of the buffer pH at 60°C the values of the equilibrium constants found by Mesmer et al. [3] for that temperature were used. The pH value found by the calculation differed by 0.004 from the value given in the literature.

The algorithm of this calculation, written as a program IONEX [4], can also be used for treatment of data obtained in the study of heterogeneous equilibria on ion-exchangers at varied temperature.

14.2 TREATMENT OF SIDE-REACTIONS IN THE CALCULATION OF MASKING EFFICIENCY

Masking of interfering metal ions by binding them as stable complexes is often used in chemical analysis. Also, in the food-stuff industry masking agents are added to prevent deterioration of the product, caused by the catalytic action of certain metal ions. In medicine, the toxic effects of some metal ions can be eliminated by suitable masking. For example, the ions Cd(II) and Pb(II) can be removed from a living organism by the action of a solution of the calcium chelate CaL^{2-n} of some of the EDTA-type reagents (Table 14.1) of formula H_nL. The calcium complex

Table 14.1
Logarithmic values of equilibrium constant K for the displacement reaction
$$CaL^{2-n} + M^{2+} \rightleftharpoons ML^{2-n} + Ca^{2+}$$
(for pH 7–9, 25°C, and $I = 0.1$)

Chelating ligand	Toxic metal ion				
	Sr^{2+}	Cd^{2+}	Pb^{2+} *		
			pH 7.0	8.0	9.0
DCTA	−2.57	6.69	7.02	6.64	5.70
DTPA	−1.07	8.25	7.84	7.46	6.52
EDTA	−1.93	5.75	7.20	6.82	5.88
EGTA	−2.43	5.64	3.61	3.23	2.29
HEDTA	−1.40	4.90	7.23	6.85	5.91
TTHA	−0.63	8.71	8.54	8.16	7.22

* $\log \alpha_{Pb(OH)}$ has the following values: 0.07 at pH 7.0, 0.45 at pH 8.0, 1.39 at pH 9.0

Sec. 14.2] Treatment of Side-Reactions in the Calculation of Masking Efficiency

reacts with the heavy metal ion by the displacement reaction

$$CaL^{2-n} + M^{2+} \rightleftharpoons ML^{2-n} + Ca^{2+}.$$

The heavy metal chelate is then passed out of the organism and the non-toxic calcium remains. With the aid of the data on the stability of the complexes of calcium, cadmium, and lead with DCTA, DTPA, EDTA, EGTA, HEDTA, and TTHA (for abbreviations and data see Table 7.3) the most suitable masking agent can be found. It is also possible to decide whether such a calcium chelate can be used to remove Sr(II) ions.

In our further consideration of the course of this displacement reaction, it will be assumed that all processes proceed at pH 7–9 and at $I = 0.1$, and that the concentration of free calcium ion Ca^{2+} is $0.01 M$. These conditions correspond approximately to the composition of human blood plasma.

The equilibrium constant K for the displacement reaction can be expressed as

$$K = \beta'_{ML}/\beta'_{CaL} \tag{14.10}$$

where β' is the conditional stability constant for the complex involved, and hence influenced by protonation of the ligand, formation of hydroxo-complexes of the metal ion, and formation of protonated and polynuclear complexes of the type MHL, MH$_2$L, M$_2$L, etc. [cf. the expressions in Eqs. (3.2)–(3.8)]. Within the given pH range, the values of $\alpha_{L(H)}$ indicate that protonation of the ligand has the greatest influence on the stability of the complex; however, the contribution of the coefficient $\alpha_{L(H)}$ cancels in Eq. (14.10). The coefficients α_{ML} are equal to 1 in the pH range 7–9 and of the values of $\alpha_{M(OH)}$ only the coefficient $\alpha_{Pb(OH)}$ attains values greater than 1 (see note to Table 14.1).

If the stability constants given in Table 7.3 are used to find the constant K for the displacement reaction, the data listed in Table 14.1 are obtained. It can be seen that for the ions Cd(II) and Pb(II), the displacement equilibrium is shifted furthest to the right if the ligand TTHA is used. The efficiency of detoxication by the use of chelating ligands decreases in the following sequences:

Cd: TTHA > DTPA > DCTA > EDTA > EGTA > HEDTA
Pb: TTHA > DTPA > HEDTA > EDTA > DCTA > EGTA.

The value of K for the reaction with strontium indicates that it is not possible to remove Sr(II) from an organism by displacement with a calcium chelate; the stability of the calcium complexes is much higher than that of the corresponding strontium complexes.

From the known value of the displacement reaction constant it is possible to calculate the minimum dosage of calcium chelate that will lower the concentration of the toxic metal ion to a given value. For given conditions, the constant K can be written in terms of equilibrium concentrations as

$$K = \frac{[Ca^{2+}][ML^{2-n}]}{[CaL^{2-n}][M']}.$$

Only for lead must a conditional concentration for the metal ion be inserted; i.e. $[Pb'] = [Pb^{2+}] + [Pb(OH)^+] + [Pb(OH)_2]$. For example, when $K = 10^{8.0}$ and $[Ca^{2+}] = 10^{-2}M$, and the initial concentration of lead $c_{Pb} = 10^{-2}M$, and if nearly all the lead is required to be transformed into the complex at equilibrium, the following expression is obtained:

$$[CaL^{2-n}][Pb'] = [Ca^{2+}][PbL^{2-n}]/K = 10^{-12}.$$

Hence, to decrease the concentration of free lead $[Pb']$ to $10^{-6}M$ an equilibrium concentration of $[CaL^{2-n}] = 10^{-6}M$ is required and the molar amount of calcium chelate to be added is the same as that of the lead present; to reduce the free lead concentration to $10^{-10}M$, a final calcium chelate concentration of $10^{-2}M$ is needed, so the dosage must be doubled.

14.3 CONSIDERATION OF THE COURSE OF A REDOX REACTION BY USE OF A PREDOMINANCE-AREA DIAGRAM

When permanganate is used to oxidize substances in acidic aqueous solution, for example in the determination of organic compounds in waste water by titration with permanganate, it is necessary to prevent oxidation of the chloride commonly present in the samples.

With complicated redox systems where the elements can exist in several oxidation states (such as manganese and chlorine), useful information can be obtained from the predominance-area diagrams. Figures 14.1 and 14.2 are diagrams showing the relative contributions of the various redox forms of the two elements as a function of pH and redox potential E.

In the reaction between chloride and permanganate in acidic medium the following equilibria have to be considered:

$$MnO_4^- + 4H^+ + 3e \rightleftharpoons MnO_{2(s)} + 2H_2O \tag{14.11}$$

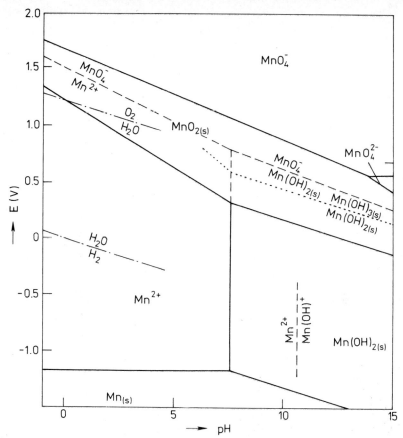

Fig. 14.1 — Predominance-area diagram for the various redox forms of manganese in aqueous solution as a function of redox potential and pH (25°C, $I = 0$; for equilibrium data see Tables 5.1, 8.1, and 9.1). Full lines for the boundaries correspond to predominant species in the system; broken or dotted lines denote the boundaries between species which do not predominate under the given conditions. Dot-and-dash lines denote conditional redox potentials for the couples O_2/H_2O and H_2O/H_2.

$$MnO_4^- + 8H^+ + 5e \rightleftharpoons Mn^{2+} + 4H_2O \qquad (14.12)$$

$$MnO_{2(s)} + 4H^+ + 2e \rightleftharpoons Mn^{2+} + 2H_2O \qquad (14.13)$$

$$HClO + H^+ + e \rightleftharpoons \tfrac{1}{2}Cl_{2(aq)} + H_2O \qquad (14.14)$$

$$HClO \rightleftharpoons H^+ + ClO^- \qquad (14.15)$$

$$HClO + H^+ + 2e \rightleftharpoons Cl^- + H_2O \qquad (14.16)$$

$$ClO^- + 2H^+ + 2e \rightleftharpoons Cl^- + H_2O \qquad (14.17)$$

$$\tfrac{1}{2}Cl_{2(aq)} + e \rightleftharpoons Cl^- \qquad (14.18)$$

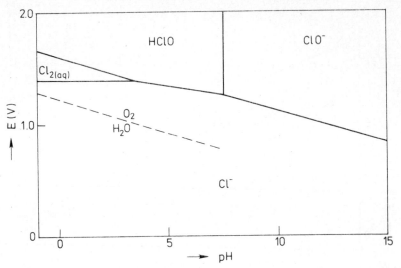

Fig. 14.2 — Predominance-area diagram for the system Cl(I)–Cl–Cl(−I) in aqueous solution as a function of redox potential and pH (25°C, $I = 0$, for equilibrium data see Tables 5.1 and 9.1). Full lines separate regions of predominance for the species given; the dotted line is for the redox potential of O_2/H_2O.

and also the reaction

$$\tfrac{1}{2}O_{2(g)} + 2H^+ + 2e \rightleftharpoons H_2O \qquad (14.19)$$

By combining the Nernst–Peters equations, the expression for the dissociation constant of hypochlorous acid and the relevant data from Tables 5.1 and 9.1, the following equations for the lines in Fig. 14.3 can be written.

Line (boundary)	Equation for the boundary line	See equation
1	$E = 1.68 - 0.08\,\text{pH}$	(14.11)
2	$E = 1.51 - 0.096\,\text{pH}$	(14.12)
2'	$E = 1.51 - 0.096\,\text{pH} - 0.12$	(14.12) $\left(\text{for } \dfrac{[Mn^{2+}]}{[MnO_4^-]} = 10\right)$
3	$E = 1.23 - 0.12\,\text{pH}$	(14.13)
4	$E = 1.30 - 0.06\,\text{pH}$	(14.19)
Cl_2/Cl^-	$E = 1.39$	(14.18)
$HClO/Cl_2$	$E = 1.60 - 0.06\,\text{pH}$	(14.14)
$HClO/Cl^-$	$E = 1.49 - 0.03\,\text{pH}$	(14.16)
$HClO/ClO^-$	$\text{pH} = 7.54$	(14.15)
ClO^-/Cl^-	$E = 1.72 - 0.06\,\text{pH}$	(14.17)

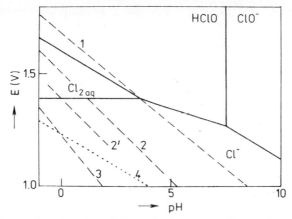

Fig. 14.3 — Enlarged section of the predominance-area diagram for the system Cl(I)–Cl–Cl(−I) (cf. Fig. 14.2) combined with broken lines showing the pH-dependence of various redox systems of manganese. Lines: 1 — $E°'(MnO_4^-, MnO_{2(s)})$; 2 — $E°'(MnO_4^-, Mn^{2+})$; 2' — E for MnO_4^-/Mn^{2+} in the presence of a 10:1 ratio of Mn(II) to Mn(VII); 3 — $E°'(MnO_{2(s)}, Mn^{2+})$; 4 — $E°'(O_2, H_2O)$.

Figure 14.3 is an enlarged part of the predominance-area diagram for the redox forms Cl(+I), Cl, and Cl(−I) in the pH range from −1 to 10 and for redox potentials between +1.0 and +1.8 V (lines drawn in full). This diagram also shows the pH-dependence of the conditional redox potentials for couples $MnO_4^-/MnO_{2(s)}$ (dashed line 1), MnO_4^-/Mn^{2+} (dashed line 2), and $MnO_{2(s)}/Mn^{2+}$ (dashed line 3). Line 2' shows the pH-dependence of the redox potential for the couple MnO_4^-/Mn^{2+} in presence of a 10:1 ratio of Mn(II) to MnO_4^-.

Figure 14.3 clearly shows the power of permanganate to oxidize chloride to chlorine in acidic medium (at pH < 3.7). However, the reaction is rather slow because the mechanism of the redox process between the anions MnO_4^- and Cl^- is complicated if no suitable catalyst is present [e.g. the Fe(III)/Fe(II) couple]. Addition of manganese(II) causes the redox potential for MnO_4^-/Mn^{2+} to be lowered (cf. line 2') and to exceed the value of $E°$ for Cl_2/Cl^- only at pH $\leq$ 0. This is utilized in the titration of iron(II) with permanganate in presence of chloride; manganese(II) sulphate is one of the components of the Zimmermann–Reinhardt reagent, which is added to the solution to be titrated.

Figures 14.1 and 14.3 also show the great thermodynamic instability of permanganate in acidic medium; it may even oxidize water to oxygen and be reduced to Mn(II) and $MnO_{2(s)}$, but the high oxygen overpotential (about +0.5 V) inhibits rapid reaction.

Figures 14.2 and 14.3 also show that a solution of chlorine in water is not thermodynamically stable even in strongly acidic medium. Even at pH -1, the redox potential for the couple $Cl_{2(aq)}/Cl^-$ is higher than that for O_2/H_2O; again it is the high oxygen overpotential that prevents the expected reaction, and chlorine in water is not reduced to chloride with evolution of oxygen. However, oxygen present in water cannot oxidize chloride to chlorine. The figures also illustrate the disproportionation of chlorine in water to yield HClO and Cl^- or, at pH $\gtrsim 8$, ClO^- and Cl^-.

14.4 HYDROLYSIS OF METAL IONS IN AQUEOUS SOLUTION

Knowledge of the actual form in which a metal ion exists in solution may often have a great influence on the accuracy attainable in an analytical determination, the assessment of the catalytic action of the metal in a reaction (this is sometimes of considerable importance for a technological process or a biochemical problem), or its toxic effect on a living organism. Even in the absence of complex-forming or precipitation reactions, metal ions in aqueous solution, especially those of metals in higher oxidation states, participate in reactions with water. These reactions may lead to formation of soluble hydroxo-complexes (mono- and polynuclear) and even to precipitation of sparingly soluble hydroxides or hydrated oxides. The present state of knowledge on the hydrolysis of metal ions is discussed by Baes and Mesmer in their outstanding monograph [5], which also includes a critical compilation of values of equilibrium constants for hydrolytic processes and precipitation of hydroxides.

An interesting application of such data to allow information to be obtained on the forms of a metal ion present in solution was presented by Kragten [6]. This approach will be illustrated here by the following examples.

14.4.1 Use of the log $\alpha_{M(OH)}$–pH relationship for the determination of the extent of hydrolysis of a metal ion in aqueous solution

A general expression for the side-reaction coefficient $\alpha_{M(OH)}$, where polynuclear hydroxo-complexes may be formed, can be written in analogy with Eq. (3.6), p. 45, as

$$\alpha_{M(OH)} = 1 + \sum_{n=1}^{N} \sum_{n=1}^{M} m\beta_{mn}[M]^{m-1}[OH^-]^n. \qquad (14.20)$$

Application of this complicated expression is difficult because the free concentration of the metal ion is unknown – it is itself the aim of the calculation. Kragten [6] advocates a simplifying approach based on the assumption that in any solution often only one species of the system of hydroxo-complexes predominates, and the particular species will depend on the pH and the total concentration of the metal. At low concentration of the metal, sometimes not above $10^{-5}M$, the presence of polynuclear complexes can be neglected; hence

$$\alpha_{M(OH)} = 1 + \sum_{n=1}^{N} \beta_n K_w^n \cdot 10^{n\text{pH}}. \qquad (14.21)$$

From this expression it can be seen that the contribution of hydroxo-complexes to the total metal concentration (in the absence of complexing agents) is negligible ($< 0.1\%$) when

$$\beta_1 K_w \cdot 10^{\text{pH}} < 0.01$$

which is attained at

$$\text{pH} < 12 - \log \beta_1. \qquad (14.22)$$

In this region, $\log \alpha_{M(OH)}$ is zero. Equation (14.22) is applicable at 25°C and for dilute solutions with $I \leq 1$, where K_w approaches 10^{-14}. At a pH value higher than that given by Eq. (14.22), simplified expressions can be written for the particular regions where each hydroxo-complex $M(OH)_n$ predominates.

$$\log \alpha_{M(OH)_n} = n\,\text{pH} + \log \beta_n - n\,pK_w. \qquad (14.23)$$

Thus, in the absence of polynuclear complexes the relationship $\log \alpha_{M(OH)} = f(\text{pH})$ can be considered to be composed of several linear sections which approximate the true function. In Fig. 14.4 these lines are shown as tangents to the curve $\log \alpha_{Pb(OH)} = f(\text{pH})$ plotted as curve a for the absence of polynuclear hydroxo-complexes.

At higher concentrations of metal ion (commonly at $[M] \geq 10^{-4}M$) polynuclear complexes become significant. Again, it is possible to find pH regions where only one polynuclear complex $M_m(OH)_n$ predominates; thus the simplified side-reaction coefficient is written as

$$\alpha_{M_m(OH)_n} = m[M]^{m-1} \beta_{nm} K_w^n \cdot 10^{n\text{pH}}. \qquad (14.24)$$

If the unknown free metal ion concentration is replaced, as proposed by Kragten [6], by $[M] = [M']/\alpha_{M_m(OH)_n}$ [cf. Eq. (3.6), p. 45)], where $[M'] = m[M_m(OH)_n]$, Eq. (14.24) can be written in logarithmic form as

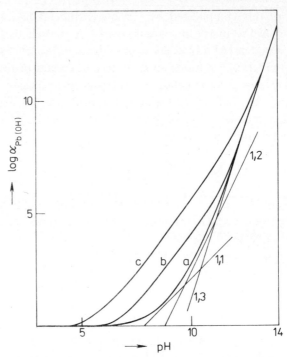

Fig. 14.4 — Side-reaction coefficient for lead(II) reacting with hydroxide ions, for different total concentrations of lead. Curve: a — $\log \alpha_{Pb(OH)}$ for $[Pb'] = 10^{-5} M$; b — $[Pb'] = 10^{-3} M$; c — $[Pb'] = 10^{-1} M$. (The lines denoted 1,1, 1,2, 1,3 correspond to the contributions of the relevant mononuclear hydroxo-complexes ($n = 1$–3) to the total value of $\alpha_{Pb(OH)}$.

$$\log \alpha_{M_m(OH)_n} = \frac{1}{m} [\log m + (m - 1) \log [M']$$
$$+ \log \beta_{nm} - n\,pK_w + n\,pH] \qquad (14.25)$$

which is again the equation of a straight line, the slope of which is given by the stoichiometry of the polynuclear complex (n/m); the intercept on the $\log \alpha$ axis depends on the stability of the polynuclear complex and on the concentration of the metal, i.e. on the value of $[M']$. Equation (14.25) has general validity; for $m = 1$ it is identical to Eq. (14.23).

If known values of the stability constants of hydroxo-complexes are used in Eq. (14.25), it is easy to find which of the metal forms becomes important within a given pH region and at a particular metal concentration. This is illustrated by the plots of particular sections of $\log \alpha_{Pb(OH)}$ shown in Figs. 14.5 and 14.6 for two different total concentrations of lead(II). The functions plotted are given as equations in Table 14.2. The

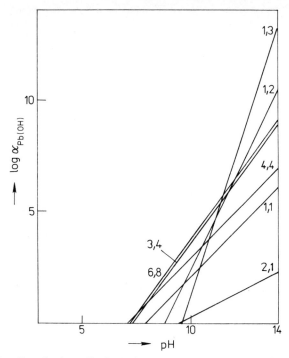

Fig. 14.5 — Contributions of individual hydroxo-complexes of lead(II) to the total value of the coefficient $\alpha_{Pb(OH)}$ at $[Pb'] = 10^{-3}M$, showing the dependence on pH. The lines are labelled with the stoichiometric coefficients of the complexes $Pb_m(OH)_n$; the corresponding equations for the lines obtained from Eq. (14.25) are given in Table 14.2.

Table 14.2

Contributions of individual lead(II) hydroxo-complexes to the side-reaction coefficient $\alpha_{Pb(OH)}$

Equilibrium species	Coefficients m, n	Equation for the straight line† $\log \alpha = f(\text{pH})$
Pb^{2+}	1, 0	$\log \alpha = 0$
$Pb(OH)^+$	1, 1	$= \text{pH} - 7.95$
$Pb(OH)_2$	1, 2	$= 2\,\text{pH} - 17.63$
$Pb(OH)_3^-$	1, 3	$= 3\,\text{pH} - 28.66$
$Pb_2(OH)^{3+}$	2, 1	$= \frac{1}{2}\text{pH} - 3.19 - 0.5\,\text{pM}'$
$Pb_3(OH)_4^{2+}$	3, 4	$= \frac{4}{3}\text{pH} - 7.69 - \frac{2}{3}\text{pM}'$
$Pb_4(OH)_4^{4+}$	4, 4	$= \text{pH} - 4.91 - \frac{3}{4}\text{pM}'$
$Pb_6(OH)_8^{4+}$	6, 8	$= \frac{4}{3}\text{pH} - 7.97 - \frac{5}{6}\text{pM}'$

† The equations are obtained from Eq. (14.25) by inserting values of the coefficients m and n and using the stability constants for the lead(II) hydroxo-complexes at $I = 1$ (Table 6.12).

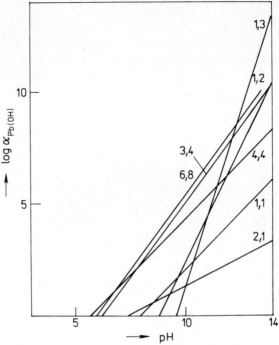

Fig. 14.6 — Contributions of individual hydroxo-complexes of lead(II) to the total value of the coefficient $\alpha_{Pb(OH)}$ at $[Pb'] = 10^{-1} M$, plotted as a function of pH. The lines are labelled as for Fig. 14.5; for the equations of the lines see Table 14.2.

resulting curves, $\log \alpha_{Pb(OH)} = f(pH)$, for three different total concentrations of lead(II) are shown in Fig. 14.4.

The fraction of the metal present in solution as a polynuclear complex often does not participate in the reaction under investigation because of the slow reactions of polynuclear species. Inert polynuclear species can often cause significant errors in analysis [7–9]. Therefore it is important to evaluate not only the overall coefficient $\alpha_{M(OH)}$ but also the contribution of the polynuclear complex to the total concentration of the metal ion in solution. Kragten [6] calculates the conditions for which this contribution amounts to 1%; that is

$$m[M_m(OH)_n] = 0.01[M'] \qquad (14.26)$$

where

$$[M'] = [M]\alpha_{M(OH)_n}. \qquad (14.27)$$

Sec. 14.4]   Hydrolysis of Metal Ions in Aqueous Solution   395

On substitution from the expression $\beta_{nm} = [M_m(OH)_n]/[M]^m[OH]^n$ for the equilibrium concentration $[M_m(OH)_n]$, the following equation is obtained

$$m\beta_{nm}[M']^{m-1}[OH^-]^n = 0.01\alpha^n_{M(OH)_n}.\tag{14.28}$$

In logarithmic form, this becomes

$$-\log[M']_{1\%} = pM'_{1\%} = \frac{1}{m-1}(2 - m\log\alpha_{M(OH)_n} + \log m$$
$$+ \log\beta_{nm} + n\log K_w + n\,pH).\tag{14.29}$$

For calculation of the values of $pM'_{1\%}$ as a function of pH from this expression it is again assumed that only one polynuclear hydroxo-complex predominates in solution. Some data for such calculations are given in Table 14.3 for the equilibria of lead(II) hydroxo-complexes. The values of $\alpha_{M(OH)}$ needed for calculation of $pM'_{1\%}$ at particular pH values, and the highest $pM'_{1\%}$ value for each pH, are underlined. The relationship $pM'_{1\%} = f(pH)$ is plotted as a dashed curve in Fig. 14.7.

Table 14.3
Values for construction of the curves† $pPb' = f(pH)$ in Fig. 14.7

pH	$\log\alpha_{M(OH)_n}$	$pPb'_{1\%}$ polynuclear complexes				pPb'_{min}	prevailing complex m, n
		2, 1	3, 4	4, 4	6, 8		
4	0	−0.08	−2.54	−0.55	−1.77	−20.72	6, 8
5	0	0.92	−0.54	0.78	−0.17	−16.72	6, 8
6	0	1.92	1.46	2.12	1.43	−12.72	6, 8
7	0	2.92	3.46	3.45	3.03	−8.72	6, 8
8	0.05	3.82	5.38	4.71	4.57	−4.72	6, 8
9	1.05	2.82	5.89	4.71	4.97	−0.72	6, 8
10	2.37	1.18	5.91	4.29	4.99	3.08	6, 8 + 3, 4 + 1, 2
11	4.67	−2.42	4.90	2.96	4.19	3.82	1, 2 + 1, 3
12	7.34		2.41	0.33	2.19	3.14	1, 3
13	10.34		−0.05	−2.34	0.19	2.14	1, 3
14	13.34		−2.47	−5.01	−1.83	1.14	1, 3

† A value for $\log\alpha_{Pb(OH)_n}$ is calculated from the expressions in Table 14.2; $pPb'_{1\%}$ is expressed by Eq. (14.29) and its highest value for each pH is taken for the graph; pPb'_{min} is calculated from Eq. (14.32).

[Ch. 14

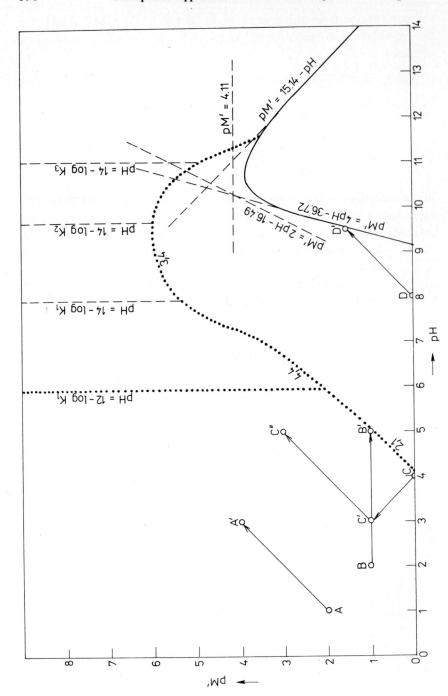

Fig. 14.7 – Diagram for the equilibria of lead(II) with hydroxide ion in aqueous solution showing the dependence on pH. Dotted line – contribution of $Pb(OH)^+$ reaches 1%. Dashed lines – boundary lines for predominance areas of Pb(II) and mononuclear complex species. Dotted curve – 1% contribution of $Pb_m(OH)_n$ [stoichiometric coefficients m and n of the predominating species are written at the relevant region; the data needed for construction of the diagram are listed in Table 14.3; Eq. (14.29) was used for the calculation]. Full line – plot of the function $pPb'_{min} = f(pH)$, which corresponds to Eq. (14.32) and represents the maximum concentration of Pb(II) in solution corresponding to non-precipitation of lead(II) hydroxide. The values of pPb'_{min} are also given in Table 14.3.

14.4.2 Precipitation of sparingly soluble hydroxides

In the presence of hydroxo-complexes, precipitation of a metal hydroxide is characterized by a conditional solubility product [cf. Eq. (3.22)]

$$K'_s = [M'][OH]^k = K_s \alpha_{M(OH)} \qquad (14.30)$$

for the hydroxide $M(OH)_k$.

Hence, the highest attainable concentration of all metal species in solution is

$$[M']_{max} = K_s \alpha_{M(OH)}/[OH]^k. \qquad (14.31)$$

If the concentration $[M']$ is increased above this value at a given pH, a precipitate of the hydroxide $M(OH)_k$ is formed.

Equation (14.31) can be used to derive the expression for the minimum value of pM' which can be attained before precipitation occurs:

$$pM'_{min} = k\,pH - \log K_s + k \log K_w - \log \alpha_{M(OH)}. \qquad (14.32)$$

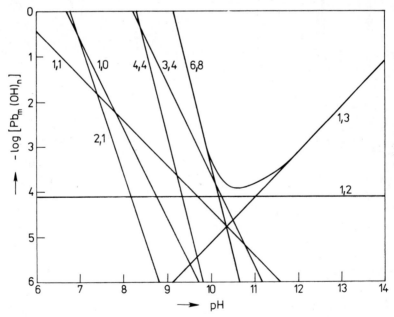

Fig. 14.8 — Logarithmic solubility diagram for lead(II) hydroxide, showing the influence of the formation of lead(II) hydroxo-complexes. (Equation for the lines are given in Table 14.2.) Lines in the diagram are labelled with the coefficients m and n which correspond to the relevant hydroxo-complexes $Pb_m(OH)_n$. The full line shows the pH-dependence of the total solubility of $Pb(OH)_2$.

If Eq. (14.20) is used to express the side-reaction coefficient, an equation for the pH-dependence of pM'_{min} is obtained; by using the approximation in Eq. (14.25) it is possible to write equations for the straight lines which approximate the function $pM'_{min} = f(pH)$ in relevant pH regions.

A curve for lead(II) hydroxide and its approximations are shown in Fig. 14.7. In the vicinity of the maximum several complexes contribute to the solubility; therefore, in this region, the error resulting from use of straight-line approximations is greater than it is for the increasing and decreasing parts of the curve, where only a single complex predominates. A plot of the curve $pM'_{min} = f(pH)$ provides information on solubility of the hydroxide; however, nothing about the presence of particular complex species can be read from such a graph. It is therefore useful to plot the function $pM'_{1\%} = f(pH)$ on the same graph, to indicate the contribution of polynuclear complexes. Also the boundary lines for the predominance areas of mononuclear complexes can be drawn (from the equation $pH = 14 - \log K_n$, where K_n denotes the consecutive stability constants of the complexes).

If the concentrations of individual metal species in solution have to be known, it is best to construct a logarithmic solubility diagram, such as given for lead(II) hydroxide in Fig. 14.8. In this graph the curve of total solubility is a mirror image of the curve $pM'_{min} = f(pH)$.

14.4.3 Use of plots of $pM' = f(pH)$

Any solution containing a particular concentration of the metal ion at a given pH is represented by a point in the diagram. When the solution is diluted or the pH value is adjusted, the location of the point changes accordingly. Thus it is possible to see immediately how the change has influenced the metal ion in solution.

For example, in Fig. 14.7 the point A corresponds to $c_{Pb} = 10^{-2}M$ and pH 1.0 and the location of the point indicates that lead is present in solution as the ion Pb^{2+}. On 100-fold dilution of the solution, the point A' is reached ($c_{Pb} = 10^{-4}M$, pH = 3), but the form of the metal ion in solution is not changed. The point B represents $c_{Pb} = 10^{-1}M$ at pH 2. When the pH is increased to 5 (point B'), no appreciable concentration of lead(II) hydroxo-complexes is formed in solution. If the concentration of lead is $1M$, about 1% of a polynuclear hydroxo-complex is present in solution at pH 4 (point C); on dilution with water the contribution of polynuclear complexes slowly increases, and in analytical manipulations, it is very important to avoid increasing this contribution

[7]. If the solution is acidified to about pH 3 (point C' is reached by a 10-fold dilution with $10^{-3} M$ $HClO_4$), it can be further diluted with water without forming polynuclear complexes (point C" is reached by a 100-fold dilution with water). A solution of lead(II) in pure water, which has a pH of approximately 8 (point D), contains lead predominantly in the form of hydroxo-complexes. It can be diluted only 30-fold; further dilution causes precipitation of lead(II) hydroxide (point D').

All these considerations about the forms of lead(II) in aqueous solution are valid provided that atmospheric carbon dioxide is excluded and oxidation by oxygen is prevented.

REFERENCES

[1] D. D. Perrin and I. G. Sayce, *Talanta*, 1967, **14**, 833.
[2] R. G. Bates, *Determination of pH*, 2nd Ed., p. 73, Wiley, New York, 1973.
[3] R. E. Mesmer, C. F. Baes and F. H. Sweeton, *Inorg. Chem.*, 1972, **11**, 536.
[4] M. Suchánek and L. Šůcha, *Sb. Vys. Šk. Chemicko-Technol. v Praze*, 1979, **H14**, 79.
[5] C. F. Baes, Jr. and R. E. Mesmer, *The Hydrolysis of Cations*, Wiley, New York, 1976.
[6] J. Kragten, *Atlas of Metal–Ligand Equilibria in Aqueous Solution*, Horwood, Chichester, 1977.
[7] J. Kragten, *Talanta*, 1977, **24**, 483.
[8] J. Kragten, *Analyst*, 1974, **99**, 43.
[9] J. Kragten, *Talanta*, 1975, **22**, 205.

Index

Abbreviation acac, see acetylacetone
Acetaldehyde, redox potentials
  for reduction to ethanol and oxidation to acetic acid, 248
Acetic acid (ac),
  acetato-complexes, 156–7
  protonation, 156
    in aqueous dioxane, 102
    log $\alpha_{L(H)}$ (No. 5), 256
Acetohydroxamic acid,
  protonation and complexes, 186–7
Acetylacetone,
  complexes, 188–9
  protonation, 188
    in mixed and non-aqueous solvents, 103
    log $\alpha_{L(H)}$ (No. 52), 258
Activity, 12
  and concentration scales, 16, 17
Activity coefficients,
  for non-electrolytes, 20–1
    electrolytes, 22–9
      mean activity coefficients, 24–6
      individual ion, estimation, 23
Adenine,
  protonation in aqueous dioxane, 103
    log $\alpha_{L(H)}$ (No. 50), 258
Adenosine,
  protonation, 98
    log $\alpha_{L(H)}$ (No. 129), 262
Adenosine-5'-diphosphoric acid,
  protonation, 98
    trihydrogen salt (ADP), log $\alpha_{L(H)}$ (No. 133), 262
Adenosine-2'-monophosphoric acid,
  protonation, 98
    dihydrogen salt (AMP-2), log $\alpha_{L(H)}$ (No. 130), 262
Adenosine-3'-monophosphoric acid,
  protonation, 98
    dihydrogen salt (AMP-3), log $\alpha_{L(H)}$ (No. 131), 262
Adenosine-5'-monophosphoric acid,
  protonation, 98
    dihydrogen salt (AMP-5), log $\alpha_{L(H)}$ (No. 132), 262
Adenosine-5'-triphosphoric acid,
  protonation, 98
    tetrahydrogen salt (ATP), log $\alpha_{L(H)}$ (No. 137), 262
Adipic acid,
  protonation, 96
    log $\alpha_{L(H)}$ (No. 85), 260
Alanine (2-aminopropionic acid),
  protonation and complexes, 167–8
    log $\alpha_{L(H)}$ values (No. 18), 256
β-Alanine (3-aminopropionic acid),
  protonation and complexes, 168
    log $\alpha_{L(H)}$ (No. 19), 256
Alizarin Red S,
  protonation and complexes, 202
Aluminium,
  compounds, solubility product, 212
  log $\alpha_{M(OH)}$, 284, log $\alpha_{M(L,OH)}$, 290, 308
  redox potential, 223
Americium,
  redox potentials, 224
Amidosulphate, $NH_2SO_3^-$,
  protonation, 91
Aminoacetic acid,
  complexes, 167
  protonation, 167
    in aqueous dioxane, 102
    log $\alpha_{L(H)}$ (No. 8), 256
3-Aminobenzenesulphonic and 4-aminobenzenesulphonic acid,
  protonation, 96
2-Aminobenzoic acid (anthranilic acid),

complexes, 174
protonation, 174
in aqueous dioxane, 104
log $\alpha_{L(H)}$ (No. 106), 260
3-Aminobenzoic and 4-aminobenzoic acid,
protonation, 97
2-Aminoethanethiol
protonation, 93
log $\alpha_{L(H)}$ (No. 11), 256
2-Aminoethanol,
protonation and complexes, 143
log $\alpha_{L(H)}$ (No. 10), 256
2-(2'-Aminoethylamino)ethanol,
protonation, 94
2-Amino-2-(hydroxymethyl)-1,3-propanediol
[tris(hydroxymethyl)aminomethane, THAM, tris],
metal complexes, 145
protonation, 145
aqueous methanol, 102
log $\alpha_{L(H)}$ (No. 45), 258
2-Aminophenol,
protonation, 104
6-Aminopurine,
protonation, 95
2-Aminopyridine and 4-aminopyridine,
protonation, 95
3-Aminopyridine,
protonation, 95
log $\alpha_{L(H)}$ (No. 51), 258
8-Aminoquinoline,
protonation, 98
Aminosuccinic acid, see aspartic acid
Ammonia,
ammine-complexes, 115
protonation, 90
in mixed and non-aqueous solvents, 99
log $\alpha_{L(H)}$, 254
Aniline,
protonation, 96
in aqueous ethanol, 103
Anthranilic acid, see 2-aminobenzoic acid
Antimonic acid,
hexahydroxoantimonate, Sb(OH)$_6^-$, protonation, 91
Antimony,
antimony(III) oxide and sulphide, solubility products, 217
compounds, redox potentials, 242–3
log $\alpha_{M(OH)}$, 286
log $\alpha_{M(L,OH)}$, 302, 330
Arginine,
protonation, 97
log $\alpha_{L(H)}$ (No. 93), 260

Arsenazo I,
protonation and complexes, 202–3
log $\alpha_{L(H)}$ (No. 173), 264
Arsenazo III,
protonation and complexes, 205
log $\alpha_{L(H)}$ (No. 185), 264
Arsenic,
compounds, redox potentials, 224
oxide, solubility product, 212
Arsenic acid,
arsenate, protonation, 91
log $\alpha_{L(H)}$, 254
Arsenous acid,
tetrahydroxoarsenite, protonation, 91
log $\alpha_{L(H)}$, 254
Ascorbic acid (asco),
protonation and complexes, 161–2
log $\alpha_{L(H)}$ (No. 78), 260
redox couple with dehydroascorbic acid, redox potential, 249
Asparagine,
protonation, 94
log $\alpha_{L(H)}$ (No. 35), 256
Aspartic acid,
protonation and complexes, 169–70
log $\alpha_{L(H)}$ (No. 37), 256
Azide,
protonation, 90
in methanol, 99
log $\alpha_{L(H)}$, 254

BAL, see 2,3-dimercaptopropanol
Barium,
compounds, solubility products, 212
Ba^{2+}, redox potential, 225
log $\alpha_{M(L,OH)}$, 290, 309
Bathophenanthroline, see 4,7-diphenyl-1,10-phenanthroline
Benzenearsonic acid,
log $\alpha_{L(H)}$ (No. 76), 258
Benzenecarboxylic acid, see benzoic acid
Benzene-1,2-dicarboxylic acid, see phthalic acid
Benzenethiol,
protonation, 96
log $\alpha_{L(H)}$ (No. 73), 258
Benzil dioxime,
protonation and complexes, 199
Benzoic acid,
complexes, 163
protonation, 163
in aqueous ethanol and dioxane, 104
Benzoylacetone,
complexes, 198

Index

protonation, 98
  in mixed and non-aqueous solvents, 105
  $\log \alpha_{L(H)}$ (No. 128), 262
Berkelium,
  redox potentials, 225
Beryllium,
  hydroxide, solubility product, 212
  $\log \alpha_{M(OH)}$, 284
  $\log \alpha_{M(L,OH)}$, 290, 310
  redox potential, 225
Biacetyl dioxime, see 2,3-butanedione dioxime
2,2'-Bipyridyl (bipy),
  complexes, 150–1
  protonation, 150
    in aqueous ethanol and dioxane, 105
  $\log \alpha_{L(H)}$ (No. 125), 262
N,N'-Bis(2-hydroxybenzyl)ethylenedinitrilo-N,N'-diacetic acid (HBED),
  $\log \alpha_{L(H)}$ (No. 181–3), 264
Bismuth,
  compounds, solubility products, 212
    redox potentials, 225
  $\log \alpha_{M(OH)}$, 284
  $\log \alpha_{M(L,OH)}$, 290, 310
Boric acid,
  borate,
    protonation, 90
      in mixed solvents, 99
  borato-complexes, 111
  calculation of the pH in presence of polyborates, 382–4
  pH-dependence solubility diagram, 345
  polyborates, 344
    protonation, 90
  solubility in water, 343
Boron,
  compounds, redox potentials, 225
Bromoacetic acid,
  protonation, 93
Bromide,
  bromo-complexes, 139–40
  protonation in non-aqueous solvents, 100
Bromine,
  oxidation states, redox potentials, 225–6
Butanedioic acid, see succinic acid
2-Butenedioic acid, see maleic acid
2,3-Butanedione dioxime,
  complexes, 188
  protonation, 188
    in aqueous dioxane, 102
  $\log \alpha_{L(H)}$ (No. 38), 256
Butanoic acid, see butyric acid

Butyric acid,
  $\log \alpha_{L(H)}$ (No. 39), 258

Cadmium,
  compounds, solubility products, 213
    redox potentials, 226
  Cd^{2+}, redox equilibria, effect of chloro-complexes, 379–80
  $\log \alpha_{M(OH)}$, 284
  $\log \alpha_{M(L,OH)}$, 292, 310–13
Calcichrome,
  protonation and complexes, 205
  $\log \alpha_{L(H)}$ (No. 191), 266
Calcium,
  compounds, solubility products, 212–3
  $\log \alpha_{M(L,OH)}$, 292, 310
  redox potential, 226
Calcon,
  protonation and complexes, 204
Californium,
  redox potential, 227
Calmagite,
  protonation and complexes, 203
Carbon,
  redox potentials for inorganic compounds, 226
  redox potential for the reduction to methane, 248
Carbonic acid
  carbonato-complexes, 111
  protonation, 90
  $\log \alpha_{L(H)}$, 254
Cerium,
  compounds, solubility products, 213
  $\log \alpha_{M(OH)}$, 284
  $\log \alpha_{M(L,OH)}$, 292, 312
  redox potentials, 226–7
Cesium,
  periodate, solubility product, 213
  redox potential, 228

Chelidamic acid,
  protonation, 97
  $\log \alpha_{L(H)}$ (No. 97), 260
Chemical potential,
  relation to activity, 12
Chloranilic acid,
  $\log \alpha_{L(H)}$ (No. 64), 258
  protonation and complexes, 190
Chlorate, ClO_3^-,
  protonation, 92
Chloride (hydrogen chloride)
  chloro-complexes, 135–8
  protonation, 92
    in mixed and non-aqueous solvents, 100

Chlorine,
   oxidation states and redox potentials, 227
Chlorite, ClO_2^-,
   protonation, 92
Chloroacetic acid (Cl-ac),
   $\log \alpha_{L(H)}$ (No. 4), 256
2-Chloro-1,10-phenanthroline,
   $\log \alpha_{L(H)}$ (No. 146), 262
   protonation and complexes, 151
5-Chloro-1,10-phenanthroline,
   $\log \alpha_{L(H)}$ (No. 147), 262
   protonation and complexes, 151
Chlorophosphonazo III,
   protonation and complexes, 204
6-Chloropurine,
   protonation in aqueous dioxane, 102
Chromate,
   protonation constants, 91
   $\log \alpha_{L(H)}$, 254
Chrome Azurol S,
   protonation and complexes, 205
Chromotropic acid,
   $\log \alpha_{L(H)}$ (No. 127), 262
   protonation and complexes, 197–8
Citric acid (citr),
   $\log \alpha_{L(H)}$ (No. 79–81), 260
   protonation and complexes, 162
Cobalt,
   compounds, solubility products, 213
   redox potentials, 228
   $\log \alpha_{M(OH)}$, 284
   $\log \alpha_{M(L,OH)}$, 292–4, 312–5
Concentration scales,
   molality and molar scale, interconversion, 17
Conditional concentrations
   of reacting species, 44–5
Conditional extraction constant, 57
   in calculation of extraction coefficient, 59
Conditional redox potentials, 53–7
   effect of formation of complexes, 55
      for different concentrations of the ligand, 56
      pH-dependence for the Fe(III)–Fe(II) couple in presence of 5-sulphosalicylate, 54
   effect of precipitation, 55–6
Conditional solubility product,
   definition, 52
   pH-dependence for metal hydroxides, 340
      metal sulphides, 342
         some sparingly soluble precipitates, 52, 343
      for presence of masking agents, 344
Conditional stability constant, 32–5
   for calculation of displacement equilibria of Ca^{2+} from EDTA-type complexes, 385–6

formulation for complicated reaction systems, 49, 50
diagram for iron(III)-EDTA reaction, in dependence on pH, 51
Copper,
   compounds, solubility products, 214
   redox potentials, 229
   $\log \alpha_{M(OH)}$, 284
   $\log \alpha_{M(L,OH)}$, 294, 314–7
CPDTA, see 1,2-cyclopentylenedinitrilotetra-acetic acid
Cresol Red,
   $\log \alpha_{L(H)}$ (No. 184), 264
Cupferron, see N-nitrosophenylhydroxylamine
Curium,
   redox potential, 227
Cyanide,
   cyano-complexes, 112
   protonation, 90
   $\log \alpha_{L(H)}$, 254
1,2-Cyclohexanedione dioxime (nioxime),
   complexes, 194
   protonation, 96
      in aqueous dioxane, 104
   $\log \alpha_{L(H)}$ (No. 84), 260
cis-1,2-Cyclohexylenedinitrilotetra-acetic acid (cis-DCTA),
   protonation and complexes, 179
trans-1,2-Cyclohexylenedinitrilotetra-acetic acid (DCTA),
   $\log \alpha_{L(H)}$ (No. 162–3), 264
   protonation and complexes, 179–80
N-Cyclohexyliminodiacetic acid,
   protonation, 99
   $\log \alpha_{L(H)}$ (No. 138), 262
trans-1,2-Cyclopentylenedinitrilotetra-acetic acid (CPDTA),
   $\log \alpha_{L(H)}$ (No. 158), 264
   protonation and complexes, 179
Cyclo-tetraphosphate, $P_4O_{12}^{4-}$,
   $\log \alpha_{L(H)}$, 254
   protonation, 91
Cyclo-triphosphate, $P_3O_9^{3-}$
   protonation, 90
   $\log \alpha_{L(H)}$, 254
L-Cysteine (cyst),
   for reduction from cystine, redox potential, 249
   protonation, 94
   $\log \alpha_{L(H)}$ (No. 21), 256

Davies equation, 25
DCTA, see trans-1,2-cyclohexylenedinitrilotetra-acetic acid

Index

DGENTA, see ethylenebis(imino-1-oxoethylene-
nitrilo)tetra-acetic acid
Debye-Hückel equations, 22–5
  additional terms by Pitzer et al., 27–9
  Davies equation, 25
  Guggenheim expression, 25
  ion-size parameters, 24
Deuterium,
  redox potential, 231
1,3-Diamino-2-propanol,
  log $\alpha_{L(H)}$ (No. 28), 256
  protonation and complexes, 145
Dibenzoylmethane,
  protonation in mixed and non-aqueous solvents, 106
Dichloroacetic acid,
  protonation, 93
2,5-Dichloro-3,6-dihydroxy-1,4-benzoquinone, see chloranilic acid
4,7-Dichloro-1,10-phenanthroline,
  protonation in aqueous dioxane, 105
Dichromate,
  formation constant, 91
Diphosphate, $P_2O_7^{4-}$,
  diphosphato-complexes, 120
  protonation, 90
  log $\alpha_{L(H)}$, 254
dien, see diethylenetriamine
Diethylamine,
  protonation, 94
N,N-Diethyldithiocarbamic acid (DDC),
  log $\alpha_{L(H)}$ (No. 63), 258
  protonation and complexes, 189
Diethylenetriamine,
  log $\alpha_{L(H)}$ (No. 47), 258
Diethylenetrinitrilopenta-acetic acid (DTPA),
  log $\alpha_{L(H)}$ (No. 166–7), 264
  protonation and complexes, 181–2
Diglycine,
  log $\alpha_{L(H)}$ (No. 41), 258
1,2-Dihydroxybenzene (pyrocatechol),
  complexes, 190–1
  protonation, 190
    in mixed and non-aqueous solvents, 103
  log $\alpha_{L(H)}$ (No. 69), 258
1,3-Dihydroxybenzene (resorcinol),
  protonation and complexes, 191
1,4-Dihydroxybenzene (hydroquinone),
  reduction from p-quinone, redox potential, 249
1,2-Dihydroxybenzene-3,5-disulphonic acid (tiron),
  log $\alpha_{L(H)}$ (No. 74), 258
  protonation and complexes, 192–3
1,2-Dihydroxybenzene-4-sulphonic acid,
  protonation and complexes, 192
3,4-Dihydroxybenzoic acid,
  protonation and complexes, 164
3,5-Dihydroxybenzoic acid,
  protonation and complexes, 164
1,8-Dihydroxynaphthalene-3,6-disulphonic acid,
  see chromotropic acid
1,3-Dihydroxy-4-(2-pyridylazo)benzene, see PAR
2,3-Dimercaptopropanol,
  protonation and complexes, 187
  log $\alpha_{L(H)}$ (No. 24), 256
Dimethylamine, protonation, 93
5-(4′-Dimethylaminobenzylidene)rhodanine,
  protonation and complexes, 198–9
Dimethylglyoxime, see 2,3-butanedione dioxime
Dimethyl-1,10-phenanthroline, log $\alpha_{L(H)}$,
  2,9- –, neocuproine, (No. 159), 264
  4,7- –, (No. 160), 264
    protonation in aqueous dioxane, 106
  5,6- –, (No. 161), 264
    protonation in aqueous dioxane, 106
  protonation and complexes, 152–3
Dimethylpyridine,
  2,3- –, 2,4- –, 2,5- –, 2,6- –, 3,4- –, 3,5- –,
  protonation, 97
1,5-Diphenylcarbazone,
  complexes, 201
  protonation, 201
    in dioxane, 106
4,7-Diphenyl-1,10-phenanthroline
(bathophenanthroline),
  complexes, 153
  protonation, 153
  log $\alpha_{L(H)}$ (No. 187), 264
1,5-Diphenylthiocarbazone, see dithizone
Dipicolinic acid, see pyridine-2,6-dicarboxylic acid
Displacement reaction
  of calcium from EDTA-type complexes, 384–6
  equilibrium constant, calculation, 385
Dissociation constant of an acid, 30
  conversion to protonation constant, 31
Distribution diagrams, 67–9
  calculation of distribution coefficients, 46
  for a consecutive protonation, 68
    stepwise complexation, 68
    presence of polynuclear species, 69
Dithio-oxamide, protonation, 93
Dithizone (1,5-diphenylthiocarbazone),
  complexes, 201
  protonation in various solvents, 106
DTPA, see diethylenetrinitrilopenta-acetic acid

EDDA, see ethylenediamine-N,N'-diacetic acid

Index

EDDG, see ethylenedi-iminodi-2-pentanedioic acid
EDDP, see ethylenediamine-N,N'-di-3-propanoic acid
EDDS, see ethylenedi-iminodibutanedioic acid
EDTA, see ethylenedinitrilotetra-acetic acid
EEDTA, see oxybis(ethylenenitrilo)tetra-acetic acid
EGTA, see ethylenebis(oxyethylenenitrilo)tetra-acetic acid
EHPG, see ethylenedi-iminobis(2-hydroxyphenyl)-acetic acid
Einsteinium, log $\alpha_{M(OH)}$, 284
en, see ethylenediamine
Enthalpy,
   change of standard enthalpy, 13
Entropy, 13
Equilibrium constant,
   conditional (effective) constant, notation, 43
   for a chemical reaction, definition, 13, 16
      effect of ionic strength, 20–31
      effect of temperature, 18–20
   for an overall redox reaction, 35
      calculation from standard redox potentials, 36
   for a redox couple, partial equilibrium constant by Sillén, 37
      and Sillén's pe quantity, 37
   for a precipitation equilibrium with formation of complexes with common ion, 41
   thermodynamic and concentration equilibrium constant, interconversion, 29–30
Eriochrome Black R, see Calcon
Eriochrome Black T,
   protonation and complexes, 203–4
   log $\alpha_{L(H)}$ (No. 180), 264
Eriochrome Cyanine R,
   protonation and complexes, 205
   log $\alpha_{L(H)}$ (No. 186), 264
Ethane-1,2-dithiol, protonation, 93
Ethylamine, protonation, 93
Ethyldi-iminodi-3-propionic acid, protonation, 98
Ethylenebis(imino-1-oxyethylenenitrilo)tetra-acetic acid, (DGENTA),
   log $\alpha_{L(H)}$ (No. 164–5), 264
Ethylenebis(oxyethylenenitrilo)tetra-acetic acid (EGTA),
   complexes, 182–3
   protonation, 182
   log $\alpha_{L(H)}$ (No. 168–9), 264
Ethylenediamine (en),
   complexes, 143–4
   protonation, 143
      in aqueous dioxane, 102
Ethylenediamine-N,N'-diacetic acid (EDDA),
   log $\alpha_{L(H)}$ (No. 89), 260

Ethylenediamine-N,N'-di-3-propanoic acid, (EDDP),
   log $\alpha_{L(H)}$ (No. 114), 260
Ethylenedi-iminobis[(2-hydroxyphenyl)acetic] acid (EHPG),
   log $\alpha_{L(H)}$ (No. 175), 264
Ethylenedi-iminodibutanedioic acid (EDDS),
   log $\alpha_{L(H)}$ (No. 136), 262
Ethylenedi-iminodi-2-pentanedioic acid (EDDG),
   log $\alpha_{L(H)}$ (No. 150–1), 262
Ethylenedinitrilotetra-acetic acid (EDTA),
   protonation and complexes, 175–7
      log $\alpha_{L(H)}$ (No. 134–5), 262
Ethylenedinitrilotetra-acetohydroxamic acid,
   protonation, 99
N-Ethyliminodimethylenediphosphonic acid,
   protonation, 94
   log $\alpha_{L(H)}$ (No. 46), 258
4-Ethylpyridine, protonation, 97
Extraction,
   constants, definitions, 38–9
   distribution coefficient, 38

Fermium, redox potential, 230
Ferrocene,
   ferricinium-ferrocene redox potential, 249
Ferrocyanide, protonation constant, 92
Ferron, see 7-iodo-8-hydroxyquinoline-5-sulphonic acid
Fluoroacetic acid, protonation, 93
Fluoride,
   complexes, 133–5
   protonation, 92
      in ethanol, 100
   log $\alpha_{L(H)}$, 254
Fluorine,
   redox potential, 229
Fluorophosphate, protonation, 91
Formaldehyde,
   formaldehyde–methanol redox potential, 248
Formic acid,
   complexes, 154
   protonation, 154
      in aqueous dioxane, 102
      log $\alpha_{L(H)}$ (No. 1), 256
   redox potential, formation from CO_2, 248
      formic acid–formaldehyde, 248
Fructose, complexes, 194
Fumaric acid, protonation, 94
   log $\alpha_{L(H)}$ (No. 31), 256
3-Furancarboxaldehyde oxime, protonation in aqueous dioxane, 103
α-Furil dioxime, protonation, 197

log $\alpha_{L(H)}$ (No. 126), 262
complexes, 197

Gallic acid, protonation, 164
   log $\alpha_{L(H)}$ (No. 103–4), 260
   complexes, 164
Gallium,
   compounds,
      redox potentials, 230
      solubility products, 214
   log $\alpha_{M(OH)}$, 284
   log $\alpha_{M(L,OH)}$, 296, 320
Germanates,
   mono- and polynuclear species, protonation, 90
Germanium,
   monoxide, solubility product, 214
   oxidation states and redox potentials, 230–1
Gibbs free energy,
   of a reaction, 12, 35
   partial molal, 12
   relation to other thermodynamic quantities, 13
Glucose, complexes, 194
Glutamic acid (glutam),
   protonation, 95
   log $\alpha_{L(H)}$ (No. 56), 258
Glutamine,
   protonation, 95
   log $\alpha_{L(H)}$ (No. 59), 258
Glutaric acid,
   log $\alpha_{L(H)}$ (No. 53), 258
Glycerol, complexes, 187–8
Glycinamide,
   protonation, 93
   log $\alpha_{L(H)}$ (No. 9), 256
Glycine (glyc), see aminoacetic acid
Glycollic acid see hydroxyacetic acid
Glycylglycine, protonation, 170
   in aqueous dioxane, 102
   complexes, 170–1
Glycylglycylglycine (triglycine), protonation, 97
   log $\alpha_{L(H)}$ (No. 88), 260
Glycylglycylglycylglycine (tetraglycine),
   protonation, 98
   log $\alpha_{L(H)}$ (No. 113), 260
Glyoxylic acid, protonation, 93
   log $\alpha_{L(H)}$ (No. 2), 256
Gold,
   compounds, redox potentials, 224
   hydroxide, solubility product, 212
   log $\alpha_{M(OH)}$, 284
   log $\alpha_{M(L,OH)}$, 290
Guanidine, protonation, 93

Hafnium,
   dioxide, solubility product, 215
   redox potential, 231
   log $\alpha_{M(OH)}$, 284
   log $\alpha_{M(L,OH)}$, 296
HBED, see N,N'-bis(2-hydroxybenzyl)ethylenedinitrilo-N,N'-diacetic acid
N-(2-Hydroxyethyl)ethylenedinitrilo-N,N',N'-triacetic acid (HEDTA),
   log $\alpha_{L(H)}$ (No. 139–40), 262
Hexafluoroacetylacetone, protonation in aqueous dioxane, 102
Histamine (histam), protonation, 95
   log $\alpha_{L(H)}$ (No. 58), 258
L-Histidine (histid), protonation, 96
   log $\alpha_{L(H)}$ (No. 83), 260
L-Homoserine, protonation, 94
   log $\alpha_{L(H)}$ (No. 42), 258
Hydrazine, protonation, 90
Hydrochloric acid, see chloride
Hydrogen,
   redox potentials, 231
Hydrogen peroxide, protonation of HO_2^-, 91
   log $\alpha_{L(H)}$, 254
   redox potential, 237
Hydrogen sulphide, see sulphide
Hydrolysis,
   of metal ions in aqueous solution, 390–7
      diagrams of alpha-coefficients, 281–2
      pH calculation for negligible contributions of hydroxo-complexes, 391
      pH region for safe dilution with water, 399
      pM'-pH diagram for lead(II), 395
         contribution of polynuclear species, 394
Hydroxyacetic (glycollic) acid,
   complexes, 157
   protonation, 157
   in aqueous dioxane, 102
   log $\alpha_{L(H)}$ (No. 7), 256
Hydroxo-complexes,
   of Fe(III) and Fe(II), 346, 360
   indium(III), 348
   lead(II), pH-dependence, 393
   stability constants for various ions, 124–9
2-Hydroxybenzaldehyde, see salicylaldehyde
2-Hydroxybenzaldehyde oxime, protonation
   in various solvents, 104
L-Hydroxybutanedioic acid, see malic acid
2-Hydroxyethylamine, protonation in ethanol, 102
N-(2-Hydroxyethyl)ethylenedinitrilo-N,N',N'-triacetic acid (HEDTA),
   protonation and complexes, 178–9
   log $\alpha_{L(H)}$ (No. 139–40), 262

N-(2-Hydroxyethyl)iminodiacetic acid,
  protonation, 96
Hydroxylamine, protonation, 90
2-Hydroxyphenylarsonic acid and 4- – –,
  protonation, 96
L-Hydroxyproline, protonation, 95
  $\log \alpha_{L(H)}$ (No. 55), 258
6-Hydroxypurine, protonation in aqueous dioxane,
102
4-Hydroxypyridine-2,6-dicarboxylic acid
(chelidamic acid),
  $\log \alpha_{L(H)}$ (No. 97), 260
8-Hydroxyquinoline (8-quinolinol, oxine),
  complexes, 149
  protonation, 149
    in various solvents, 104
    $\log \alpha_{L(H)}$ (No. 117), 262
8-Hydroxyquinoline-5-sulphonic acid (sulfoxine),
  complexes, 150
  protonation, 150
    in various solvents, 105
    $\log \alpha_{L(H)}$ (No. 118), 262
Hypochlorite, protonation, 92
Hypoiodite, protonation, 92
Hyponitrite, $N_2O_2^{2-}$, protonation, 90

IDA, see iminodiacetic acid
Imidazole, protonation, 93
  $\log \alpha_{L(H)}$ (No. 13), 256
Iminobis(2-ethylamine) (dien),
  protonation and complexes, 146
  $\log \alpha_{L(H)}$ (No. 47), 258
Iminodiacetic acid (IDA),
  protonation and complexes, 168–9
  $\log \alpha_{L(H)}$ (No. 36), 256
2,2'-Iminodiethanol, protonation, 94
  in aqueous dioxane, 102
Indicators
  see Cresol Red,
  metallochromic indicators, 200–7
    $\log \alpha_{L(H)}$, 271–7
  see Methylene Blue
Indium,
  compounds, solubility products, 215
    redox potentials, 232
  indium(III) hydroxo-complexes, 128
    predominance regions, 376
    and redox equilibria, 370–6
  indium(III) mixed hydroxo-chloro-complexes,
348
    predominance-area diagram, 351–5
    $\log \alpha_{M(OH)}$, 284–7
    $\log \alpha_{M(L,OH)}$, 298, 322

Iodate, protonation, 92
Iodide, protonation, 92
  in pyridine, 100
  iodo-complexes, 141
Iodine,
  redox potentials for various compounds, 232
7-Iodo-8-hydroxyquinoline-5-sulphonic acid
(ferron),
  complexes, 148–9
  protonation, 148
    in aqueous dioxane, 105
    $\log \alpha_{L(H)}$ (No. 116), 262
Ionic product of water for varied ionic strength,
91, 124
Iridium,
  compounds, redox potentials, 233
Iron,
  compounds, solubility products, 214
    redox potentials, 229–30
    solubility of iron(III) hydroxide, 346
  iron(III) hydroxo-complexes, 126–7, 346
  logarithmic activity-ratio diagram for the
    Fe(VI)–Fe(III)–Fe(II)–Fe equilibrium, 357–9
  influence of hydrolysis, 359–70
  $\log \alpha_{M(OH)}$, 284, $\log \alpha_{M(L,OH)}$, 294–7, 316–21
Isobutyric acid,
  $\log \alpha_{L(H)}$ (No. 40), 258
L-Isoleucine, protonation, 97
  $\log \alpha_{L(H)}$ (No. 91), 260
DL-Isoserine, protonation, 94
  $\log \alpha_{L(H)}$ (No. 23), 256

Kojic acid,
  complexes, 191–2
  protonation, 103, 191
    $\log \alpha_{L(H)}$ (No. 71), 258

Lactic acid,
  complexes, 158–9
  protonation, 158
    $\log \alpha_{L(H)}$ (No. 17), 256
  pyruvic–lactic acid, redox potential, 249
Lanthanum,
  redox potential, 233
  compounds, solubility products, 215
  $\log \alpha_{M(OH)}$, 286
  $\log \alpha_{M(L,OH)}$, 298, 322
Lanthanides (Ln),
  complexes with inorganic ligands, 111–41
    organic ligands, 143–207
  redox potentials, 233
  solubility products of some salts, 215

Index

Lead,
    compounds, solubility products, 216–7
    hydroxo-complexes, 128, 393
    masking and removal from a living organism, 384
L-Leucine, protonation, 97
    log $\alpha_{L(H)}$ (No. 90), 260
Lithium,
    redox potential, 233
    fluoride, solubility product, 215
Logarithmic activity-ratio diagram,
    example of a simple redox system, 78–9
    for redox system of iron, simplified, 357
    and hydrolysis, 366–8
Logarithmic concentration diagrams for protonation equilibria,
    construction, 70
        for close values of protonation constants, 73
        reading concentrations of species, 72
Logarithmic solubility diagram,
    effect of complexation with common ion, 74–7
    example of construction, 74
    for boric acid, 345
        indium(III) hydroxide, presence of hydroxo- and chloro-complexes, 349–50
        iron(III) hydroxide, 347
        lead(II) hydroxide, 398
L-Lysine, protonation, 97
    log $\alpha_{L(H)}$ (No. 92), 260

Magnesium,
    compounds, solubility products, 216
    log $\alpha_{M(OH)}$, 286
    log $\alpha_{M(L,OH)}$, 298–301, 324
    redox potential, 233
Manganese,
    compounds, solubility products, 216
    redox equilibria and redox potentials, 233–4
        in presence of chloride 386–90
Maleic acid,
    protonation and complexes, 159
    log $\alpha_{L(H)}$ (No. 30), 256
L-Malic acid,
    protonation and complexes, 160
    log $\alpha_{L(H)}$ (No. 33), 256
Malonic acid,
    protonation and complexes, 157–8
    log $\alpha_{L(H)}$ (No. 15), 256
Mandelic acid,
    protonation and complexes, 166
    log $\alpha_{L(H)}$ (No. 111), 260
Mannitol, complexes, 194
Masking,
    efficiency, effect of side-reactions, 385

Mercaptoacetic acid,
    log $\alpha_{L(H)}$ (No. 6), 256
2-Mercaptobenzo-1,3-thiazole,
    protonation and complexes, 195
    log $\alpha_{L(H)}$ (No. 99), 260
2-Mercaptoethanol, complexes, 187
N-(2-Mercaptoethyl)iminodiacetic acid,
    protonation, 96
    log $\alpha_{L(H)}$ (No. 86), 260
2-Mercapto-N-2-naphthylacetamide, see thionalide
3-Mercaptopropionic acid, protonation, 94
6-Mercaptopurine, protonation in aqueous dioxane, 103
8-Mercaptoquinoline, protonation in aqueous dioxane, 104
Mercury,
    compounds, solubility products, 215
    log $\alpha_{M(OH)}$, 284
    log $\alpha_{M(L,OH)}$, 296–9, 320
    redox potentials, 231
Metalphthalein, see phthaleincomplexone
DL-Methionine, protonation, 95
Methylamine, protonation, 93
2- –, 4- Methylaniline, protonation, 97
Methylene Blue, redox potential, 249
DL-(Methylene)dinitrilotetra-acetic acid,
    log $\alpha_{L(H)}$ (No. 143), 262
1-Methylethylenediamine,
    protonation and complexes, 144
2- –, 4-Methyl-8-hydroxyquinoline,
    protonation in aqueous dioxane, 105
N-Methyliminodiacetic acid (MIDA),
    log $\alpha_{L(H)}$ (No. 57), 258
    protonation and complexes, 171
5-Methyl-1,10-phenanthroline,
    log $\alpha_{L(H)}$ (No. 156), 264
    protonation and complexes, 152
3- –, 4-Methylphenol, protonation, 97
2-Methylpropanoic acid,
    log $\alpha_{L(H)}$ (No. 40), 258
4-Methylpyridine,
    protonation in aqueous dioxane, 103
Methylthymol Blue,
    log $\alpha_{L(H)}$ (No. 197–9), 266
    protonation and complexes, 207
Molybdate,
    and polymolybdates, protonation, 91
    log $\alpha_{L(H)}$, 254
Monochloroacetic acid,
    log $\alpha_{L(H)}$ (No. 4), 256
    protonation and complexes, 155–6
        in aqueous dioxane, 102
Murexide, see purpuric acid

1- –, 2-Naphthol, protonation, 98
Neocuproine, see 2,9-dimethyl-1,10-phenanthroline
Neptunium,
  compounds, solubility products, 216
    redox potentials, 237
  $\log \alpha_{M(OH)}$, 286
  $\log \alpha_{M(L,OH)}$, 300, 328
Nernst-Peters equation,
  for a redox half-reaction, 36
Nickel,
  compounds, solubility products, 216
    redox potentials, 236
  $\log \alpha_{M(OH)}$, 286
  $\log \alpha_{M(L,OH)}$, 300, 326
Nicotinic acid, protonation, 95
  $\log \alpha_{L(H)}$ (No. 67), 258
Nioxime, see 1,2-cyclohexanedione dioxime
Nitrate, protonation, 90
  in various solvents, 100
  nitrato-complexes, 116
Nitrolotriacetic acid (NTA),
  $\log \alpha_{L(H)}$ (No. 82), 260
  protonation and complexes, 104, 172–3
Nitrilotris(2-ethylamine) (tren),
  $\log \alpha_{L(H)}$ (No. 96), 260
  protonation and complexes, 148
Nitrilotris(methylenephosphonic) acid,
  protonation and complexes, 188
Nitrite, protonation, 90
  $\log \alpha_{L(H)}$, 254
5-Nitro-1,10-phenanthroline,
  $\log \alpha_{L(H)}$ (No. 148), 262
  protonation and complexes, 105, 151
2-Nitroso-1-naphthol,
  $\log \alpha_{L(H)}$ (No. 122), 262
  protonation and complexes, 105, 196
1-Nitroso-2-naphthol,
  $\log \alpha_{L(H)}$ (No. 123), 262
  protonation and complexes, 105, 196
2-Nitroso-1-naphthol-4-sulphonic acid,
  protonation and complexes, 196–7
1-Nitroso-2-naphthol-3,6-disulphonic acid
  (Nitroso-R acid),
    protonation and complexes, 197
N-Nitrosophenylhydroxylamine (cupferron),
  protonation, 96
  $\log \alpha_{L(H)}$ (No. 72), 258
DL-Norvaline, protonation, 95
  $\log \alpha_{L(H)}$ (No. 62), 258
NTA, see nitrilotriacetic acid

L-Ornithine, protonation, 95
Oxalic acid (ox),

$\log \alpha_{L(H)}$ (No. 3), 256
  protonation and complexes, 154–5
  redox potentials,
    carbon dioxide–oxalic acid,
    carbonic–oxalic acid
    oxalic–formic acid, 248
Oxidation-reduction (redox),
  equilibrium constant, calculation from standard
  redox potentials, 36
Oxidation-reduction (redox) potential,
  conditional redox potential, 53–7
  definition, 36
Oxoacetic acid,
  $\log \alpha_{L(H)}$ (No. 2), 256
2-Oxopropionic acid, protonation, 93
  $\log \alpha_{L(H)}$ (No. 14), 256
Oxybis(ethylenenitrilo)tetra-acetic acid (EEDTA),
  $\log \alpha_{L(H)}$ (No. 154–5), 264

Palladium,
  compounds, solubility products, 217
    redox potentials, 239
  $\log \alpha_{M(OH)}$, 286
  $\log \alpha_{M(L,OH)}$, 302, 328
PAN [1-(2-pyridylazo)-2-naphthol],
  complexes, 202
  $\log \alpha_{L(H)}$ (No. 170), 264
  protonation in aqueous dioxane, 106
p-PAN [1-(2-pyridylazo)-4-naphthol],
  complexes, 202
  $\log \alpha_{L(H)}$ (No. 171), 264
  protonation in aqueous dioxane, 106
PAR [4-(2-pyridylazo)resorcinol, 1,3-dihydroxy-4-
  (2-pyridylazo)-benzene]
    complexes, 200–1
    $\log \alpha_{L(H)}$ (No. 141), 262
    protonation in aqueous dioxane, 105
Partition constant
  for a liquid–liquid distribution, 38
2,4-Pentanedione, see acetylacetone,
Perchloric acid, protonation, 92
  in acetic acid, 100
Permanganate, protonation, 92
Peroxophosphate, PO_5^{3-}, protonation, 91
Peroxosulphate, HSO_5^-, protonation, 91
pH,
  calculation for the equilibrium of boric acid and
  polyborates, 382–4
  concentration scale, standardization, 88
1,10-Phenanthroline (phen),
  complexes, 151–2
  $\log \alpha_{L(H)}$ (No. 149), 262
  protonation in various solvents, 105

Phenol, protonation, 95
   $\log \alpha_{L(H)}$ (No. 68), 258
DL-Phenylalanine, protonation, 98
Phenylarsonic acid, protonation, 96
Phenyl-1,3-butanedione, see benzoylacetone
N-Phenylglycine, protonation, 98
   $\log \alpha_{L(H)}$ (No. 112), 260
L-Phenylhydroxyacetic acid, see mandelic acid
5-Phenyl-1,10-phenanthroline,
   $\log \alpha_{L(H)}$ (No. 174), 264
(Phenylthio)acetic acid,
   protonation and complexes, 166
Phosphate, protonation, 90
   in various solvents, 100
   $\log \alpha_{L(H)}$, 254
   phosphato-complexes, 118–9
L-Phosphoserine, protonation, 94
   $\log \alpha_{L(H)}$ (No. 25), 256
Phthaleincomplexone,
   $\log \alpha_{L(H)}$ (No. 195–6), 266
   protonation and complexes, 206
Phthalic acid, complexes, 165–6,
   protonation, 98
   $\log \alpha_{L(H)}$ (No. 110), 260
2- –, 3-Picoline, protonation, 96
4-Picoline, protonation, 96
   $\log \alpha_{L(H)}$ (No. 77), 260
Picolinic acid, see pyridine-2-carboxylic acid
Pimelic acid, protonation, 97
   $\log \alpha_{L(H)}$ (No. 107), 260
Piperazine, protonation, 94
   $\log \alpha_{L(H)}$ (No. 44), 258
Piperidine, protonation, 95
   $\log \alpha_{L(H)}$ (No. 60), 258
Plutonium,
   compounds, solubility products, 217
      redox potentials, 239–40
   $\log \alpha_{M(OH)}$, 286
      diagram of pH-dependence, 280
   $\log \alpha_{M(L,OH)}$, 302, 328–31
pM'–pH diagram for hydrolysis of lead(II), 396
   application, 399
Polymetaphosphato-complexes, 123
Polyphosphato-complexes, 122
Predominance-area diagrams,
   for variable concentrations of two reactants, 80
      example of construction, 80–3
      for a saturated solution and formation of complexes, 85
         indium(III) hydroxo- and chloro-complexes, 352
         redox system In(III)–In(I)–In in presence of hydroxo-complexes, 371–2

         redox equilibria of indium in presence of solid indium(III) hydroxide, 372–6
   for redox and protolytic equilibria of manganese, 387
   for redox system Cl(I)–Cl–Cl(–I), 388–9
   for redox system Cd(II)–Cd in presence of chloride, 379
   for redox system O_2–H_2O_2–H_2O as function of pH, 377
Programs for computers,
   COMICS, 60–2
   DH-LETAG, 33
   HALTAFALL, 62–6
   IONEX, 384
   LETAGROP, 33
L-Proline, protonation, 95
   $\log \alpha_{L(H)}$ (No. 54), 258
Propane-1,2,3-tricarboxylic acid, protonation, 96
Propionic acid, protonation, 93, 102
   $\log \alpha_{L(H)}$ (No. 16), 256
1,2-Propylenediamine (pn), see 1-methylethylenediamine
Protonation (acid-base) equilibria, 30–4
Protonation constants,
   and pH scales, 87
   for a consecutive equilibrium, 32
      overall constant, 32
   thermodynamic constant, 31
      and mixed or concentration constants, 32–4
      calculation with the aid of computer programs, 33
Purine, protonation, 95
   $\log \alpha_{L(H)}$ (No. 48), 258
Purpuric acid, ammonium salt (murexide),
   $\log \alpha_{L(H)}$ (No. 109), 260
   protonation and complexes, 200
Pyridine (py),
   $\log \alpha_{L(H)}$ (No. 49), 258
   protonation and complexes, 146–7
Pyridine-2-carboxylic acid,
   protonation and complexes, 103, 171–2
Pyridine-2,6-dicarboxylic acid,
   $\log \alpha_{L(H)}$ (No. 98), 260
   protonation and complexes, 173–4
1-(2'-Pyridylazo)-2-naphthol, see PAN
1-(2'-Pyridylazo)-4-naphthol, see p-PAN
2-(2'-Pyridylazo)phenol and 4- –, protonation in aqueous methanol, 105
4-(2-Pyridylazo)resorcinol, see PAR
N-(2-Pyridylmethyl)iminodiacetic acid, protonation, 98
Pyrocatechol, see 1,2-dihydroxybenzene
Pyrocatechol Violet,

log $\alpha_{L(H)}$ (No. 179), 264
protonation and complexes, 203
Pyrogallol, *see* 1,2,3-trihydroxybenzene
Pyrophosphoric acid, *see* diphosphate
Pyrrolidine, protonation, 94
Pyruvic acid,
log $\alpha_{L(H)}$ (No. 14), 256

Quinoline, protonation, 98
Quinoline-2-carboxylic acid (quinaldic, quinaldinic acid),
log $\alpha_{L(H)}$ values (No. 124), 262
protonation and complexes, 175
Quinoline-8-carboxylic acid,
protonation and complexes, 175
8-Quinolinol, *see* 8-hydroxyquinoline

Resorcinol (1,3-dihydroxybenzene),
log $\alpha_{L(H)}$ (No. 70), 258
protonation and complexes, 191
Rhodanine, protonation, 93
Rhodizonic acid, protonation, 95
log $\alpha_{L(H)}$ (No. 65), 258

Salicylaldehyde (2-hydroxybenzaldehyde),
log $\alpha_{L(H)}$ (No. 100), 260
protonation and complexes, 104, 195
Salicylic acid (sal),
log $\alpha_{L(H)}$ (No. 102), 260
protonation and complexes, 104, 163
Saligenine, protonation in aqueous dioxane, 104
Sarcosine, protonation, 94
log $\alpha_{L(H)}$ (No. 20), 256
Scandium,
hydroxide, solubility product, 217
log $\alpha_{M(OH)}$, 286
log $\alpha_{M(L,OH)}$, 302, 330
redox potential, 243
Selectivity coefficient,
for equilibrium on an ion-exchanger, 39
relation to the distribution coefficient for an ion, 40
Selenate, protonation, 91
Selenide, protonation, 91
Selenite, protonation, 91
Semicarbazide,
protonation and complexes, 186
Semixylenol Orange,
log $\alpha_{L(H)}$ (No. 188–90), 264
L-Serine, protonation, 94
log $\alpha_{L(H)}$ (No. 22), 256
Side-reaction coefficients,
calculation, 45–9

for zinc-ammine complexes, 46
definitions, 45
for complexation of metals in dependence on pH, 47
logarithmic values for some metal ions,
inorganic ligands, *see* Table 11.2
organic ligands, *see* Table 11.3
diagram for zinc-ammine complexes at different concentrations of ammonia, 283
diagrams for the hydrolysis of lead(II), 392–4
for extraction distribution of 8-hydroxyquinoline, calculation, 58
for hydrolysis of metals, log $\alpha_{M(OH)}$, 45, 49
logarithmic values for some metal ions, 284–9
diagrams of pH-dependence, 280–2
for protonation of ligands,
definition, 48, 251
logarithmic values, log $\alpha_{L(H)}$
for some inorganic ligands, 254–5
for some organic ligands, 256–65
Silicic acid,
protonation of mono- and polynuclear species, 90
Silver,
compounds, solubility products, 211
redox potentials, 223
log $\alpha_{M(OH)}$, 284
log $\alpha_{M(L,OH)}$, 290, 306
Solochrome Violet R
log $\alpha_{L(H)}$ (No. 172), 264
Solubility,
calculation with the use of conditional solubility product, 52, 339
common ion effect and formation of complexes, 52–3, 339
of indium(III) hydroxide in presence of chloride, 350–5
minimum solubility, 346
Solubility product,
conditional, 52, 339
definition, 40, 339
values for some sparingly soluble compounds, 211–9
Succinic acid,
protonation and complexes, 159–60
log $\alpha_{L(H)}$ (No. 32), 256
Sulphate, protonation, 91
in various solvents, 100
log $\alpha_{L(H)}$, 254
sulphato-complexes, 130–2
Sulphide
and polysulphides, protonation, 91
log $\alpha_{L(H)}$, 254
sulphido-complexes, 129

Sulphite, protonation, 91
  in aqueous methanol, 100
  log $\alpha_{L(H)}$, 254
  sulphito-complexes, 130
Sulphonazo III,
  protonation and complexes, 204
5-Sulphosalicylaldehyde,
  protonation and complexes, 196
5-Sulphosalicylic acid (sulfosal),
  complexes, 164–5
  protonation, 104
  log $\alpha_{L(H)}$ (No. 105), 260
Stability constant of a complex
  overall constant, definition, 35
  stepwise (consecutive) constant, 34
Standard state,
  and the reference state, 17
  for solutes, 16
  of a substance, 12

TAN [1-(1′,3′-thiazol-2′-ylazo)-2-naphthol]
  protonation and complexes, 201
  log $\alpha_{L(H)}$ (No. 157), 264
TAR [4-(2′-thiazolylazo)-1,3-dihydroxybenzene]
  protonation in various solvents, 105
Tartaric acid (tart),
  protonation and complexes, 102, 160–1
  log $\alpha_{L(H)}$ (No. 34), 256
TEDTA, see thiobis(ethylenenitrilo)tetra-acetic acid
Tetra(aminomethyl)methane, protonation, 95
Tetraethylenepentamine (tetren),
  protonation and metal complexes, 148
  log $\alpha_{L(H)}$ (No. 115), 262
Tetraglycine, protonation, 98
  log $\alpha_{L(H)}$ (No. 113), 260
Tetrametaphosphate, see cyclo-tetraphosphate
Tetraphosphate, $P_4O_{13}^{6-}$, protonation, 90
  log $\alpha_{L(H)}$, 254
Thallium,
  compounds, solubility products, 218
    redox potentials, 244
  log $\alpha_{M(OH)}$, 288
  log $\alpha_{M(L,OH)}$, 304, 332–5
  logarithmic activity-ratio diagram for the system Tl(III)–Tl(I)–Tl, 78–9
3-Thenoylacetone,
  protonation in aqueous dioxane, 104
2-Thenoyltrifluoroacetone (TTA),
  protonation and complexes, 104, 196
Thermodynamics,
  of chemical equilibrium, 12–3
Thiobis(2-ethylamine), protonation, 94
Thiobis(ethylenenitrilo)tetra-acetic acid (TEDTA),
  log $\alpha_{L(H)}$ (No. 152–3), 262–4
Thiocyanate, protonation, 90
  log $\alpha_{L(H)}$, 254
  thiocyanato-complexes, 113–4
Thioglycollic acid, see mercaptoacetic acid
Thionalide [2-mercapto-N-(2′-naphthyl)acetamide],
  protonation and complexes, 105, 198
Thiophene-2-carboxylic acid,
  protonation in aqueous dioxane, 103
Thiosalicylic acid,
  protonation in aqueous dioxane, 104
Thiosemicarbazide,
  protonation and complexes, 186
Thiosulphate, protonation, 91
  log $\alpha_{L(H)}$, 254
  thiosulphato-complexes, 132
Thiourea
  complexes, 185
Thorium,
  compounds, solubility products, 218
    redox potentials, 244
  log $\alpha_{M(OH)}$, 288
  log $\alpha_{M(L,OH)}$, 304, 332
L-Threonine, protonation, 94
  log $\alpha_{L(H)}$ (No. 43), 258
Thymolphthalexone,
  protonation and complexes, 207
  log $\alpha_{L(H)}$ (No. 200), 266
Tin,
  compounds, solubility products, 217
    redox potentials, 243
  log $\alpha_{M(OH)}$, 288
  log $\alpha_{M(L,OH)}$, 302, 330
Tiron, see 1,2-dihydroxybenzene-3,5-disulphonic acid
Titanium,
  compounds, solubility products, 218
    redox potentials, 244
  log $\alpha_{M(OH)}$, 288
  log $\alpha_{M(L,OH)}$, 304, 332
TMDTA, see trimethylenedinitrilotetra-acetic acid
1,2,3-Triaminopropane,
  protonation and complexes, 145
  log $\alpha_{L(H)}$ (No. 29), 256
Trichloroacetic acid, protonation, 93
Triethanolamine (TEA),
  protonation and complexes, 147
  log $\alpha_{L(H)}$ (No. 94), 260
Triethylenetetramine (trien),
  protonation and metal complexes, 147
  log $\alpha_{L(H)}$ (No. 95), 260
Triethylenetetranitrilohexa-acetic acid (TTHA),
  protonation and complexes, 183–4

$\log \alpha_{L(H)}$ (No. 176–8), 264
1,1,1-Trifluoroacetylacetone,
  protonation in aqueous dioxane, 103
1,2,3-Trihydroxybenzene (pyrogallol),
  protonation and complexes, 191
  $\log \alpha_{L(H)}$ (No. 75), 258
Trimetaphosphate, see cyclo-triphosphate
Trimethylenediamine,
  protonation and complexes, 144
  $\log \alpha_{L(H)}$ (No. 27), 256
Trimethylenedinitrilotetra-acetic acid (TMDTA),
  $\log \alpha_{L(H)}$ (No. 144–5), 262
Triphosphate, $P_3O_{10}^{5-}$,
  protonation, 90
  $\log \alpha_{L(H)}$, 254
  triphosphato-complexes, 121
Tris(aminomethyl)methane, protonation, 94
Tropolone,
  protonation and complexes, 195–6
  $\log \alpha_{L(H)}$ (No. 101), 260
L-Tryptophan (tryp),
  protonation, 99
  $\log \alpha_{L(H)}$ (No. 142), 262
TTHA, see triethylenetetranitrilohexa-acetic acid
L-Tyrosine, protonation, 98
  $\log \alpha_{L(H)}$ (No. 121), 262
DL-o-Tyrosine, protonation, 98
  $\log \alpha_{L(H)}$ (No. 120), 262

Uranium,
  compounds, solubility products, 218
  redox potentials, 245
  $\log \alpha_{M(OH)}$, 288
  $\log \alpha_{M(L,OH)}$, 304, 334
Urea, protonation, 93

L-Valine, protonation, 95
  $\log \alpha_{L(H)}$ (No. 61), 258
Vanadium,
  compounds, solubility products, 218–9
  redox potentials, 245–6
  $\log \alpha_{M(OH)}$, 288
  $\log \alpha_{M(L,OH)}$, 304, 334
vant'Hoff equation, 18

Water,
  protonation in non-aqueous media, 100

Xylenol Orange,
  protonation and complexes, 206
  $\log \alpha_{L(H)}$ (No. 192–4), 266

Yttrium,
  redox potentials, 246
  $\log \alpha_{M(OH)}$, 288
  $\log \alpha_{M(L,OH)}$, 304, 336

Zinc,
  compounds, solubility products, 219
  redox potentials, 247
  $\log \alpha_{M(OH)}$, 288
  $\log \alpha_{M(L,OH)}$, 304–7, 336
Zincon,
  protonation and complexes, 204
Zirconium,
  compounds, solubility products, 219
  redox potentials, 247
  $\log \alpha_{M(OH)}$, 288
  $\log \alpha_{M(L,OH)}$, 306, 336